VOLUME SEVENTY FOUR

ADVANCES IN PHARMACOLOGY

VOLUME SEVENTY FOUR

ADVANCES IN PHARMACOLOGY

Cytochrome P450 Function and Pharmacological Roles in Inflammation and Cancer

Edited by

JAMES P. HARDWICK

Biochemistry and Molecular Pathology in Department of Integrative Medical Sciences, Northeast Ohio Medical University, Rootstown, Ohio, USA

Serial Editor

S. J. ENNA

Department of Molecular and Integrative Physiology, Department of Pharmacology, Toxicology and Therapeutics, University of Kansas Medical Center, Kansas City, Kansas, USA

Managing Editor

LYNN LECOUNT

University of Kansas Medical Center School of Medicine, Kansas City, Kansas, USA

ELSEVIER

AMSTERDAM • BOSTON • HEIDELBERG • LONDON
NEW YORK • OXFORD • PARIS • SAN DIEGO
SAN FRANCISCO • SINGAPORE • SYDNEY • TOKYO

Academic Press is an imprint of Elsevier

Academic Press is an imprint of Elsevier
225 Wyman Street, Waltham, MA 02451, USA
525 B Street, Suite 1800, San Diego, CA 92101-4495, USA
125 London Wall, London, EC2Y 5AS, UK
The Boulevard, Langford Lane, Kidlington, Oxford OX5 1GB, UK

First edition 2015

ISBN: 978-0-12-803119-3
ISSN: 1054-3589

For information on all Academic Press publications visit our website at http://store.elsevier.com/

CONTENTS

CONTRIBUTORS

Mohamed A. Abdelmegeed
Section of Molecular Pharmacology and Toxicology, Laboratory of Membrane Biochemistry and Biophysics, National Institute on Alcohol Abuse and Alcoholism, Bethesda, Maryland, USA

Mohammed Akbar
Section of Molecular Pharmacology and Toxicology, Laboratory of Membrane Biochemistry and Biophysics, National Institute on Alcohol Abuse and Alcoholism, Bethesda, Maryland, USA

Udayan Apte
Department of Pharmacology, Toxicology and Therapeutics, The University of Kansas Medical Center, Kansas City, Kansas, USA

Stelvio M. Bandiera
Faculty of Pharmaceutical Sciences, The University of British Columbia, Vancouver, British Columbia, Canada

K. Cameron Falkner
Division of Gastroenterology, Hepatology, and Nutrition, Department of Medicine, University of Louisville, Louisville, Kentucky, USA

Matt C. Cave
Division of Gastroenterology, Hepatology, and Nutrition, Department of Medicine, and Department of Biochemistry & Molecular Biology, University of Louisville, Louisville, Kentucky, USA

Chen Chen
Department of Internal Medicine and Gene Therapy Center, Tongji Hospital, Tongji Medical College, Huazhong University of Science and Technology, Wuhan, PR China

Peter Christmas
Department of Biology, Radford University, Radford, Virginia, USA

Ann K. Daly
Institute of Cellular Medicine, Newcastle University, Medical School, Newcastle upon Tyne, United Kingdom

Katheryne Z. Edson
Department of Medicinal Chemistry, School of Pharmacy, University of Washington, Seattle, Washington, and Amgen Inc., Thousand Oaks, California, USA

Giovanna Gentile
NESMOS Department, Faculty of Medicine and Psychology, Sapienza University of Rome, and Advanced Molecular Diagnostics Unit, Sant'Andrea Hospital, Rome, Italy

Charlotte Gröschel
Department of Pathophysiology and Allergy Research, Center of Pathophysiology, Infectiology & Immunology, Medical University of Vienna, Währinger Gürtel, Vienna, Austria

James P. Hardwick
Biochemistry and Molecular Pathology in Department of Integrative Medical Sciences, Northeast Ohio Medical University, Rootstown, Ohio, USA

Eugene G. Hrycay
Faculty of Pharmaceutical Sciences, The University of British Columbia, Vancouver, British Columbia, Canada

Nina Isoherranen
Department of Pharmaceutics, School of Pharmacy, University of Washington, Seattle, Washington, USA

Hartmut Jaeschke
Department of Pharmacology, Toxicology and Therapeutics, Kansas University Medical Center, Kansas City, Kansas, USA

Jing Jing
Department of Pharmaceutics, School of Pharmacy, University of Washington, Seattle, Washington, USA

Inho Jo
Department of Molecular Medicine, Ewha Womans University School of Medicine, Seoul, South Korea

Amanda L. Johnson
Department of Medicinal Chemistry, School of Pharmacy, University of Washington, Seattle, Washington, USA

Enikö Kállay
Department of Pathophysiology and Allergy Research, Center of Pathophysiology, Infectiology & Immunology, Medical University of Vienna, Währinger Gürtel, Vienna, Austria

Tiangang Li
Department of Pharmacology, Toxicology and Therapeutics, The University of Kansas Medical Center, Kansas City, Kansas, USA

Robert Preissner
Structural Bioinformatics Group, Charité—Universitätsmedizin Berlin, Berlin, Germany

Saskia Preissner
Dental, Oral and Maxillary Medicine, Charité—Universitätsmedizin Berlin, Berlin, Germany

Russell A. Prough
Department of Biochemistry & Molecular Biology, University of Louisville, Louisville, Kentucky, USA

Allan E. Rettie
Department of Medicinal Chemistry, School of Pharmacy, University of Washington, Seattle, Washington, USA

Maurizio Simmaco
NESMOS Department, Faculty of Medicine and Psychology, Sapienza University of Rome, and Advanced Molecular Diagnostics Unit, Sant'Andrea Hospital, Rome, Italy

Byoung-Joon Song
Section of Molecular Pharmacology and Toxicology, Laboratory of Membrane Biochemistry and Biophysics, National Institute on Alcohol Abuse and Alcoholism, Bethesda, Maryland, USA

Faith Stevison
Department of Pharmaceutics, School of Pharmacy, University of Washington, Seattle, Washington, USA

Samawansha Tennakoon
Department of Pathophysiology and Allergy Research, Center of Pathophysiology, Infectiology & Immunology, Medical University of Vienna, Währinger Gürtel, Vienna, Austria

Rheem A. Totah
Department of Medicinal Chemistry, School of Pharmacy, University of Washington, Seattle, Washington, USA

Sasmita Tripathy
Department of Pharmaceutics, School of Pharmacy, University of Washington, Seattle, Washington, USA

B. Wahlang
Division of Gastroenterology, Hepatology, and Nutrition, Department of Medicine, University of Louisville, Louisville, Kentucky, USA

Dao Wen Wang
Department of Internal Medicine and Gene Therapy Center, Tongji Hospital, Tongji Medical College, Huazhong University of Science and Technology, Wuhan, PR China

Benjamin L. Woolbright
Department of Pharmacology, Toxicology and Therapeutics, Kansas University Medical Center, Kansas City, Kansas, USA

PREFACE

ABBREVIATIONS

ALD alcoholic liver disease
CYP cytochrome P450
EETs epoxyeicosatrienoic acids
ER endoplasmic reticulum
GWAS genome-wide association studies
LTB_4 leukotriene B_4
NF-κB nuclear factor kappa-light-chain-enhancer of activated B cells
RA retinoic acid
ROS reactive oxygen species
UDCA ursodeoxycholic acid

Cytochrome P450s (CYPs) are critical monooxygenase enzymes in the synthesis and degradation of both endogenous and exogenous substrates that influence cellular homeostasis through regulation of cellular metabolism, growth, differentiation, apoptosis, and neuroendocrine function. To date, over 23 human diseases are associated with mutations in members of the 18 CYP families (Nebert, Wikvall, & Miller, 2013). Furthermore, variants of human CYP genes have a profound effect on drug and endogenous metabolite levels by altering drug therapeutic efficacy or drug toxicity.

Unfortunately, the functional role of CYP isoforms and genetic variants in complex human diseases through altered metabolism of endogenous substrates has been largely unexplored. Several genome-wide association studies (GWAS) have revealed that genetic variations of CYP1, CYP2, CYP3, CYP4, and CYP7 families are associated with significant individual risk for complex polygenic human disease. This is apparent when one considers the fact that CYP members participate in the metabolism of fatty acid, cholesterol, and bile acid pathways, in addition to modulating vitamin A, vitamin D, vitamin K, vitamin E, and eicosanoid metabolic pathways. The association of an individual's P450 profile to complex human diseases becomes a formidable scientific task because of the broad substrate specificity of individual P450s, differential regulation of CYP genes by multiple nuclear hormone receptors, and the role of polymorphic CYP variants in the metabolism of endogenous and exogenous compounds. However, the system biology approaches with new technologies to measure CYP copy number variation (CNV), the level of variant CYP expression by digital PCR, quantitative proteomic P450 analysis (MacLeod et al., 2015; Sakamoto et al.,

2011), and metabolomics analysis of endogenous substrate metabolism by mass spectrometry (Guengerich, Tang, Cheng, & Salamanca-Pinzon, 2011; Xiao & Guengerich, 2012) have new avenues to associate CYP variants with altered endogenous metabolites to complex polygenetic human disease.

This *Advances in Pharmacology* volume on the role of CYP in inflammation and cancer compiles a comprehensive collection of monographs on the role of different CYP gene families in human susceptibility, initiation, and progression of inflammation and cancer. Extraordinary progress has been made by the authors within this volume on the identification of and the role of the 57 human P450s from 18 CYP gene families, including the correlation of CYP polymorphic variants with inflammatory disorders and initiation and progression of cancer. Although variants of the CYP1, CYP2C, CYP2A6, CYP2A6, CYP2B6, CYP2D6, CYP2E, CYP3A, and CYP7A families have been associated with individual increased risk of cancer through alteration in metabolism and drug susceptibility from animal studies, the association of human CYP variant alleles with these diseases lacks proof because of subpar statistical power from GWAS. Therefore, it is difficult to conclusively associate human studies with transgenic and knockout animal data, even though many published GWAS have found positive correlations between several CYP family members and several complex human diseases. It is refreshing that the contributors to this volume not only provide a comprehensive detailed analysis of the functional and pharmacological role of different CYP families in inflammation and cancer but also offer insightful theories and experimental plans to associate an individual's CYP genotype with disease phenotype in complex human diseases.

In review one by Bruce Wahlang, Cameron Falkner, Matt Cave, and Russell Prough, the authors discuss the historical significance and background of CYP in the activation of procarcinogens that are genotoxic agents forming DNA adducts that lead to DNA mutations. They also provide insight into how chemotherapeutic prodrugs are used in cancer chemotherapy through metabolism by selective members of the CYP family. The authors offer a personalized medical approach in the treatment of cancer. By understanding the expression pattern of an individual's CYP genes in different tissue tumors, the authors suggest efficacy of many chemotherapeutic drugs can be increased if we increase the expression of the tumor CYP through activation of nuclear hormone receptors with coadministration of chemotherapeutic prodrugs metabolized by these induced CYP. By using this theoretical approach, one can design therapies with increased chemotherapeutic efficacy and reduced drug toxicity. Lastly, the authors discuss the role of P450 polymorphic variants in cancer susceptibility and how informatics approaches, combining CYP genomic, proteomics, and metabolomics data, are being used to translate our basic understanding of CYP expression and drug metabolism into personalized cancer treatments.

In the second review, Eugene G. Hrycay and Stelvio M. Bandiera provide us with a solid foundation to understand the P450 catalytic cycle and how P450 functions as a monooxygenase or an oxidase and peroxygenase under different conditions of cellular stress can increase drug concentration or cause adverse drug–drug interactions. The outcome of this P450-related stress is increased formation of reactive oxygen species (ROS) through coupling of the P450 catalytic cycle, which eventually results in increased endoplasmic reticulum (ER) lipid peroxidation and initiation of ER unfolded response to reestablish cellular homeostasis. The uncoupling of the P450 cycle is also discussed by the authors extensively, citing three models of uncoupling: (1) substrate-dependent access of water that destabilizes the ferric-superoxo intermediate, (2) reduction of the ferric-peroxo intermediate, and (3) dissociation of the hyperoxide anion from ferric iron leading to generation of hydrogen peroxide. The uncoupling of electron transport to substrate oxidation by different members of the CYP gene families varies significantly, ranging from 0.5% to greater than 50%, indicating that a specific P450 isozyme can play a significant role in the generation of ROS. It is presently unknown whether these P450s have the functional role in providing an oxidizing environment necessary for ER protein unfolding. In any case, CYP enzymes are quantitatively one of the most important cellular sources of ROS and thus have an important role in the initiation and propagation of lipid peroxidation. The ultimate consequence of increased polyunsaturated fatty acid lipid peroxidation is the formation of fatty acid peroxides, which can function as chemotactic agents to attract immune cells to the site of injury and initiate the inflammatory response. Thus, the persistent uncoupling of selective P450 isozymes by drugs or through increased expression can contribute to chronic inflammation with the consequences of an observable pathology in numerous disease processes. It is very apparent that CYP isoforms have a pivotal role in initiating the acute inflammatory response by endogenous or exogenous substrates capable of uncoupling the P450 catalytic cycle. To ameliorate adverse drug reaction and the consequence of idiosyncratic drug interaction would require choosing either a drug not metabolized by the same CYP or a drug with the ability to act as a reducing agent (e.g., antioxidant) and thus prevent CYP catalytic cycle uncoupling and ROS formation. Of course, in the design of chemotherapeutic agents the reverse scenario may increase the efficacy of combination chemotherapeutic drug therapy.

The third review by Ann Daly details our current understanding of the relevance of polymorphic CYP variants in disease initiation, progression, and individual susceptibility to cancer. Present-day dogma generated solely from genomic data suggests that dominance variation at common SNPs explains only a small fraction of phenotypic variation for complex human diseases or the role of CYP variants in cancer susceptibility. The formation of either sporadic or familial inherited SNP plays a substantial role in inflammation and cancer only if the metabolite has a critical central role in complex human diseases. This is very evident in arginine mutations of isocitrate dehydrogenases (IDH1 and IDH2) in colorectal, glioma, and acute myeloid leukemia. These mutations lead to conversion of α-ketoglutarate to R-2-hydroxyglutarate metabolite that inhibits α-ketoglutarate-dependent enzymes of the tricarboxylic cycle and mitochondrial respiratory chain inhibition of cytochrome *c* oxidase (complex IV) and ATP synthase (complex V). This results in the induction of antiapoptotic BCL-2 that antagonizes cell death signaling. Until we can discern the phenotype caused by variant CYP in producing new drug and/or endogenous biochemical metabolite, the associate of variant CYP with individual susceptibility to complex human disease will not be realized. This is demonstrated in the highly polymorphic human CYP2D6 subfamily by which dopamine can be synthesized from the false neurotransmitter tyramine, and in humanized CYP2D6-transgenic mice 5-methoxytryptamine, a precursor to melatonin, can be O-demethylated by CYP2D6 to serotonin. It will be of interest to determine whether debrisoquine high metabolizers who display a Type A personality phenotype because of increased serotonin blood levels are more susceptible to the detrimental effects of serotonin uptake inhibitors. The association of CYP polymorphisms appears to be relevant to susceptibility to increased cancer risk, which is largely dependent upon the type of xenobiotic exposure. GWAS and meta-analysis revealed that increased risk of lung cancer is associated with CYP2A6 high metabolizers of nicotine and nitrosamine activation. In addition, several studies have found correlations between CYP3A4 polymorphism, aflatoxin exposure, and risk of hepatocellular carcinoma (HCC). Similarly, CYP1A2 polymorphism and heterocyclic aryl amine exposure increase risk of colon cancer and nitrosamine-related nasopharyngeal cancer in individuals with increased activity or expression of the ethanol-metabolizing CYP2E1. Unfortunately, many of these GWAS almost exclusively perform genotype analysis with no correlative studies on the levels of insulting metabolites produced by the particular CYP or whether the identified CYP has a functional role in the metabolism of a biochemical metabolite that regulates cell differentiation, growth, or proliferation. This is apparent from Dr. Daly's discussion of CYP2D6 polymorphism where poor metabolizers show a significantly decreased risk of lung cancer while genotyping studies failed to find a significant association of CYP2D6 alleles with lung cancer susceptibility. The pharmacologically successful use of CYP19 aromatase inhibitors in postmenopause women to prevent conversion of testosterone to estradiol in fat tissues has been shown to be an exceedingly successful target, reducing the risk of breast cancer by 47%. It is apparent that obese women with increased fat deposits would most likely have increased risk of breast cancer because of increased expression of CYP19 or a polymorphic variant with increased activity to produce growth-activating estradiol. It is also apparent that GWAS have identified a possible weak involvement of CYP polymorphisms with increased risk of cancer; however, without confirmatory phenotype studies, the power of these genotype studies will not be fully realized until additional metabolomics and tissue-specific proteomic studies are performed.

In review four, Saskia Preissner, Maurizio Simmaco, Giovanna Gentile, and Robert Preissner discuss how CYP variability can influence the effectiveness of cancer therapy, and present the preemptive challenge to clinical oncologists such as to why CYPs and drug transport expression in cancer tissues are not evaluated in providing the most effective personalized cancer chemotherapy. The authors state the simple dictum that we should genotype patients because over 2000 drug-metabolizing reactions are performed by 57 human CYPs and that CYP3A4, 2D6, 1A2, 2C9, and 2C19 are responsible for 60% of all drug metabolisms. The association of CYP SNPs can have dramatic effects on activation of chemotherapeutic prodrugs or lead to adverse drug reactions with side effects and toxicities. This review summarizes the influence of common CYP SNPs in cancer therapy and establishes a strong case to underscore personalized medicine in cancer chemotherapy. The authors also present the case for analyzing CYP variant expression in tumors. They suggest that tissue-specific CYP profile can be used to design effective therapies with fewer adverse side effects. The field of cancer chemotherapy is slowly but surely moving toward personalized therapies that, hopefully, will increase drug therapy efficiency and reduce overt severe side effects. The utility of this approach to personalized cancer therapy is seen in the use of Tegafur (a prodrug for 5-fluorouracil) in patients with increased tumor-specific overexpression of dihydropyrimidine dehydrogenase (DDP), which assists in the conversion to inactive 5-fluoro-β-alanine. It is readily evident that tumors expressing low levels of DPP would be effectively treated by this chemotherapeutic drug, while the same patient with high expression of DDP in normal highly proliferative tissues (hematopoietic and intestine) would experience fewer adverse side effects from this therapy. Similarly, the topoisomerase inhibitor irinotecan used in colorectal cancer therapy is activated by carboxylesterases to a potent inhibitor of DNA replication and transcription SN-38, which is conjugated by UDP-glucuronosyltransferase and removed from the cell by ABCC2 transporters, or irinotecan can be metabolized and inactivated by CYP3A4 and CYP3A5. The toxicity or efficacy of irinotecan chemotherapy is dependent on the genetic polymorphism of CYP3A, UDP-glucuronosyltransferase, and ABC transporters and environmental factors that control expression of these gene family members. Thus, genotyping with serum metabolomics of irinotecan metabolites will lead to optimal chemotherapeutic efficacy and limitation of adverse drug side effects characterized by myelotoxicity, neutropenia, and most prominently diarrhea. This personalized medical approach is being established in the United States by Vanderbilt University for genotyping transplant patients, and in the near future, the European Commission's Horizon 2020 program will implement these techniques in clinical practice.

Having established the utility of a comprehensive genotypic and phenotypic characterization of CYP alleles in cancer susceptibility and optimized chemotherapy, review five by Benjamin Woolbright and Hartmut Jaeschke discusses the relationship between the CYP system and inflammatory response in the liver. This topic has particular relevance in the treatment of sepsis patients who display a hypercatabolic state with persistent hepatic synthesis of acute-phase proteins takes persistence to maintain cellular homeostasis at the expense of drug metabolism. This presents a therapeutic challenge for physicians to treat these patients. The authors of this timely review discuss the divergent roles of the CYP in the liver hepatocytes in xenobiotic and endobiotic metabolism and their functional roles in modulating the innate immune response (Kupffer cells, natural killer T cell, and blood-derived neutrophils) to orchestrate a complementary response to maintain homeostasis. When these complementary systems are improperly balanced to combat foreign insults, then chronic inflammation and tissue injury ensue leading to liver dysfunction, fibrosis, cirrhosis, and cancer.

A complete discussion on how particular CYP and drugs can produce idiosyncratic drug-induced liver injury through activation of the liver innate system enlightens us on how the endogenous metabolites mediate this cross talk between the CYP system and activates the innate immune response in the modulation of the inflammation in cholestasis and sepsis. Although immune cells have low levels of CYP enzymes, their functional role in immune cell activity through metabolism of endogenous substrates represents an unexplored field that might have profound significance in modulating the immune response and inflammation. This argument was recently substantiated in triple knockout of CYP1a1/1a2/1b1$^{(-/-)}$ mice that showed an increased proinflammatory response due to elevated leukotriene B_4 (LTB_4) production in exudates of Zymosan-stimulated mice with a reduction in anti-inflammatory eicosanoids (Divanovic et al., 2013). Future studies with both tissue-specific CYP-humanized transgenic and knockout mice with the analysis of eicosanoid metabolism will offer new insights into the functional role of tissue-specific CYP gene family members in the control of inflammation.

In review six, Peter Christmas discusses the role of specific CYPs in inflammation and how their modulation may have clinical significance in the treatment of chronic inflammatory disorders. Dr. Christmas makes the case that a patient's response to acute, chronic, or the resolution phase of inflammation is a clinical problem for the physician and pharmacists since the intensity and time course are difficult to determine. Identification of a patient's CYP profile during these stages may provide a valuable diagnostic tool in both the choice of therapy and drug dosage. To support this clinical scenario, the regulation of CYPs in animal models of inflammation was determined since it is well established that the activity of hepatic CYPs varies depending on the insulting agent (chemical injury, lipopolysaccharide, live bacterial infection, and sterile activation of inflammation). These studies reveal a selective regulation of different CYPs, which are dependent on disease state rather than the uniform downregulation of CYPs in response to global activation of cytokine and chemokine production in immune cells. Because blood leukocytes' expression pattern of CYPs closely mimics hepatic CYP expression, a clinical opportunity exists in monitoring the progression of the inflammatory response and disease state through the induction or suppression of different CYPs. Dr. Christmas extends the functional role of CYPs in modulating inflammation by discussing the largely opposing role of the eicosanoid CYPs of the epoxygenase (CYP2C) and omega hydroxylase (CYP4) family members. The epoxygenase CYP2C family members produce four bioactive regioisomeric epoxyeicosatrienoic acids (EETs) from arachidonic acid that can be inactivated by soluble epoxide hydratase to less active dihydroxyeicosatrienoic acid. The CYP2C family members are expressed in the vascular system where EET can cause vasodilation and inhibit inflammation through inactivation of NF-κB (nuclear factor kappa-light-chain-enhancer of activated B cells) signaling in immune cells. It is currently believed that loss of CYP2C expression or the expression of CYP2C alleles with reduced ability to metabolize arachidonic acid assist in driving inflammation to chronic stage. Drug companies have come to realize the beneficial local effects of EETs in the amelioration of several disease states and thus have prudently designed specific soluble epoxide hydratase inhibitors to maintain elevated tissue EET levels. Because EETs accelerate wound healing and vascularization as well as stimulation of tumor growth and metastasis, drugs to increase EET warrant caution and further investigation of both the tissue-specific biological function of different EETs and the CYP2 variant alleles associated with different disease states. Again, a detailed genomic, proteomic, and tissue-specific lipidomic approach will provide important information on CYP2C-produced EETs in the amelioration of different disease states and progression of inflammation. Lastly, Dr. Christmas provides a discussion of the CYP4 family members in the production of 20-hydroxyeicosatetraenoic acid (20-HETE) from arachidonic acid, which largely has an opposing role compared to EETs by causing vasoconstriction and activation of NF-κB. It is interesting that in humans, 20-HETE production is largely mediated by CYP4A11 while CYP4F members metabolize and inactivate the potent proinflammatory chemotactic LTB_4. In humans, CYP4F3A is expressed in neutrophils and mediates the inactivation of LTB_4. The generation of knockout mice to the human homologue of CYP4F3A *CYP4F18* abolishes LTB_4 hydroxylase activity in neutrophils, but does not impact inflammation in a mouse model of renal ischemia-reperfusion injury. The mechanisms linking CYP activity to inflammatory disease will require further refined approaches using CRISPR/Cas9-derived RNA-guided nuclease DNA targeting system to target multiple CYPs in disease models with lipidomics and metabolomics approaches.

In review seven, Chen Chen and Dao Wen Wang discuss the importance of the CYP2 gene subfamilies and the role of epoxygenase P450s in inflammation and cancer. They describe how pharmaceutical companies are targeting these CYP to increase the beneficial effects of EETs in inflammatory-mediated cardiovascular disease where their potent antihypertensive, vasodilatory effects are cardioprotective and antiatherosclerosis.

In this review, the authors discuss recent progress in identification and association of CYP2 polymorphism in disease susceptibility as well as their tissue distribution in inflammation and cancer. To truly understand the function of CYP2C epoxygenase in inflammation, one must understand at what step in inflammation, rubor-redness, calor-heat, tumor-swelling, and dolar-pain EETs have a functional role in preventing chronic inflammation and induction of acute inflammation that leads to resolution and return of tissue homeostasis. In addition, the authors of this topical review categorize the biological steps of human tumor development, proliferative signaling, evading growth suppressors, resistance to cell death, replicative immortality, angiogenesis, and metastasis and tissue invasion while they describe the role of EETs in these processes. Further research will be necessary to fully understand how different regioisomers control the process of inflammation and cancer through activation of tissue-specific EET receptor subtypes. This is apparent when we consider the reason why 11,12 EET has a broad spectrum of anti-inflammatory effects against both acute and chronic inflammation, but other regioisomers do not. Many questions remain unresolved. These include what EET receptor does 11,12 EET activate, or does it impart its anti-inflammatory effect through activation of specific PPARs or by inhibition of NF-κβ? Similar questions arise on the role of different CYP genes in the progression of cancer. Designing flox knockout mice for each of CYP genes contributing to EET production and transgenic mouse model expressing CYP2 variant alleles will greatly enhance our understanding of the functional and pharmacological role of the epoxygenase CYP2 P450s in the progression of inflammation and cancer.

In review eight of this volume, Amanda L. Johnson, Katheryne Z. Edson, Rheem A. Totah, and Allan E. Rettie extend the previous introduction and functional role of the CYP4 members in inflammation and cancer through their metabolism of endogenous substrates, which include a variety of different chain-length fatty acids, eicosanoids, and vitamins. The CYP4 family consists of several subfamilies (CYP4, CYP4B, CYP4F, CYP4Z, CYP4V, and CYP4X) whose members preferentially metabolize fatty acids with different chain length. By far, the most studied are members of the CYP4A and CYP4F gene subfamilies because of their role in the metabolism of the important eicosanoid, arachidonic acid, and proinflammatory LTB_4. The functional role of these members in inflammation is not completely understood even though the human CYP4F3A has a pivotal role in the metabolism and inactivation of the potent chemotactic LTB_4. As described by Dr. Christmas in this volume, knockout of the mouse orthologue

CYP4F18 did not increase inflammation in a mouse model of ischemia-reperfusion injury. These results strongly suggest that the functional role of CYP4F P450 in the process of acute, chronic, and resolution phases of inflammation is unknown. Observational evidence of patients with celiac disease suggests that increased level of CYP4F2 mRNA correlates with resolution of inflammation. It is likely that different CYPA and CYP4F members have different functional roles in modulating inflammation in various phases of the inflammatory response through expression in diverse tissues and immune cells that orchestrate this complex process. It is possible that CYP4A11, which metabolizes arachidonic acid to the vasoconstrictive 20-HETE, may function like thromboxane A_2 to limit tissue involvement, while expression of CYP4F may be involved in the resolution phase by limiting the chemotactic potential of proinflammatory LTB_4. It is imperative that a detailed understanding of the expression of critical eicosanoids metabolizing CYP4A and CYP4F members must be determined in the different phases of the initiation and resolution of inflammation. Dr. Rettie forwards this idea by detailing chemical tools for dissecting CYP4 P450 enzymology and biological mechanisms in both inflammation and cancer. In contrast, the CYP4A and CYP4F subfamily members and members of the CYP4Z, CYP4B, and CYP4V subfamilies have been strongly associated with several human cancers where they promote tumor angiogenesis and proliferation. The development of specific CYP4 inhibitors may provide vital tools to treat inflammation as well as targeted therapies for tissue-specific cancers.

Review nine by Tiangang Li and Udayan Apte explores the medical implication of the bile acid metabolizing CYP7A1, CYP8B1, CYP27A1, CYP7B1, and CYP3A4 in cholestasis, initiation of inflammation, and their role in of cancer. Cholesterol metabolism and the production of bile acids in human health are a complex multistage process where perturbations at any stage can have physiological consequences. Bile acids function as physiological detergents that facilitate hepatic excretion of endogenous metabolites (e.g., cholesterol, phospholipids, and bilirubin), xenobiotics, and drugs while assisting in the absorption of fat and vitamins in the intestine. Bile acids also function as signaling molecules to regulate cellular response by activating nuclear hormone receptors (FXR, PXR, VDR, SHP) and G-coupled receptors (TGR5) that tightly regulate metabolism. In contrast, increased levels of hepatic–biliary bile acids are highly toxic contributing to several human diseases, biliary cirrhosis, primary sclerosing cholangitis, gallstones, and cholestasis, resulting in chronic inflammation, fibrosis, cirrhosis, and HCC. Unlike the previous eight reviews where polymorphic variants may have a significantly undefined role in individual susceptibility to inflammation and cancer through an altered catalytic activity and metabolite production, the cholesterol-metabolizing CYPs association with human diseases is due to their altered regulation by bile acid-mediated activation of nuclear hormone receptors. This tightly regulated physiological system has become a pharmaceutical target in the treatment of a variety of human diseases. For instance, ursodeoxycholic acid (UDCA) and vitamin D_3 are used for primary biliary cirrhosis, norUDCA for primary sclerosing cholangitis, cholestyramine for hypercholesterolemia, rifampicin for cholestasis, chenodeoxycholic acid (CDCA) and UDCA for gallstones, and FXR agonist obeticholic acid for primary biliary cirrhosis, primary sclerosing cholangitis, and cholestasis-induced fibrosis. Further studies are required to determine whether targeted therapies to modulate the cholesterol-metabolizing CYPs and the ratiocholic acid:chenodeoxycholic (CA: CDCA) can be of therapeutic value in treating these disorders where hydrophobic bile acids induce inflammation and cancer progression.

In review ten by Byoung-Joon Song, Mohammed Akbar, Inho Jo, James P. Hardwick, and Mohamed A. Abdelmegeed, the authors detail the role of alcohol-metabolizing enzymes and CYP2E and CYP4 family members in the initiation and progression of alcoholic and nonalcoholic liver disease. This comprehensive review provides a basis for translating research results into the development of new therapies based on our present understanding of how CYPs function in alcohol and fatty acid-induced progression of liver disease. The association of CYP2E1 with liver disease is due to uncoupling of the P450 catalytic cycle and existence of P4502E1 in a high-spin state, offering a continuous source of ROS that activates the redox-sensitive transcription factor NF-κB. Dr. Song describes in detail the role of CYP2E1 in inflammation and initiation of cancer through the metabolism and activation of several carcinogenic amines, leading to drug-induced liver injury. The functional role of CYP2E1 in the initiation and progression of alcoholic liver disease (ALD) largely stems from CYP2E1-related oxidative stress. This oxidative stress inhibits autophagy and mitophagy, thereby increasing intracellular levels of misfolded protein and dysfunctional organelles that contribute to ER stress and liver injury. CYP2E1 also activates the redox-sensitive NF-κB transcription factor that activates iNOS, leading to increase in reactive nitrogen species (RNS). This results in ROS/RNS-mediated release of damage-associated molecular pattern molecules, such as high-mobility group box-1, which activate hepatic stellate and endothelial cells resulting in cell injury. In addition, the leakiness of intestine caused by binge ethanol increases the liver exposure to pathogen-associated molecular pattern molecules that can activate the immune response leading to severe hepatitis. It is of particular interest that release of vitamin A from activated hepatic stellate cells initiates the fibrotic pathway, and that vitamin A is metabolized by CYP2E1. It has recently been shown that hematopoietic stem cell self-renewal is controlled by retinoic acid (RA) signaling, and elimination of RA by CYP26 P450 members prevents differentiation of hematopoietic stem cells. It will be of interest to determine whether CYP2E1-mediated metabolism of RA from activated stellate cells accounts for the proliferation of biliary ductal cells in the progression of fibrosis to cirrhosis and alcohol-induced liver cancer. Dr. Soon concludes his discussion of ethanol-metabolizing enzymes by providing a detailed discussion of current clinical management of ALD with glucocorticosteroids, pentoxifylline, and PPAR agonists. Promising new areas of drug development focus on reducing nitroxidative stress through the use of SOD–catalase mimetic and mitochondria-targeted ubiquinone (Mito-Q) as well as small-molecule activators of aldehyde dehydrogenase (ALDH2) to prevent the toxic effects of acetaldehyde, which is in contrast to ALDH inhibitor disulfiram in that is used to increase alcoholics' aversion to ethanol intake.

Article eleven is a timely significant review by Faith Stevison, Jing Jing, Sasmita Tripathy, and Nina Isoherranen on the role of RA-metabolizing cytochrome, CYP26 family member's role in inflammation and cancer. The critical function of RA in numerous physiological processes in vision, epithelia, bone and immune cell differentiation, and ligands for nuclear hormone receptors identifies RA-metabolizing CYP26 members as sentinels in maintaining RA homeostasis in diverse biological systems. The critical function of CYP26 isozyme in balancing the cellular retinoic levels necessary for retinoic signaling is evident since knockout mice have server developmental malformations. CYP26 regulation of RA metabolism plays an important role in modulating immune response and inflammation. Thus, CYP26 regulation of RA level is critical in inflammatory, autoimmune, and alcoholic and nonalcoholic liver disease. This is apparent in that mice fed a high-fat diet and retinoic do not develop hepatic steatohepatitis, suggesting that RA may have utility in the treatment of diseases associated with metabolic syndrome. Unfortunately, the activity of ALDH1A in the synthesis of RA from reinaldehyde or metabolism of RA by CYP26 family members has not been studied in many of these disease models. In inflammatory bowel disease, RA reduces symptoms of inflammation by decreasing neutrophil infiltration, reducing NF-κB, TNFα, and IL17a levels while increasing the level of Treg cells. Furthermore, a tentative relationship exists between a high RA-metabolizing CYP26 polymorphic variants and increased human risk of developing Crohn's disease. Whether increased intestinal RA concentrations can prevent alcohol or fatty acid-induced gut leakiness and thus pathogen-associated molecular pattern molecules' induction of hepatic inflammation and steatohepatitis is a vitally important research area. Dr. Isoherranen and others provide a thoughtful and complete discussion of CYP26 role in development of psoriasis through increased expression of CYP2S1 in psoriatic plaques and the use of CYP16 inhibitors to potentiate cell differentiation effects of RA. Because active RA metabolites have a central role in regulating cell proliferation, differentiation, and apoptosis, the regulation of RA levels has been an important therapeutic target in the treatment of cancer. Elevated CYP26 levels in numerous human cancers lead to RA deficiency, which has been linked in epidemiological studies to carcinogenesis. Thus, the synthesis of synthetic retinoid and CYP26 inhibitors has been employed to increase RA levels and thus cell differentiation. The importance of homeostatic regulation of RA levels by RA-mediated CYP26 members is seen in patients with relapsed acute promyelocytic leukemia where CYP26 inhibitors may initiate RA-induced promyelocyte differentiation. The functional role and modulation of CYP26 levels have amazing therapeutic promise in numerous inflammatory diseases and the treatment of cancer. Hopefully, future studies will not only characterize the expression and proteomic levels of ALDH1 RA-synthesizing and CYP26 catabolic enzymes with the metabolic levels of retinoic metabolites to obtain an overall view on the importance of RA in cell proliferation and differentiation.

The twelfth review by Charlotte Gröschel, Samawansha Tennakoon, and Enikö Kállay brings awareness to our understanding of vitamin D insufficiency with respect to the synthesis and catabolism of active vitamin D metabolites and the functional role of vitamin D hydroxylase in maintaining active vitamin D ($1,25(OH)_2D_3$) calcitriol levels in tissue homeostasis. Dr. Kállay also discusses the close connection between vitamin D and bile acid metabolism previously discussed by Drs. Li and Apte in their review on bile acid-metabolizing P450. The tissue-specific paracrine and autocrine influence of locally synthesized $1,25(OH)_2D_3$ are diverse and include activation of innate immune response and suppression of autoimmune disease through activation of the adaptive immune system, thus suppressing inflammation and exerting antitumorigenic activity. Although it was long thought that liver and kidney are the primary sites of 25-hydroxylation

(CYP2R1, CYP27A1) and 1α-hydroxylation (CYP27B1), respectively, to produce the active metabolite $1,25(OH)_2D_3$, subsequent studies have shown that the CYPs that activate and inactivate vitamin D_3 are expressed in many tissues. Active vitamin D_3 levels are largely controlled by vitamin D_3 metabolite activation or inactivation of tissue-specific vitamin D_3-metabolizing CYP genes. It is known that $1,25(OH)_2D_3$ induces its own catabolism through vitamin D receptor (VDR)-mediated CYP24A1 gene activation, and reduced synthesis by VDR-mediated decrease in CYP27B1 expression through CYP24A1 conversion of calcitriol to $1,25(OH)_2D$-26,23 lactone that inhibits VDR activation. The intimate relationship between cholesterol and bile acids with vitamin D_3 metabolism is elegantly discussed in the context of the rate-limiting step in steroid biosynthesis by cholesterol side-chain cleavage enzyme (CYP11A1) converts vitamin D_3 to several metabolites with unknown cellular functions. A thoughtful and enlightening discussion on the role of CYP in the control of calcitriol levels provides an explanation of calcitriol anti-inflammatory action. Calcitriol inhibits inflammation through suppression of innate immune response, downregulation of Toll-like receptors, inhibition of NF-κB activation, prostaglandin synthesis, and increased synthesis of anti-inflammatory cytokines. In contrast, calcitriol also activates the adaptive immune response through an increase in Th2 cells and the induction of regulatory T-cells. The diverse role of calcitriol in modulation of both the innate and adaptive arms of the immune response identifies vitamin D_3 hydroxylase as important drug targets for the treatment of inflammatory disorders. Vitamin D deficiency increases the severity of several autoimmune diseases and has been closely associated with the inflammatory bowel disease, ulcerative colitis, and Crohn's disease where increased CYP27B1 expression reduces intestinal calcitriol levels. Dr. Kállay discusses the role of calcitriol in suppressing cancer through inhibition of cell proliferation, angiogenesis, and activation of cell differentiation and apoptosis. A complete discussion of vitamin D metabolites in the prevention of cancer development indicates that increasing tissue levels of calcitriol in organs with sporadic cancer can prevent neoplasia development. These data suggest that vitamin D-metabolizing P450 polymorphisms may influence synthesis or degradation of vitamin D, which correlates with tissue- and organ-specific risk, incidence, and development of different cancers. Numerous studies on diverse tissue cancers reveal that deficiency of calcitriol in tumors is either due to increased CYP24A1 expression or reduced CYP27B1 expression. Such results indicate that specific CYP24A1 inhibitors may be a therapeutic avenue to maintain or increase tissue calcitriol levels. It is of vital importance that future studies of tumor vitamin D metabolism should be correlated with proteomic levels of vitamin D-metabolizing enzyme and metabolomics determination of tissue vitamin D metabolites. Cumulatively, these detailed studies will allow determination of relative risk association of CYP variants with vitamin D metabolites and cancer.

A number of themes are revealed from these superb reviews on the role of different CYP gene family members and variants in human susceptibility to inflammatory disorders and cancer. First, GWAS indicate that there is a weak association of CYP variant alleles with complex human disease. Second, the true endogenous biological substrate for members of the CYP1, CYP2, and CYP3 is largely unknown, thereby correlating CYP allelic variants with GWAS of complex human disease remains a biological uncertainty. Third, altered expression of CYP alleles through epigenetic modification by xenobiotic exposure may be more strongly correlated with the functional significance of CYPs in individual susceptibility, initiation, and progression of inflammatory disease and cancer. The scientific and clinically relevant question is how can the results from CYP mouse knockout and transgenic models be extrapolated to human studies? Although the mouse has more CYP genes (102) than humans (57), there are 36 orthologous pairs that potentially have similar profiles of endogenous substrates, while nonorthologous members either metabolize a unique species-specific endogenous substrate or a persistent environmental xenobiotic. Therefore, the phenotype of orthologous gene knockouts may more closely predict the human phenotype rather than the knockouts of nonorthologous gene of the expanded mouse subfamily. To obtain useful information applicable to a human disease phenotype, multiple nonorthologous CYP genes in the mouse need to be disrupted simultaneously. This has been shown in triple knockout mice of *Cyp1a1/1a2/1b1*$^{(-/-)}$ where increased inflammation due to increased LTB_4, but decreased 5-HETE, 15S-HEPE, 12S-HEPE are necessary for transcellular synthesis of anti-inflammatory lipoxins (Divanovic et al., 2013). The investigators present a strong case for this method since all these genes have human orthologs, suggesting that knock-in of human-specific CYPs may provide a useful tool to understand the role of human CYP in inflammation through biosynthesis and catabolism of bioactive lipids. Although these mouse-based strategies will yield insightful information on the role of human CYPs in complex diseases, confounding participation of nonorthologous solute ligand carrier proteins, conjugating enzymes, and ATP-binding cassette transports can skew the resulting phenotype seen in humans. A possible solution to this dichotomy is to use chimeric mice with humanized liver in which the human hepatocytes can be engineered to express critical human CYP variant that may confirm the human GWAS. Another critical endeavor to compare mouse and human studies is to identify the true endogenous metabolic substrate for orthologous orphan CYP pairs. A powerful method developed by Dr. Guengerich's laboratory has been used in the identification of endogenous substrates in tissue extracts using metabolomics and isotopic labeling approaches (Guengerich et al., 2011). The validity of this approach led to the identification of lysophospholipids as substrates for human CYP2W1, which is expressed at high levels in malignant colorectal cancer extracts (Xiao & Guengerich, 2012). It will be of interest to see if the mouse orthologue *CYP2W1* P450 displays a similar substrate specificity. Although human GWAS provide tantalizing clues of which genes have strong odd ratio with disease, the lack of statistical power in the identification of contributing genes associated with complex human diseases of inflammation and cancer most likely will not be realized without additional supporting experimental evidence. To increase the power of human GWAS of complex disease, it is critical that further methodologies be employed, such as digital PCR to determine quantitative allele-specific CYP expression with CNV, coupled with proteomic characterization of allele-specific P450 levels and metabolomics or lipidomic metabolites cumulatively will provide a substantial platform to correlate the genotype of CYP orthologous pairs with the phenotype of inflammatory disease and cancer. Confirmation of the genomic, proteomic, and metabolomics approaches in identifying the functional role of variant CYPs in susceptibility to inflammation and cancer can only be conclusively confirmed in human patients. Therefore, patients that have variant CYP alleles and display clinical symptoms and phenotypes would be of substantial value in validating this methodology and its use in identification of CYP role in polygenic complex disease. An ideal CYP system to associate variant CYP alleles with disease would be to use this approach on steroid-metabolizing CYPs, which have well-characterized mutations in several CYPs that closely associate with alteration in clinical values of mineralocorticoids, glucocorticoids, androgen, and estrogens. To date, there are over 120 variant alleles of CYP11A1, CYP11B2, CYP21A2, CYP17A1, CYP19A1 that may increase individual susceptibility to inflammatory disorders and cancer.

I wish to thank all the contributors for their excellent reviews on the functional and pharmacological role of different CYP family members in inflammation and cancer. I also wish to thank Lynn LeCount for her dedication and expertise in handling the submissions. I also thank Dr. Enna for his vision and recognition of the importance of publishing this collection of reviews at an exciting time in identifying the pathophysiological roles of CYPs in human diseases. Finally, special thanks to Elsevier B.V. for recognizing the timeliness of publishing this collection of monographs. I sincerely hope that these thoughtful and comprehensive reviews will further scientific discussion and generate new ideas and excitement among investigators defining the pharmacological and metabolic role of CYP in human diseases.

James P. Hardwick
Department of Integrative Medical Sciences, Northeast Medical University, Rootstown, Ohio, USA

REFERENCES

Divanovic, S., Dalli, J., Jorge-Nebert, L. F., Flick, L. M., Galvez-Peralta, M., Boespflug, N. D., et al. (2013). Contributions of the three CYP1 monooxygenases to pro-inflammatory and inflammation-resolution lipid mediator pathways. *The Journal of Immunology*, *191*(6), 3347–3357. http://dx.doi.org/10.4049/jimmunol.1300699.

Guengerich, F. P., Tang, Z., Cheng, Q., & Salamanca-Pinzon, S. G. (2011). Approaches to deorphanization of human and microbial cytochrome P450 enzymes. *Biochimica et Biophysica Acta*, *1814*(1), 139–145. http://dx.doi.org/10.1016/j.bbapap.2010.05.005.

MacLeod, A. K., Fallon, P. G., Sharp, S., Henderson, C. J., Wolf, C. R., & Huang, J. T. (2015). An enhanced in vivo stable isotope labeling by amino acids in cell culture (SILAC) model for quantification of drug metabolism enzymes. *Molecular and Cellular Proteomics*, *14*(3), 750–760. http://dx.doi.org/10.1074/mcp.M114.043661.

Nebert, D. W., Wikvall, K., & Miller, W. L. (2013). Human cytochromes P450 in health and disease. *Philosophical Transactions of the Royal Society of London. Series B, Biological Sciences*, *368*(1612), 20120431. http://dx.doi.org/10.1098/rstb.2012.0431.

Sakamoto, A., Matsumaru, T., Ishiguro, N., Schaefer, O., Ohtsuki, S., Inoue, T., et al. (2011). Reliability and robustness of simultaneous absolute quantification of drug transporters, cytochrome P450 enzymes, and Udp-glucuronosyltransferases in human liver tissue by multiplexed MRM/selected reaction monitoring mode tandem mass spectrometry with nano-liquid chromatography. *Journal of Pharmaceutical Sciences*, *100*(9), 4037–4043. http://dx.doi.org/10.1002/jps.22591.

Xiao, Y., & Guengerich, F. P. (2012). Metabolomic analysis and identification of a role for the orphan human cytochrome P450 2W1 in selective oxidation of lysophospholipids. *The Journal of Lipid Research*, *53*(8), 1610–1617. http://dx.doi.org/10.1194/jlr.M027185.

CHAPTER ONE

Role of Cytochrome P450 Monooxygenase in Carcinogen and Chemotherapeutic Drug Metabolism

B. Wahlang*, K. Cameron Falkner*, Matt C. Cave*,†, Russell A. Prough†,1

*Division of Gastroenterology, Hepatology, and Nutrition, Department of Medicine, University of Louisville, Louisville, Kentucky, USA

†Department of Biochemistry & Molecular Biology, University of Louisville, Louisville, Kentucky, USA

[1]Corresponding author: e-mail address: raprou01@louisville.edu

Contents

Advances in Pharmacology, Volume 74
ISSN 1054-3589
http://dx.doi.org/10.1016/bs.apha.2015.04.004

Abstract

The purpose of this chapter is to provide insight into which human cytochromes P450 (CYPs) may be involved in metabolism of chemical carcinogens and anticancer drugs. A historical overview of this field and the development of literature using relevant animal models and expressed human CYPs have provided information about which specific CYPs may be involved in carcinogen metabolism. Definition of the biochemical properties of CYP activity came from several groups who studied the reaction stoichiometry of butter yellow and benzo[*α*]pyrene, including their role in induction of these enzyme systems. This chapter will list as much as is known about the human CYPs involved in carcinogen and anticancer drug metabolism, as well as summarize studies with rodent CYPs. A review of three major classes of anticancer drugs and their metabolism in humans is covered for cyclophosphamide, procarbazine, and anthracycline antibiotics, cancer chemotherapeutic compounds extensively metabolized by CYPs. The emerging information about human CYP gene polymorphisms as well as other enzymes involved in foreign compound metabolism provides considerable information about how these genetic variants affect carcinogen and anticancer drug metabolism. With information available from individual's genomic sequences, consideration of populations who may be at risk due to environmental exposure to carcinogens or how to optimize their cancer therapy regimens to enhance efficacy of the anticancer drugs appears to be an important field of study to benefit individuals in the future.

ABBREVIATIONS

AB 4-aminoazobenzene
AHR arylhydrocarbon receptor
B[*α*]P benzo[*α*]pyrene
CAR constitutive androstane receptor
CYP cytochrome P450
DAB *N,N*-dimethylaminoazobenzene
GR glucocorticoid receptor
MAB *N*-methyl-aminoazobenzene
MC methylcholanthrene
P450 cytochrome P450
PAH polycyclic aromatic hydrocarbon
PCB polychlorinated biphenyl
PPAR peroxisome proliferator-activated receptor
PXR pregnane X receptor

1. INTRODUCTION

The major focus of this chapter is to address the role of specific cytochromes P450 in carcinogen and chemotherapeutic drug metabolism. The chapter will cover some historical aspects of the early biochemical studies of carcinogen and chemotherapeutic drug metabolism and provide details defining the role of specific cytochromes P450 in these reactions. While much has been published about specific CYPs in the metabolism of an individual drugs designed by Pharma (Michaels & Wang, 2014; Paine et al., 2006) in liver and intestine, no compilation of the CYPs involved in carcinogen and anticancer drug metabolism exists. Several of the major workers in CYPs and carcinogen metabolism are now deceased and the authors wish to acknowledge James A. Miller, Elizabeth C. Miller, and Ronald W. Estabrook, all whom were helpful to us during our careers.

2. HISTORICAL ASPECTS OF CHEMICAL CARCINOGENESIS LEADING TO DISCOVERY OF THE CYTOCHROME P450 MONOOXYGENASE

Early studies focusing on the enzymologic role of the cytochrome P450 monooxygenase in metabolism of carcinogens, drugs, steroids, and other lipids came from a number of groups, including James A. Miller and Elizabeth C. Miller (McArdle Institute, University of Wisconsin–Madison), Bert N. LaDu, James R. Gillette, and Bernard B. Brodie (NIH), Julius Axelrod (NIH), and David O. Cooper, Ronald W. Estabrook, and Otto Rosenthal (University of Pennsylvania) who defined the biochemical properties of hepatic enzymes involved in foreign compound metabolism. In reviewing the literature in this area, we recommend the article provided by James A. Miller in *Drug Metabolism Reviews* in 1998 (Miller, 1998), due to its coverage of the history and perspectives of Dr. Miller. In this chapter, we will review some of the more salient points in azo dye and polycyclic aromatic hydrocarbon (PAH) metabolism discovered at that time and then define major metabolic routes for anticancer drugs. The work performed in the 1940 and 1950s was focused on biochemical knowledge that was available at that time when only enzymes and proteins were being extensively studied. This was a time prior to the discovery of DNA by Watson and Crick (1953).

2.1 Cytochromes P450

The cytochrome P450 monooxygenases are a heme-thiolate proteins (cysteine serving as a 5-coordination ligand to the heme group of these proteins) that serve as a powerful oxidizing catalysis for many fat-soluble compounds (Coon, 2005). The reaction catalyzed by these heme proteins is that of a monooxygenase (Eq. 1). In some CYPs, such as the well-characterized bacterial CYP101, the stoichiometry is strictly maintained, while in

$$SH + O_2 + NADPH + H^+ \rightarrow S - OH + NADP + H_2O \qquad (1)$$

others, such as CYP3A, a mixed-function oxidase reaction is observed, in which approximately 60% of the electrons in the reaction obey the stoichiometry above, while the remaining electrons go to unproductive catalysis in the form of superoxide anion radical or hydrogen peroxide production. The CYPs are part of a superfamily of heme-thiolate monooxygenases, and more than 21,000 CYP genes have been identified in the biome to date (Gonzalez & Yu, 2006). The human genome contains 57 CYP genes, some with defined reactions involving endogenous compounds and many metabolizing foreign compounds. The catalytic mechanism of these oxidizing enzymes involves higher oxidation states of the heme iron, cycling from its ferric form, to the ferryl form, to the perferryl form, and back to the ferric form (Coon, 2005). These reactions involve oxidation of aliphatic and aromatic compounds, as well as nitrogen and sulfur functional groups. Due to the importance of this heme protein superfamily to medicine, the environment, and pharmacological sciences, one will observe over 80,780 references in PubMed for this enzyme on March 31, 2015 on the topic "cytochrome P450."

2.2 Butter Yellow Carcinogenesis

Butter yellow (*N,N*-dimethylaminoazobenzene or DAB) was first noted to be carcinogenic in rats by Japanese researchers (Sasaki & Yoshida, 1935), and an initial study of its metabolism by rat liver homogenates was performed by Stevenson, Dobriner, and Rhoad (1942). Because this compound had been used in Europe to color lipid extracts in consideration of its production as a coloring for edible fats (butter yellow color), the Millers' first joint research project was focused on DAB so that the carcinogenicity of such colored dyes would be better understood (Miller, 1998).

After repeating the carcinogenesis study with DAB (Sasaki & Yoshida, 1935), the Millers noted that the liver tissue from rats treated with DAB displayed a yellow color and extensive perfusion and organic extraction of the proteins did not remove the yellow color. They set out to characterize the protein covalent adducts they deduced were formed *in vivo*, using enzymatic and alcohol/base hydrolysis to free the metabolites, *N*-methylaminoazobenzene (MAB) and 4-aminoazobenzene (AB) (Miller, 1998) from protein. This result supported the suggestion of Stevenson et al. (1942) that DAB can be sequentially demethylated in liver homogenates to MAB and AB. The report from the Millers documented protein covalent binding during metabolism of the chemical carcinogen based upon retention of the yellow color of the parent molecule with the protein fraction and subsequently, they demonstrated this same residue in the chemical adduct bound to DNA (Lin, Miller, & Miller, 1975). Returning to carcinogenicity testing, they were surprised that while DAB and MAB were equally carcinogenic in rats, AB was not a carcinogen. Their subsequent studies demonstrated that covalent binding of DAB and MAB to hepatic protein *in vivo*, while AB did not appreciably bind to hepatic protein. They also observed that there were few DAB/MAB protein adducts in the tumor tissue itself, and in 1947 they initially proposed a protein adduct-based mechanism of cancer formation (Miller & Miller, 1952). In their characterization of the protein adduct, they used a reducing reagent followed by steam distillation to demonstrate that aniline was a degradation product, suggesting that the intact aniline portion of the DAB molecule existed in the protein adduct. At this time, many others were reporting protein–carcinogen adducts, and the interpretation of protein–carcinogenic adducts of DAB (Lin et al., 1975), benzo[*α*]pyrene (Koreeda et al., 1978), dibenz[*ah*]anthracene (Mlcoch, Fuchs, Oesch, & Platt, 1993; Wiest & Heidelberger, 1953), 2-acetylaminofluorene (Evans, Miller, & Beland, 1980; Weisburger, Weisburger, & Morris, 1953), and napthylamine (Kadlubar, Unruh, Beland, Straub, & Evans, 1980) were important observations for understanding the mechanism of chemical carcinogenesis at that time (Table 1).

As James Miller pointed out in his *Drug Metabolism Reviews* article (Miller, 1998), the focus on protein–carcinogen adducts was a matter of the technology available to study carcinogen metabolism and reactivity. McCarty and Avery (1946) provided strong data that nucleic acids encode genetic material, but the lack of radiolabeled substrates and sophisticated analytic methods prevented further study of carcinogen metabolism to DNA-binding reactive intermediates. The Millers and others did eventually perform experiments leading to the realization that covalent binding to DNA was also an endpoint of carcinogen metabolism (Table 1). The

Table 1 Covalent Adducts of Chemicals Documented First as DNA–Carcinogen Adducts *In Vivo*

Carcinogen	DNA Adduct	Reference
N,*N*-Dimethylaminoazobenzene	*N*-(Guanosin-8-yl)-4-monomethyl-aminoazobenzene	Lin et al. (1975)
Benzo[*a*]pyrene	(+)-7β,8α-Dihydroxy-9α,10α-epoxy-7,8,9,10-tetrahydro-B[α]P-deoxyguanosine	Koreeda et al. (1978)
Dibenz[*a*,*h*]anthracene	(*anti*)-3*S*,4*R*-Dihydroxy-1*R*,2*S*-epoxy-1,2,3,4-tetrahydrodibenz[*a*,*h*]anthracene-deoxyguanosine	Mlcoch et al. (1993), Wiest and Heidelberger (1953)
2-Acetylaminofluorene	*N*-(Deoxyguanosin-8-yl)-2-acetylaminofluorene and *N*-(deoxyguanosin-8-yl)-2-aminofluorene	Evans et al. (1980), Weisburger et al. (1953)
Naphthylamine (NA)	1-(Deoxyguanosin-*N*2-yl)-2-NA, 1-(deoxyadenosin-*N*6-yl)-2-NA, and 1-[5-(2-6-diamino-4-oxopyrimidinyl-*N*6-deoxyriboside)]-2-(2-naphthyl) urea	Kadlubar et al. (1980)

availability of isotopic-labeled substrates has allowed the rapid expansion of the study of covalent DNA adducts as seen by the work of Lin et al. (1975).

With the advent of radiolabeled carcinogen substrates from commercial or readily available sources, the door opened to more sophisticated biochemical studies. In addition, the availability of commercially available ultracentrifuges and high-speed centrifuges allowed the separation of various organelles from tissues, such as liver. With these tools, the focus of biochemists moved from looking at proteins to their catalytic reactions. This effort provided a puzzling feature of the CYPs, in that they were polysubstrate enzymes, having many possible substrates upon which each act catalytically, relative to other metabolic enzymes with rigid substrate specificity. Within a few years, groups working on chemical carcinogens (Millers, Heidelburger, Weisburger), drugs (Brody, LaDu, Gillette), neurotransmitters (Axelrod), and sterols (Cooper, Estabrook, Rosenthal, Boyd) all noted that the metabolism of each of their compound of interest had a defined stoichiometry, namely, that of a mixed-function oxidase or oxygenase reaction and likely utilized CYPs as a catalyst. One of the earliest reports on the stoichiometry of foreign compound metabolism is that of the Millers looking at DAB metabolism in rat liver homogenates. The conversion of DAB to MAB, AB, and a ring-hydroxylated metabolite required an atmosphere of molecular oxygen and a source of reducing equivalents (NADH or NADPH) (Mueller & Miller, 1949). In a subsequent experiment, they demonstrated that carbon monoxide could inhibit this reaction catalyzed by rat liver microsomes (Mueller & Miller, 1950). This inhibitory effect of carbon monoxide was later linked to the heme protein, cytochrome P450, by Cooper, Levin, Narasimhulu, and Rosenthal (1965), who showed that CO served as an inhibitor by coordinating with the 6-coordination position of heme in P450 and its inhibitory effect could be reversed by illumination with light at 450 nm. In this manner, similar studies were performed by many others on their favorite carcinogen, drug, or sterol substrates providing the literature that defined the role of CYPs in carcinogen metabolism (Simpson & Boyd, 1967).

2.3 PAH Carcinogenesis: Protein Adducts and Induction of Metabolism

In her independent research as a junior faculty member, Elizabeth Miller showed that benzo[α]pyrene was also metabolized in the liver using the fluorescence of the benzo[α]pyrene molecule (Miller, 1951). Her initial studies lead the Millers to begin study of benzo[α]pyrene (B[α]P) metabolism and its role in carcinogenesis. Their interest was simplified by the fact that to study B[α]P metabolites (Miller, 1951), one could easily follow the increased fluorescence of its metabolites. This led Alan Conney, their graduate student, to observe that pretreating rats with B[α]P several days prior to preparing liver homogenates also resulted in higher rates of product formation of fluorescent metabolites (Conney, Miller, & Miller, 1957) in treated versus untreated liver. This observation led them to study the effects of pretreatment of rats with PAHs on metabolism of both *N,N*-dimethylaminoazo-benzene and benzo[α]pyrene (Conney, Brown, Miller, & Miller, 1957; Conney, Miller, & Miller, 1956; Conney, Miller, et al., 1957). These studies were among the earliest addressing the possibility that foreign chemicals, including chemical carcinogens, may induce their own metabolism. Conney, the Millers, and many others spent most of the 1960s cataloguing whether or not animal pretreatment with PAHs, barbiturates, and many other drugs caused the metabolism of xenobiotics to be increased. Similar studies were applied to the enzymes that conjugate drug/carcinogen metabolites to glucuronic acid and sulfate. These observations were of keen interest to academics like the Millers and pharmaceutical scientists like Conney and the group at Hoffmann-La Roche in Nutley, NJ. The process of induction was masterfully reviewed by Conney in *Pharmacological Reviews* in 1970 (Conney, 1967) and provided the impetus for the pharmacological companies world-wide to pursue research on drug metabolism and disposition of their compounds.

Another facet of this area is the superb chemical work of several groups to deduce the structure of major adducts of carcinogens (Koreeda et al., 1978; Yagi et al., 1977). As shown in Table 1, the chemical nature of major carcinogen–DNA adducts has been characterized and well documented in the literature, adding to the understanding of the molecular mechanism of somatic cell mutations in rodent and human carcinogenesis. A final area to study has been the enzymes that catalyze these deleterious reactions. With the advent of procedures allowing purification of the cytochromes P450 first described by Lu and Coon (1968), one can relatively easily determine which form of the cytochrome metabolizes a given drug, carcinogen, or sterol using recombinant proteins that can be obtained commercially. Many of the studies elucidating the role of specific CYPs in metabolism that were published in the period between 1980 and 1990 utilized *in vitro* experiments to ascertain the relative catalytic activity of newly purified CYPs reconstituted with NADPH:cytochrome P450 oxidoreductase and form-specific inhibitors have been reported for major CYPs. Subsequently, bacterial or yeast cells expressing specific recombinant CYPs, plus NADPH:cytochrome P450 oxidoreductase, have been used, after the gene sequences for CYPs were determined. While much was done with CYPs purified from native membranes, this approach has been particularly fruitful for study of transporters and other enzymes of foreign compound metabolism.

3. CYTOCHROME P450-MONOOXYGENASE-MEDIATED ACTIVATION OF PROCARCINOGEN AND CHEMOTHERAPEUTIC PRODRUGS

The next section will review the form-specific metabolism of various carcinogens and chemotherapeutic compounds. In this review, we will focus on a limited set of carcinogens that provide the background for bioactivation of chemicals as chemical carcinogens and toxicants. We will also focus on specific CYPs involved in specific chemicals metabolism to carcinogens (Table 2), their ability to be induced by drugs or other compounds, and the cancer process.

3.1 Aromatic Azo Compounds

Through the work of the Millers, it became clear that treatment of animals with 3-methylcholanthrene (3MC) resulted in increased DAB metabolism by some enzyme induction process. We now know that members of the CYP1A family

Table 2 Human CYPs Believed To Be Associated with Metabolism of Chemical Carcinogens and Anticancer Agents

Carcinogen/Anticancer Agent	Metabolite Formed	Rodent CYPs	Human CYPs
N,N-Dimethyl-aminoazobenzene	DAB MAB	CYP1A1, CYP2B1	CYP1A1, CYP2B6
Benzo[α]pyrene	Phenols/quinones Epoxides/dihydrodiols	Many CYP2B1 CYP1A1	CYP2B6 CYP1A1
Polychlorinated biphenyls	Low chlorine content High chlorine content	CYP1A1, CYP2B1	CYP1A1, CYP2B6
Vinyl chloride	Chloroethylene oxide	CYP2E1	CYP2E1
Nitrosamines	Dimethylnitrosamine Tobacco-specific nitrosamines	CYP2E1, CYP2B1 CYP2B1, CYP3A, CYP2A	CYP2E1, CYP2B6 CYP2B6, CYP3A4, CYP2A6
Furans	Furan Aflatoxin B1	CYP2E1 CYP3A, 1A2, 2A13	CYP2E1 CYP3A, 1A2, 2A13
Cyclophosphamide	Phosphoramide mustard	CYP2B1, 2C6, 2C11	CYP2B6, 2C9, 3A4
Procarbazine	Azoprocarbazine Azoxyprocarbazine	CYP1A1, CYP2B1, CYP2C6 CYP1A1, CYP2B1	
Anthracycline antibiotics	Alkylaminoanthraquinone di-*N*-oxide	CYP3A1	CYP3A4

are induced through arylhydrocarbon receptor (AHR) to increase metabolism. Raza and Levine demonstrated that both β-naphthoflavone (an AHR agonist) and phenobarbital (PB) (a constitutive androstane receptor or CAR agonist) induced DAB demethylation reactions (Raza & Levine, 1986b). In light of our current knowledge, these observations implicated members of the CYP1 and CYP2 or three families in DAB and MAB demethylation. In a subsequent study, Levine and Lu (1982) demonstrated that partially purified CYP fraction from livers of 3MC-treated rats displayed high levels of demethylase activity. Their studies measured the demethylation of DAB to MAB and MAB to AB, as well as 4′-hydroxylation of the phenyl ring. The partially purified CYP fraction from rat liver microsomal protein after 3MC induction was shown to contain CYP1A1 as the catalyst, which displayed both greater specific enzyme activity toward demethylation of DAB than MAB and significant 4′-hydroxylation of DAB. The major enzyme of P450 in liver microsomal fractions from PB-treated rats was noted by Levine and Lu to demethylate DAB more rapidly than MAB. Based upon our current knowledge, the fact that it was induced over fourfold by PB treatment and was highly inhibited by metyrapone suggested it may have utilized the CYP2B1 enzyme. This CYP also catalyzed ring hydroxylation of DAB. The activity seen in liver microsomal fractions from untreated rats must apparently be due to a constitutive cytochrome P450, such as rat CYP2C11 or CYP3A1.

The elegant work of Kadlubar and the Millers provided a definitive mechanism for the bioactivation of MAB, namely MAB is N-hydroxylated and subsequently sulfated to form a powerful electrophilic intermediate (Kadlubar, Miller, & Miller, 1976; Kadlubar et al., 1980). This reactive intermediate accounts for the potent protein and DNA alkylation caused by the MAB reactive electrophile. Because the CYPs in rats often have similar enzymatic properties as the orthologous human CYPs, we would anticipate that CYP1A1 and CYP2B6 would be major enzymes in human liver metabolizing DAB and MAB, respectively (Table 2), and their associated activities should be inducible with AHR and CAR ligands *in vivo*, respectively. In human, CYP3A4/3A5, CYP2C9, and CYP2C19 may also metabolize MAB and DAB, because there is appreciable constitutive demethylase enzyme activity due to CYPs, but no experiments have been performed to directly define which of these human P450s are involved in DAB and MAB metabolism.

Another route of metabolism for aminoazobenzene compounds is azo reduction as shown in 1957 by Fouts et al. in which neoprontosil and many other azo drugs are metabolized microsomal enzymes and required NADPH for the reaction (Fouts, Kamm, & Brodie, 1957). Levine and Raza (1988) and Raza and Levine (1986a) demonstrated that this reaction azo reduction pathway does not lead to reactive intermediates for disubstituted azo compounds, like DAB. They did note that probably CYP4A1 in rat, and likely CYP4A11 in humans from our current understanding, catalyzed the heme protein-dependent reduction of DAB. This reaction is also catalyzed by NADPH:cytochrome P450 oxidoreductase alone in a nonheme protein-dependent reaction. Most of these studies did not provide evidence for specific cytochromes P450 involved in these reduction reactions, except those cases where liver microsomes from PB- or 3MC-pretreated rodents could be associated with the known CYP1A or CYP2B activities (Mallett, King, & Walker, 1985). These reactions have been considered to lead to detoxification of these compounds, not bioactivation, although toxicity can result if redox cycling occurs with flavoproteins like NADPH:cytochrome P450 oxidoreductase.

3.2 Polycyclic Aromatic Hydrocarbons

The metabolism of PAHs by purified P450s has been extensively studied by the group of researchers who worked with Alan Conney at Hoffmann-La Roche in Nutley, NJ (Holder et al., 1974, 1975; Thakker et al., 1977; Wood et al., 1976) or in the collaboration between M. Judd Coon and Harry Gelboin (Wiebel et al., 1975). These investigators utilized high-pressure liquid chromatography to separate classes of B[α]P metabolites (Selkirk, Croy, & Gelboin, 1974; Thakker et al., 1977; Wood et al., 1976), such as phenols, quinones, or dihydrodiols, namely, the 7,8- and 9,10-dihydrodiol. The elegant chemical studies on the stereospecific metabolism of B[α]P to DNA reactive species by the Jerina group (Holder et al., 1974; Thakker et al., 1977) served as a guiding factor in definition of an ultimate carcinogenic form of the PAH and guided research in the study of many other PAH compounds.

Initially, the work documented the fact that the CYP1 family of enzymes and epoxide hydratase are critically involved in production of the position-specific oxidative reaction products leading to the dihydrodiol epoxide intermediates of benzo[α]pyrene. These were subsequently shown to lead to protein and DNA adducts (Kapitulnik, Levin, Conney, Yagi, & Jerina, 1977; Koreeda et al., 1978). This family of CYPs is induced by PAH, thereby leading to induction of their own metabolism when an organism is exposed to high levels of these compounds. Members of the CYP Family

2 and 3 do not preferentially form dihydrodiols and dihydrodiol epoxide derivatives at high levels (Table 2), but they do metabolize B[α]P to a series of phenolic and quinone products, some of which are carcinogenic (Kapitulnik, Levin, Yagi, Jerina, & Conney, 1976; Wood et al., 1976; Yang, Selkirk, Plotkin, & Gelboin, 1975), because they can be further metabolized to reactive intermediates (Capdevila, Estabrook, & Prough, 1978b; Prough, Saeki, & Capdevila, 1981). Because B[α]P quinones can be reduced by CYPs to diphenols (Capdevila et al., 1978a), these products, like the phenolic products formed, can undergo sulfation and glucuronidation for elimination. Therefore, induction of this second set of CYPs by drugs may lead to detoxification by forming phenol and quinone conjugates for excretion.

3.3 Polychlorinated Biphenyls

3.3.1 Polychlorinated Biphenyls with Low Chlorine Content

Polychlorinated biphenyls (PCBs) with low chlorine content have been observed to be readily metabolized by CYPs, while those with higher chlorine content are poorly metabolized. This is believed to be due to the position of the chlorine atom in the structure and its possible steric hindrance leads to diminished occupancy of the heme protein active site, preventing productive oxygenation of the biphenyl ring (Grimm et al., 2015). For this reason, the more chlorinated PCBs bioaccumulate in fatty tissues, like adipose, while the less chlorinated PCBs are cleared by metabolism and result in much lower body burdens for such PCB compounds. For the low chlorine content PCBs, both rat CYP1A1 and 2B1 have been shown to metabolize these compounds (Table 2). Rat CYP2B1 preferentially metabolizes noncoplanar dichlorobiphenyls, while rat CYP1A1 metabolizes coplanar dichlorobiphenyls. The studies of Kaminsky, Kennedy, Adams, and Guengerich (1981) showed that these two P450s exhibit differential regioselectivity for dichlorinated biphenyls, with CYP2B1 enzyme preferring unchlorinated phenyl rings and CYP1A1 enzyme preferring chlorinated rings. Both CYPs metabolized monochlorinated biphenyls. The conclusion of Kaminsky et al. (1981) was that the rates of metabolism and the regioselectivity of metabolism are dependent upon which CYP enzyme is induced, and that the toxicity and carcinogenicity may also correlate with CYP enzyme involved. In the case of human enzymes, CYP1A1 and CYP2B6 have been shown to metabolize 2,4,5,2′,4′,5′-hexachlorobiphenyl and other PCBs (Ariyoshi, Oguri, Koga, Yoshimura, & Funae, 1995; Grimm et al., 2015; Warner, Martin, & Wong, 2009), suggesting humans metabolize PCBs in a similar manner as the rat (Table 2). Recently, a role for the CYP2C family enzymes in PCB metabolism has been shown with rat, dog, and human P450s (Ariyoshi, Ito, et al., 1995), suggesting that some of the constitutively expressed P450s of the CYP2C family may also metabolize PCBs.

3.3.2 Highly Chlorinated Biphenyls

As noted earlier, more highly chlorinated PCBs tend to bioaccumulate in fatty tissues because they are not readily metabolized by CYPs. It is known that significant differences in toxicity exists as a function of chlorination level, and the mechanism of toxic action may diverge depending on whether the metabolites are chemically reactive intermediates for low chlorinated PCBs, or whether the highly chlorinated PCBs may cause specific receptor-mediated toxicities. Because PCBs have been considered to be high-affinity agonists for receptors like the AHR, this bioaccumulation may result in sufficiently high levels of PCB to increase metabolism through AHR or other nuclear receptors. In our recent studies, we have shown that some highly chlorinated PCBs activate CAR and pregnane X receptor (PXR) (Wahlang, Falkner, et al., 2014; Wahlang, Song, et al., 2014). Because no reactive intermediates are apparently formed by these PCBs, different mechanisms of toxicity may exist, such as receptor activation through ionotrophic receptors, G-coupled receptors, kinase-linked receptors, or nuclear receptors.

3.4 Vinyl Chloride

Vinyl chloride is high production volume industrial chemical produced for manufacturing synthetic rubber and plastics. It has been shown that CYP2E1 is involved in vinyl chloride metabolism to chloroethylene oxide, a known DNA-alkylating agent (el Ghissassi, Barbin, & Bartsch, 1998), and liver cancer and toxicity including hemangiosarcoma incidents were reported in vinyl chloride workers in Louisville, KT (Creech & Johnson, 1974; Falk, Creech, Heath, Johnson, & Key, 1974; Makk, Creech, Whelan, & Johnson, 1974). Noteworthy from the study by el Ghissassi et al. (1998) was the demonstration that this compound which has an extremely low boiling point/vapor pressure serves as a substrate for microsomal cytochrome P450 and exhibits a low Michaelis constant for the enzyme-mediated reaction. The low K_m allows gaseous vinyl chloride with limited solubility to be dissolved at sufficient concentrations to allow the CYP2E1 to catalyze the oxidation reaction in solution. A second reactive metabolite is formed during the decomposition of chloroethylene oxide to 2-chloroacetaldehyde, which is a protein-alkylating agent, but not as effective as chloroethylene oxide in forming DNA adducts (Bolt, 1986). As a human carcinogen, it is known that individuals with polymorphisms in CYP2E1, glutathione *S*-transferase tau, and aldehyde dehydrogenase 2 all have higher odds ratio for cancer than wild-type individuals (Wong et al., 2002; Wong, Wang, Hsieh, & Cheng, 2003), suggesting that not only is CYP2E1 important for vinyl chloride bioactivation (Table 2), but lack of two other detoxification enzymes, glutathione *S*-transferase tau and aldehyde dehydrogenase 2, are critical for clearance of the two metabolites, chloroethylene oxide and 2-chloroacetaldehyde. Studies by Yang, Yoo, Ishizaki, and Hong (1990) have provided support that the human CYP2E1 also catalyzes this epoxidation reaction and accounts for the bioactivation of this human carcinogen.

3.5 Nitrosamines

Early studies on metabolism of *N,N*-dimethylnitrosamine documented a role for CYP2E1 and CYP2B1 in the bioactivation of this compound (Lorr, Tu, & Yang, 1982; Yamazaki, Oda, et al., 1992; Yang et al., 1990). As the array of purified CYPs available for research became larger and the spectrum of nitrosamines increased (Table 2), other P450s were shown to be involved (Kokkinakis, Koop, Scarpelli, Coon, & Hollenberg, 1985; Shu & Hollenberg, 1996; Yamazaki, Inui, Yun, Guengerich, & Shimada, 1992). Small molecule nitrosamines were excellent substrates for CYP2E1, but less so for nitrosamines containing larger substituents, such as, *N*-nitrosodialkyl amines and *N*-nitroso-2,6-dimethylmorpholine, the more complex tobacco-related nitrosamines (Kokkinakis et al., 1985; Shu & Hollenberg, 1996; Yamazaki, Oda, et al., 1992). For the larger nitrosamines, CYP2B1, CYP3A4, and CYP2A6 were found to play important roles in nitrosamine metabolism (Patten et al., 1992, 1997). CYP2B1 is a PB-inducible P450 and may not play a role in metabolism in uninduced animals, but CYP2A enzymes have increasingly been shown to be important in nitrosamine metabolism and carcinogenesis. More recently, the importance of human P450s has been shown, namely CYP2A6 (Raunio, Rautio, Gullsten, & Pelkonen, 2001) and CYP2A13 (He, Shen, Ding, Lu, & Hong, 2004; Megaraj et al., 2014; Zhou, D'Agostino, Xie, & Ding, 2012). The tissue-specific expression of these CYP2A enzymes in nasal epithelium most likely accounts for nitrosamine carcinogenesis in those tissues.

3.6 Furans

Furan and related compounds are known to be carcinogenic in rodents and it is known that CYPs are involved in their bioactivation (Peterson, 2006). These compounds are hepatoxic and furan's route of metabolism by CYP2E1 is believed to proceed via a reactive α,β-unsaturated dialdehyde, *cis*-2-butene-1,4-dial (Gates, Lu, & Peterson, 2012). Other CYPs were not as active a catalyst as CYP2E1, but for larger molecular with a furan ring, such as aflatoxin B1, their metabolism has been shown to require CYP3A (Ueng, Shimada, Yamazaki, & Guengerich, 1995), CYP1A2 (Ueng et al., 1995), and CYP2A13 (Zhang et al., 2014) as a major catalyst for oxidation (Table 2). Because lung is a major site of tumorigenesis for aflatoxin compounds, the role of CYP2A13 is highly important in that tissue (Zhang et al., 2014; Zhu et al., 2006), while in liver CYP3A4 may be important for carcinogenesis.

3.7 Chemotherapeutic Prodrugs

3.7.1 Cyclophosphamide

Cyclophosphamide is a drug that upon bioactivation becomes an alkylating agent with a high therapeutic index and a broad spectrum of activity against human tumors (Sladek, 1988). The metabolism of cyclophosphamide has been extensively studied. The compound is deactivated in a CYP-catalyzed reaction by N-dechloroethylation to yield dechloroethylated oxazaphosphorine, and activated by 4-hydroxylation to form phosphoramide mustard, the ultimate anticancer agent, and acrolein. While not therapeutically active, some of the side toxicities of the anticancer drug apparently may be ascribed to dechlorinated oxazaphosphorine and acrolein. Waxman and coworkers have provided a detailed analysis of the specific CYPs involved in cyclophosphamide metabolism in the rat (Yu & Waxman, 1996), using inducing agents and inhibitors specific for subfamilies of P450. For the 4-hydroxylation pathway, CYP2C6 and CYP2C11 appeared to be the most catalytic active CYP in liver microsomes from untreated rats, while PB treatment allowed association of CYP2B1 and increased cyclophosphamide metabolism (Table 2). Dexamethasone pretreatment of rats also increased liver microsomal metabolism of cyclophosphamide and this activity was most sensitive to inhibitors of CYP2B and CYP2C6/2C11. For the N-dechloroethylation pathway, CYP3A was the major CYP isozyme in untreated rat or rats treated with PB or dexamethasone. Similar experiments were performed with human CYPs or isolated human hepatocytes. Cyclophosphamide metabolism by expressed human CYP2A6, -2B6, -2C8, -2C9, and -3A4 in *E. coli* was noted with CYP2C8/2C9 displaying a much lower K_m for cyclophosphamide than the others (Chang, Weber, Crespi, & Waxman, 1993). This observation explained the complex biphasic kinetic properties seen for cyclophosphamide using human liver microsomal fractions. However, CYP1A1, -1A2, -2D6, and -2E1 did not display enzyme activity at higher substrate concentrations for cyclophosphamide metabolism. An interesting observation for this anticancer drug has been the fact that cyclophosphamide itself serves as an inducer for cyclophosphamide metabolism and CYP2B6, CYP2C9, and CYP3A4 enzyme activities (Chang, Yu, Maurel, & Waxman, 1997; Lindley et al., 2002). The role of cyclophosphamide serving as a CAR ligand activator was not appreciated until after this observation in humans. Because cyclophosphamide has been used in treatment of a number of cancers and autoimmune diseases, its success as a drug depends both on its metabolism to a toxic intermediate and its ability to induce its own metabolism. Factors regulating basal expression of CYP2B6 expression, such as obesity (Zannikos, Bandyopadhyay, Robertson, & Blouin, 1994), may alter cyclophosphamide metabolism and possibly therapeutic efficacy.

3.7.2 Procarbazine and Other Hydrazine Derivatives

Like cyclophosphamide, procarbazine (Matulane) was discovered in a screening activity for high-dose chemotherapy by the National Cancer Institute in the 1960s. This compound, *N*-isopropyl-4-[(2-methylhydrazino)methyl] benzamide, has been used for treatment of Hodgkin's lymphoma and brain cancers, such as glioblastoma multiforme. Its mechanism of action appears to be that of a methylating agent and as such, procarbazine causes strand breaks in DNA. Its mechanism of action depends upon its oxidation to an azoxy derivative via the azo intermediate. Both reactions utilize a CYP enzyme, although procarbazine itself is somewhat easily air oxidized. Oxidation of procarbazine is catalyzed by liver microsomes from untreated and PB-treated rats, although the level of induction by PB treatment was only 1.8-fold for procarbazine oxidation, while metabolism of a CYP2B substrate would have been around 8-fold (Dunn, Lubet, & Prough, 1979). The reaction was inhibited by metyrapone and SKF-525A, indicating that the CYP enzymes involved were most likely a member of the CYP2 or CYP3 family. Under the conditions of this enzyme assay, chemical oxidation was not appreciable, unless EDTA was omitted from the reaction mixture.

The second oxidative step in procarbazine metabolism to two isomeric azoxy metabolites *N*-isopropyl-α-(2-methyl-ONN-azoxy)-*p*-toluamide (methyl-azoxy) and *N*-isopropyl-α-(2-methyl-NNO-azoxy)-*p*-toluamide (benzyl-azoxy) has been extensively studied by Yost and our group (Swaffar et al., 1989; Wiebkin & Prough, 1980). In our hands, the methyl-azoxy metabolite was formed stereoselectively and preferentially by control liver microsomal fractions or microsomal fractions from β-naphthoflavone- (CYP1A inducer) treated rats. Liver microsomes from PB-treated rats also produced the benzyl-azoxy metabolite. The azo derivative of procarbazine displayed a K_m value of around 50 μM, indicating that it is a good substrate for the constitutive and β-naphthoflavone-induced CYPs (Table 2). When PB was used as an inducing agent, both methyl and benzyl-azoxy metabolites were formed in a 5:1 ratio in a carbon monoxide-sensitive reaction requiring NADPH and molecular oxygen. Inhibitors of CYPs, chemicals and anti-NADPH:cytochrome P450 oxidoreductase globulin, blocked the enzymatic reaction. Purified rat CYP2C6 appears to be the major constitutive P450 involved, while CYP2B1 was a minor contributor (Prough, Brown, Dannan, & Guengerich, 1984). An antibody against rat CYP1A1 inhibited the reaction catalyzed by the major β-naphthoflavone-induced CYP. Both groups reported that the methyl-azoxy derivative could be bioactivated to cytotoxic and protein-alkylating agents. Methane and *p*-formyl-*N*-isopropylbenzamide were also shown to be metabolites formed during CYP metabolism (Moloney & Prough, 1983; Moloney, Wiebkin, Cummings, & Prough, 1985). Using sensitive mass spectral analysis (Swaffar, Harker, Pomerantz, Lim, & Yost, 1992), Yost's group also noted that chemical oxidation could convert procarbazine and azoxy-procarbazine to reactive intermediates, suggesting that chemical oxidation as well as enzyme-mediated reactions may produce reactive chemical species, perhaps accounting for procarbazine toxicity.

3.7.3 Anthracycline Antibiotics

A subtle role of CYPs has been noted in the metabolism and disposition of these nonspecific, cell-cycle inhibitor drugs used in cancer chemotherapy. These compounds are derived from *Streptomyces bacterium*, including daunorubicin (daunomycin), doxorubicin (adriamycin), epirubicin, idarubicin, valrubicin, and mitoxantrone. Their utility as anticancer drugs is considered by some to be limited due to their cytotoxicity. However, doxorubicin is used as a standard in several cancers, such as B cell lymphoma, leukemias, and others. A common chemical feature of these compounds is the anthracycline ring and their mechanism of action appears to involve at least four or more modes of action: (1) Inhibition of DNA and RNA synthesis by intercalating between base pairs of the DNA/RNA strand, (2) inhibition of topoisomerase II enzyme, (3) iron-mediated generation of free oxygen radicals that damage the DNA, proteins, and cell membranes, and (4) induction of histone displacement from chromatin altering transcriptional regulation.

Being compounds with a quinone structure, researchers presumed that anthracycline antibiotics may undergo redox cycling for their mode of action. Some like methoxymorpholinyl doxorubicin required CYP3A bioactivation to its reactive form, a fact also noted for adriamycin (Goeptar, te Koppele, Lamme, Pique, & Vermeulen, 1993), mitomycin C (Goeptar et al., 1993), methoxymorpholinyl doxorubicin (Lewis, Lau, Duran, Wolf, & Sikic, 1992; Lu & Waxman, 2005), and alkylaminoanthraquinone di-*N*-oxide (AQ4N) (Raleigh, Wanogho, Burke, & Patterson, 1999). In the case of these four anthracycline compounds, conversion of a prodrug form to a more reactive intermediate required NADPH, anaerobic conditions, and CYP2B or CYP3A to potentiate the anticancer activity, i.e., converting the prodrug to the active drug. Inhibitors of CYP2B and 3A all prevented the conversion of the prodrug to its active form (Table 2). Raleigh, Wanogho, Burke, McKeown, and Patterson (1998) demonstrated a 2- and 4-electron reduced form of the *N*-oxide functional groups on AQ4N allowing the formation of a more easily redox cycled cytostatic form of the drug. They subsequently demonstrated that it was a reductive substrate for CYP3A4 in the human as well. One can envision similar reduction reactions for adriamycin and doxorubicin derivatives as part of the mechanism of action. A role for NADPH:cytochrome P450 oxidoreductase is also implicated in the case of some anthracycline compounds, suggesting that CYP3A is critical for the first reduction step of bioactivation and reductive enzymes (such as NADPH:cytochrome P450 oxidoreductase, NADPH oxidase, and xanthine oxidase) for subsequent reduction reactions, in order to sustain a robust redox cycling process leading to DNA strand breaks, depletion of GSH, and other redox-sensitive processes (Pawlowska, Tarasiuk, Wolf, Paine, & Borowski, 2003).

Interestingly, CYP3A family gene products appear to be able to serve as reductants for other quinones and α,β-unsaturated compounds. For example, Capdevila et al., 1978a, 1978b showed that benzo[*α*]pyrene-3,6-quinone is reduced by rat liver microsomes in a reaction that is sensitive to CYP inhibitors including anti-NADPH:cytochrome P450 oxidoreductase globulin, but not CO, dicoumarol (NAD(P)H:quinone oxidoreductase inhibitor), molecular oxygen, or strict anaerobiosis. The product formed is the 3,6-diphenol which can serve as a substrate for UDP glycosyltransferases. In addition, we have shown that 4-hydroxynonenal can be reduced to 1,4-dihydroxynonene (Amunom et al., 2011) and oxidized to 4-hydroxynonenoic acid (Amunom et al., 2007). Both reactions were catalyzed by human CYP3A4, human CYP2B6, and murine CYP2c29. As seen with the anthracycline derivatives, it appears that a 2-electron reduction occurs to form the stable dialcohol, an unusual reaction for a CYP enzyme.

4. FACTORS THAT REGULATE CHANGES IN GENE EXPRESSION: DIFFERENCES BETWEEN RODENTS AND HUMANS

4.1 Hepatic Nuclear Receptors and Enzyme Induction

Regulation of cytochrome P450 and many other enzymes involved in foreign compound metabolism is known to involve a variety of cellular receptors. Many are members of the steroid hormone nuclear receptor family (Conney, 2003; Evans & Mangelsdorf, 2014), while others are members of the basic helix-loop-helix transcription family (AHR) (Denison & Nagy, 2003) or basic leucine zipper (bZIP) protein family (nuclear factor (erythroid-derived 2)-like 2, Nrf2) (Mukaigasa et al., 2012; Taguchi, Motohashi, & Yamamoto, 2011). Table 3 summarizes these receptors involved in the induction of CYPs, including a discussion of their target genes. These transcription factors play an important role in regulating the expression of many enzymes acting on exogenous and endogenous compounds and have been the focus of intense research in pharmacological and biochemical sciences. Some of the unique differences between rodents and animals that regulate differential gene expression in rodents and man are due to differences in expression or function of these receptors as will be seen below.

4.2 Differences Between Rodent and Human Receptors

An example of the difference between rodent and human receptors is the observation that rodents utilize pregnenolone-16α-carbonitrile as a preferred ligand activator for PXR, while the human receptor utilizes the drug rifampicin (Kliewer, Goodwin, & Willson, 2002). Their respective response to the alternative ligand is far less robust than the ligand first described for each of them. Almost as striking is the fact that the mouse AhR has a higher

Table 3 Transcription Factors That Regulate Expression of Various CYPs and Other Genes Encoding Enzymes of Foreign Compound Metabolism

Transcription Factor		Biochemical Feature	Ligand	Target
Constitutive androstane receptor	NR1I3 CAR	Zinc finger nuclear receptor	1,4-bis[2-(3,5-Dichloropyridyloxy)] benzene (TCPOBOP)	CYP2B
Pregnane X receptor	NR1I2 PXR	Zinc finger nuclear receptor	Murine pregnenolone-16α-carbonitrile Human rifampicin	CYP3A
Glucocorticoid receptor	NR3C1 GR	Zinc finger nuclear receptor	Cortisol and dexamethasone	CYP1A1 and glutathione *S*-transferase A2
Aryl hydrocarbon receptor	AHR	Basic helix-loop-helix	2,3,7,8-Tetrachloro-dibenzo-*p*-dioxin, benzo[*α*]pyrene	CYP1A1, NAD(P)H: quinone oxidoreductase, UGT1A1
Nuclear factor (erythroid-derived 2)-like 2	Nrf2	bZIP domain	*t*-Butylhydroquinone	NAD(P)H: quinone oxidoreductase and glutathione *S*-transferase A2
Fos/Jun	Fos/Jun	bZIP domain	Butylated hydroxy-anisole	Cyp2c29, CYP2C9, ALDH1A1

affinity by ~10-fold for 2,3,7,8-tetrachlorodibenzo-*p*-dioxin compared to the human AHR (Flaveny, Murray, Chiaro, & Perdew, 2009). Yet other ligands, such as indirubin and quercetin, have a higher affinity for the human receptor than the murine receptor. These differences must be due to the changes in primary amino acid sequence seen between the two proteins (only about 85% sequence identity).

In addition, there are many differences between the downstream targets of the AHR in mice compared to humans. Carlson et al. (2009) showed that only five responsive orthologous genes are shared in liver between rat and human AHR, suggesting very distinct differences between the respective inducible genes. Clearly, both the specific genes that respond to AHR and the agonist potency vary. With this variation in receptor function and response, it is clear that animal models may be particularly limited in studying mechanisms of human cancer and treatment. The advent of "humanized" mice will provide better models, but the differences in downstream targets may still obviate a direct comparison between species (Shen, Jiang, Gonzalez, & Yu, 2011).

4.3 CAR Splice Variants and Ligand Specificity

A third variance between murine and human receptors is seen with the number of different splice variants of CAR in the human and the single receptor seen in mice (Auerbach et al., 2003). There are three major CAR variants expressed in human liver, termed hCAR1, hCAR2, and hCAR3. hCAR1 is constitutively active in transient transfection assays, as is the murine form of the receptor that has been more highly characterized. However, the other two are not constitutively active *in vitro* and are ligand activated. In our studies with Omiecinski (Wahlang, Falkner, et al., 2014; Wahlang, Song, et al., 2014), we showed that hCAR3 was activated by some PCB congeners in a ligand-dependent manner. PCB138, 149, 151, 183, and 187 all activated CAR3, while PXR was activated only by PCB126, 138, and 183 and AHR was activated only by PCB126 and PCB138. These results demonstrate that these three receptors display very different ligand specificity for receptor activation.

In vivo, murine CAR1 and human CAR1 are not constitutively active, because they are sequestered in the cytoplasm of most cells (Mutoh et al., 2009). A role for EGF receptor in activation of CAR1 has been proposed by Mutoh and coworkers and protein phosphorylation prevents nuclear translocation, accounting for the low level of gene activation in naïve animals (Mutoh et al., 2009, 2013). Their studies demonstrate clearly that EGF receptor binds PB and that this prevents the phosphorylation of CAR. Unphosphorylated CAR then easily translocates to the nucleus of cells like liver cells.

4.4 Exogenous Compounds Acting on the PPAR

Species differences can also arise when the level of receptor expression differs between an animal model and humans. This is seen for PPARα because mice are a peroxisome proliferator responsive species, while humans are not (Mukherjee, Jow, Noonan, & McDonnell, 1994); that is, treatment with a compound like fibric acid analogues used in human therapy causes hepatomegaly and induction of lipid oxidation enzymes in mouse liver, but similar fibrates do not cause this hepatomegaly in human liver (Gonzalez, Peters, & Cattley, 1998). It is also noteworthy that treatment with high doses of fibrates caused hepatocarcinoma formation in mice, but not in humans. However, the latter event can be attributed to a miRNA formed from murine PPARα 3′-flanking region that is lacking in the human PPARα gene (Shah et al., 2007).

Recently, perchloro- and perfluoroalkyl carboxylic acid, developed as industrial surfactant agents, have also been noted to be bioaccumulated and as they attain sufficient concentration in the body, they have been noted to activate a number of nuclear receptors and display more subtle adverse health effects (Betts, 2008). As we develop compounds (plasticizers, detergents, etc.) that seem to be less toxic than industrial compounds of the past, like vinyl chloride or PCBs, we must be mindful that more subtle forms of toxicity like immunotoxicity or enhanced steatosis may occur (Betts, 2008; Wahlang, Falkner, et al., 2014; Wahlang, Song, et al., 2014).

4.5 Underlying Regulation by Other Hormones and Transcription Factors

As anticipated, variation in a regulated pathway leading to CYP gene expression can be caused by many factors, such as the existence of multiple receptors (such as CAR isoforms) or differences in receptor-ligand affinity, or differences in downstream targets. One factor that has not been mentioned is the regulation of the receptors controlling target gene expression, i.e., what regulates AHR, CAR, PXR, etc. Maurel and his group have demonstrated that many of the hepatic nuclear receptors (CAR, PPAR, and PXR) are all regulated by a common ligand-activated receptor, namely glucocorticoid receptor (GR) (Pascussi, Drocourt, Fabre, Maurel, & Vilarem, 2000; Pascussi, Gerbal-Chaloin, Fabre, Maurel, & Vilarem, 2000). In addition, one can have specific gene regulation by both a nuclear receptor and the GR, as seen for CYP2C9, which has both a glucocorticoid-responsive element and a CAR-responsive element (Gerbal-Chaloin et al., 2002). In addition, many genes have tissue- or developmental-stage-specific transcription factors that also regulate basal or inducible expression, as is the case for HNF4α regulation of CYP2A6 (Onica et al., 2008). Many genes utilize multiple transcription factors activity for basal, inducible, development-specific, and tissue-specific expression.

4.6 Effects of Environmental Agents

Exposure to environmental agents and drugs can result in induction of various CYPs as has been discussed earlier. Some compounds have been developed as a therapeutic agent, like cyclophosphamide, that are not only substrates for bioactivation but they also induce their own metabolism. In the case of CAR and PXR, their receptor's ligand-binding cavity is large and plastic, accommodating many different molecular structures as ligands (Xiao, Cui, Madison, White, & Cheng, 2002). Therefore, whenever a new chemical or drug is developed or discovered, one must consider the possibility of autoinduction of metabolism, as was shown with PAH for AHR and cyclophosphamide for CAR. In addition, some lipophilic compounds may bioaccumulate in fatty depots of the body, if they cannot be cleared by metabolism. An example of these would be the highly chlorinated PCBs that we ingest through our diet. If stored in adipose, conditions like starvation or weight loss may release these compounds to the liver, where they may induce CYP expression through action of CAR and PXR (Wahlang, Falkner, et al., 2014; Wahlang, Song, et al., 2014). This possibility requires physicians to know more about their patient's exposures, because drug efficacy may be altered through this internal exposure process.

5. POLYMORPHIC CYTOCHROME P450 VARIANTS IMPACTING EXPRESSION AND CANCER SUSCEPTIBILITY

As the human genome sequences became available for individuals, increasingly we have learned that there are many gene polymorphisms in our DNA and researchers have sought to see if subpopulations of humans are more or less at risk to disease as a result of these polymorphisms. This topic is covered in two chapters of this Series (Ann Daley, "Polymorphic variants of cytochrome P450: Relevance to cancer and other diseases" and Saskia Preissner et al., "Personalized cancer therapy considering Cytochrome P450 variability"). Therefore, the potential variability in P450 function and expression will be briefly mentioned in this section. However, it is well known that gene polymorphisms have been observed in nearly every P450 gene studied (see http://cypalleles.ki.se). A few of these polymorphisms associated with the carcinogen metabolism discussed in this chapter will be mentioned to highlight the possibility that physicians or environmental consultants may use this clinical information to ascertain populations that may be affected by these polymorphisms. Sound counseling or avoidance of chemical exposures by some groups at higher risk may practically be applied to human populations.

5.1 Cytochrome P4501A1 and Glutathione *S*-Transferase M1

Hayashi, Watanabe, Nakachi, and Kawajiri (1991) discovered a polymorphism in CYP1A1 in a small population of Japanese individuals that displayed a polymorphism in Exon 7 at residue 462 of that gene product leading to an amino acid replacement (Ile to Val) in the region that encodes the binding site for the catalytic heme group. They had also defined a human population who displayed high genetic risk to cigarette smoking and linked the *Msp*I restriction site polymorphism to the Exon 7 mutation (Hayashi et al., 1991). In their studies, the odds ratio for heavy smokers and lung cancer was around 2–3, relative to nonsmokers. Subsequently, a combined polymorphism of CYP1A1 and glutathione *S*-transferase M1 leads to higher odds ratios (16–40) than either alone (Hayashi, Watanabe, & Kawajiri, 1992). The accumulation of such deleterious polymorphisms may cause small groups of individuals be at high risk to environmental hazards.

5.2 Cytochrome P4502D6 and Cytochrome P4502E1

As reviewed by Meyer and Zanger (1997) and Meyer, Skoda, and Zanger (1990), CYP2D6 polymorphisms have been shown in human populations leading to the identification of a small group (10–15%) in the population who are poor metabolizers of debrisoquine. CYP2E1 polymorphisms have also been discovered and may have impacted the human exposures to vinyl chloride seen in factories producing vinyl chloride for rubber and plastic products (Raunio et al., 1995, 2001). Some of the subsequent studies on CYP2D6 have shown different responses in which poor metabolizers' genotypes have been associated with increased cancer incidents (Smith, Moss, Gough, Spurr, & Wolf, 1992; Wolf et al., 1994) and the role of these polymorphism in human disease is still under study. However, these examples of polymorphisms of foreign compound-metabolizing genes demonstrate the importance of understanding these differences between extensive and poor metabolism phenotypes in therapy, cancer initiation, and chemical toxicity. Consideration of these differences is required as we move toward use of personalized medicine in our society.

6. CONCLUSION

As clinical medicine begins to apply all the information available through analysis of an individual human's genome to polydrug pharmacy, it is envisioned that clinicians will be able to fine-tune an individual's drug regimen to optimize their treatment (e.g., pharmacogenomics and personalized medicine). This same set of genomic data may also allow us to continue to define populations who may be more sensitive to chemical carcinogens or toxins, as well as anticancer drugs, and alter their susceptibility to carcinogenesis or toxicity by other methods. For example, use of compounds such as rifampin as therapeutic approaches may enhance mechanisms of elimination to protect against accumulation of chemicals or even hasten elimination. Alternatives may also include counseling to limit exposure to industrial or other sources of chemical exposure, including tobacco smoking or wood burning. Development of new anticancer drugs should be coupled with animal pharmacokinetic studies to deduce potential drug–drug interactions or autoinduction at the same time they are applied in clinical trials for efficacy. Medical ethical issues must be applied to both normal pharmaceutical treatment of disease as well as consideration of who will be more sensitive to carcinogen/anticancer drug treatment. These phenomena have implications for occupational health because sensitive individuals may need to further reduce their environmental exposure if they are poor metabolizers of the chemicals they accumulate. However, with informatics methods being developed, clinicians, toxicologists, and oncologists should be able to improve outcomes to both groups, protecting individuals with no polymorphic genes and individuals who are most susceptible to chemical toxicity and carcinogenesis due to polymorphic genes.

CONFLICT OF INTEREST

The authors have no conflict of interest to declare.

ACKNOWLEDGMENT

This work was supported in part by NIEHS grant 1 R01 ES021375 (M.C.C.) and 1 T35 ES 014559 (R.A.P.)

REFERENCES

Amunom, I., Dieter, L. J., Tamasi, V., Cai, J., Conklin, D. J., Srivastava, S., et al. (2011). Cytochromes P450 catalyze the reduction of alpha, beta-unsaturated aldehydes. *Chemical Research in Toxicology*, *24*, 1223–1230.

Amunom, I., Stephens, L. J., Tamasi, V., Cai, J., Pierce, W. M., Jr., Conklin, D. J., et al. (2007). Cytochromes P450 catalyze oxidation of alpha, beta-unsaturated aldehydes. *Archives of Biochemistry and Biophysics*, *464*, 187–196.

Ariyoshi, N., Ito, S., Okudaira, A., Mise, M., Matsusue, K., Yamada, H., et al. (1995). Studies on PCB toxicity involving 2C subfamily cytochrome P450. *Fukuoka Igaku Zasshi*, *86*, 153–162.

Ariyoshi, N., Oguri, K., Koga, N., Yoshimura, H., & Funae, Y. (1995). Metabolism of highly persistent PCB congener, 2,4,5,2',4',5'-hexachlorobiphenyl, by human CYP2B6. *Biochemical and Biophysical Research Communications*, *212*, 455–460.

Auerbach, S. S., Ramsden, R., Stoner, M. A., Verlinde, C., Hassett, C., & Omiecinski, C. J. (2003). Alternatively spliced isoforms of the human constitutive androstane receptor. *Nucleic Acids Research*, *31*, 3194–3207.

Betts, K. S. (2008). Not immune to PFOS effects? *Environmental Health Perspectives*, *116*, A290.

Bolt, H. M. (1986). Metabolic activation of vinyl chloride, formation of nucleic acid adducts and relevance to carcinogenesis. *IARC Scientific Publications*, 261–268.

Capdevila, J., Estabrook, R. W., & Prough, R. A. (1978a). The existence of a benzo(a) pyrene-3,6-quinone reductase in rat liver microsomal fractions. *Biochemical and Biophysical Research Communications*, *83*, 1291–1298.

Capdevila, J., Estabrook, R. W., & Prough, R. A. (1978b). The microsomal metabolism of benzo(a) pyrene phenols. *Biochemical and Biophysical Research Communications*, *82*, 518–525.

Carlson, E. A., McCulloch, C., Koganti, A., Goodwin, S. B., Sutter, T. R., & Silkworth, J. B. (2009). Divergent transcriptomic responses to aryl hydrocarbon receptor agonists between rat and human primary hepatocytes. *Toxicological Sciences*, *112*, 257–272.

Chang, T. K., Weber, G. F., Crespi, C. L., & Waxman, D. J. (1993). Differential activation of cyclophosphamide and ifosphamide by cytochromes P-450 2B and 3A in human liver microsomes. *Cancer Research*, *53*, 5629–5637.

Chang, T. K., Yu, L., Maurel, P., & Waxman, D. J. (1997). Enhanced cyclophosphamide and ifosfamide activation in primary human hepatocyte cultures: Response to cytochrome P-450 inducers and autoinduction by oxazaphosphorines. *Cancer Research*, *57*, 1946–1954.

Conney, A. H. (1967). Pharmacological implications of microsomal enzyme induction. *Pharmacological Reviews*, *19*, 317–366.

Conney, A. H. (2003). Induction of drug-metabolizing enzymes: A path to the discovery of multiple cytochromes P450. *Annual Review of Pharmacology and Toxicology*, *43*, 1–30.

Conney, A. H., Brown, R. R., Miller, J. A., & Miller, E. C. (1957). The metabolism of methylated aminoazo dyes. VI. Intracellular distribution and properties of the demethylase system. *Cancer Research*, *17*, 628–633.

Conney, A. H., Miller, E. C., & Miller, J. A. (1956). The metabolism of methylated aminoazo dyes. V. Evidence for induction of enzyme synthesis in the rat by 3-methylcholanthrene. *Cancer Research*, *16*, 450–459.

Conney, A. H., Miller, E. C., & Miller, J. A. (1957). Substrate-induced synthesis and other properties of benzpyrene hydroxylase in rat liver. *Journal of Biological Chemistry*, *228*, 753–766.

Coon, M. J. (2005). Cytochrome P450: Nature's most versatile biological catalyst. *Annual Review of Pharmacology and Toxicology*, *45*, 1–25.

Cooper, D. Y., Levin, S., Narasimhulu, S., & Rosenthal, O. (1965). Photochemical action spectrum of the terminal oxidase of mixed function oxidase systems. *Science*, *147*, 400–402.

Creech, J. L., Jr., & Johnson, M. N. (1974). Angiosarcoma of liver in the manufacture of polyvinyl chloride. *Journal of Occupational Medicine*, *16*, 150–151.

Denison, M. S., & Nagy, S. R. (2003). Activation of the aryl hydrocarbon receptor by structurally diverse exogenous and endogenous chemicals. *Annual Review of Pharmacology and Toxicology*, *43*, 309–334.

Dunn, D. L., Lubet, R. A., & Prough, R. A. (1979). Oxidative metabolism of N-isopropyl-alpha-(2-methylhydrazino)-p-toluamide hydrochloride (procarbazine) by rat liver microsomes. *Cancer Research*, *39*, 4555–4563.

el Ghissassi, F., Barbin, A., & Bartsch, H. (1998). Metabolic activation of vinyl chloride by rat liver microsomes: Low-dose kinetics and involvement of cytochrome P450 2E1. *Biochemical Pharmacology*, *55*, 1445–1452.

Evans, R. M., & Mangelsdorf, D. J. (2014). Nuclear receptors, RXR, and the Big Bang. *Cell*, *157*, 255–266.

Evans, F. E., Miller, D. W., & Beland, F. A. (1980). Sensitivity of the conformation of deoxyguanosine to binding at the C-8 position by N-acetylated and unacetylated 2-aminofluorene. *Carcinogenesis*, *1*, 955–959.

Falk, H., Creech, J. L., Jr., Heath, C. W., Jr., Johnson, M. N., & Key, M. M. (1974). Hepatic disease among workers at a vinyl chloride polymerization plant. *JAMA*, *230*, 59–63.

Flaveny, C. A., Murray, I. A., Chiaro, C. R., & Perdew, G. H. (2009). Ligand selectivity and gene regulation by the human aryl hydrocarbon receptor in transgenic mice. *Molecular Pharmacology*, *75*, 1412–1420.

Fouts, J. R., Kamm, J. J., & Brodie, B. B. (1957). Enzymatic reduction of prontosil and other azo dyes. *Journal of Pharmacology and Experimental Therapeutics*, *120*, 291–300.

Gates, L. A., Lu, D., & Peterson, L. A. (2012). Trapping of cis-2-butene-1,4-dial to measure furan metabolism in human liver microsomes by cytochrome P450 enzymes. *Drug Metabolism and Disposition*, *40*, 596–601.

Gerbal-Chaloin, S., Daujat, M., Pascussi, J. M., Pichard-Garcia, L., Vilarem, M. J., & Maurel, P. (2002). Transcriptional regulation of CYP2C9 gene. Role of glucocorticoid receptor and constitutive androstane receptor. *The Journal of Biological Chemistry*, *277*, 209–217.

Goeptar, A. R., te Koppele, J. M., Lamme, E. K., Pique, J. M., & Vermeulen, N. P. (1993). Cytochrome P450 2B1-mediated one-electron reduction of adriamycin: A study with rat liver microsomes and purified enzymes. *Molecular Pharmacology*, *44*, 1267–1277.

Gonzalez, F. J., Peters, J. M., & Cattley, R. C. (1998). Mechanism of action of the non-genotoxic peroxisome proliferators: Role of the peroxisome proliferator-activator receptor alpha. *Journal of the National Cancer Institute*, *90*, 1702–1709.

Gonzalez, F. J., & Yu, A. M. (2006). Cytochrome P450 and xenobiotic receptor humanized mice. *Annual Review of Pharmacology and Toxicology*, *46*, 41–64.

Grimm, F. A., Hu, D., Kania-Korwel, I., Lehmler, H. J., Ludewig, G., Hornbuckle, K. C., et al. (2015). Metabolism and metabolites of polychlorinated biphenyls. *Critical Reviews in Toxicology*, *45*, 245–272.

Hayashi, S., Watanabe, J., & Kawajiri, K. (1992). High susceptibility to lung cancer analyzed in terms of combined genotypes of P450IA1 and Mu-class glutathione S-transferase genes. *Japanese Journal of Cancer Research*, *83*, 866–870.

Hayashi, S., Watanabe, J., Nakachi, K., & Kawajiri, K. (1991). Genetic linkage of lung cancer-associated MspI polymorphisms with amino acid replacement in the heme binding region of the human cytochrome P450IA1 gene. *Journal of Biochemistry*, *110*, 407–411.

He, X. Y., Shen, J., Ding, X., Lu, A. Y., & Hong, J. Y. (2004). Identification of critical amino acid residues of human CYP2A13 for the metabolic activation of 4-(methylnitrosamino)-1-(3-pyridyl)-1-butanone, a tobacco-specific carcinogen. *Drug Metabolism and Disposition*, *32*, 1516–1521.

Holder, G., Yagi, H., Dansette, P., Jerina, D. M., Levin, W., Lu, A. Y., et al. (1974). Effects of inducers and epoxide hydrase on the metabolism of benzo(a)pyrene by liver microsomes and a reconstituted system: Analysis by high pressure liquid chromatography. *Proceedings of the National academy of Sciences of the United States of America*, *71*, 4356–4360.

Holder, G. M., Yagi, H., Jerina, D. M., Levin, W., Lu, A. Y., & Conney, A. H. (1975). Metabolism of benzo[a]pyrene. Effect of substrate concentration and 3-methylcholanthrene pretreatment on hepatic metabolism by microsomes from rats and mice. *Archives of Biochemistry and Biophysics*, *170*, 557–566.

Kadlubar, F. F., Miller, J. A., & Miller, E. C. (1976). Microsomal N-oxidation of the hepatocarcinogen N-methyl-4-aminoazobenzene and the reactivity of N-hydroxy-N-methyl-4-aminoazobenzene. *Cancer Research*, *36*, 1196–1206.

Kadlubar, F. F., Unruh, L. E., Beland, F. A., Straub, K. M., & Evans, F. E. (1980). In vitro reaction of the carcinogen, N-hydroxy-2-naphthylamine, with DNA at the C-8 and N2 atoms of guanine and at the N6 atom of adenine. *Carcinogenesis*, *1*, 139–150.

Kaminsky, L. S., Kennedy, M. W., Adams, S. M., & Guengerich, F. P. (1981). Metabolism of dichlorobiphenyls by highly purified isozymes of rat liver cytochrome P-450. *Biochemistry*, *20*, 7379–7384.

Kapitulnik, J., Levin, W., Conney, A. H., Yagi, H., & Jerina, D. M. (1977). Benzo[a]pyrene 7,8-dihydrodiol is more carcinogenic than benzo[a]pyrene in newborn mice. *Nature*, *266*, 378–380.

Kapitulnik, J., Levin, W., Yagi, H., Jerina, D. M., & Conney, A. H. (1976). Lack of carcinogenicity of 4-, 5-, 6-, 7-, 8-, 9-, and 10-hydroxybenzo(a)pyrene on mouse skin. *Cancer Research*, *36*, 3625–3628.

Kliewer, S. A., Goodwin, B., & Willson, T. M. (2002). The nuclear pregnane X receptor: A key regulator of xenobiotic metabolism. *Endocrine Reviews*, *23*, 687–702.

Kokkinakis, D. M., Koop, D. R., Scarpelli, D. G., Coon, M. J., & Hollenberg, P. F. (1985). Metabolism of N-nitroso-2,6-dimethylmorpholine by isozymes of rabbit liver microsomal cytochrome P-450. *Cancer Research*, *45*, 619–624.

Koreeda, M., Moore, P. D., Wislocki, P. G., Levin, W., Yagi, H., & Jerina, D. M. (1978). Binding of benzo[a]pyrene 7,8-diol-9,10-epoxides to DNA, RNA, and protein of mouse skin occurs with high stereoselectivity. *Science*, *199*, 778–781.

Levine, W. G., & Lu, A. Y. (1982). Role of isozymes of cytochrome P-450 in the metabolism of N, N-dimethyl-4-aminoazobenzene in the rat. *Drug Metabolism and Disposition*, *10*, 102–109.

Levine, W. G., & Raza, H. (1988). Mechanism of azoreduction of dimethylaminoazobenzene by rat liver NADPH-cytochrome P-450 reductase and partially purified cytochrome P-450. Oxygen and carbon monoxide sensitivity and stimulation by FAD and FMN. *Drug Metabolism and Disposition*, *16*, 441–448.

Lewis, A. D., Lau, D. H., Duran, G. E., Wolf, C. R., & Sikic, B. I. (1992). Role of cytochrome P-450 from the human CYP3A gene family in the potentiation of morpholino doxorubicin by human liver microsomes. *Cancer Research*, *52*, 4379–4384.

Lin, J. K., Miller, J. A., & Miller, E. C. (1975). Structures of hepatic nucleic acid-bound dyes in rats given the carcinogen N-methyl-4-aminoazobenzene. *Cancer Research*, *35*, 844–850.

Lindley, C., Hamilton, G., McCune, J. S., Faucette, S., Shord, S. S., Hawke, R. L., et al. (2002). The effect of cyclophosphamide with and without dexamethasone on cytochrome P450 3A4 and 2B6 in human hepatocytes. *Drug Metabolism and Disposition*, *30*, 814–822.

Lorr, N. A., Tu, Y. Y., & Yang, C. S. (1982). The nature of nitrosamine denitrosation by rat liver microsomes. *Carcinogenesis*, *3*, 1039–1043.

Lu, A. Y., & Coon, M. J. (1968). Role of hemoprotein P-450 in fatty acid omega-hydroxylation in a soluble enzyme system from liver microsomes. *Journal of Biological Chemistry*, *243*, 1331–1332.

Lu, H., & Waxman, D. J. (2005). Antitumor activity of methoxymorpholinyl doxorubicin: Potentiation by cytochrome P450 3A metabolism. *Molecular Pharmacology*, *67*, 212–219.

Makk, L., Creech, J. L., Whelan, J. G., Jr., & Johnson, M. N. (1974). Liver damage and angiosarcoma in vinyl chloride workers. A systematic detection program. *JAMA*, *230*, 64–68.

Mallett, A. K., King, L. J., & Walker, R. (1985). Solubilisation, purification and reconstitution of hepatic microsomal azoreductase activity. *Biochemical Pharmacology*, *34*, 337–342.

McCarty, M., & Avery, O. T. (1946). Studies on the chemical nature of the substance inducing transformation of pneumococcal types: III. An improved method for the isolation of the transforming substance and its application to pneumococcus types II, III, and VI. *The Journal of Experimental Medicine*, *83*, 97–104.

Megaraj, V., Zhou, X., Xie, F., Liu, Z., Yang, W., & Ding, X. (2014). Role of CYP2A13 in the bioactivation and lung tumorigenicity of the tobacco-specific lung procarcinogen 4-(methylnitrosamino)-1-(3-pyridyl)-1-butanone: In vivo studies using a CYP2A13-humanized mouse model. *Carcinogenesis*, *35*, 131–137.

Meyer, U. A., Skoda, R. C., & Zanger, U. M. (1990). The genetic polymorphism of debrisoquine/sparteine metabolism-molecular mechanisms. *Pharmacology and Therapeutics*, *46*, 297–308.

Meyer, U. A., & Zanger, U. M. (1997). Molecular mechanisms of genetic polymorphisms of drug metabolism. *Annual Review of Pharmacology and Toxicology*, *37*, 269–296.

Michaels, S., & Wang, M. Z. (2014). The revised human liver cytochrome P450 "pie": Absolute protein quantification of CYP4F and CYP3A enzymes using targeted quantitative proteomics. *Drug Metabolism and Disposition*, *42*, 1241–1251.

Miller, E. C. (1951). Studies on the formation of protein-bound derivatives of 3,4-benzpyrene in the epidermal fraction of mouse skin. *Cancer Research*, *11*, 100–108.

Miller, J. A. (1998). The metabolism of xenobiotics to reactive electrophiles in chemical carcinogenesis and mutagenesis: A collaboration with Elizabeth Cavert Miller and our associates. *Drug Metabolism Reviews*, *30*, 645–674.

Miller, E. C., & Miller, J. A. (1952). In vivo combinations between carcinogens and tissue constituents and their possible role in carcinogenesis. *Cancer Research*, *12*, 547–556.

Mlcoch, J., Fuchs, J., Oesch, F., & Platt, K. L. (1993). Characterization of DNA adducts at the bay region of dibenz[a,h]anthracene formed in vitro. *Carcinogenesis*, *14*, 469–473.

Moloney, S. J., & Prough, R. A. (1983). Studies on the pathway of methane formation from procarbazine, a 2-methylbenzylhydrazine derivative, by rat liver microsomes. *Archives of Biochemistry and Biophysics*, *221*, 577–584.

Moloney, S. J., Wiebkin, P., Cummings, S. W., & Prough, R. A. (1985). Metabolic activation of the terminal N-methyl group of N-isopropyl-alpha-(2-methylhydrazino)-p-toluamide hydrochloride (procarbazine). *Carcinogenesis*, *6*, 397–401.

Mueller, G. C., & Miller, J. A. (1949). The reductive cleavage of 4-dimethylaminoazobenzene by rat liver; the intracellular distribution of the enzyme system and its requirement for triphosphopyridine nucleotide. *Journal of Biological Chemistry*, *180*, 1125–1136.

Mueller, G. C., & Miller, J. A. (1950). The reductive cleavage of 4-dimethylaminoazobenzene by rat liver: Reactivation of carbon dioxide-treated homogenates by riboflavin-adenine dinucleotide. *Journal of Biological Chemistry*, *185*, 145–154.

Mukaigasa, K., Nguyen, L. T., Li, L., Nakajima, H., Yamamoto, M., & Kobayashi, M. (2012). Genetic evidence of an evolutionarily conserved role for Nrf2 in the protection against oxidative stress. *Molecular Cell. Biology*, *32*, 4455–4461.

Mukherjee, R., Jow, L., Noonan, D., & McDonnell, D. P. (1994). Human and rat peroxisome proliferator activated receptors (PPARs) demonstrate similar tissue distribution but different responsiveness to PPAR activators. *Journal of Steroid Biochemistry and Molecular Biology*, *51*, 157–166.
Mutoh, S., Osabe, M., Inoue, K., Moore, R., Pedersen, L., Perera, L., et al. (2009). Dephosphorylation of threonine 38 is required for nuclear translocation and activation of human xenobiotic receptor CAR (NR1I3). *Journal of Biological Chemistry*, *284*, 34785–34792.
Mutoh, S., Sobhany, M., Moore, R., Perera, L., Pedersen, L., Sueyoshi, T., et al. (2013). Phenobarbital indirectly activates the constitutive active androstane receptor (CAR) by inhibition of epidermal growth factor receptor signaling. *Science Signaling*, *6*, ra31.
Onica, T., Nichols, K., Larin, M., Ng, L., Maslen, A., Dvorak, Z., et al. (2008). Dexamethasone-mediated up-regulation of human CYP2A6 involves the glucocorticoid receptor and increased binding of hepatic nuclear factor 4 alpha to the proximal promoter. *Molecular Pharmacology*, *73*, 451–460.
Paine, M. F., Hart, H. L., Ludington, S. S., Haining, R. L., Rettie, A. E., & Zeldin, D. C. (2006). The human intestinal cytochrome P450 "pie" *Drug Metabolism and Disposition*, *34*, 880–886.
Pascussi, J. M., Drocourt, L., Fabre, J. M., Maurel, P., & Vilarem, M. J. (2000). Dexamethasone induces pregnane X receptor and retinoid X receptor-alpha expression in human hepatocytes: Synergistic increase of CYP3A4 induction by pregnane X receptor activators. *Molecular Pharmacology*, *58*, 361–372.
Pascussi, J. M., Gerbal-Chaloin, S., Fabre, J. M., Maurel, P., & Vilarem, M. J. (2000). Dexamethasone enhances constitutive androstane receptor expression in human hepatocytes: Consequences on cytochrome P450 gene regulation. *Molecular Pharmacology*, *58*, 1441–1450.
Patten, C. J., Ishizaki, H., Aoyama, T., Lee, M., Ning, S. M., Huang, W., et al. (1992). Catalytic properties of the human cytochrome P450 2E1 produced by cDNA expression in mammalian cells. *Archives of Biochemistry and Biophysics*, *299*, 163–171.
Patten, C. J., Smith, T. J., Friesen, M. J., Tynes, R. E., Yang, C. S., & Murphy, S. E. (1997). Evidence for cytochrome P450 2A6 and 3A4 as major catalysts for N'-nitrosonornicotine alpha-hydroxylation by human liver microsomes. *Carcinogenesis*, *18*, 1623–1630.
Pawlowska, J., Tarasiuk, J., Wolf, C. R., Paine, M. J., & Borowski, E. (2003). Differential ability of cytostatics from anthraquinone group to generate free radicals in three enzymatic systems: NADH dehydrogenase, NADPH cytochrome P450 reductase, and xanthine oxidase. *Oncology Research*, *13*, 245–252.
Peterson, L. A. (2006). Electrophilic intermediates produced by bioactivation of furan. *Drug Metabolism Reviews*, *38*, 615–626.
Prough, R. A., Brown, M. I., Dannan, G. A., & Guengerich, F. P. (1984). Major isozymes of rat liver microsomal cytochrome P-450 involved in the N-oxidation of N-isopropyl-alpha-(2-methylazo)-p-toluamide, the azo derivative of procarbazine. *Cancer Research*, *44*, 543–548.
Prough, R. A., Saeki, Y., & Capdevila, J. (1981). The metabolism of benzo[a]pyrene phenols by rat liver microsomal fractions. *Archives of Biochemistry and Biophysics*, *212*, 136–146.
Raleigh, S. M., Wanogho, E., Burke, M. D., McKeown, S. R., & Patterson, L. H. (1998). Involvement of human cytochromes P450 (CYP) in the reductive metabolism of AQ4N, a hypoxia activated anthraquinone di-N-oxide prodrug. *International Journal of Radiation Oncology, Biology, and Physics*, *42*, 763–767.
Raleigh, S. M., Wanogho, E., Burke, M. D., & Patterson, L. H. (1999). Rat cytochromes P450 (CYP) specifically contribute to the reductive bioactivation of AQ4N, an alkylaminoanthraquinone-di-N-oxide anticancer prodrug. *Xenobiotica*, *29*, 1115–1122.

Raunio, H., Husgafvel-Pursiainen, K., Anttila, S., Hietanen, E., Hirvonen, A., & Pelkonen, O. (1995). Diagnosis of polymorphisms in carcinogen-activating and inactivating enzymes and cancer susceptibility—A review. *Gene*, *159*, 113–121.
Raunio, H., Rautio, A., Gullsten, H., & Pelkonen, O. (2001). Polymorphisms of CYP2A6 and its practical consequences. *British Journal of Clinical Pharmacology*, *52*, 357–363.
Raza, H., & Levine, W. G. (1986a). Azoreduction of N, N-dimethyl-4-aminoazobenzene (DAB) by rat hepatic microsomes. Selective induction by clofibrate. *Drug Metabolism and Disposition*, *14*, 19–24.
Raza, H., & Levine, W. G. (1986b). Effect of phenobarbital and beta-naphthoflavone on oxidative metabolism of N, N-dimethyl-4-aminoazobenzene by regenerating rat-liver microsomes and its response to sulphydryl compounds. *Xenobiotica*, *16*, 827–837.
Sasaki, T., & Yoshida, T. (1935). Experimentelle Erzeugung des Lebercarcinomas durch Futterung mit o-Amidoazotoluol. *Virchows Archiv für pathologische Anatomie und Physiologie und für klinische Medizin*, *295(2)*, 175–200.
Selkirk, J. K., Croy, R. G., & Gelboin, H. V. (1974). Benzo(a)pyrene metabolites: Efficient and rapid separation by high-pressure liquid chromatography. *Science*, *184*, 169–171.
Shah, Y. M., Morimura, K., Yang, Q., Tanabe, T., Takagi, M., & Gonzalez, F. J. (2007). Peroxisome proliferator-activated receptor alpha regulates a microRNA-mediated signaling cascade responsible for hepatocellular proliferation. *Molecular Cell. Biology*, *27*, 4238–4247.
Shen, H. W., Jiang, X. L., Gonzalez, F. J., & Yu, A. M. (2011). Humanized transgenic mouse models for drug metabolism and pharmacokinetic research. *Current Drug Metabolism*, *12*, 997–1006.
Shu, L., & Hollenberg, P. F. (1996). Identification of the cytochrome P450 isozymes involved in the metabolism of N-nitrosodipropyl-, N-nitrosodibutyl- and N-nitroso-n-butyl-n-propylamine. *Carcinogenesis*, *17*, 839–848.
Simpson, E. R., & Boyd, G. S. (1967). The cholesterol side-chain cleavage system of bovine adrenal cortex. *European Journal of Biochemistry*, *2*, 275–285.
Sladek, N. E. (1988). Metabolism of oxazaphosphorines. *Pharmacology and Therapeutics*, *37*, 301–355.
Smith, C. A., Moss, J. E., Gough, A. C., Spurr, N. K., & Wolf, C. R. (1992). Molecular genetic analysis of the cytochrome P450-debrisoquine hydroxylase locus and association with cancer susceptibility. *Environmental Health Perspectives*, *98*, 107–112.
Stevenson, E. S., Dobriner, K., & Rhoad, C. P. (1942). The metabolism of dimethylaminoazobenzene (butter yellow) in rats. *Cancer Research*, *2*, 160–167.
Swaffar, D. S., Harker, W. G., Pomerantz, S. C., Lim, H. K., & Yost, G. S. (1992). Isolation, purification, and characterization of two new chemical decomposition products of methylazoxyprocarbazine. *Drug Metabolism and Disposition*, *20*, 632–642.
Swaffar, D. S., Horstman, M. G., Jaw, J. Y., Thrall, B. D., Meadows, G. G., Harker, W. G., et al. (1989). Methylazoxyprocarbazine, the active metabolite responsible for the anticancer activity of procarbazine against L1210 leukemia. *Cancer Research*, *49*, 2442–2447.
Taguchi, K., Motohashi, H., & Yamamoto, M. (2011). Molecular mechanisms of the Keap1-Nrf2 pathway in stress response and cancer evolution. *Genes to Cells*, *16*, 123–140.
Thakker, D. R., Yagi, H., Akagi, H., Koreeda, M., Lu, A. H., Levin, W., et al. (1977). Metabolism of benzo[a]pyrene. VI. Stereoselective metabolism of benzo[a]pyrene and benzo[a]pyrene 7,8-dihydrodiol to diol epoxides. *Chemico-Biological Interactions*, *16*, 281–300.
Ueng, Y. F., Shimada, T., Yamazaki, H., & Guengerich, F. P. (1995). Oxidation of aflatoxin B1 by bacterial recombinant human cytochrome P450 enzymes. *Chemical Research in Toxicology*, *8*, 218–225.

Wahlang, B., Falkner, K. C., Clair, H. B., Al-Eryani, L., Prough, R. A., States, J. C., et al. (2014). Human receptor activation by aroclor 1260, a polychlorinated biphenyl mixture. *Toxicological Sciences*, *140*, 283–297.

Wahlang, B., Song, M., Beier, J. I., Cameron, F. K., Al-Eryani, L., Clair, H. B., et al. (2014). Evaluation of Aroclor 1260 exposure in a mouse model of diet-induced obesity and non-alcoholic fatty liver disease. *Toxicology and Applied Pharmacology*, *279*, 380–390.

Warner, N. A., Martin, J. W., & Wong, C. S. (2009). Chiral polychlorinated biphenyls are biotransformed enantioselectively by mammalian cytochrome P-450 isozymes to form hydroxylated metabolites. *Environmental Science and Technology*, *43*, 114–121.

Watson, J. D., & Crick, F. H. (1953). The structure of DNA. *Cold Spring Harbor Symposia on Quantitative Biology*, *18*, 123–131.

Weisburger, E. K., Weisburger, J. H., & Morris, H. P. (1953). Studies on the metabolism of 2-acetylaminofluorene-9-C14. *Archives of Biochemistry and Biophysics*, *43*, 474–484.

Wiebel, F. J., Selkirk, J. K., Gelboin, H. V., Haugen, D. A., van der Hoeven, T. A., & Coon, M. J. (1975). Position-specific oxygenation of benzo(a)pyrene by different forms of purified cytochrome P-450 from rabbit liver. *Proceedings of the National academy of Sciences of the United States of America*, *72*, 3917–3920.

Wiebkin, P., & Prough, R. A. (1980). Oxidative metabolism of N-isopropyl-alpha-(2-methylazo)-p-toluamide (azoprocarbazine) by rodent liver microsomes. *Cancer Research*, *40*, 3524–3529.

Wiest, W. G., & Heidelberger, C. (1953). The interaction of carcinogenic hydrocarbons with tissue constituents. II. 1,2,5,6-Dibenzanthracene-9,10-C14 in skin. *Cancer Research*, *13*, 250–254.

Wolf, C. R., Smith, C. A., Bishop, T., Forman, D., Gough, A. C., & Spurr, N. K. (1994). CYP2D6 genotyping and the association with lung cancer susceptibility. *Pharmacogenetics*, *4*, 104–106.

Wong, R. H., Du, C. L., Wang, J. D., Chan, C. C., Luo, J. C., & Cheng, T. J. (2002). XRCC1 and CYP2E1 polymorphisms as susceptibility factors of plasma mutant p53 protein and anti-p53 antibody expression in vinyl chloride monomer-exposed polyvinyl chloride workers. *Cancer Epidemiology, Biomarkers and Prevention*, *11*, 475–482.

Wong, R. H., Wang, J. D., Hsieh, L. L., & Cheng, T. J. (2003). XRCC1, CYP2E1 and ALDH2 genetic polymorphisms and sister chromatid exchange frequency alterations amongst vinyl chloride monomer-exposed polyvinyl chloride workers. *Archives of Toxicology*, *77*, 433–440.

Wood, A. W., Levin, W., Lu, A. Y., Yagi, H., Hernandez, O., Jerina, D. M., et al. (1976). Metabolism of benzo(a)pyrene and benzo(a)pyrene derivatives to mutagenic products by highly purified hepatic microsomal enzymes. *Journal of Biological Chemistry*, *251*, 4882–4890.

Xiao, L., Cui, X., Madison, V., White, R. E., & Cheng, K. C. (2002). Insights from a three-dimensional model into ligand binding to constitutive active receptor. *Drug Metabolism and Disposition*, *30*, 951–956.

Yagi, H., Akagi, H., Thakker, D. R., Mah, H. D., Koreeda, M., & Jerina, D. M. (1977). Absolute sterochemistry of the highly mutagenic 7,8-diol 9,10-epoxides derived from the potent carcinogen trans-7,8-dihydroxy-7,8-dihydrobenzol[a]pyrene. *Journal of the American Chemical Society*, *99*, 2358–2359.

Yamazaki, H., Inui, Y., Yun, C. H., Guengerich, F. P., & Shimada, T. (1992). Cytochrome P450 2E1 and 2A6 enzymes as major catalysts for metabolic activation of N-nitrosodialkylamines and tobacco-related nitrosamines in human liver microsomes. *Carcinogenesis*, *13*, 1789–1794.

Yamazaki, H., Oda, Y., Funae, Y., Imaoka, S., Inui, Y., Guengerich, F. P., et al. (1992). Participation of rat liver cytochrome P450 2E1 in the activation of N-nitrosodimethylamine and N-nitrosodiethylamine to products genotoxic in an acetyltransferase-overexpressing Salmonella typhimurium strain (NM2009). *Carcinogenesis*, *13*, 979–985.

Yang, S. K., Selkirk, J. K., Plotkin, E. V., & Gelboin, H. V. (1975). Kinetic analysis of the metabolism of benzo(a)pyrene to phenols, dihydrodiols, and quinones by high-pressure chromatography compared to analysis by aryl hydrocarbon hydroxylase assay, and the effect of enzyme induction. *Cancer Research*, *35*, 3642–3650.

Yang, C. S., Yoo, J. S., Ishizaki, H., & Hong, J. Y. (1990). Cytochrome P450IIE1: Roles in nitrosamine metabolism and mechanisms of regulation. *Drug Metabolism Reviews*, *22*, 147–159.

Yu, L., & Waxman, D. J. (1996). Role of cytochrome P450 in oxazaphosphorine metabolism. Deactivation via N-dechloroethylation and activation via 4-hydroxylation catalyzed by distinct subsets of rat liver cytochromes P450. *Drug Metabolism and Disposition*, *24*, 1254–1262.

Zannikos, P. N., Bandyopadhyay, A. M., Robertson, L. W., & Blouin, R. A. (1994). Cytochrome P450 2B enzyme induction defect after 2,2',4,4',5,5'-hexachlorobiphenyl treatment in the fa/fa Zucker rat. *Journal of Pharmacology and Experimental Therapeutics*, *268*, 1565–1570.

Zhang, Z., Lu, H., Huan, F., Meghan, C., Yang, X., Wang, Y., et al. (2014). Cytochrome P450 2A13 mediates the neoplastic transformation of human bronchial epithelial cells at a low concentration of aflatoxin B1. *International Journal of Cancer*, *134*, 1539–1548.

Zhou, X., D'Agostino, J., Xie, F., & Ding, X. (2012). Role of CYP2A5 in the bioactivation of the lung carcinogen 4-(methylnitrosamino)-1-(3-pyridyl)-1-butanone in mice. *Journal of Pharmacology and Experimental Therapeutics*, *341*, 233–241.

Zhu, L. R., Thomas, P. E., Lu, G., Reuhl, K. R., Yang, G. Y., Wang, L. D., et al. (2006). CYP2A13 in human respiratory tissues and lung cancers: An immunohistochemical study with a new peptide-specific antibody. *Drug Metabolism and Disposition*, *34*, 1672–1676.

CHAPTER TWO

Involvement of Cytochrome P450 in Reactive Oxygen Species Formation and Cancer

Eugene G. Hrycay[1,2], Stelvio M. Bandiera
Faculty of Pharmaceutical Sciences, The University of British Columbia, Vancouver, British Columbia, Canada
[1]Corresponding author: e-mail address: eugenehrycay@netscape.net

Contents

Abstract

This review examines the involvement of cytochrome P450 (CYP) enzymes in the formation of reactive oxygen species in biological systems and discusses the possible involvement of reactive oxygen species and CYP enzymes in cancer. Reactive oxygen species are formed in biological systems as byproducts of the reduction of molecular

[2] Dr. Eugene Hrycay is presently retired.

☆Members of the CYP superfamily are called enzymes and should not be referred to as isoforms or iso(en)-zymes. All CYP members are different enzymes and are products of different genes of the CYP superfamily (Dr. R. Feyereisen, CNRS, Univ. Nice Sophia Antipolis, France, personal communication).

Advances in Pharmacology, Volume 74
ISSN 1054-3589
http://dx.doi.org/10.1016/bs.apha.2015.03.003

oxygen and include the superoxide radical anion ($^{\bullet}O_2^-$), hydrogen peroxide (H_2O_2), hydroxyl radical ($^{\bullet}OH$), hydroperoxyl radical ($HOO^{\bullet}$), singlet oxygen ($^1O_2$), and peroxyl radical ($ROO^{\bullet}$). Two endogenous sources of reactive oxygen species are the mammalian CYP-dependent microsomal electron transport system and the mitochondrial electron transport chain. CYP enzymes catalyze the oxygenation of an organic substrate and the simultaneous reduction of molecular oxygen. If the transfer of oxygen to a substrate is not tightly controlled, uncoupling occurs and leads to the formation of reactive oxygen species. Reactive oxygen species are capable of causing oxidative damage to cellular membranes and macromolecules that can lead to the development of human diseases such as cancer. In normal cells, intracellular levels of reactive oxygen species are maintained in balance with intracellular biochemical antioxidants to prevent cellular damage. Oxidative stress occurs when this critical balance is disrupted. Topics covered in this review include the role of reactive oxygen species in intracellular cell signaling and the relationship between CYP enzymes and cancer. Outlines of CYP expression in neoplastic tissues, CYP enzyme polymorphism and cancer risk, CYP enzymes in cancer therapy and the metabolic activation of chemical procarcinogens by CYP enzymes are also provided.

ABBREVIATIONS

AFB_1 aflatoxin B_1
Cpd I Compound I
CYP cytochrome P450
DMBA 7,12-dimethylbenz[*a*]anthracene
Eq. equation
Por porphyrin
Redox reduction-oxidation

1. INTRODUCTION

Biological life is driven primarily by energy, which is derived mainly from the biological oxidation of a variety of compounds (Szent-Györgyi, 1982). The primary oxidant in this process is molecular oxygen (O_2). Oxygen is a highly abundant element that is found in the atmosphere and terrestrially as the diatomic molecule (O_2), chemically bound with hydrogen in water (H_2O), or in anionic form as hydroxide (OH^-), carbonate (CO_3^{2-}), nitrate (NO_3^-), and sulfate (SO_4^{2-}) (Lewis, 2002). In biological systems, oxygen is found in macromolecules such as lipids, proteins, nucleic acids, and carbohydrates and in small organic molecules including amino acids, fatty acids, nucleotides, sugars, steroid hormones, and vitamins.

Cytochrome P450 (CYP) enzymes and an electron transfer partner, the nonheme iron–sulfur protein (ferredoxin) (Davis, 2002; Wächtershäuser, 1988; Wickramasinghe & Villee, 1975), are postulated to have arisen in an anaerobic prokaryote some 3.5 billion years ago in a time that predated atmospheric oxygen and the combustion of organic matter (Hrycay & Bandiera, 2015; Nelson et al., 1993; Wickramasinghe & Villee, 1975). CYP enzymes and nonheme iron–sulfur proteins display low iron reduction-oxidation (redox) potentials (Lewis, 2001; Wickramasinghe & Villee, 1975) and could have functioned in anaerobic reductive reactions carried out by ancient prokaryotes in the early anoxic terrestrial environment. Even today, anaerobic reductive reactions driven by mammalian CYP enzyme systems continue to play a vital role in the reduction of O_2, hydroperoxides, arene- and *N*-oxides, azido and azo compounds, halogen and nitro compounds, hydroxylamines, and other xenobiotics (Goeptar, Scheerens, & Vermeulen, 1995; Hrycay & Bandiera, 2008, 2012). A crucial reductive activity of the CYP enzyme system, discovered by Hrycay & O'Brien (1971, 1972), is its peroxidase function whereby CYP enzymes reduce biological hydroperoxides such as hydrogen peroxide, lipid hydroperoxides, steroid hydroperoxides and exogenous organic hydroperoxides (Hrycay & Bandiera, 2012, 2015; Hrycay, Gustafsson, Ingelman-Sundberg, & Ernster, 1975; Hrycay & O'Brien, 1971, 1972, 1973; Hrycay & Prough, 1974). The ancient anoxic environment could have been relatively rich in H_2O_2 and peroxygenated organic chemicals (Halliwell, 2006a; Joo, Lin, & Arnold, 1999; Lane, 2002; McKay & Hartman, 1991; Samuilov, 1997), raising the interesting possibility that CYP enzymes of ancient prokaryotes used H_2O_2 and other peroxy compounds as oxygen atom donors for monooxygenase reactions (Hrycay & Bandiera, 2012, 2015). Under global anaerobic conditions, primordial CYP enzymes could have functioned as peroxidases and peroxygenases well before their monooxygenase function developed.

Around 2.45–2.32 billion years ago, primitive anaerobic photosynthesizing cyanobacteria are believed to have heralded the process of oxygenating the Earth's atmosphere (Schirrmeister, de Voss, Antonelli, & Bagheri, 2013). Steadily increasing atmospheric O_2 levels, coupled with the reductive metabolism of O_2 by prokaryotic CYP enzymes, produced reactive oxygen species such as the superoxide radical anion ($^{\bullet}O_2^-$), H_2O_2 and the hydroxyl radical ($^{\bullet}OH$) (Table 1) that are toxic to anaerobic and aerobic life-forms (Wickramasinghe & Villee, 1975). Anaerobic organisms evolved antioxidant defense systems to protect against O_2 cytotoxicity. Moreover, the inexorable rise in atmospheric O_2 levels was advantageous

Table 1 Major Reactive Oxygen Species Formed in Biological Systems[a]

Reactive Oxygen Species	Description and/or Reactivity
Free radicals	
Superoxide radical anion, $^{\cdot}O_2^-$	One-electron reduction state of O_2; formed as a byproduct of the CYP-dependent microsomal electron transport chain; also formed in reactions catalyzed by NADPH oxidases and xanthine oxidase
Hydroperoxyl radical, $HOO^{\cdot}$	Protonated form of $^{\cdot}O_2^-$ and more lipid soluble
Hydroxyl radical, $^{\cdot}OH$	Extremely reactive; three-electron reduction state of O_2; primarily formed by the Fenton reaction
Peroxyl radical, $ROO^{\cdot}$	Generated by hydrogen atom radical abstraction from a polyunsaturated fatty acid by a reactive oxygen radical species (e.g., $^{\cdot}OH$), generating a lipid radical ($R^{\cdot}$) that then reacts with O_2
Alkoxyl radical, $RO^{\cdot}$	Generated by the reaction of a lipid hydroperoxide with ferrous iron or purified CYP1A2
Nonradicals	
Peroxide dianion, O_2^{2-}	Two-electron reduction state of O_2; primarily formed by the one-electron reduction of $^{\cdot}O_2^-$
Hydrogen peroxide, H_2O_2	Protonated form of O_2^{2-}; generated by dismutation of superoxide anion catalyzed by superoxide dismutase; H_2O_2 is converted to water by catalase and is utilized for generation of the more highly reactive $^{\cdot}OH$ radical
Singlet molecular oxygen (known simply as singlet oxygen), $^1\Delta gO_2$, abbreviated as $^1O_2$	Highly reactive; red or infrared photoemission; nonradical singlet oxygen ($^1\Delta gO_2$) is generated from the radical singlet oxygen state ($^1\sum g^+O_2$) initially formed from triplet $^3O_2$
Ozone, O_3	Toxic oxidizing species formed in air

[a]Reactive oxygen species of a radical and nonradical nature are listed in the table. R signifies an organic substituent. Information shown in the table was compiled from references Halliwell (2006a, 2006b), Halliwell and Gutteridge (1984), Manda, Nechifor, and Neagu (2009), Rhee, Chae, and Kim (2005), Rota et al. (1997), Sies (1986), and Winterbourne, Hampton, Livesey, and Kettle (2006).

in that it led to the formation of the ozone (O_3) layer in the stratosphere that protected living organisms from solar UV radiation, and removed ferrous iron (Fe^{2+}) from aqueous environments by forming insoluble ferric complexes (Halliwell, 2006a). Removing Fe^{2+} was advantageous because

Fe^{2+} reacts with H_2O_2 to produce the extremely toxic hydroxyl radical ($^{\bullet}OH$) via the Fenton reaction, as shown in equation Eq. (1).

$$Fe^{2+} + H_2O_2 \rightarrow [Fe(II)H_2O_2] \rightarrow Fe^{3+} + {}^{\bullet}OH + HO^- \quad (1)$$

Increased atmospheric O_2 levels triggered the evolution of multicellular life-forms containing specific CYP enzymes that recognized distinct structural classes of chemicals (Lewis & Sheridan, 2001). These eukaryotic CYP enzymes developed critical catalytic functions that were used to metabolize endogenous compounds and xenobiotics (Nelson et al., 1993; Wickramasinghe & Villee, 1975). One vital CYP activity that evolved in an oxygen-rich environment was the monooxygenase function, which is retained to this day in contemporary organisms because of its importance in anabolic and catabolic processes.

Reactive oxygen species, which are a normal attribute of aerobic life today, are capable of oxidatively damaging cellular membranes and macromolecules such as lipids, proteins, nucleic acids, and carbohydrates, leading to the development and progression of numerous human diseases including cancer (Table 2). CYP enzymes are key players in the formation of reactive oxygen species and may be involved in cancer pathogenesis (Rodriguez-Antona & Ingelman-Sundberg, 2006). This review examines the involvement of CYP enzymes in reactive oxygen species formation in biological systems, describes the antioxidant systems used by humans to prevent oxidative damage to cellular components by reactive oxygen species, and discusses the vital roles played by reactive oxygen species and CYP enzymes in the development and progression of cancer.

2. REACTIVE OXYGEN SPECIES FORMATION IN BIOLOGICAL SYSTEMS

2.1 Oxygen Reactivity and Formation of Reactive Oxygen Species

In its most common and stable form, oxygen exists in the triplet ground state ($^3\sum g^- O_2$, abbreviated as $^3O_2$), which occurs due to the presence of two unpaired electrons that have parallel spins. The presence of unpaired electrons residing in two antibonding molecular orbitals signifies that molecular oxygen is a diradical and paramagnetic species (Lewis, 2002). In contrast, the majority of organic molecules in living organisms exist in singlet state and contain paired electrons in their molecular orbitals. Reactions between triplet and singlet molecules are termed "spin forbidden" and are not

Table 2 Selected Human Diseases/Disorders Associated with Reactive Oxygen Species

Human Diseases/Disorders	References
Aging disorders	Afanas'ev (2011), Alfadda and Sallam (2012), Ashok and Ali (1999), Cadenas and Davies (2000), El Assar, Angulo, and Rodríguez-Mañas (2013), Gorenne and Bennett (2010), Halliwell (1987), Kidd (1997), Krause (2007), Kregel and Zhang (2007), Repetto, Semprine, and Boveris (2012), and Serra et al. (2009)
Cancer (e.g., cancer of the bladder, breast, colon, kidney, liver, lung, ovary, pancreas, prostate, skin, small intestine, stomach)	Afanas'ev (2011), Agbor (2011), Bickers and Athar (2006), Brieger, Schiavone, Miller, and Krause (2012), Handa, Naito, and Yoshikawa (2011), Hecht (1999), Klaunig and Kamendulis (2004), Manda et al. (2009), Marks and Fürstenberger (1985), Petros et al. (2005), Repetto et al. (2012), Valavanidis, Vlachogianne, Fiotakis, and Loridas (2013), Waris and Ahsan (2006), and Wiseman and Halliwell (1996)
Heart diseases (e.g., angina pectoris, aortic valve stenosis, chronic heart failure, ischemia, myocardial infarction)	Alfadda and Sallam (2012), Birukov (2006), Gottlieb (2003), Gorenne and Bennett (2010), Halliwell (1987), Peña-Silva, Miller, Chu, and Heistad (2009), Rahman, Hosen, Islam, and Shekhar (2012), Rowland and Mangoni (2014), Sato, Yokoyama, Fujita, Okumura, and Ishikawa (2011), and Singh et al. (1995)
Inflammatory diseases/disorders (e.g., asthma, atherosclerosis, inflammatory bowel disease, lupus, rheumatoid arthritis, sepsis)	Alfadda and Sallam (2012), Alvarez, Evelson, and Cimolai (2008), Birukov (2006), Fearon and Faux (2009), Gorenne and Bennett (2010), Halliwell (1987), Halliwell (2007), Harris et al. (1992), Kaneto, Katakami, Matsuhisa, and Matsuoka (2010), Paoletti, Gotto, and Hajjar (2004), Parke, Ioannides, Lewis, and Parke (1991), Ramos de Andrade Jr., Becco de Souza, Alves dos Santos, and Ramos de Andrade (2005), Rosenfeld (1998), Rossi (2012), Singh et al. (1995), Wiseman and Halliwell (1996), Wood, Gibson, and Garg (2003), Woodruff et al. (1986), and Young and Woodside (2001)

Liver diseases/injuries (e.g., alcoholic and nonalcoholic steatohepatitis, CCl_4 liver injury, endotoxin liver injury, hepatosteatosis)	Abdelmegeed, Moon, Hardwick, Gonzalez, and Song (2009), Abdelmegeed et al. (2012), Abdelmegeed et al. (2013), Al-Omar, Beedham, and Alsarra (2004), Comporti (1985), Halliwell (1987), Ibrahim et al. (2008), Pohl, Schulick, Highet, and George (1984), Tsuji et al. (2005), and Uzma, Kumar, and Anees (2011)
Lung diseases/disorders (e.g., acute respiratory distress syndrome, chronic obstructive pulmonary disease, emphysema, pulmonary fibrosis, sleep apnea syndrome)	Halliwell (1987), Kondrikov, Caldwell, Dong, and Su (2011), Rahman et al. (2012), and Ramos de Andrade et al. (2005)
Neurodegenerative diseases/disorders (e.g., Alzheimer's disease, amyotrophic lateral sclerosis, ataxia, bipolar disorder, Down's syndrome, epilepsy, Huntington's disease, multiple sclerosis, muscular dystrophy, Parkinson's disease, prion diseases, schizophrenia, Tardive Dyskinesia)	Al-Omar et al. (2004), Barnham, Masters, and Bush (2004), Behl (1999), Brieger et al. (2012), de Vries, Schreibelt, and van Horssen (2011), Farooqui and Farooqui (2011), Halliwell (2006b), Kovacic and Somanathan (2012), Murphy and Kehrer (1989), Pohanka (2013), Rahman et al. (2012), Repetto et al. (2012), and Valko et al. (2007)

energetically favorable, requiring a high activation energy (Lewis, 2002). Thus, the chemical reactivity of triplet molecular oxygen with organic compounds at ambient temperatures is intrinsically low, which is a highly desirable situation for aerobic life-forms that rely on oxygen for respiration. If this were not the case, oxygen would spontaneously oxidize organic molecules, which would be extremely detrimental to living organisms (Hrycay & Bandiera, 2012; Sono, Roach, Coulter, & Dawson, 1996).

The O_2 molecule is utilized in a multitude of synthetic biochemical reactions that are fundamental to most aspects of cell growth, development, and reproduction of aerobic organisms. As the ultimate electron acceptor in cellular metabolism, O_2 is a four-electron oxidant that can be enzymatically activated and reduced to H_2O by the stepwise sequential addition of four electrons (Goeptar et al., 1995; Halliwell, 2006a, 2006b; Kappus, 1993). This reduction process generates reactive oxygen species such as the superoxide radical anion ($^{\cdot}O_2^{-}$), peroxide dianion (O_2^{2-}) (which is protonated to form H_2O_2), and the hydroxyl radical ($^{\cdot}OH$) (Bast, 1986). Generation of reactive oxygen species occurs at a controlled rate in healthy cells of aerobic organisms. Reactive oxygen species can oxidatively damage membranes of organelles and cells, as well as macromolecules such as lipids, proteins, nucleic acids, and carbohydrates (Kappus & Sies, 1981; Lewis, 2002; Sies, 1986; Wiseman & Halliwell, 1996), resulting in a condition known as oxidative stress. As currently defined by Sies and Jones, oxidative stress refers to an imbalance between oxidants and antioxidants in favor of the oxidants, leading to a disruption of cellular redox signaling and control and/or to molecular damage (Sies, 2014; Sies & Jones, 2007). Severe oxidative stress can result in cellular apoptosis or necrosis and cell death (Al-Omar et al., 2004; Lennon, Martin, & Cotter, 1991).

Reactive oxygen species can be produced by both endogenous and exogenous sources (Cash, Pan, & Simon, 2007). A potential endogenous source of reactive oxygen species is the mitochondrial electron transport chain (Alfadda & Sallam, 2012). The process of oxidative phosphorylation within mitochondria generates ATP as an energy source for mammalian cells. Mitochondria in human cells reduce about 95% of the oxygen they take up to water, a process that is achieved by mitochondrial cytochrome oxidase. Cytochrome oxidase is a large membrane-associated multiprotein assembly containing transition metal ions (iron and copper) at its active site, which allows transfer of single electrons to facilitate redox reactions (Babcock, 1999; Halliwell, 2006a; Montezano & Touyz, 2012). The terminal cytochrome oxidase complex catalyzes several single-electron reduction steps by which four electrons are added sequentially to each O_2 molecule at normal oxygen levels. Cytochrome oxidase keeps the partially reduced oxygen intermediates formed in the reduction process safely bound until they can be fully reduced to water, without releasing the superoxide radical or other reactive oxygen radicals from the mitochondrial membrane (Halliwell, 2006a; Sies, 1997). Only 0.15% of the total molecular oxygen consumed by mitochondria in a healthy cell is converted into superoxide, due to incomplete reduction of O_2 and "electron leakage" during the electron transport process (Han, Williams, & Cadenas, 2001; Kagan et al., 2009; St-Pierre, Buckingham, Roebuck, & Brand, 2002).

Another major endogenous source of reactive oxygen species is the mammalian CYP-dependent microsomal electron transport system (see Section 4) (Halliwell, 2006b; Hrycay & Bandiera, 2012, 2015; Lewis, 2002; Zangar, Davydov, & Verma, 2004). Additional sources include peroxisomes (Rahman et al., 2012); cardiac and vascular cells (Montezano & Touyz, 2012); brain cells (Barnham et al., 2004; Behl, 1999); phagocytic cells such as leukocytes, macrophages, monocytes, neutrophils, and eosinophils (Sies, 1986; Vagula & Konieczko, 2012); and various oxidases such as NADPH oxidases, aldehyde oxidase, glucose oxidase, and xanthine oxidase (Al-Omar et al., 2004; El-Aal, 2012; Halliwell & Gutteridge, 1984; Manda et al., 2009; Maraldi, 2013; Montezano & Touyz, 2012).

2.2 Types of Reactive Oxygen Species and Their Reactivities

The most common reactive oxygen species generated in cells of aerobic organisms include the superoxide radical anion ($^{\bullet}O_2^-$), hydrogen peroxide (H_2O_2), hydroxyl radical ($^{\bullet}OH$), peroxyl radical ($ROO^{\bullet}$), alkoxyl radical ($RO^{\bullet}$) (R signifies an organic substituent), and singlet oxygen ($^1\Delta gO_2$, abbreviated as $^1O_2$) (see Table 1).

2.2.1 Superoxide Radical Anion

As mentioned above, the superoxide radical anion ($^{\bullet}O_2^-$) is formed by a one-electron reduction of O_2 (Bast, 1986; Halliwell, 2006a, 2006b; Halliwell & Gutteridge, 1984). The electron enters one of the two π^* 2p antibonding orbitals of O_2, leaving an ionic species with a single unpaired electron and a negative charge (Halliwell, 2006a). As is the case with molecular oxygen, the reaction of $^{\bullet}O_2^-$ with singlet molecules is spin forbidden. Therefore, $^{\bullet}O_2^-$ generally displays low reactivity in biological systems (Manda et al., 2009), is short-lived (Montezano & Touyz, 2012), and is deactivated by dismutation in a disproportionation reaction. Dismutation of superoxide can be spontaneous, or enzymatic via a rapid reaction catalyzed by superoxide dismutase (Al-Omar et al., 2004), leading to the formation of H_2O_2 and O_2, as shown in Eq. (2).

$$2^{\bullet}O_2^{-} + 2H^{+} \rightarrow H_2O_2 + O_2 \quad (2)$$

The negative charge on $^{\bullet}O_2^{-}$ makes it unable to cross cellular membranes except via transmembrane anion channels (Kellogg & Fridovich, 1977). Superoxide is not a strong oxidant but it is a precursor of other reactive oxygen species such as hydrogen peroxide, the hydroxyl radical, and peroxynitrite.

As mentioned above, superoxide is generated during the sequential reduction of O_2 in the mitochondrial electron transport chain and in CYP-catalyzed microsomal oxidative reactions. Superoxide is utilized by the immune system and is produced in large quantities by NADPH oxidases in phagocytes, as part of the oxygen-dependent mechanism used to destroy invading microorganisms (Maraldi, 2013).

2.2.2 Hydrogen Peroxide

Addition of two electrons to O_2 produces the peroxide dianion (O_2^{2-}), which is protonated to form H_2O_2. H_2O_2 is generated mainly by spontaneous dismutation of $^{\bullet}O_2^{-}$ or in a reaction catalyzed by superoxide dismutase. H_2O_2 can also be generated by the mammalian CYP-dependent microsomal electron transport system (see Section 4).

H_2O_2 is unstable and a strong oxidizing agent, and when present in sufficient quantities can interfere with normal cellular metabolism. Unlike the superoxide anion, H_2O_2 is lipid soluble and, due to its structural similarity to water and low innate reactivity, can diffuse freely in and out of cells and through tissues to other sites where it can exert its oxidative actions (Manda et al., 2009; Montezano & Touyz, 2012). H_2O_2 can also serve as the source of more toxic reactive oxygen species such as the extremely reactive hydroxyl radical ($^{\bullet}OH$), formed primarily by the Fenton reaction (Koppenol, 2001), and hypochlorous acid (HOCl), formed by the myeloperoxidase reaction (Manda et al., 2009). For this reason, H_2O_2 levels are tightly regulated by intracellular and extracellular enzymes, including catalase, glutathione peroxidases, and peroxiredoxins (Manda et al., 2009; Park, Lee, Lee, & Kang, 2014; Sies, 2014). H_2O_2 is rapidly converted to H_2O and O_2 by catalase. Glutathione peroxidases reduce H_2O_2 by transferring the energy of the reactive peroxides to reduced glutathione (GSH) (Table 3). Mammalian peroxiredoxins catalyze the NADPH-dependent

Table 3 Major Enzymatic and Nonenzymatic Antioxidants Used in Mammalian Systems[a]

Antioxidants	Description/Reaction Catalyzed
Enzymatic	
Superoxide dismutases (SOD[b])	The three SODs are CuZn-SOD (SOD1), Mn-SOD (SOD2), EC-SOD (SOD3); $2O_2^- + 2H^+ \rightarrow H_2O_2 + O_2$
Glutathione peroxidases	There are five mammalian GSH peroxidases, which are mostly selenoenzymes that can act on H_2O_2 and other hydroperoxides; $H_2O_2 + 2GSH \rightarrow GSSG + 2H_2O$
Catalase	Heme enzyme localized mainly in peroxisomes; $H_2O_2 + H_2O \rightarrow 2H_2O + O_2$
Glutathione reductase	Key enzyme of GSSG reduction; $GSSG + NADPH + H^+ \rightarrow 2GSH + NADP^+$
Glutathione-*S*-transferases	Involved in the conjugation (with GSH) of 4-hydroxy-nonenal (a product of lipid peroxidation), endogenous substrates including H_2O_2 and lipid hydroperoxides, and xenobiotics such as carcinogens, mutagens, and anticancer drugs
Peroxiredoxins	In mitochondria, cytosol and nucleus, mammalian peroxiredoxins catalyze the NADPH-mediated reduction of H_2O_2 and other peroxides by coupling with thioredoxin and thioredoxin reductase
Nonenzymatic	
Glutathione (GSH)	Potent water-soluble reducing agent and radical scavenger; also acts as a cofactor for glutathione peroxidases and glutathione-*S*-transferases
β-Carotene (provitamin A)	Lipid-soluble radical scavenger and singlet oxygen quencher
Riboflavin (vitamin B2)	Radical scavenger involved in the renewal of GSH
Ascorbic acid (vitamin C)	Water-soluble reducing agent and radical scavenger; also functions in vitamin E regeneration

Continued

Table 3 Major Enzymatic and Nonenzymatic Antioxidants Used in Mammalian Systems—cont'd

Antioxidants	Description/Reaction Catalyzed
α-Tocopherol (vitamin E)	Lipid-soluble radical scavenger that inhibits lipid peroxidation
Lycopene	Lipid-soluble carotenoid peroxyl radical scavenger and singlet oxygen quencher
Melatonin	Lipid-soluble compound that reacts with H_2O_2 to produce less reactive species; potent peroxyl radical and hydroxyl radical scavenger that inhibits lipid peroxidation; stimulates antioxidant enzymes to maintain normal ROS levels
Chelators of free iron or copper (e.g., desferrioxamine, penicillamine)	Desferrioxamine acts as a chelator of free iron and a hydroxyl radical scavenger; penicillamine acts as a chelator of copper and a scavenger of ROS; both inhibit reactions that produce peroxides and hydroxyl radicals
Punicalagin	Most abundant antioxidant found in pomegranate juice
Plant flavonoids (e.g., capsaicin, daidzein, garlic, genistein, kaempferol, lutein, quercetin, resveratrol, rottlerin, rutin)	Many flavonoids display anti-inflammatory and antioxidant properties
Herbs (e.g., bilberry, blackberry, ginkgo biloba, grape seed, turmeric)	Certain compounds found in herbs display anti-inflammatory and antioxidant properties

[a]Information shown in the table was compiled from references Al-Omar et al. (2004), Alfadda and Sallam (2012), Cámara et al. (2013), El-Aal (2012), Grune et al. (2010), Gutman (2002), Halliwell (2006a, 2006b), Hofmann, Hecht, and Flohé (2005), Ishikawa, Esterbauer, and Sies (1986), Kidd (1997), Lewis (2002), Manda et al. (2009), Mortensen and Skibsted (1998), Park et al. (2014), Pieri, Marra, Moroni, Recchioni, and Marcheselli (1994), Reiter (1997), Reiter, Carneiro, and Oh (1997), Sies (1986, 2014), Stahl and Sies (1996), Vagula and Konieczko (2012), Young and Woodside (2001), and Zhou, Jing, Styblo, Chen, and Waxman (2005).
[b]*Abbreviations*: GSSG, oxidized glutathione; ROS, reactive oxygen species; SOD, superoxide dismutase.

reduction of H_2O_2 and other peroxides by coupling with thioredoxin and thioredoxin reductase.

2.2.3 Hydroxyl Radical

The hydroxyl radical ($^{\bullet}OH$), the three-electron reduction state of O_2, can be generated *in vivo* by the iron II (or copper I)-catalyzed decomposition of H_2O_2 via the Fenton reaction. In addition, stepwise univalent reduction of O_2 leads to the initial formation of $^{\bullet}O_2^-$, which can then act as an electron donor in the generation of the hydroxyl radical by the Haber–Weiss reaction, as shown in Eq. (3).

$$^{\bullet}O_2^- + H^+ + H_2O_2 \rightarrow O_2 + {^{\bullet}OH} + H_2O \quad (3)$$

It has since been shown that the Haber–Weiss reaction has a limited role in the production of $^{\bullet}OH$, and superoxide has been established as the source of H_2O_2, which generates hydroxyl radicals via the Fenton reaction (see Eq. 1) (Koppenol, 2001). The hydroxyl radical can also be produced during CYP-catalyzed hepatic microsomal oxidative reactions (see Section 4).

Although the hydroxyl radical has a very short half-life (10^{-9} s) (Sies, 1993), it is one of the most reactive oxygen species due to its high reduction potential, compared to other oxygen species (Al-Omar et al., 2004). The hydroxyl radical is carcinogenic (El-Aal, 2012) and reacts indiscriminately, with extremely high rate constants, with almost every type of molecule found in living cells, including lipids, amino acids, nucleotides, and sugars (Halliwell, 2006a, 2006b; Young & Woodside, 2001). The hydroxyl radical is so reactive that no enzyme system can use it as a substrate. Its activity is minimized by the removal of H_2O_2 and transition metals from the cell. Unlike superoxide and H_2O_2, the hydroxyl radical cannot be eliminated by an enzymatic reaction, as this would require its diffusion to the enzyme's active site. As diffusion is slower than the radical half-life, the hydroxyl radical is neutralized by harmful reactions with any oxidizable molecule in its vicinity (Halliwell, 2006a, 2006b; Manda et al., 2009; Young & Woodside, 2001). The major means of protecting cellular structures and molecular components from hydroxyl radical-mediated damage are effective repair systems (Halliwell, 2006a; Halliwell & Gutteridge, 2007; Manda et al., 2009; Sies, 1986, 1993, 1997).

2.2.4 Peroxyl Radical

The peroxyl radical ($ROO^{\bullet}$) (R signifies an organic substituent) can be generated in an initiation reaction involving the hydroxyl radical. Thus, $^{\bullet}OH$

initially abstracts a hydrogen atom from the hydrocarbon side-chain of a membrane polyunsaturated fatty acid residue to produce a carbon-centered fatty acid radical ($R^{\bullet}$), which reacts quickly with O_2 to generate the reactive fatty acid peroxyl radical (Halliwell, 2006b). Peroxyl radicals can oxidize membrane proteins and cholesterol, as well as fatty acid side-chains.

2.2.5 Singlet Oxygen

Singlet oxygen is produced by photochemical, thermal, chemical, or enzymatic activation of O_2. Singlet oxygen (represented as $^1\Delta gO_2$, abbreviated as $^1O_2$) is not a radical and represents an excited state of O_2 in which the spin of one of the unpaired electrons is changed to yield two electrons with opposite spins. The newly paired electrons can exist in separate antibonding orbitals or in the same orbital. Singlet oxygen ($^1O_2$) has a half-life of 10^{-5} s (Sies, 1993) and is occasionally encountered in biological systems (Halliwell, 2006a; Sies, 1986). Singlet oxygen is a very strong oxidant and readily oxidizes cellular components such as lipids, proteins, and nucleic acids. Singlet oxygen can be quite long-lived in a cell and can diffuse over appreciable distances, including across cellular membranes into extracellular compartments (Manda et al., 2009). Most of the singlet oxygen in biology is generated by photosensitization reactions (Halliwell, 2006b) in which an appropriate sensitizer is electronically excited and then transfers energy to oxygen (Sies, 1986). In photodynamic therapy, a photosensitizer is excited by irradiation to produce singlet oxygen, which is then used to destroy cancer cells (Manda et al., 2009). Singlet oxygen can also be generated in CYP-catalyzed hepatic microsomal oxidative reactions (see Section 4).

3. PROTECTIVE AGENTS TO COMBAT REACTIVE OXYGEN SPECIES

Rapid detoxification of reactive oxygen species is critical for the survival of all aerobic life-forms. To reduce the damaging effects of physiologically generated reactive oxygen species and maintain a state of homeostasis, aerobic life-forms strive to keep these highly reactive oxygen species under tight control with the aid of a diverse system of antioxidants (Alfadda & Sallam, 2012; Al-Omar et al., 2004; El-Aal, 2012; Halliwell & Gutteridge, 2007; Ilie & Margină, 2012; Veskoukis, Tsatsakis, & Kouretas, 2012). Normally, intracellular levels and deleterious effects of reactive oxygen species are maintained in balance with intracellular biochemical antioxidants to prevent cellular damage. In healthy cells, the balance between production of reactive oxygen species and their removal by antioxidants is sometimes referred to as the redox state of a cell. Oxidative stress occurs when this critical balance is disrupted by the excessive production of reactive oxygen species or depletion of antioxidants, or both (Waris & Ahsan, 2006). A number of antioxidant defense systems have evolved to counteract the destructive effects of reactive oxygen species and minimize their potential to cause tissue damage (Bickers & Athar, 2006). Today, levels of reactive oxygen species in biological cells and extracellular body fluids are controlled by the concerted actions of enzymatic and nonenzymatic antioxidants (Steinbrenner & Sies, 2009). Table 3 outlines some of the reactions catalyzed by antioxidant enzymes and briefly describes the roles played by antioxidant enzymes and nonenzymatic antioxidants to combat the cytotoxic effects of reactive oxygen species.

3.1 Antioxidants

3.1.1 Antioxidant Enzymes

Cellular antioxidant enzymes are the primary defense against reactive oxygen species, with the most important enzymes being superoxide dismutases, catalase, glutathione peroxidases, glutathione reductase, glutathione-*S*-transferases, and peroxiredoxins. The major function of antioxidant enzymes is averting reactive oxygen radical-induced tissue damage by preventing formation of reactive oxygen radical species (Sies, 1997; Young & Woodside, 2001), or by scavenging the highly reactive oxygen radical species and neutralizing them to inactive compounds (Al-Omar et al., 2004). There are three mammalian forms of superoxide dismutase that catalyze the dismutation of superoxide anion to H_2O_2 and O_2 (see Table 3) (Antonyuk, Strange, Marklund, & Hasnain, 2009). The importance of superoxide dismutase as the body's first line of defense against reactive oxygen species is highlighted by studies showing that genetic inactivation of superoxide dismutase produces severely compromised phenotypes in organisms ranging from bacteria to mice. Mice lacking mitochondrial manganese superoxide dismutase die at approximately 21 days after birth due to neurodegeneration, cardiomyopathy and lactic acidosis, while mice lacking cytosolic copper-zinc superoxide dismutase suffer multiple pathologies including reduced lifespan, liver cancer, muscle atrophy, thymic involution, anemia, and decreased female fertility (Muller, Lustgarten, Jang, Richardson, & Van Remmen, 2007).

Catalase rapidly converts H_2O_2 to O_2 and H_2O, whereas glutathione peroxidases convert H_2O_2 only to H_2O. Glutathione reductase maintains

GSH in the reduced state, so that it can be used by glutathione-*S*-transferases to neutralize diverse electrophilic and reactive compounds by conjugating them with GSH. Glutathione-*S*-transferases also metabolize H_2O_2 and many other endogenous hydroperoxides (Al-Omar et al., 2004; Vagula & Konieczko, 2012; Zhou et al., 2005). Mammalian peroxiredoxins are the most recently identified family of antioxidant enzymes that act as peroxidases and catalyze the NADPH-dependent reduction of H_2O_2 and other peroxides by coupling with thioredoxin and thioredoxin reductase (Halliwell & Gutteridge, 2007; Hofmann et al., 2005; Park et al., 2014; Rhee et al., 2005). As a cofactor for peroxiredoxins, human thioredoxin-1 is a powerful singlet oxygen quencher, hydroxyl radical scavenger, and reducer of hydroperoxides (Das & Das, 2000; Manda et al., 2009). Thioredoxin reductase-1 is a selenoprotein with oxidoreductase activity and is involved in the reduction of cellular peroxides and reactive oxygen species (Nordberg & Arner, 2001). The essential trace element selenium exerts its antioxidant function primarily in the form of selenocysteine residues as an integral constituent of selenoenzymes such as glutathione peroxidases, thioredoxin reductases, and selenoprotein P, which efficiently detoxify reactive oxygen species (Steinbrenner & Sies, 2009). Some of the reactions catalyzed by antioxidant enzymes are shown in the scheme below.

$$2^{\cdot}O_2^- + 2H^+ \xrightarrow[\text{Superoxide dismutase}]{} H_2O_2 + O_2 \qquad (4)$$

$$2H_2O_2 \xrightarrow[\text{Catalase}]{} 2H_2O + O_2 \qquad (5)$$

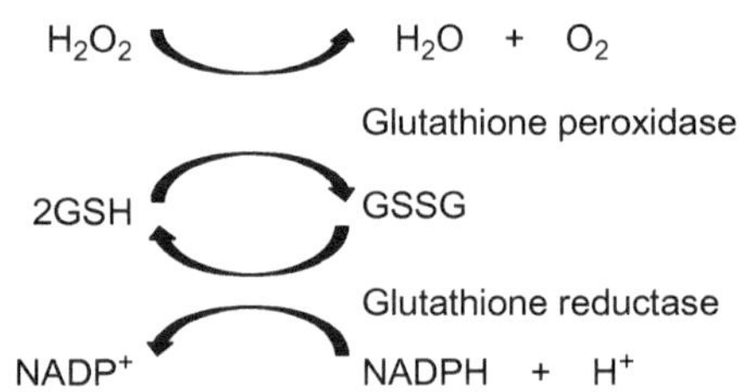

(6)

3.1.2 Nonenzymatic Antioxidants

Low-molecular-weight nonenzymatic antioxidants add significantly to the prolific defense provided by antioxidant enzymes. Nonenzymatic antioxidants include endogenous agents such as GSH, hydroquinone, melatonin and NADPH, and dietary antioxidants such as β-carotene (provitamin A), riboflavin (vitamin B2), ascorbic acid (vitamin C), α-tocopherol (vitamin E), lycopene, and flavonoids, all of which can protect against the deleterious effects of reactive oxygen species (see Table 3 for references). The characteristics and reactivities of the major nonenzymatic antioxidants are briefly described below.

3.1.2.1 Glutathione

GSH is one of the most important cellular antioxidants and is the first line of defense against reactive oxygen species. GSH is present in mammalian cells in reduced (GSH) and oxidized (GSSG) forms (Meister & Anderson, 1983). The cellular concentration of the reduced form is normally 10–100 times greater than that of the oxidized form and the GSH/GSSG ratio is closely regulated and can serve as an indicator of the health of a cell (Kidd, 1997). GSH displays high electron-donating capacity, linked to its sulfhydryl (–SH) group, as indicated by the high negative redox potential of the GSH/GSSG redox couple $\left(E_0' = -0.33\text{V}\right)$. This renders GSH a potent antioxidant, a powerful reducing agent, and a suitable cofactor for antioxidant enzymes such as glutathione peroxidases and glutathione-*S*-transferases that require readily available reducing equivalents. In addition to serving as a cofactor for antioxidant enzymes, GSH can function directly as an antioxidant, scavenging a variety of free-radical species (Young & Woodside, 2001).

The considerable reducing power of GSH is a measure of its electron-donating, sulfhydryl-donating, and free-radical-scavenging capacity and is key to the multiple actions of the thiol at molecular, cellular and tissue levels. GSH also plays a vital role in the regeneration of the antioxidant vitamins C and E (Veskoukis et al., 2012). Moreover, GSH functions as the primary thiol in cellular redox reactions and is a major determinant of, and contributor to, the overall cellular redox balance (El-Aal, 2012; Schafer & Buettner, 2001). Appropriately, GSH has been termed "the body's most powerful protector" (Gutman, 2002).

3.1.2.2 Vitamin E, Vitamin C, and β-Carotene

Vitamin E and β-carotene are lipid-soluble antioxidants, while vitamin C is a water-soluble antioxidant. Vitamin E is the collective name for eight related tocopherols and tocotrienols, which display antioxidant properties (Roberts, Oates, & Linton, 2007). Of these, only α-tocopherol is functional in humans

(Vagula & Konieczko, 2012) as it has the highest bioavailability and is the form of vitamin E that is absorbed (El-Aal, 2012). There is little difference *in vitro* in antioxidant activity between the different tocopherols and tocotrienols (Halliwell & Gutteridge, 2007).

In human biological systems, vitamin C can reduce and neutralize peroxyl, alkoxyl, and hydroxyl radicals as well as reactive oxygen nonradical species such as hydrogen peroxide (El-Aal, 2012). Vitamin C also functions in the regeneration of vitamin E (Chan, 1993). The all-*trans*-β-carotene E-isomer, abbreviated as β-carotene, is also known as provitamin A and is the most important carotenoid precursor for vitamin A (Grune et al., 2010). β-Carotene can act as a lipid radical scavenger that neutralizes lipid peroxyl radicals (Mortensen & Skibsted, 1998). Carotenoids including β-carotene can neutralize reactive oxygen radical species that damage DNA and cause diseases such as macular degeneration, coronary artery disease, and cancer (Vagula & Konieczko, 2012). Carotenoids such as β-carotene and lycopene act as singlet oxygen quenchers (Grune et al., 2010; Stahl & Sies, 1996). Studies have shown that individuals who ingest a diet rich in carotenoids including β-carotene have a much lower risk of developing cardiovascular disease, diabetes mellitus, and cancers of the breast, colon, lung, and prostate (www.cancer.org; www.aminomics.com). The major dietary sources of β-carotene include carrots, sweet potatoes, spinach, squash, lettuce, and cabbage (www.cancer.org).

3.1.2.3 Lycopene

Lycopene is a carotenoid that has attracted attention due to its antioxidant and biological activities. Its antioxidant activities include singlet oxygen quenching, peroxyl radical scavenging, and reaction of its radical anion with O_2. Lycopene appears to be the most efficient carotenoid singlet oxygen quencher *in vitro* (Di Mascio, Kaiser, & Sies, 1989; Halliwell & Gutteridge, 2007; Stahl & Sies, 1996). Biological activities include induction of gap junctional communication, growth inhibition of chemically transformed cells, and suppression of cell proliferation (Stahl & Sies, 1996). Human epidemiological studies suggest that lycopene has protective effects against cardiovascular disease, atherosclerosis, skin damage, and cancer of the prostate, breast, and lung (Cámara et al., 2013; Halliwell & Gutteridge, 2007; Stahl & Sies, 1996). Diets rich in fruits, grain, and vegetables, which contain carotenoids and other antioxidants, appear to be protective against several human diseases, especially cardiovascular disease, diabetes, and some types of cancer (Halliwell & Gutteridge, 2007). The major dietary sources of lycopene include sun-dried tomatoes, processed tomato products, watermelon, guava, and papaya (Halliwell & Gutteridge, 2007).

3.1.2.4 Melatonin

Melatonin is synthesized from serotonin in the mammalian pineal gland and released during the dark phase of the circadian cycle (El-Aal, 2012). Being highly lipid-soluble as well as relatively water-soluble, melatonin is readily taken up by cells. Melatonin easily traverses the blood brain barrier as well as other morphophysiological barriers and, therefore, has ready access to every cell in the body (Reiter, 1997). Melatonin is a highly potent free-radical scavenger. The indole protects a variety of tissues from damage by the highly toxic hydroxyl radical produced by the three-electron reduction of O_2 (Reiter, 1997; Reiter et al., 1997). In addition, melatonin scavenges the lipid peroxyl radical, generated during the oxidation of unsaturated fatty acids, with an efficiency greater than that of vitamin E (Pieri et al., 1994). Also, melatonin stimulates antioxidant enzymes to maintain normal levels of reactive oxygen species (El-Aal, 2012; Reiter et al., 1997).

4. CYP CATALYTIC CYCLE AND INVOLVEMENT OF CYP ENZYMES IN REACTIVE OXYGEN SPECIES FORMATION

CYP enzymes function as part of the NADPH/O_2-dependent microsomal electron transport system and are one of the most important cellular sources of reactive oxygen species (Al-Omar et al., 2004; Halliwell, 2006b; Hrycay & Bandiera, 2012; Lewis, 2002; Zangar et al., 2004). The ability of CYP enzymes to incorporate an oxygen atom from O_2 into numerous organic substrates (monooxygenase activity), to utilize H_2O_2, cumene hydroperoxide, and other hydroperoxides as oxygen atom donors to oxygenate substrates (peroxygenase activity), and to use H_2O_2 and other peroxides in the one-electron oxidation of substrates (peroxidase activity) demonstrates the catalytic versatility of CYP enzymes (Hrycay & Bandiera, 2012, 2015). CYP enzyme systems are the predominant biotransformation pathway in the body for exogenous and endogenous compounds. Exogenous substrates include drugs, carcinogens, halogenated biphenyls and diphenyl ethers, polycyclic aromatic hydrocarbons, organic solvents, pesticides, and plant products, as well as physiological substrates such as amino acids, bile acids, cholesterol, fatty acids, leukotrienes, prostaglandins, steroid hormones, and vitamins (Hrycay & Bandiera, 2008, 2012, 2015).

CYP enzymes are localized primarily in the smooth endoplasmic reticulum of liver and other organs and function as the terminal oxidases of an electron transport chain in which one atom of oxygen from molecular oxygen is incorporated into the substrate and the other atom is reduced to water. The electron transport chain also includes a flavoprotein, NADPH-CYP oxidoreductase, for the microsomal CYP enzyme system. CYP as the terminal oxidase of the chain is the enzyme active site, the oxygen- and substrate-binding site and the determinant of substrate specificity.

CYP enzymes contain a heme prosthetic group in the form of iron protoporphyrin IX, which is essential for enzymatic activity. The iron atom in protoporphyrin IX can exist in two oxidation states, Fe^{II+} (reduced) and Fe^{III+} (oxidized). The heme prosthetic group is connected to a single polypeptide chain composed of approximately 500 amino acids. Protoporphyrin IX is bound noncovalently, at the fifth coordination site of iron, to the sulfur atom of a cysteine residue located close to the C-terminus of the polypeptide chain. During catalysis by microsomal CYP enzymes, two electrons are acquired from NADPH and migrate from the flavin adenine dinucleotide domain of the flavoprotein reductase to the CYP heme group. A water molecule normally occupies the sixth coordination site of heme iron but is replaced by molecular oxygen when the catalytic reaction begins (Hrycay & Bandiera, 2012, 2015).

The sequential steps of the CYP monooxygenase cycle (Fig. 1) and the systematic generation of CYP intermediates were originally documented using the bacterial CYP101A1 system (Mueller, Loida, & Sligar, 1995) and the mammalian hepatic microsomal CYP system (Estabrook, Martinez-Zedillo, Young, Peterson, & McCarthy, 1975). The CYP monooxygenase cycle commences with the iron of the heme group in the oxidized ferric state (Fe^{III}) (**A**) and continues according to the following sequence. In step 1, the substrate (signified by A) binds to the substrate-binding site on the polypeptide chain of CYP and forms a high-spin ferric CYP-substrate complex, which then undergoes a one-electron reduction to the ferrous state (Fe^{II}) (**B**). However, substrate binding is not always a prerequisite for reduction of the heme iron, as shown for some human CYP enzymes such as CYP1A2 and CYP2E1 (Guengerich & Johnson, 1997). In step 2, O_2 binds to ferrous CYP to produce a relatively stable ferric-superoxo anion complex $[Cys-Fe^{III}-O_2^{-}]$ (**C**), which is the resonance hybrid of the ferrous–dioxygen complex $[Cys\text{–}Fe^{II}\text{–}O_2]$. In step 3, transfer of a second electron produces the ferric-peroxo anion intermediate $^{-}[Cys-Fe^{III}-O_2^{-}]$ (**D**), which carries a negative charge on the oxygen and a second negative charge

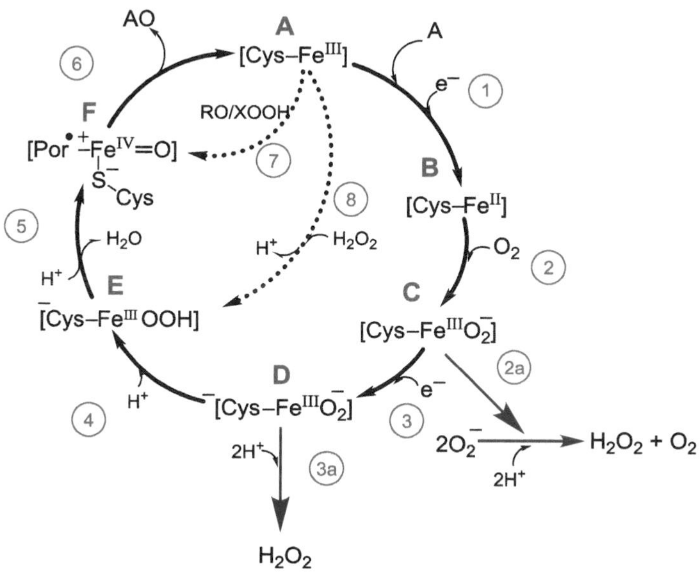

Figure 1 The CYP monooxygenase cycle. The sequential steps by which CYP enzymes catalyze the NAD(P)H/O_2-mediated oxidation of a substrate (signified by A) to a monooxygenated substrate (signified by AO) are illustrated numerically in steps 1–6, indicated by encircled numbers shown in purple (light gray in the print version). CYP intermediary states are depicted by capital letters shown in red (dark gray in the print version). The substrate (A) is bound to states **A** through **F** but is only shown in step 1 and step 6 as a monooxygenated substrate (AO) to simplify the scheme. States **A** through **F** contain a cysteine thiolate ligand (Cys) that displays a delocalized negative charge in states **D** and **E**. H_2O_2-forming pathways in uncoupled reactions are shown in steps 2A and 3A. Ferric CYP (Fe^{III}) can react with oxotransfer agents (RO) and peroxy compounds (XOOH) to generate the Cpd I species [$Por^{\cdot +}Fe^{IV}{=}O$] (**F**) via the shunt pathway (step 7). Ferric CYP enzymes such as CYPcam, CYP2B4, and CYP3A4 can react with H_2O_2 to form the ferric-hydroperoxo species $^{-}$[Cys–Fe^{III}–OOH] (**E**) (step 8), which is believed to rapidly dissociate, upon protonation, to produce Cpd I (**F**) and H_2O (step 5). Due to spatial considerations, the line representing the chemical bond between the iron and oxygen atom (e.g., in Fe^{III}OOH) is omitted. The + charges on the iron atom are also omitted. All intermediates (**A** through **F**) contain the porphyrin (Por) moiety that is shown as a $Por^{\cdot +}$ radical cation only for Cpd I (**F**). Briefly, the CYP monooxygenase cycle follows the order of substrate (signified by A) binding to ferric CYP, a one-electron reduction of ferric CYP, O_2 binding to ferrous CYP to produce the ferric-superoxo anion complex [Cys – Fe^{III} – O_2^{-}] (**C**), which is the resonance hybrid of the ferrous-dioxygen complex [Cys–Fe^{II}–O_2], a second one-electron reduction to produce the ferric-peroxo anion intermediate $^{-}$[Cys – Fe^{III} – O_2^{-}] (**D**), an initial protonation to form the ferric-hydroperoxo intermediate $^{-}$[Cys–Fe^{III}–OOH] (**E**), a second protonation followed by loss of a water molecule to generate the Por π radical ferryl intermediate [$Por^{\cdot +}Fe^{IV}{=}O$] (**F**) known as Compound I (Cpd I), and transfer of the ferryl oxygen atom to the substrate, forming the monooxygenated substrate (signified by AO) and regenerated ferric CYP (Hrycay & Bandiera, 2012, 2015). *Adapted from Hrycay and Bandiera (2015).*

delocalized over the cysteine ligand. In step 4, protonation of the distal oxygen atom produces the ferric-hydroperoxo intermediate $^{-}$[Cys–Fe^{III}–OOH] (**E**). In step 5, a second protonation followed by loss of a water molecule generates the porphyrin (Por) π radical ferryl intermediate [Por$^{\cdot+}$Fe^{IV}=O] (**F**) known as Compound I (Cpd I). In step 6, the [Por$^{\cdot+}$Fe^{IV}=O] intermediate transfers the ferryl oxygen atom to the substrate, forming the monooxygenated substrate (signified by AO) and regenerated ferric CYP (Hrycay & Bandiera, 2012, 2015). The precise electronic configuration of Cpd I formed in NADPH/O_2-mediated reactions has not been defined due to its unstable nature and high reactivity (Hrycay & Bandiera, 2015). Other reactive intermediary species such as the ferric oxenoid complex [PorFe^{III}=O], ferryl radical resonance form [PorFe^{IV}–O$^{\cdot}$] of Cpd I, perferryl entity [PorFe^{V}=O], and ferric-hydroperoxo intermediate $^{-}$[Cys–Fe^{III}–OOH] have also been proposed to act as transitory oxygenating species in monooxygenase reactions (Hrycay & Bandiera, 2015).

During substrate monooxygenation reactions catalyzed by mammalian hepatic CYP enzymes, if the transfer of an oxygen atom to a substrate is not tightly coupled to NADPH utilization, then one or more electron equivalents derived from NADPH are not utilized to oxygenate substrates but instead are unproductively deployed in the reduction of intermediary CYP–oxygen complexes of the CYP catalytic cycle. The complexes then dissociate and liberate reactive oxygen species such as O_2^-, H_2O_2, and $^{\cdot}$OH rather than monooxygenated substrate in a process known as uncoupling (Hrycay & Bandiera, 2012, 2015; Kappus, 1993; Lewis, 2002). One likely reason for uncoupling is the substrate-dependent access of water to the CYP active site (Denisov, Makris, Sligar, & Schlichting, 2005), which destabilizes the ferric-superoxo intermediate [Cys – Fe^{III} – O_2^-] (**C**) of the CYP cycle and results in dissociation of the complex to form O_2^- and ferric CYP. Subsequent dismutation of O_2^- generates H_2O_2 (see Fig. 1, step 2A). A second mode of uncoupling involves reduction of the ferric-peroxo intermediate $^{-}$[Cys – Fe^{III} – O_2^-] (**D**) and subsequent dissociation to produce H_2O_2 (see Fig. 1, step 3A). A third mode of uncoupling involves dissociation of the hydroperoxide anion from ferric iron (**E**) to generate H_2O_2.

Hydrogen peroxide is produced at various steps of the CYP cycle and can be converted to the hydroxyl radical in the presence of ferrous iron via the Fenton reaction (see Eq. 1). Hydroxyl radicals diverted from the CYP cycle can then oxidatively biotransform xenobiotic substrates such as aniline, benzene, ethanol, and safrole (Cederbaum & Berl, 1982; Ingelman-Sundberg & Ekström, 1982; Ingelman-Sundberg & Hagbjörk, 1982; Ioannides, Delaforge, & Parke, 1981). Singlet oxygen ($^1O_2$) is an additional reactive oxygen species formed in the CYP cycle and can oxygenate substrates such as coumarin, 7-ethoxyresorufin, nifedipine, and tolbutamide in human liver microsomes (Yasui et al., 2002), and aminopyrine, aniline, 7-ethoxycoumarin, lauric acid, *p*-nitroanisole, and *p*-nitrophenol in rat liver microsomes (Hayashi, Yasui, & Sakurai, 2005; Lewis, Ioannides, & Parke, 1989; Osada, Ogura, Yasui, & Sakurai, 1999; Yasui et al., 2002; Yasui, Hayashi, & Sakurai, 2005).

The degree of coupling in mammalian microsomal systems varies among the CYP enzymes. It is usually less than 50% and sometimes can be as low as 0.5–3% (Gruenke, Konopka, Cadieu, & Waskell, 1995; Yasui et al., 2005). CYP2E1 is notable in that the enzyme leaks electrons readily during the course of its catalytic cycle, producing more reactive oxygen species than other CYP enzymes. During the process of uncoupling, reactive oxygen species such as superoxide anion and H_2O_2 escape from the CYP2E1 catalytic cycle. Significantly, induction of CYP2E1 by compounds that are poor substrates leads to uncoupling and increased production of reactive oxygen species, which exceeds the capacity of the natural defense system to cope with the high levels produced (Lewis, 2002). Interestingly, even in the absence of substrates, expression of CYP2E1 can generate reactive oxygen species (Caro & Cederbaum, 2004; Gonzalez, 2005).

Puntarulo and Cederbaum (1998) conducted experiments to evaluate the ability of individually expressed human CYP1A1, CYP1A2, CYP2B6, or CYP3A4 enzymes to consume NADPH and produce reactive oxygen species. Microsomes enriched in the individual CYP proteins were prepared from the human AHH-1 TK$\pm$cell line (β-lymphoblastoid cells) that had been engineered to stably express human CYP1A1, CYP1A2, CYP2B6, or CYP3A4. On a per nmole CYP basis, CYP3A4 was the most active in catalyzing NADPH oxidation and superoxide production, while CYP1A1 was the next most active, and CYP1A2 and CYP2B6 displayed lower activity. Because CYP3A4 is highly expressed in human liver microsomes (Shimada, Yamazaki, Mimura, Inui, & Guengerich, 1994) and because it was found to be active in catalyzing production of superoxide, there is a high expectation that CYP3A4 can be an important contributor for the generation of reactive oxygen species by human liver microsomes. In contrast, human CYP1A1 showed high activity in consuming NADPH and generating superoxide, but the enzyme is not present in significant amounts in human liver microsomes. CYP1A1 mRNA and protein have been detected in lungs of cigarette smokers (McLemore et al., 1990; Shimada et al., 1992) and under such circumstances, oxygen radical generation catalyzed by CYP1A1 in lung tissue may be of toxicological significance.

5. ROLES OF REACTIVE OXYGEN SPECIES IN CANCER

Reactive oxygen species are known to be cytotoxic and have been implicated in the etiology of a variety of human diseases including cancer. Reactive oxygen species are considered potential carcinogens as they were shown to facilitate mutagenesis, cancer promotion, and progression (Manda et al., 2009; Waris & Ahsan, 2006). Cancer can result from damage to, or from the activation, suppression or mutation of, a large number of genes. The association between reactive oxygen species and cancer is complex and not fully understood. Increased levels of reactive oxygen species have been implicated in cancer initiation, in cancer progression and paradoxically, in the killing of cancer cells.

According to the mitochondrial theory of cancer (Wallace, 2005), mutations in nuclear or mitochondrial genes encoding components of the mitochondrial electron transport chain can lead to an increase in generation of reactive oxygen species if electron transfer is partially inhibited. This leads to an accumulation of electrons at sites along the electron transport chain, where they can be captured by O_2, leading to the formation of superoxide, which is rapidly dismutated by superoxide dismutase to yield hydrogen peroxide. Hydrogen peroxide can diffuse into the nucleus and react with DNA, contributing to genetic damage.

Whether reactive oxygen species are generated by the mitochondrial electron transport chain or by CYP enzymes, they can produce single- or double-stranded DNA breaks, purine, pyrimidine or deoxyribose modifications, and DNA cross links (Halliwell & Gutteridge, 2007). These changes in DNA are usually repaired by DNA repair enzymes but persistent DNA damage can lead to replication errors, activation of oncogenes, inactivation of tumor suppressor genes and genomic instability, all of which are associated with initiation of various cancers. An example has been described in prostate cancer models, where mutations in mitochondrial DNA, which encodes specific subunits of the electron transport chain, have been linked to increased production of reactive oxygen species and tumor progression (Petros et al., 2005).

Hydrogen peroxide is not very reactive toward DNA. Most of the damaging effects on DNA are due to hydroxyl radicals, which are generated by the Fenton and Haber–Weiss reactions. Hydroxyl radicals, if produced in close proximity to DNA, can interact with DNA to produce oxidized bases and single- and double-strand breaks. The most extensively studied DNA adduct is 8-hydroxyguanine, which is formed by the oxidation of guanine at the C8 position. 8-Hydroxyguanine is mutagenic to mammalian cells and results in G to T transversions, which are widely found in mutated oncogenes and tumor suppressor genes. Other oxidative DNA lesions include 8-oxo-adenine, thymine glycol, 5-hydroxy-deoxycytidine, and several oxidized uracil compounds, which are mutagenic. DNA damage can lead to mutations when cells undergo division. Oxidative modification of proteins can occur and result in altered protein function. For example, irreversible oxidation of phosphatases within the catalytic site hinders enzymatic activity. Enzymes that may be affected include DNA repair enzymes and DNA polymerases.

Reactive oxygen species influence cellular processes such as proliferation, apoptosis, and senescence, which play a role in the development of cancer. Understanding the role of reactive oxygen species as key mediators in signaling cascades provides opportunities for understanding cancer initiation and progression and provides, paradoxically, insight into possible avenues for pharmacological intervention.

5.1 Reactive Oxygen Species and Cell Signaling

The redox state of a cell is kept within a narrow range to allow for normal cell functioning. An imbalance in the redox state of a cell can lead to altered cell signaling. Stress-responsive signaling pathways triggered by disruption of the intracellular redox state include the mitogen-activated protein kinase (MAPK) pathway, the PI3K/Akt pathway and the Keap 1/Nrf2/ARE pathway, and represent mechanisms to restore redox homeostasis. In addition, fluctuations in the steady-state concentrations of intracellular reactive oxygen species can activate and inactivate nuclear transcription factors and protein tyrosine phosphatases.

MAPK are a family of serine/threonine kinases that modulate gene expression, mitosis, proliferation, motility, and programmed cell death. MAPK modulate gene expression through phosphorylation of nuclear transcription factors such as cAMP response element binding protein (CREB) and AP-1 (Kyriakis & Avruch, 2001). Activation of the MAPK/extracellular signal-regulated kinase (ERK) pathway, also known as the Raf/MEK/ERK signaling cascade, by reactive oxygen species occurs through several mechanisms and leads to activation of transcription factors, including c-Jun, c-Myc, and CREB, that are important for cell proliferation and survival (McCubrey et al., 2007). Reactive oxygen species can act directly on growth factor receptors such as EGFR, in a ligand-independent fashion, and activate Ras and ERK signaling (McCubrey et al., 2007). Reactive oxygen species can also activate Ras and ERK via upstream regulators that function independently of growth factor receptors. In breast cancer cells, for example, estrogen-stimulated cell proliferation results from reactive oxygen species-mediated activation of the ERK signaling pathway and the transcription factor, CREB (Felty, Singh, & Roy, 2005; Roy, Cai, Felty, & Narayan, 2007). ERK signaling is also increased in ovarian cancer cells in the presence of high concentrations of reactive oxygen species (Chan et al., 2008). Activation of the ERK pathway by reactive oxygen species increases cell survival, anchorage-independent growth, and motility in multiple cancers (McCubrey et al., 2007; Roberts & Der, 2007).

Hydrogen peroxide and singlet oxygen can activate the c-Jun NH_2-terminal kinase (JNK) and p38 kinase pathways, leading to increased apoptosis (Chang & Karin, 2001). Activation of the JNK pathway likely occurs through the apoptosis signal-related kinase (ASK1) and MEKK1 (McCubrey et al., 2007). Under basal conditions, reduced thioredoxin binds to ASK1 and inhibits ASK1 activity. Under conditions of oxidative stress, thioredoxin is oxidized and dissociates from ASK1, which allows ASK1 to autophosphorylate and become activated. Once activated, JNK proteins translocate to the nucleus and phosphorylate c-Jun and ATF2, potent transcriptional regulators, and thereby induce AP-1. The effect of AP-1 activation is increased cell proliferation and modulation of regulators of the cell cycle and apoptosis. Activation of the p38 pathway likely occurs via mechanisms similar to the JNK pathway and involving ASK1 (McCubrey et al., 2007). The balance between ERK and JNK activation is a key factor for cell survival as a decrease in ERK and an increase in JNK are required for apoptosis (Xia, Dickens, Raingeaud, Davis, & Greenberg, 1995).

Another kinase pathway that is activated by reactive oxygen species is the PI3K/Akt pathway, also known as the PI3K/Akt/mTOR signaling cascade. Akt, also called protein kinase B, is a serine/threonine kinase. Akt plays a role in cell survival through phosphorylation and inactivation of proapoptotic substrates and activation of transcription factors such as CREB. The Akt pathway is tightly controlled by a cascade that includes the kinases, phosphatidylinositol-dependent kinase 1 (PDK-1), mammalian target of rapamycin (mTOR) and phosphoinositide 3-kinase (PI3K), and by the phosphatase, PTEN. PTEN, which is a negative regulator of the PI3K/Akt pathway, is reversibly inactivated by hydrogen peroxide leading to activation of Akt activity (Liou & Storz, 2010). Activation of Akt activity stimulates mTOR, an important regulator of protein synthesis, and activates CREB, a transcription factor that promotes cell proliferation.

The Keap 1/Nrf2/ARE signaling cascade plays a protective role in response to environmental stresses including oxidative stress. The transcription factor, Nrf2 (nuclear factor erythroid 2-related factor, NFE2L2) is a basic leucine zipper protein that directly regulates the redox homeostatic gene regulatory network (Jaiswal, 2004). Nrf2 acts in the nucleus to induce expression of more than 500 cytoprotective genes including NADPH quinone oxidoreductase, sulfiredoxin 1, heme oxygenase, and glutathione-*S*-transferases (Hayes et al., 2000; Jarmi & Agarwal, 2009; Neumann, Cao, & Manevich, 2009; Venugopal & Jaiswal, 1996). Under basal conditions, Nrf2 is retained in the cytoplasm by a cluster of proteins that include Kelch-like ECH-associated protein 1 (Keap 1) and Cullin 3, which degrade Nrf2 by ubiquitination and results in low cellular levels of Nrf2 (Itoh et al., 1999; Kobayashi et al., 2004). Under conditions of oxidative stress, Nrf2 is not degraded and translocates to the nucleus. An increase in intracellular levels of reactive oxygen species promotes oxidation of critical cysteine residues in Keap 1, thereby disrupting the Keap 1-Cul 3 ubiquitination system and allowing cytosolic Nrf2 levels to rise (Yamamoto et al., 2008). Stabilized NRF2 is more easily translocated and retained in the nucleus (Jain, Bloom, & Jaiswal, 2005). In the nucleus, Nrf2 forms a heteromeric transcriptional complex with Maf protein, which binds to antioxidant response elements on the 5′ untranslated regions of target cytoprotective genes (Itoh et al., 1997). Thus, oxidative stress activates Nrf2 and activation of Nrf2 protects the cell from oxidative damage.

Protein tyrosine phosphatases are one of the best-characterized direct targets of reactive oxygen species (Denu & Tanner, 1998; Valko et al., 2007). Protein tyrosine phosphatases catalyze the dephosphorylation, while protein tyrosine kinases catalyze the phosphorylation of tyrosine residues on diverse protein substrates including epidermal growth factor receptors such as HER2/Neu and HER1/EGFR. Because dephosphorylation typically leads to deactivation of the target proteins, protein tyrosine phosphatases can positively and negatively regulate signaling pathways. For example, inhibition of phosphatase activity results in activation of ERK signaling (McCubrey et al., 2007). The effects of reactive oxygen species occur through oxidation of an essential cysteine residue in the active sites of tyrosine phosphatases, producing sulfenic acid intermediates that can further react with thiols to form disulfides (Salmeen & Barford, 2005). As a result, the protein tyrosine phosphatases are rendered catalytically inactive. Reactive oxygen species can also upregulate mRNA levels of cyclins, including cyclin B2, cyclin D3, cyclin E1, and cyclin E2, which participate in the cell cycle to expedite the transition from G1 to S phases (Sarsour, Venkataraman, Kalen, Oberley, & Goswami, 2008).

Some studies have suggested that cancer cells produce increased levels of reactive oxygen species compared to normal cells. For example, immortalized epithelial cells that had been transfected (with H-RASV or BCR-ABL) exhibited greater levels of intracellular oxidant stress than nontransformed cells (Trachootham et al., 2006). This suggests that increased oxidative stress may contribute to the survival and proliferation of cancer cells. Conversely, diverse chemotherapeutic agents are toxic to cancer cells because they augment oxidant stress in the cells. Neoplastic cells may be more vulnerable because basal levels of reactive oxygen species are already higher in neoplastic cells. The addition of an agent that increases levels of reactive oxygen species could push neoplastic cells beyond a limit of oxidant stress that the cells are able to absorb. If cancer cells have decreased levels of antioxidants or antioxidant scavenging capacity, due to gene mutations, they may be more susceptible to reactive oxygen species stress. A disproportionate increase in intracellular reactive oxygen species can result in cancer cell cycle arrest, senescence, and apoptosis. For example, pancreatic cancer cells generally show high basal levels of endogenous oxidative stress as compared to normal cells (Lewis et al., 2005). The increased reactive oxygen species levels have been linked to increased proliferation in these and other cancers (Kumar, Koul, Khandrika, Meacham, & Koul, 2008; Lewis et al., 2005). Combination therapy of gemcitabine with trichostatin A, epigallocate-3-galate, capsaicin, and benzyl isothiocyanate is one of the few effective treatments for pancreatic cancer (Shankar, Suthakar, & Srivastava, 2007; Srivastava & Singh, 2004; Zhang, Humphreys, Sahu, Shi, & Srivastava, 2008). These drugs share a common mechanism, namely, to elevate intracellular levels of reactive oxygen species and trigger apoptosis (Zhang et al., 2008). Similarly, aminoflavone induces cell death in MCF-7 and

MDA-MD-468 breast cancer cells by increasing intracellular reactive oxygen species and activating caspase 3 activity and apoptosis (McLean et al., 2008).

6. RELATIONSHIP BETWEEN CYP ENZYMES AND CANCER

There is little experimental evidence to support the premise that CYP enzymes play a direct causative role in cancer. CYP enzymes are, however, involved in various types of cancer by several mechanisms. First, CYP enzymes catalyze the bioactivation of chemical procarcinogens into highly reactive conversion products. Second, they participate in the activation of drugs that are used to treat cancers. Third, CYP enzymes serve as drug targets in cancer therapy. Fourth and most applicable to this chapter, CYP-mediated metabolism can lead to formation of reactive oxygen species, which can promote carcinogenesis. Thus, the potential for CYP enzymes to contribute to the etiology of cancer exists.

6.1 CYP Expression in Neoplastic Tissues

There are a large number of CYP genes coding for CYP proteins that are expressed in normal human organs. In humans, 57 genes belonging to 18 families and 43 subfamilies have been identified (http://drnelson.uthsc.edu/P450.statsfile.html). One approach used to explore the relationship between CYP expression and cancer has been to measure tissue-specific expression of CYPs in neoplastic and normal tissues. Table 4 lists human CYPs that are expressed in specific types of malignant primary or cell tumors. Significantly, most of the CYPs shown in Table 4 were overexpressed in various malignant tumor tissues, compared to adjacent normal tissues.

Murray and coworkers carried out a series of studies regarding expression of extrahepatic human CYPs in various tissues from cancerous and normal origin. The most significant finding was the identification of CYP1B1 as a tumor marker (Murray et al., 1997). Notably, the protein was absent in the corresponding normal tissue samples. Similar associations were drawn regarding the differential overexpression of CYP1A1 in cancerous and non-cancerous tissues. Significantly, CYP1A1 is present to a greater extent in malignant than in normal breast tissues (Androutsopolous, Tsatsakis, & Spandidos, 2009).

Kumarakulasingham and coworkers (2005) carried out studies to define the expression profile of human CYP in colorectal cancer and identify which CYPs are overexpressed, and also establish their prognostic significance. In primary colorectal cancer, several CYPs (1B1, 2S1, 2U1, 3A5, 51) were

Table 4 Expression of Human CYPs in Malignant Tumor Tissues[a]

Type of Primary Tumor	CYPs Expressed	References
Adrenal	2W1[b], 19A1	Karlgren et al. (2006) and Moreau et al. (2008)
Bone	3A4, 3A5	Dhaini et al. (2003)
Brain	1A1, 1B1	Murray et al. (1997) and Wahid, Mahjabeen, Baig, and Kayani (2013)
Breast	1A1, 1B1, 2B6, 2C9, 2D6, 2E1, 2J2, 2W1, 3A4, 3A5, 4F11, 4Z1, 4Z2P, 8A1	Huang, Fasco, Figge, Keyomarsi, and Kaminsky (1996), Hellmold et al. (1998), Jiang et al. (2005), Jiang, Fu, Chen, and Wang (2009), Leung et al. (2013), Rieger et al. (2004), Schmidt et al. (2004), and Yu et al. (2012)
Colon	1A1, 1B1, 2J2, 2W1, 3A4, 24A1	Anderson, Nakane, Ruan, Kroeger, and Wu-Wong (2006), Androutsopolous et al. (2013), Gibson et al. (2003), Jiang et al. (2005), Jiang et al. (2009), Murray (2000), Karlgren et al. (2006), Murray et al. (1997), and Peters et al. (1992)
Colorectal	1B1, 2A6, 2S1, 2U1, 3A5, 51	Kumarakulasingham et al. (2005) and Matsuda et al. (2007)
Endometrial	19A1	Sasano et al. (1996)
Esophageal	1A, 1B1, 2A6, 2A7, 2C, 2E1, 2J2, 3A, 24A1	Jiang et al. (2005), Jiang et al. (2009), Mimori et al. (2004), Murray et al. (1994), Murray et al. (1997), and Godoy et al. (2002)
Gastric	2J2	Jiang et al. (2005) and Jiang et al. (2009)
Glioblastoma multiforme	27B1	Maas et al. (2001)
Liver	2J2, 3A4	Fritz, Behrele, Beaune, Eichelbaum, and Kroemer (1993), Jiang et al. (2005), and Jiang et al. (2009)

Lung	1A1, 1B1, 2A6, 2J2, 3A4, 3A5, 4B1, 24A1	Anderson et al. (2006), Czerwinski, McLemore, Gelboin, and Gonzalez (1994), Jiang et al. (2005), Jiang et al. (2009), Kivistö et al. (1995), Murray (2000), Murray et al. (1997), Ariyoshi et al. (2002), and Spivack et al. (2001)
Lymph node	1B1, 2A, 2B, 2C, 2F1, 4V2, 39	Kumarakulasingham et al. (2005) and Murray et al. (1997)
Oral squamous cell carcinoma	2R1	Sundaram, Sambandam, Tsuruga, Wagner, and Reddy (2014)
Ovarian	1B1, 2A, 2B, 2F1, 2R1, 2U1, 3A5, 3A7, 3A43, 4Z1, 24A1, 26A1, 51	Anderson et al. (2006) and Downie et al. (2005)
Prostate	1B1	Carnell et al. (2004)
Renal	1B1, 3A4, 3A5, 3A7	Murray et al. (1999) and Murray et al. (1997)
Skin	1A1, 1B1	Murray et al. (1997)
Testis	1B1	Murray et al. (1997)
Urinary bladder	1A1, 1B1	Androutsopolous et al. (2013)
Uterine	1B1	Murray et al. (1997)

[a]The CYPs listed in the table represent those expressed in human primary or cell tumors. Expression of CYPs in tumors or adjacent normal tissues was determined using RT-PCR, Western blot analysis, or immunohistochemistry.
[b]Due to spatial considerations, the CYP designation is not shown for the CYPs.

present at significantly higher levels compared to normal colon. Notably, CYP2S1 and CYP51 show prognostic significance and CYP51 is an independent marker of prognosis in colorectal cancer.

Rieger and coworkers (2004) investigated the tissue distribution of human CYP4Z1 to evaluate its overexpression in breast cancer tissue, in view of its potential use as a breast cancer-related antigen. Preferential expression of CYP4Z1 cDNA was observed in breast carcinoma tissue and mammary gland, and protein expression was verified in breast tissue specimens. Even in liver, known as the organ with the highest level of CYP proteins, the expression of hepatic CYP4Z1 was 86-fold lower than in breast carcinoma. This high expression of CYP4Z1 is a unique feature among the CYP superfamily. Studies investigating the expression of a broad range of CYPs showed the existence of CYPs (1A1, 1B1, 2A6, 2C, 2B6, 2E1, 3A4, 4A11, 4B1) in different breast tissues including breast cancer (Hellmold et al., 1998; Rieger et al., 2004).

A recent study examined the CYP gene expression profile in tumor and paired normal tissues of 13 child patients with rhabdomyosarcoma (Molina-Ortiz et al., 2014). Higher CYP2W1, CYP3A4, and CYP3A5 mRNA expression was found in tumor samples compared to adjacent normal tissues for several, but not all, of the patients. Previous studies that examined CYP1B1 expression using immunohistochemical analysis found higher CYP1B1 protein expression in samples of breast, ovarian, prostate, testis, and uterine cancer (Sissung, Price, Sparreboom, & Figg, 2006). Similar studies have been carried out with other CYP enzymes. Differential overexpression of a CYP enzyme in a tumor has been interpreted to suggest a functional role for the enzyme in initiation or pathogenesis of the specific cancer (Jiang et al., 2009; Yu et al., 2012). However, these findings must be interpreted cautiously and increased CYP enzyme expression may be no more than a consequence of the cancer.

6.2 CYP Polymorphism and Cancer Risk

Genetic polymorphism is a common phenomenon in the CYP superfamily. Mutations leading to a single base change or multiple base changes in the coding and/or noncoding regions of a CYP gene occur frequently and produce variant forms of the CYP enzyme with activity that is equal to less than, or greater than, that of the native enzyme, or with decreased or increased enzyme expression in tissues. Not all CYP enzymes are associated with identifiable polymorphic forms, but well-studied functional polymorphism has been described for CYP2A6, CYP2B6, CYP2C9, CYP2C19, and CYP2D6. The CYP enzymes involved in bioactivation of chemical procarcinogens do not exhibit functional polymorphism, and those that do are not generally involved in these reactions. For this reason, consistent and strong association between variant CYP alleles and risk for various types of cancer has not been reported (Rodriguez-Antona & Ingelman-Sundberg, 2006).

Numerous studies examined the association between CYP1B1 expression and estrogen-related cancers because of the role of CYP1B1 in catalyzing 17β-estradiol hydroxylation. These case-only and case–control studies assessed the association between CYP1B1 polymorphisms and breast, endometrial, ovarian, and prostate cancer in different human populations. No strong or consistent associations between CYP1B1 genotype and cancer risk were found in most of the studies (Sissung et al., 2006). Similar outcomes were obtained with studies investigating polymorphism of other CYP enzymes and various types of cancer (Rodriguez-Antona & Ingelman-Sundberg, 2006). In other words, the inheritance of genes coding for weakly active or inactive CYP enzymes that participate in bioactivation of chemical procarcinogens has not been shown to correlate with increased cancer risk in humans. Possible explanations are that the CYP enzymes do not play a significant role in the etiology of the cancers studied, or that other endogenous and environmental factors that were not controlled for in the studies are more important contributors to cancer risk.

6.3 CYP Enzymes and Chemical Carcinogenesis

The deleterious effects of the chemical procarcinogens encountered in the environment are attributed to metabolic bioactivation by CYP enzymes to highly reactive conversion products. Such reactive metabolites cause carcinogenicity in experimental animals and humans, whereas their corresponding parent compounds are chemically inactive. Table 5 lists the major procarcinogens that are bioactivated by CYP enzymes. Among the CYP enzymes studied in relation to chemical carcinogenesis, CYP1A1, CYP1A2, CYP1B1, and CYP3A4 have been the most thoroughly investigated. CYP1A1 and CYP1B1 catalyze the bioactivation of polycyclic aromatic hydrocarbons such as benzo[*a*]pyrene and 7,12-dimethylbenz[*a*]anthracene (DMBA) (see Table 5). Benzo[*a*]pyrene is converted to its ultimate carcinogen, benzo[*a*]pyrene-7,8-diol-9,10-epoxide, through the action of CYP1A enzymes. Hepatic CYP1A1 and CYP1A2 initially oxidize benzo[*a*]pyrene

Table 5 Bioactivation of Selected Chemical Procarcinogens by Human CYP Enzymes[a]

CYP Enzyme	Chemical Procarcinogen
CYP1A1	*o*-Aminoazotoluene, 4-aminobiphenyl, 6-aminochrysene, benzo[*b*]-fluoranthrene-9,10-diol, benzo[*a*]pyrene, benzo[*a*]pyrene-7,8-diol, dibenzo[*a*,*l*]pyrene, DMBA[b], 3-methoxy-4-aminoazobenzene, PhIP, Trp-P-1
CYP1A2	2-Acetylaminofluorene, AFB1, 2-aminoanthracene; *o*-aminoazotoluene; 4-aminobiphenyl, 2-aminofluorene, Glu-P-1, Glu-P-2, IQ, 2-naphthylamine, *N*-nitrosodimethylamine, PhIP, Trp-P-1, Trp-P-2
CYP1B1	2-Aminoanthracene, 6-aminochrysene, benz[*a*]anthracene, benzo[*g*]-chrysene-11,12-diol, benzo[*c*]phenanthrene-3,4-diol, benzo[*a*]pyrene, benzo[*a*]pyrene-4,5-diol, benzo[*a*]pyrene-7,8-diol, DMBA, 17β-estradiol, 3-MC, 3-methoxy-4-aminoazobenzene, 2-nitropyrene, Trp-P-1, Trp-P-2
CYP2A6	AFB1, DEN, DMN, NATB, NDEA, NNAL, NNK, NNN, butadiene, 2,6-dichlorobenzonitrile
CYP2B6	AFB1, 4-(methylnitrosamino)-1-(3-pyridyl)-1-butanone
CYP2E1	Acrylonitrile, benzene, butadiene, carbon tetrachloride, chloroform, ethyl carbamate, *N*-nitrosodimethylamine, NNK, NNN, styrene, trichloroethylene, vinyl chloride
CYP3A4	AFB1, aflatoxin G1, 6-aminochrysene, benzo[*a*]pyrene-7,8-diol, 17β-estradiol, MOCA, 1-nitropyrene, NNN, senesionine, sterigmatocystin, tris(2,3-dibromopropyl)phosphate

[a]This Table shows the bioactivation of selected procarcinogens to reactive carcinogenic metabolites by human CYP enzymes. Information contained in the table was compiled from the following references Guengerich (1991), Guengerich (1993), Guengerich and Shimada (1998), Lewis (2001), and Rodriguez-Antona and Ingelman-Sundberg (2006).

[b]*Abbreviations*: AFB1, aflatoxin B1; DEN, *N*,*N*-diethylnitrosamine; DMN, *N*,*N*-dimethylnitrosamine; DMBA, 7,12-dimethylbenz[*a*]anthracene; Glu-P-1,2-amino-6-methyldipyrido[1,2-*a*:3,2′-*d*]imidazole; IQ, 2-amino-3-methylimidazo[4,5-*f*]quinolone; 3-MC, 3-methylcholanthrene; MOCA, 4,4′-methylene bis(2-chloroaniline); NATB, *N*-nitrosoanatabine; NDEA, *N*-nitrosodiethylamine; NNAL, 4-(methylnitrosoamino)-1-(3-pyridyl)-1-butanol; NNK, 4-(methylnitrosoamino)-1-(3-pyridyl)-1-butanone; NNN, nornitrosonicotine; PhIP, 2-amino-1-methyl-6-phenylimidazo[4,5-*b*]pyridine; Trp-P-1, 3-amino-1,4-dimethyl-5*H*-pyrido[4,3-*b*]indole; Trp-P-2, 3-amino-1-methyl-5*H*-pyrido[4,3-*b*]indole.

to the 7,8-epoxide metabolite, which is then hydrolyzed by epoxide hydrolase to form the 7,8-diol metabolite (Conney, Chang, Jerina, & Wei, 1994). Further oxidation of the diol metabolite by CYP1A1 and CYP1A2 convert it to the 7,8-diol-9,10-epoxide metabolite, which is highly reactive and can react with guanine in DNA to form DNA adducts and G to T transversion mutations (Conney et al., 1994; Hodek et al., 2013). CYP1A2, and to a lesser extent CYP1A1, have also been shown to bioactivate aristolochic acid, a herbal drug and human carcinogen that produces nephropathy and urothelial malignancies (Levová et al., 2011; Stiborová, Frei, & Schmeiser, 2008). The major metabolic activation pathway involves nitroreduction of aristolochic acid leading to formation of a reactive aristolactam I nitrenium ion species and DNA-adduct formation (to adenine and guanosine), which leads to A-T transversion mutations of the *p*53 tumor suppressor gene (Levová et al., 2011). Similarly, CYP1B1, which is expressed in extrahepatic tissues, can bioactivate DMBA to tumorigenic metabolites that bind to DNA and lead to mutations and ultimately carcinogenesis (Buters et al., 1999). Experimental evidence for the role of CYP1B1 in mediating the carcinogenicity of DMBA resulted from studies by Buters and coworkers using CYP1B1 null and wild-type mice treated with DMBA (Buters et al., 1999, 2003). DNA adducts of DMBA metabolites were found in tissues of wild-type mice but not in tissues of CYP1B1 null mice, and tumors developed in ovary, skin, spleen, thymus, and uterus of wild-type mice but not CYP1B1 null mice. Moreover, human recombinant CYP1B1 was shown to be active in the formation of DMBA metabolite-DNA adducts in Chinese hamster V79MZ cells (Buters et al., 2003).

CYP enzymes also bioactivate aromatic amines and aflatoxin B_1 (AFB_1) to carcinogenic metabolites (Code et al., 1997; He et al., 2006). CYP1A2 and CYP3A4 play a major role in the bioactivation in human liver of AFB_1, a mycotoxin produced by *Aspergillus flavus* and *Aspergillus parasiticus* that is a potent human and animal carcinogen (Gallagher, Kunze, Stapleton, & Eaton, 1996). CYP-mediated oxidation results in the formation of the carcinogenic and mutagenic metabolites, AFB_1-8,9-epoxide and AFM_1-8,9-epoxide, and leads to inactivation of the *p*53 tumor suppressor gene and activation of the K-*ras* oncogene, as the major mechanism in AFB_1-induced cancers (Bujons, Hsieh, Kado, & Messeguer, 1995; Donnelly et al., 1996). CYP2A13, an enzyme predominantly expressed in human respiratory tract, has been identified as an efficient catalyst in the bioactivation of AFB_1 in human lung, suggesting that it plays a critical role in human lung carcinogenesis related to inhalation exposure to AFB_1 (He et al., 2006). CYP2A13 was previously shown to be involved in the metabolic activation of a major tobacco carcinogen, 4-(methylnitrosamino)-1-(3-pyridyl)-1-butanone (Su et al., 2000). The role of CYP enzymes in bioactivation reactions has been covered extensively in several excellent reviews (Guengerich, 2001, 2003, 2008).

6.4 CYP Enzymes and Cancer Therapy

CYP enzymes participate in the activation of several antineoplastic agents including cyclophosphamide, ifosfamide, docetaxel, paclitaxel, irinotecan, and tamoxifen. CYP polymorphism and interindividual variability of CYP expression influences the effectiveness of treatment with these chemotherapeutic agents so that the effectiveness of these compounds is reduced in individuals who have weakly active or inactive forms, or who express low levels of the individual CYP enzymes that activate cyclophosphamide, ifosfamide, docetaxel, paclitaxel, irinotecan, and tamoxifen (Rodriguez-Antona & Ingelman-Sundberg, 2006). CYP2B6 is more effective in activating cyclophosphamide and ifosfamide (Chang, Weber, Crespi, & Waxman, 1993).

The steroidogenic CYP enzymes (CYP11A1, CYP11B1, CYP11B2, CYP17A1, CYP19A1, CYP21A2) play important roles in steroid hormone biosynthesis. These enzymes can serve as drug targets in cancer therapy. For example, inhibitors of CYP19A1 (aromatase) such as anastrozole and letrozole are used in the treatment of postmenopausal women with hormone-receptor positive breast cancer. CYP19A1 is a viable target because it converts androstenedione and testosterone to estrone and 17β-estradiol, respectively. Inhibition of CYP19A1 limits production of estrogens that promote the survival and growth of hormone-receptor positive breast cancer. Inhibitors of CYP17A1 (steroid 17α-hydroxylase) such as abiraterone are being used in the treatment of hormone-receptor positive prostate cancer. CYP17A1 catalyzes the biosynthesis of 17α-hydroxypregnenolone and dehydroepiandrosterone, which are precursors of androstenedione and testosterone. Inhibitors of this enzyme deprive prostate tumors of androgens.

Thus, the involvement of CYP enzymes in cancer is complex. As mediators of the bioactivation of chemical procarcinogens and as generators of reactive oxygen species, CYP enzymes can be considered as playing a role in cancer formation. As targets for drug therapy and as catalysts of the metabolism of anticancer drugs, CYP enzymes function as determinants of cancer therapy.

7. CONCLUSION

Reactive oxygen species, which are derived from the metabolism of molecular oxygen and include the superoxide radical anion ($^{\bullet}O_2^{-}$), singlet oxygen ($^1O_2$), hydroperoxyl radical ($HOO^{\bullet}$), hydrogen peroxide (H_2O_2), and hydroxyl radical ($^{\bullet}OH$), are found in all aerobic cells. Reactive oxygen species are known to be cytotoxic and have been implicated in the etiology of a variety of human diseases including cancer. Normally, intracellular levels and effects of reactive oxygen species are maintained in balance with intracellular biochemical antioxidants to prevent cellular damage. Oxidative stress occurs when this critical balance is disrupted because of excessive production of reactive oxygen species or depletion of antioxidants, or both. A potentially important endogenous source of reactive oxygen species is the mammalian CYP microsomal electron transport system. CYP enzymes are primarily localized in the endoplasmic reticulum and catalyze the oxygenation of an organic substrate and the simultaneous reduction of molecular oxygen. Reduction of molecular oxygen is generally tightly controlled by components of the system, namely NADPH and NADPH-CYP oxidoreductase. When the redox reaction is not tightly controlled, uncoupling occurs and leads to the formation of reactive oxygen species instead of the oxygenated substrate.

CYP enzymes can be indirectly involved in cancer at various levels. First, several CYP enzymes catalyze the bioactivation of chemical procarcinogens into reactive carcinogens. Second, they participate in the bioactivation of drugs that are used to treat cancers. Third, CYP enzymes serve as drug targets in cancer therapy. For example, inhibitors of CYP19A1 (aromatase) are used in the treatment of breast cancer, and inhibitors of CYP17A1 are used in the treatment of prostate cancer. Thus, CYP enzymes are involved in the etiology of cancer and as a determinant of cancer therapy. Fourth and perhaps most importantly, CYP-mediated metabolism can lead to formation of reactive oxygen species, which can promote carcinogenesis. Thus, if CYP enzymes are involved in the generation of reactive oxygen species and if reactive oxygen species play a role in the development and progression of cancer, then it follows that CYP enzymes are involved in the development and progression of cancer.

CONFLICT OF INTEREST

The authors have no conflicts of interest to declare.

ACKNOWLEDGMENTS

The authors acknowledge current and previous support from the Natural Sciences and Engineering Research Council of Canada and the Canadian Institutes of Health Research.

REFERENCES

Abdelmegeed, M. A., Banerjee, A., Jang, S., Yoo, S. H., Yun, J. W., Gonzalez, F. J., et al. (2013). CYP2E1 potentiates binge alcohol-induced gut leakiness, steatohepatitis, and apoptosis. *Free Radical Biology and Medicine*, *65*, 1238–1245.

Abdelmegeed, M. A., Banerjee, A., Yoo, S. H., Jang, S., Gonzalez, F. J., & Song, B. J. (2012). Critical role of cytochrome P450 2E1 (CYP2E1) in the development of high fat-induced non-alcoholic steatohepatitis. *Journal of Hepatology*, *57*, 860–866.

Abdelmegeed, M. A., Moon, K. H., Hardwick, J. P., Gonzalez, F. J., & Song, B. J. (2009). Role of peroxisome proliferator-activated receptor-α in fasting-mediated oxidative stress. *Free Radical Biology and Medicine*, *47*, 767–778.

Afanas'ev, I. (2011). Reactive oxygen species signaling in cancer: Comparison with aging. *Aging and Disease*, *2*, 219–230.

Agbor, A. A. (2011). Reactive oxygen species: Their role in severe clinical conditions and regulation of 8 redox-sensitive cell signaling transduction events. *Annals of Biological Research*, *2*, 599–609.

Alfadda, A. A., & Sallam, R. M. (2012). Reactive oxygen species in health and disease. *Journal of Biomedicine and Biotechnology*, *2012*, 1–14.

Al-Omar, M. A., Beedham, C., & Alsarra, I. A. (2004). Pathological roles of reactive oxygen species and their defence mechanisms. *Saudi Pharmaceutical Journal*, *12*, 1–18.

Alvarez, S., Evelson, P., & Cimolai, M. C. (2008). Oxygen and nitric oxide metabolism in sepsis. In S. Alvarez, P. Evelson, & A. Boveris (Eds.), *Free radical pathophysiology* (pp. 1–14): Kerala, India: Transworld Research Network, ISBN: 978-81-7895-311-3.

Anderson, M. G., Nakane, M., Ruan, X., Kroeger, P. E., & Wu-Wong, J. R. (2006). Expression of VDR and CYP24A1 mRNA in human tumors. *Cancer Chemotherapy and Pharmacology*, *57*, 234–240.

Androutsopolous, V. P., Spyrou, I., Ploumidis, A., Papalampros, A. E., Kyriakakis, M., Delakas, D., et al. (2013). Expression profile of CYP1A1 and CYP1B1 enzymes in colon and bladder tumors. *PLoS One*. *8*. http://dx.doi.org/10.1371/journal.pone.0082487, e82487.

Androutsopolous, V. P., Tsatsakis, A. M., & Spandidos, D. A. (2009). Cytochrome P450 CYP1A1: Wider roles in cancer progression and prevention. *BMC Cancer*, *9*, 187. http://dx.doi.org/10.1186/1471-2407-9-187.

Antonyuk, S. V., Strange, R. W., Marklund, S. L., & Hasnain, S. S. (2009). The structure of human extracellular copper–zinc superoxide dismutase at 1.7 Å resolution: Insights into heparin and collagen binding. *Journal of Molecular Biology*, *388*, 310–326.

Ariyoshi, N., Miyamoto, M., Umetsu, Y., Kunitoh, H., Dosaka-Akita, H., Sawamura, Y., et al. (2002). Genetic polymorphism of CYP2A6 gene and tobacco-induced lung cancer risk in male smokers. Cancer Epidemiology. *Biomarkers and Prevention*, *11*, 890–894.

Ashok, B. T., & Ali, R. (1999). The aging paradox: Free radical theory of aging. *Experimental Gerontology*, *34*, 293–303.

Babcock, G. T. (1999). How oxygen is activated and reduced in respiration. *Proceedings of the National Academy of Sciences of the United States of America*, *96*, 12971–12973.

Barnham, K. J., Masters, C. L., & Bush, A. I. (2004). Neurodegenerative diseases and oxidative stress. *Nature Reviews. Drug Discovery*, *3*, 205–214.

Bast, A. (1986). Is formation of reactive oxygen by cytochrome P-450 perilous and predictable? *Trends in Pharmacological Sciences*, *7*, 266–270.

Behl, C. (1999). Alzheimer's disease and oxidative stress: Implications for novel therapeutic approaches. *Progress in Neurobiology*, *57*, 301–323.

Bickers, D. R., & Athar, M. (2006). Oxidative stress in the pathogenesis of skin disease. *Journal of Investigative Dermatology*, *126*, 2565–2575.

Birukov, K. G. (2006). Oxidized lipids: The two faces of vascular inflammation. *Current Atherosclerosis Reports*, *8*, 223–231.

Brieger, K., Schiavone, S., Miller, F. J., Jr., & Krause, K. H. (2012). Reactive oxygen species: From health to disease. *Swiss Medical Weekly*, *142*, w13659.

Bujons, J., Hsieh, D. P., Kado, N. Y., & Messeguer, A. (1995). Aflatoxin M1 8,9-epoxide: Preparation and mutagenic activity. *Chemical Research in Toxicology*, *8*, 328–332.

Buters, J., Quintanilla-Martinez, L., Schober, W., Soballa, V. J., Hintermair, J., Wolff, T., et al. (2003). CYP1B1 determines susceptibility to low doses of 7,12-dimethylbenz[*a*]anthracene-induced ovarian cancers in mice: Correlation of CYP1B1-mediated DNA adducts with carcinogenicity. *Carcinogenesis*, *24*, 327–334.

Buters, J. T., Sakai, S., Richter, T., Pineau, T., Alexander, D. L., Savas, U., et al. (1999). Cytochrome P450 1B1 determines susceptibility to 7,12-dimethylbenz[*a*]anthracene-induced lymphomas. *Proceedings of the National Academy of Sciences of the United States of America*, *96*, 1977–1982.

Cadenas, E., & Davies, K. J. A. (2000). Mitochondrial free radical generation, oxidative stress and aging. *Free Radical Biology and Medicine*, *29*, 222–230.

Cámara, M., de Cortez Sánchez-Mata, M., Fernández-Ruiz, V., Cámara, R. M., Manzoor, S., & Caceres, J. O. (2013). Chapter 11—Lycopene: A review of chemical and biological activity related to beneficial health effects. *Studies in Natural Products Chemistry*, *40*, 383–426.

Carnell, D. M., Smith, R. E., Path, M. R. C., Daley, F. M., Barber, P. R., Hoskin, P. J., et al. (2004). Target validation of cytochrome P450 CYP1B1 in prostate carcinoma with protein expression in associated hyperplastic and premalignant tissue. *International Journal of Radiation Oncology, Biology, Physics*, *58*, 500–509.

Caro, A. A., & Cederbaum, A. I. (2004). Oxidative stress, toxicology, and pharmacology of CYP2E1. *Annual Review of Pharmacology and Toxicology*, *44*, 27–42.

Cash, T. P., Pan, Y., & Simon, M. C. (2007). Reactive oxygen species and cellular oxygen sensing. *Free Radical Biology and Medicine*, *43*, 1219–1225.

Cederbaum, A. I., & Berl, L. (1982). Pyrazole and 4-methylpyrazole inhibit oxidation of ethanol and dimethyl sulfoxide by hydroxyl radicals generated from ascorbate, xanthine oxidase, and rat liver microsomes. *Archives of Biochemistry and Biophysics*, *216*, 530–543.

Chan, A. C. (1993). Partners in defense, vitamin E and vitamin C. *Canadian Journal of Physiology and Pharmacology*, *71*, 725–731.

Chan, D. W., Liu, V. W., Tsao, G. S., Yao, K. M., Furukawa, T., Chan, K. K., et al. (2008). Loss of MKP3 mediated by oxidative stress enhances tumorigenicity and chemoresistance of ovarian cancer cells. *Carcinogenesis*, *29*, 1742–1750.

Chang, L., & Karin, M. (2001). Mammalian MAP kinase signaling cascades. *Nature*, *410*, 37–40.

Chang, T. K. H., Weber, G. F., Crespi, C. L., & Waxman, D. J. (1993). Differential activation of cyclophosphamide and ifosphamide by cytochromes P-450 2B and 3A in human liver microsomes. *Cancer Research*, *53*, 5629–5637.

Code, E. L., Crespi, C. L., Penman, B. W., Gonzalez, F. J., Chang, T. K. H., & Waxman, D. J. (1997). Interindividual hepatic expression, substrate specificity, and role in procarcinogen activation. *Drug Metabolism and Disposition*, *25*, 985–993.

Comporti, M. (1985). Lipid peroxidation and cellular damage in toxic liver injury. *Laboratory Investigation*, *53*, 599–623.

Conney, A. H., Chang, R. L., Jerina, D. M., & Wei, S. J. C. (1994). Studies on the metabolism of benzo[a]pyrene and dose-dependent differences in the mutagenic profile of its ultimate carcinogenic metabolite. *Drug Metabolism Reviews*, *26*, 125–163.

Czerwinski, M., McLemore, T. L., Gelboin, H. V., & Gonzalez, F. J. (1994). Quantification of CYP2B7, CYP4B1, and CYPOR messenger RNAs in normal human lung and lung tumours. *Cancer Research*, *54*, 1085–1091.

Das, K. C., & Das, C. K. (2000). Thioredoxin: A singlet oxygen quencher and hydroxyl radical scavenger: Redox independent functions. *Biochemical and Biophysical Research Communications*, *277*, 443–447.

Davis, B. K. (2002). Molecular evolution before the origin of species. *Progress in Biophysics & Molecular Biology*, *79*, 77–133.
de Vries, H. E., Schreibelt, G., & van Horssen, J. (2011). Oxidative stress in multiple sclerosis pathology and therapeutic potential of Nrf2 activation. In N. Gadoth & H. H. Göbel (Eds.), *Oxidative stress and free radical damage in neurology, oxidative stress in applied basic research and clinical practice* (pp. 65–77). New York/Heidelberg/London: Humana Press, Springer Science+Business Media. http://dx.doi.org/10.1007/978-1-60327-514-9.
Denisov, I. G., Makris, T. M., Sligar, S. G., & Schlichting, I. (2005). Structure and chemistry of cytochrome P450. *Chemical Reviews*, *105*, 2253–2277.
Denu, J. M., & Tanner, K. G. (1998). Specific and reversible inactivation of protein tyrosine phosphatases by hydrogen peroxide: Evidence for a sulfenic acid intermediate and implications for redox regulation. *Biochemistry*, *37*, 5633–5642.
Dhaini, H. R., Thomas, D. G., Giordano, T. J., Johnson, T. D., Biermann, J. S., Leu, K., et al. (2003). Cytochrome P450 CYP3A4/5 expression as a biomarker of outcome in osteosarcoma. *Journal of Clinical Oncology*, *21*, 2481–2485.
Di Mascio, P., Kaiser, S., & Sies, H. (1989). Lycopene as the most efficient biological carotenoid singlet oxygen quencher. *Archives of Biochemistry and Biophysics*, *274*, 532–538.
Donnelly, P. J., Devereux, T. R., Foley, J. F., Maronpot, R. R., Anderson, M. W., & Massey, T. E. (1996). Activation of K-ras in aflatoxin B1-induced lung tumors from AC3F1 (A/J x C3H/HeJ) mice. *Carcinogenesis*, *17*, 1735–1740.
Downie, D., McFadyen, M. C. E., Rooney, P. H., Cruickshank, M. E., Parkin, D. E., Miller, I. D., et al. (2005). Profiling cytochrome *P450* expression in ovarian cancer: Identification of prognostic markers. *Clinical Cancer Research*, *11*, 7369–7375.
El Assar, M., Angulo, J., & Rodríguez-Mañas, L. (2013). Oxidative stress and vascular inflammation in aging. *Free Radical Biology and Medicine*, *65*, 380–401.
El-Aal, H. A. H. M. A. (2012). Lipid peroxidation end-products as a key of oxidative stress: Effect of antioxidant on their production and transfer of free radicals. In A. Catala (Ed.), *Lipid peroxidation* (pp. 63–88). Rijeka, Croatia: InTech. http://dx.doi.org/10.5772/45944.
Estabrook, R. W., Martinez-Zedillo, G., Young, S., Peterson, J. A., & McCarthy, J. (1975). The interaction of steroids with liver microsomal cytochrome P-450—A general hypothesis. *Journal of Steroid Biochemistry*, *6*, 419–425.
Farooqui, T., & Farooqui, A. A. (2011). Lipid-mediated oxidative stress and inflammation in the pathogenesis of Parkinson's disease. *Parkinson's Disease*, *2011*, 1–9.
Fearon, I. M., & Faux, S. P. (2009). Oxidative stress and cardiovascular disease: Novel tools give (free) radical insight. *Journal of Molecular and Cellular Cardiology*, *47*, 372–381.
Felty, Q., Singh, K. P., & Roy, D. (2005). Estrogen-induced G_1/S transition of G_0-arrested estrogen-dependent breast cancer cells is regulated by mitochondrial oxidant signaling. *Oncogene*, *24*, 4883–4893.
Fritz, P., Behrele, E., Beaune, P., Eichelbaum, M., & Kroemer, H. K. (1993). Differential expression of drug metabolizing enzymes in primary and secondary liver neoplasm: Immunohistochemical characterization of cytochrome P4503A and glutathione-S-transferase. *Histochemistry*, *99*, 443–451.
Gallagher, E. P., Kunze, K. L., Stapleton, P. L., & Eaton, D. L. (1996). The kinetics of aflatoxin B1 oxidation by human cDNA-expressed and human liver microsomal cytochromes P450 1A2 and 3A4. *Toxicology and Applied Pharmacology*, *141*, 595–606.
Gibson, P., Gill, J. H., Khan, P. A., Seargent, J. M., Martin, S. W., Batman, P. A., et al. (2003). Cytochrome P450 1B1 (CYP1B1) is overexpressed in human colon adenocarcinomas relative to normal colon: Implications for drug development. *Molecular Cancer Therapeutics*, *2*, 527–534.
Godoy, W., Albano, R. M., Moraes, E. G., Alves Pinho, P. R., Nunes, R. A., Saito, E. H., et al. (2002). CYP2A6/2A7 and CYP2E1 expression in human esophageal mucosa:

Regional and inter-individual variation in expression and relevance to nitrosamine metabolism. *Carcinogenesis, 23*, 611–616.
Goeptar, A. R., Scheerens, H., & Vermeulen, N. P. E. (1995). Oxygen and xenobiotic reductase activities of cytochrome P450. *Critical Review of Toxicology, 25*, 25–65.
Gonzalez, F. J. (2005). Role of cytochrome P450 in chemical toxicity and oxidative stress: Studies with CYP2E1. *Mutation Research, 569*, 101–110.
Gorenne, I., & Bennett, M. R. (2010). Oxidative stress in vascular disease. In S. Bondy & K. Maiese (Eds.), *Aging and age-related disorders, oxidative stress in applied basic research and clinical practice* (pp. 211–235). New York/Heidelberg/London: Humana Press, Springer Science+Business Media. ISBN: 978-1-60761-601-6, http://dx.doi.org/10.1007/978-1-60761-602-3.
Gottlieb, R. A. (2003). Cytochrome P450: Major player in reperfusion injury. *Archives of Biochemistry and Biophysics, 420*, 262–267.
Gruenke, L. D., Konopka, K., Cadieu, M., & Waskell, L. (1995). The stoichiometry of the cytochrome P-450-catalyzed metabolism of methoxyflurane and benzphetamine in the presence and absence of cytochrome b_5. *The Journal of Biological Chemistry, 270*, 24707–24718.
Grune, T., Lietz, G., Palou, A., Ross, A. C., Stahl, W., Tang, G., et al. (2010). β-Carotene is an important vitamin A source for humans. *The Journal of Nutrition, 140*, 2268S–2285S.
Guengerich, F. P. (1991). Oxidation of toxic and carcinogenic chemicals by human cytochrome P-450 enzymes. *Chemical Research in Toxicology, 4*, 391–407.
Guengerich, F. P. (1993). The 1992 Bernard B. Brodie award lecture. Bioactivation and detoxication of toxic and carcinogenic chemicals. *Drug Metabolism and Disposition, 21*, 1–6.
Guengerich, F. P. (2001). Common and uncommon cytochrome P450 reactions related to metabolism and chemical toxicity. *Chemical Research in Toxicology, 14*, 611–650.
Guengerich, F. P. (2003). Cytochrome P450 oxidations in the generation of reactive electrophiles: Epoxidation and related reactions. *Archives of Biochemistry and Biophysics, 409*, 59–71.
Guengerich, F. P. (2008). Cytochrome P450 and chemical toxicology. *Chemical Research in Toxicology, 21*, 70–83.
Guengerich, F. P., & Johnson, W. W. (1997). Kinetics of ferric cytochrome P450 reduction by NADPH–cytochrome P450 reductase: Rapid reduction in the absence of substrate and variations among cytochrome P450 systems. *Biochemistry, 36*, 14741–14750.
Guengerich, F. P., & Shimada, T. (1998). Activation of procarcinogens by human cytochrome P450 enzymes. *Mutation Research, 400*, 201–213.
Gutman, J. (2002). GSH. Your body's most powerful protector. Glutathione. (3rd ed.). Montreal, Canada: Communications Kudo ca Inc.
Halliwell, B. (1987). Oxidants and human disease: Some new concepts. *The FASEB Journal, 1*, 358–364.
Halliwell, B. (2006a). Reactive species and antioxidants: Redox biology is a fundamental theme of aerobic life. *Plant Physiology, 141*, 312–322.
Halliwell, B. (2006b). Oxidative stress and neurodegeneration: Where are we now? *Journal of Neurochemistry, 97*, 1634–1658.
Halliwell, B. (2007). Oxidative stress and cancer: Have we moved forward? *Biochemical Journal, 401*, 1–11.
Halliwell, B., & Gutteridge, J. M. C. (1984). Oxygen toxicity, oxygen radicals, transition metals and disease. *Biochemical Journal, 219*, 1–14.
Halliwell, B., & Gutteridge, J. M. C. (2007). *Free radicals in biology and medicine* (4th ed.). Oxford, UK: Oxford University Press.
Han, D., Williams, E., & Cadenas, E. (2001). Mitochondrial respiratory chain-dependent generation of superoxide anion and its release into the intermembrane space. *Biochemical Journal, 353*, 411–416.

Handa, O., Naito, Y., & Yoshikawa, T. (2011). Redox biology and gastric carcinogenesis: The role of *Helicobacter pyroli*. *Redox Report*, *16*, 1–7.
Harris, M. L., Schiller, H. J., Reilly, P. M., Donowitz, M., Grisham, M. B., & Bulkley, G. B. (1992). Free radicals and other reactive oxygen metabolites in inflammatory bowel disease: Cause, consequence or epiphenomenon? *Pharmacology & Therapeutics*, *53*, 375–408.
Hayashi, S., Yasui, H., & Sakurai, H. (2005). Essential role of singlet oxygen species in cytochrome P450-dependent substrate oxygenation by rat liver microsomes. *Drug Metabolism and Pharmacokinetics*, *20*, 14–23.
Hayes, J. D., Chanas, S. A., Henderson, C. J., McMahon, M., Sun, C., Moffat, G. J., et al. (2000). The Nrf2 transcription factor contributes both to the basal expression of glutathione S-transferases in mouse liver and to their induction by the chemopreventive synthetic antioxidants, butylated hydroxyanisole and ethoxyquin. *Biochemical Society Transactions*, *28*, 33–41.
He, X. Y., Tang, L., Wang, S. L., Cai, Q. S., Wang, J. S., & Hong, J. Y. (2006). Efficient activation of aflatoxin B_1 by cytochrome P450 2A13, an enzyme predominantly expressed in human respiratory tract. *International Journal of Cancer*, *118*, 2665–2671.
Hecht, S. S. (1999). Tobacco smoke carcinogens and lung cancer. *Journal of the National Cancer Institute*, *91*, 1194–1210.
Hellmold, H., Rylander, T., Magnusson, M., Reiher, E., Warner, M., & Gustafsson, J. Å. (1998). Characterization of cytochrome P450 enzymes in human breast tissue from reduction mammaplasties. *Journal of Clinical Endocrinology and Metabolism*, *843*, 886–895.
Hodek, P., Koblihova, J., Kizek, R., Frei, E., Artl, V. M., & Stiborova, M. (2013). The relationship between DNA adduct formation by benzo[a]pyrene and expression of its activation enzyme cytochrome P450 1A1 in rat. *Environmental Toxicology and Pharmacology*, *36*, 989–996.
Hofmann, B., Hecht, H. J., & Flohé, L. (2005). Peroxiredoxins. *Biological Chemistry*, *383*, 347–364.
Hrycay, E. G., & Bandiera, S. M. (2008). Cytochrome P450 enzymes. In S. C. Gad (Ed.), *Preclinical development handbook: ADME and biopharmaceutical properties* (pp. 627–696): Hoboken, NJ: John Wiley & Sons Inc.
Hrycay, E. G., & Bandiera, S. M. (2012). The monooxygenase, peroxidase, and peroxygenase properties of cytochrome P450. *Archives of Biochemistry and Biophysics*, *522*, 71–89.
Hrycay, E. G., & Bandiera, S. M. (2015). Monooxygenase, peroxidase and peroxygenase properties and reaction mechanisms of cytochrome P450 enzymes. *Advances in Experimental Medicine and Biology*, *851*, in press.
Hrycay, E. G., Gustafsson, J. Å., Ingelman-Sundberg, M., & Ernster, L. (1975). Sodium periodate, sodium chlorite, organic hydroperoxides and H_2O_2 as hydroxylating agents in steroid hydroxylation reactions catalyzed by partially purified cytochrome P-450. *Biochemical and Biophysical Research Communications*, *66*, 209–216.
Hrycay, E. G., & O'Brien, P. J. (1971). Cytochrome P-450 as a microsomal peroxidase utilizing a lipid peroxide substrate. *Archives of Biochemistry and Biophysics*, *147*, 14–27.
Hrycay, E. G., & O'Brien, P. J. (1972). Cytochrome P-450 as a microsomal peroxidase in steroid hydroperoxide reduction. *Archives of Biochemistry and Biophysics*, *153*, 480–494.
Hrycay, E. G., & O'Brien, P. J. (1973). Microsomal electron transport: I. Reduced nicotinamide adenine dinucleotide phosphate—cytochrome *c* reductase and cytochrome P-450 as electron carriers in microsomal NADPH—peroxidase activity. *Archives of Biochemistry and Biophysics*, *157*, 7–22.
Hrycay, E. G., & Prough, R. A. (1974). Reduced nicotinamide adenine dinucleotide-cytochrome b_5 reductase and cytochrome b_5 as electron carriers in NADH-supported cytochrome *P*-450-dependent enzyme activities in liver microsomes. *Archives of Biochemistry and Biophysics*, *165*, 331–339.

Huang, Z., Fasco, M. J., Figge, H. L., Keyomarsi, K., & Kaminsky, L. S. (1996). Expression of cytochromes P450 in human breast tissue and tumors. *Drug Metabolism and Disposition, 24*, 899–905.

Ibrahim, Z. S., Ishizuka, M., Soliman, M., El Bohi, K., Sobhy, W., Muzandu, K., et al. (2008). Protection by *Nigella sativa* against carbon tetrachloride-induced downregulation of hepatic cytochrome P450 isozymes in rats. *Japanese Journal of Veterinary Research, 56*, 119–128.

Ilie, M., & Margină, D. (2012). Trends in the evaluation of lipid peroxidation processes. In A. Catala (Ed.), *Lipid peroxidation* (pp. 111–130). Rijeka, Croatia: InTech. http://dx.doi.org/10.5772/46075.

Ingelman-Sundberg, M., & Ekström, G. (1982). Aniline is hydroxylated by the cytochrome P-450-dependent hydroxyl radical-mediated oxygenation mechanism. *Biochemical and Biophysical Research Communications, 106*, 625–631.

Ingelman-Sundberg, M., & Hagbjörk, A. L. (1982). On the significance of the cytochrome P-450-dependent hydroxyl radical-mediated oxygenation mechanism. *Xenobiotica, 12*, 673–686.

Ioannides, C., Delaforge, M., & Parke, D. V. (1981). Safrole: Its metabolism, carcinogenicity and interactions with cytochrome P-450. *Food and Cosmetics Toxicology, 19*, 657–666.

Ishikawa, T., Esterbauer, H., & Sies, H. (1986). Role of cardiac glutathione transferase and of the glutathione-*S*-conjugate export system in biotransformation of 4-hydroxynonenal in the heart. *The Journal of Biological Chemistry, 261*, 1576–1581.

Itoh, K., Chiba, T., Takahashi, S., Ishi, T., Igarashi, K., Katoh, Y., et al. (1997). An Nrf2/small Maf heterodimer mediates the induction of phase II detoxifying enzyme genes through antioxidant response elements. *Biochemical and Biophysical Research Communications, 236*, 313–322.

Itoh, K., Wakabayashi, N., Katoh, Y., Ishii, T., Igarashi, K., Engel, J. D., et al. (1999). Keap 1 represses nuclear activation of antioxidant responsive elements by Nrf2 through binding to the amino-terminal Neh2 domain. *Genes and Development, 13*, 76–86.

Jain, A. K., Bloom, D. A., & Jaiswal, A. K. (2005). Nuclear import and export signals in control of Nrf2. *The Journal of Biological Chemistry, 280*, 29158–29168.

Jaiswal, A. K. (2004). Nrf2 signaling in coordinated activation of antioxidant gene expression. *Free Radical Biology & Medicine, 36*, 1199–1207.

Jarmi, T., & Agarwal, A. (2009). Heme oxygenase and renal disease. *Current Hypertension Reports, 11*, 56–62.

Jiang, J. G., Chen, C. L., Card, J. W., Yang, S., Chen, J. X., Fu, X. N., et al. (2005). Cytochrome P450 2J2 promotes the neoplastic phenotype of carcinoma cells and is up-regulated in human tumors. *Cancer Research, 65*, 4707–4715.

Jiang, J. G., Fu, X. N., Chen, C. L., & Wang, D. W. (2009). Expression of cytochrome P450 arachidonic acid epoxygenase 2J2 in human tumor tissues and cell lines. *Chinese Journal of Cancer, 28*, 93–96.

Joo, H., Lin, Z., & Arnold, F. H. (1999). Laboratory evolution of peroxide-mediated cytochrome P450 hydroxylation. *Nature, 399*, 670–673.

Kagan, V. E., Wipf, P., Stoyanovsky, D., Greenberger, J. S., Borisenko, G., Belikova, N. A., et al. (2009). Mitochondrial targeting of electron scavenging antioxidants: Regulation of selective oxidation vs random chain reactions. *Advanced Drug Delivery Reviews, 61*, 1375–1385.

Kaneto, H., Katakami, N., Matsuhisa, M., & Matsuoka, T. A. (2010). Role of reactive oxygen species in the progression of type 2 diabetes and atherosclerosis. *Mediators of Inflammation, 2010*, 1–11.

Kappus, H. (1993). Metabolic reactions: Role of cytochrome P-450 in the formation of reactive oxygen species. In J. B. Schenkman & H. Greim (Eds.), *Cytochrome P450* (pp. 145–154): Berlin/Heidelberg, Germany: Springer-Verlag.

Kappus, H., & Sies, H. (1981). Toxic drug effects associated with oxygen metabolism: Redox cycling and lipid peroxidation. *Experientia*, *37*, 1233–1241.

Karlgren, M., Gomez, A., Stark, K., Svärd, J., Rodriguez-Antona, C., Oliw, E., et al. (2006). Tumor-specific expression of the novel cytochrome P450 enzyme, CYP2W1. *Biochemical and Biophysical Research Communications*, *341*, 451–458.

Kellogg, E. W., III, & Fridovich, I. (1977). Liposome oxidation and erythrocyte lysis by enzymatically generated superoxide and hydrogen peroxide. *The Journal of Biological Chemistry*, *252*, 6721–6728.

Kidd, P. M. (1997). Glutathione: Systemic protectant against oxidative and free radical damage. *Alternative Medicine Review*, *2*, 155–176.

Kivistö, K. T., Fritz, P., Linder, A., Friedel, G., Beaune, P., & Kromer, H. K. (1995). Immunohistochemical localization of cytochrome P450 3A in human pulmonary carcinomas and normal bronchial tissue. *Histochemistry*, *103*, 25–29.

Klaunig, J. E., & Kamendulis, L. M. (2004). The role of oxidative stress in carcinogenesis. *Annual Review of Pharmacology and Toxicology*, *44*, 239–267.

Kobayashi, M., Kang, M. I., Okawa, H., Ohtsuji, M., Zenke, Y., Chiba, T., et al. (2004). Oxidative stress sensor Keap 1 functions as an adaptor for Cul3-based E3 ligase to regulate proteasomal degradation of Nrf2. *Molecular and Cellular Biology*, *24*, 7130–7139.

Kondrikov, D., Caldwell, R. B., Dong, Z., & Su, Y. (2011). Reactive oxygen species-dependent RhoA activation mediated collagen synthesis in hyperoxic lung fibrosis. *Free Radical Biology and Medicine*, *50*, 1689–1698.

Koppenol, W. H. (2001). The Haber-Weiss cycle—70 years later. *Redox Report*, *6*, 229–234.

Kovacic, P., & Somanathan, R. (2012). Redox processes in neurodegenerative disease involving reactive oxygen species. *Current Neuropharmacology*, *10*, 289–302.

Krause, K. H. (2007). Aging: A revisited theory based on free radicals generated by NOX family. *Experimental Gerontology*, *42*, 256–262.

Kregel, K. C., & Zhang, H. J. (2007). An integrated view of oxidative stress in aging: Basic mechanisms, functional effects, and pathological considerations. *American Journal of Physiology. Regulatory, Integrative and Comparative Physiology*, *292*, 18–36.

Kumar, B., Koul, S., Khandrika, L., Meacham, R. B., & Koul, H. K. (2008). Oxidative stress is inherent in prostate cancer cells and is required for aggressive phenotype. *Cancer Research*, *68*, 1777–1785.

Kumarakulasingham, M., Rooney, P. H., Dundas, S. R., Telfer, C., Melvin, W. T., Curran, S., et al. (2005). Cytochrome P450 profile of colorectal cancer: Identification of markers of prognosis. *Clinical Cancer Research*, *11*, 3758–3765.

Kyriakis, J. M., & Avruch, J. (2001). Mammalian mitogen-activated protein kinase signal transduction pathways activated by stress and inflammation. *Physiological Reviews*, *81*, 807–869.

Lane, N. (2002). *Oxygen, the molecule that made the world*. Oxford, UK: Oxford University Press.

Lennon, S. V., Martin, S. J., & Cotter, T. G. (1991). Dose-dependent induction of apoptosis in human tumour cell lines by widely diverging stimuli. *Cell Proliferation*, *24*, 203–214.

Lenová, K., Moserová, M., Kortrbová, V., Šulc, M., Henderson, C. J., Wolf, C. R., et al. (2011). Role of cytochromes P450 1A1/2 in detoxication and activation of carcinogenic aristolochic acid I: Studies with the hepatic NADPH:Cytochrome P450 reductase null (HRN) mouse model. *Toxicological Sciences*, *121*, 43–56.

Leung, T., Rajendran, R., Singh, S., Garva, R., Krstic-Demonacos, M., & Demonacos, C. (2013). Cytochrome P450 2E1 (CYP2E1) regulates the response to oxidative stress and migration of breast cancer cells. *Breast Cancer Research*, *15*, R107.

Lewis, D. F. V. (2001). *Guide to cytochromes P450: Structure and function*. London, UK: Taylor & Francis.

Lewis, D. F. V. (2002). Oxidative stress: The role of cytochromes P450 in oxygen activation. *Journal of Chemical Technology and Biotechnology*, 77, 1095–1100.
Lewis, A., Du, J., Liu, J., Ritchie, J. M., Oberley, L. W., & Cullen, J. J. (2005). Metastatic progression of pancreatic cancer: Changes in antioxidant enzymes and cell growth. *Clinical & Experimental Metastasis*, *22*, 523–532.
Lewis, D. F. V., Ioannides, C., & Parke, D. V. (1989). Molecular orbital studies of oxygen activation and mechanisms of cytochromes P-450-mediated oxidative metabolism of xenobiotics. *Chemico-Biological Interactions*, *70*, 263–280.
Lewis, D. F. V., & Sheridan, G. (2001). Cytochromes P450, oxygen, and evolution. *The Scientific World*, *1*, 151–167.
Liou, G.-Y., & Storz, P. (2010). Reactive oxygen species in cancer. *Free Radical Research*, *44*, 470–496.
Maas, R. M., Reus, K., Diesel, B., Steudel, W. I., Feiden, W., Fischer, U., et al. (2001). Amplification and expression of splice variants of the gene encoding the P450 cytochrome 25-hydroxyvitamin S_3 1,α-hydroxylase (*CYP27B1)* in human malignant glioma. *Clinical Cancer Research*, 7, 868–875.
Manda, G., Nechifor, M. T., & Neagu, T. M. (2009). Reactive oxygen species, cancer and anti-cancer therapies. *Current Chemical Biology*, *3*, 342–366.
Maraldi, T. (2013). Natural compounds as modulators of NADPH oxidases. *Oxidative Medicine and Cellular Longevity*, *Vol. 2013*, Article ID 271602, 10 pages.
Marks, F., & Fürstenberger, G. (1985). Tumor formation in skin: Are active oxygen species involved? In H. Sies (Ed.), *Oxidative stress* (pp. 437–475): London, UK/New York, NY: Academic Press.
Matsuda, Y., Saoo, K., Yamakawa, K., Yokohira, M., Suzuki, S., Kuno, T., et al. (2007). Overexpression of CYP2A6 in human colorectal tumors. *Cancer Science*, *98*, 1582–1585.
McCubrey, J. A., Steelman, L. S., Chappell, W. H., Abrams, S. L., Wong, E. W., Chang, F., et al. (2007). Roles of the Raf/MEK/ERK pathway in cell growth, malignant transformation and drug resistance. *Biochimica et Biophysica Acta*, *1773*, 1263–1284.
McKay, C. P., & Hartman, H. (1991). Hydrogen peroxide and the evolution of oxygenic photosynthesis. *Origins of Life and Evolution of the Biosphere*, *21*, 157–163.
McLean, L., Soto, U., Agama, K., Francis, J., Jimenez, R., Pommier, Y., et al. (2008). Aminoflavone induces oxidative DNA damage and reactive oxidative species-mediated apoptosis in breast cancer cells. *International Journal of Cancer*, *122*, 1665–1674.
McLemore, T. L., Adelberg, S., Liu, M. C., McMahon, N. A., Yu, S. J., Hubbard, W. C., et al. (1990). Expression of CYP1A1 gene in patients with lung cancer: Evidence for cigarette smoke-induced gene expression in normal lung tissue and for altered gene regulation in primary pulmonary carcinomas. *Journal of the National Cancer Institute*, *82*, 1333–1339.
Meister, A., & Anderson, M. E. (1983). Glutathione. *Annual Review of Biochemistry*, *52*, 711–760.
Mimori, K., Tanaka, Y., Yoshinaga, K., Masuda, T., Yamashita, K., Okamoto, M., et al. (2004). Clinical significance of the overexpression of the candidate oncogene *CYP24* in esophageal cancer. *Annals of Oncology*, *15*, 236–241.
Molina-Ortiz, D., Camacho-Carranza, R., González-Zamora, J. F., Shalkow-Kalincovstein, J., Cárdenas-Cardós, R., Nosti-Palacios, R., et al. (2014). Differential expression of cytochrome P450 enzymes in normal and tumor tissues from childhood rhabdomyosarcoma. *PLoS One*, *9*, e93261. http://dx.doi.org/10.1371/journal.pone.0093261.
Montezano, A. C., & Touyz, R. M. (2012). *Reactive oxygen species and the cardiovascular system*. San Rafael, CA: Morgan & Claypool Publishers.
Moreau, F., Mittre, H., Benhaim, A., Bois, C., Bertherat, J., Carreau, S., et al. (2008). Aromatase expression in the normal human adult adrenal and in adrenocortical tumors:

Biochemical, immunohistochemical, and molecular studies. *European Journal of Endocrinology, 160*, 93–99.
Mortensen, A., & Skibsted, L. H. (1998). Reactivity of β-carotene towards peroxyl radicals studied by laser flash and steady-state photolysis. *FEBS Letters, 426*, 392–396.
Mueller, E. J., Loida, P. J., & Sligar, S. G. (1995). Twenty-five years of P450cam research: Mechanistic insights into oxygenase catalysis. In P. R. Ortiz de Montellano (Ed.), *Cytochrome P450: Structure, mechanism, and biochemistry* (2nd ed., pp. 83–124). New York, NY: Plenum Press.
Muller, F. L., Lustgarten, M. S., Jang, Y., Richardson, A., & Van Remmen, H. (2007). Trends in oxidative aging theories. *Free Radical Biology and Medicine, 43*, 477–503.
Murphy, M. E., & Kehrer, J. P. (1989). Oxidative stress and muscular dystrophy. *Chemico-Biological Interactions, 69*, 101–173.
Murray, G. I. (2000). The role of cytochrome P450 in tumour development and progression and its potential in therapy. *The Journal of Pathology, 192*, 419–426.
Murray, G. I., McFadyen, M. C. E., Mitchell, R. T., Cheung, Y. L., Kerr, A. C., & Melvin, W. T. (1999). Cytochrome P450 CYP3A in human renal cell cancer. *British Journal of Cancer, 79*, 1836–1842.
Murray, G. I., Shaw, D., Weaver, R. J., McKay, J. A., Ewen, S. W. B., Melvin, W. T., et al. (1994). Cytochrome P450 expression in oesophageal cancer. *Gut, 35*, 599–603.
Murray, G. I., Taylor, M. C., McFadyen, M. C. E., McKay, J. A., Greenlee, W. F., Burke, M. D., et al. (1997). Tumor-specific expression of cytochrome P450 CYP1B1. *Cancer Research, 57*, 3026–3031.
Nelson, D. R., Kamataki, T., Waxman, D. J., Guengerich, F. P., Estabrook, R. W., Feyereisen, R., et al. (1993). The P450 superfamily: Update on new sequences, gene mapping, accession numbers, early trivial names of enzymes, and nomenclature. *DNA and Cell Biology, 12*, 1–51.
Neumann, C. A., Cao, J., & Manevich, Y. (2009). Peroxiredoxin 1 and its role in cell signaling. *Cell Cycle, 8*, 4072–4078.
Nordberg, J., & Arner, E. S. J. (2001). Reactive oxygen species, antioxidants, and thioredoxin system. *Free Radical Biology and Medicine, 31*, 1287–1312.
Osada, M., Ogura, Y., Yasui, H., & Sakurai, H. (1999). Involvement of singlet oxygen in cytochrome P450-dependent substrate oxidations. *Biochemical and Biophysical Research Communications, 263*, 392–397.
Paoletti, R., Gotto, A. M., Jr., & Hajjar, D. P. (2004). Inflammation in atherosclerosis and implications for therapy. *Circulation, 109*(Suppl. 3), 20–26.
Park, J., Lee, S., Lee, S., & Kang, S. W. (2014). 2-Cys peroxiredoxins: Emerging hubs determining redox dependency of mammalian signaling networks. *International Journal of Cell Biology, 2014*, 10. Article ID 715867, http//dx.doi.org/10.1155/2014/715867.
Parke, A. L., Ioannides, C., Lewis, D. F. V., & Parke, D. V. (1991). Molecular pathology of drug-disease interactions in chronic autoimmune inflammatory diseases. *Inflammopharmacology, 1*, 3–36.
Peña-Silva, R. A., Miller, J. D., Chu, Y., & Heistad, D. D. (2009). Serotonin produces monoamine oxidase-dependent oxidative stress in human heart valves. *American Journal of Physiology. Heart and Circulatory Physiology, 297*, 1354–1360.
Peters, W. H. M., Boon, C. E. W., Roelofs, H. M. J., Wobbes, T., Nagengast, F. M., & Kremers, P. G. (1992). Expression of drug-metabolizing enzymes and P-170 glycoprotein in colorectal carcinoma and normal mucosa. *Gastroenterology, 103*, 448–455.
Petros, J. A., Baumann, A. K., Ruiz-Pesini, E., Amin, M. B., Sun, C. Q., Hall, J., et al. (2005). mtDNA mutations increase tumorigenicity in prostate cancer. *Proceedings of the National Academy of Sciences of the United States of America, 102*, 719–727.
Pieri, C., Marra, M., Moroni, F., Recchioni, R., & Marcheselli, F. (1994). Melatonin: Peroxyl radical scavenger more effective than vitamin E. *Life Sciences, 55*, L271–L276.

Pohanka, M. (2013). Alzheimer's disease and oxidative stress: A review. *Current Medicinal Chemistry*, *21*, 356–364.

Pohl, L. R., Schulick, R. D., Highet, R. J., & George, J. W. (1984). Reductive-oxygenation mechanism of metabolism of carbon tetrachloride to phosgene by cytochrome P-450. *Molecular Pharmacology*, *25*, 318–321.

Puntarulo, S., & Cederbaum, A. I. (1998). Production of reactive oxygen species by microsomes enriched in specific human cytochrome P450 enzymes. *Free Radical Biology and Medicine*, *24*, 1324–1330.

Rahman, T., Hosen, I., Islam, M. M. T., & Shekhar, H. U. (2012). Oxidative stress and human health. *Advances in Bioscience and Biotechnology*, *3*, 997–1019.

Ramos de Andrade, D., Jr., Becco de Souza, R., Alves dos Santos, S., & Ramos de Andrade, D. (2005). Oxygen free radicals and pulmonary disease. *Journal of Brazilian Pneumology*, *31*, 60–68.

Reiter, R. J. (1997). Antioxidant actions of melatonin. *Advances in Pharmacology*, *38*, 103–117.

Reiter, R. J., Carneiro, R. C., & Oh, C. S. (1997). Melatonin in relation to cellular antioxidative defense mechanisms. *Hormone and Metabolic Research*, *29*, 363–372.

Repetto, M., Semprine, J., & Boveris, A. (2012). Lipid peroxidation: Chemical mechanism, biological implications and analytical determination. In A. Catala (Ed.), *Lipid peroxidation* (pp. 3–30). Rijeka, Croatia: InTech. http://dx.doi.org/10.5772/45943.

Rhee, S. G., Chae, H. Z., & Kim, K. (2005). Peroxiredoxins: A historical overview and speculative preview of novel mechanisms and emerging concepts in cell signaling. *Free Radical Biology and Medicine*, *38*, 1543–1552.

Rieger, M. A., Ebner, R., Bell, D. R., Kiessling, A., Rohayem, J., Schmitz, M., et al. (2004). Identification of a novel mammary-restricted cytochrome P450, CYP4Z1, with overexpression in breast carcinoma. *Cancer Research*, *64*, 2357–2364.

Roberts, P. J., & Der, C. J. (2007). Targeting the Raf-MEK-ERK mitogen-activated protein kinase cascade for the treatment of cancer. *Oncogene*, *26*, 3291–3310.

Roberts, L. J., Oates, J. A., & Linton, M. F. (2007). The relationship between dose of vitamin E and suppression of oxidative stress in humans. *Free Radical Biology and Medicine*, *43*, 1388–1393.

Rodriguez-Antona, C., & Ingelman-Sundberg, M. (2006). Cytochrome *P450* pharmacogenetics and cancer. *Oncogene*, *25*, 1679–1691.

Rosenfeld, M. E. (1998). Inflammation, lipids, and free radicals: Lessons learned from the atherogenic process. *Seminars in Reproductive Endocrinology*, *16*, 249–261.

Rossi, M. A. (2012). Tissue occurrence of carbonyl products of lipid peroxidation and their role in inflammatory disease. In A. Catala (Ed.), *Lipid peroxidation* (pp. 279–292). Rijeka, Croatia: InTech. http://dx.doi.org/10.5772/51343.

Rota, C., Barr, D. P., Martin, M. V., Guengerich, F. P., Tomasi, A., & Mason, R. P. (1997). Detection of free radicals produced from the reaction of cytochrome P-450 with linoleic acid hydroperoxide. *Biochemical Journal*, *328*, 565–571.

Rowland, A., & Mangoni, A. A. (2014). Cytochrome P450 and ischemic heart disease: Current concepts and future directions. *Expert Opinion on Drug Metabolism & Toxicology*, *10*, 191–213.

Roy, D., Cai, Q., Felty, Q., & Narayan, S. (2007). Estrogen-induced generation of reactive oxygen and nitrogen species, gene damage, and estrogen-dependent cancers. *Journal of Toxicology and Environmental Health B Critical Reviews*, *10*, 235–257.

Salmeen, A., & Barford, D. (2005). Functions and mechanisms of redox regulation of cysteine-based phosphatases. *Antioxidants & Redox Signaling*, *7*, 560–577.

Samuilov, V. D. (1997). Photosynthetic oxygen: The role of H_2O_2. A review. *Biochemistry (Moscow)*, *62*, 451–454.

Sarsour, E. H., Venkataraman, S., Kalen, A. L., Oberley, L. W., & Goswami, P. C. (2008). Manganese superoxide dismutase activity regulates transitions between quiescent and proliferative growth. *Aging Cell*, 7, 405–417.

Sasano, H., Kaga, K., Sato, S., Yajima, A., Nagura, H., & Harada, N. (1996). Aromatase cytochrome P450 gene expression in endometrial carcinoma. *British Journal of Cancer*, *74*, 1541–1544.

Sato, M., Yokoyama, U., Fujita, T., Okumura, S., & Ishikawa, Y. (2011). The roles of cytochrome P450 in ischemic heart disease. *Current Drug Metabolism*, *12*, 526–532.

Schafer, F. Q., & Buettner, G. R. (2001). Redox environment of the cell as viewed through the redox state of the glutathione disulfide/glutathione couple. *Free Radical Biology and Medicine*, *30*, 1191–1212.

Schirrmeister, B. E., de Voss, J. M., Antonelli, A., & Bagheri, H. C. (2013). Evolution of multicellularity coincided with increased diversification of cyanobacteria and the Great Oxidation Event. *Proceedings of the National Academy of Sciences of the United States of America*, *110*, 1791–1796.

Schmidt, R., Baumann, F., Knüpfer, H., Brauckhoff, M., Horn, L. C., Schönfelder, M., et al. (2004). CYP3A4, CYP2C9 and CYP2B6 expression and ifosfamide turnover in breast cancer tissue microsomes. *British Journal of Cancer*, *90*, 911–916.

Serra, J. A., Domínguez, R. O., Marschoff, E. R., Guareschi, E. M., Famulari, A. L., & Boveris, A. (2009). Systemic oxidative stress associated with the neurological diseases of aging. *Neurochemical Research*, *34*, 2122–2132.

Shankar, S., Suthakar, G., & Srivastava, S. K. (2007). Epigallocatechin-3-gallate inhibits cell cycle and induces apoptosis in pancreatic cancer. *Frontiers in Bioscience*, *12*, 5039–5051.

Shimada, T., Yamazaki, H., Mimura, M., Inui, Y., & Guengerich, F. P. (1994). Interindividual variations in human liver cytochrome P450 enzymes involved in the oxidation of drugs, carcinogens and toxic chemicals. *The Journal of Pharmacology and Experimental Therapeutics*, *270*, 414–423.

Shimada, T., Yun, C. H., Yamazaki, H., Gautier, J. C., Beaune, P. H., & Guengerich, F. P. (1992). Characterization of human lung microsomal cytochrome P-4501A1 and its role in the oxidation of chemical carcinogens. *Molecular Pharmacology*, *41*, 856–864.

Sies, H. (1986). Biochemistry of oxidative stress. *Angewandte Chemie International Edition in English*, *25*, 1058–1071.

Sies, H. (1993). Strategies of antioxidant defense. *European Journal of Biochemistry*, *215*, 213–219.

Sies, H. (1997). Oxidative stress: Oxidants and antioxidants. *Experimental Physiology*, *82*, 291–295.

Sies, H. (2014). Role of metabolic H_2O_2 generation: Redox signaling and oxidative stress. *The Journal of Biological Chemistry*, *289*, 8735–8741.

Sies, H., & Jones, D. (2007). Oxidative stress. In G. Fink (Ed.), *Encyclopedia of stress: Vol. 3* (2nd ed., pp. 45–48). San Diego, CA: Academic Press.

Singh, R. B., Niaz, A. M., Ghosh, S., Agarwal, P., Ahmad, S., Begum, R., et al. (1995). Randomized, controlled trial of antioxidant vitamins and cardioprotective diet on hyperlipidemia, oxidative stress, and development of experimental atherosclerosis: The diet and antioxidant trial on atherosclerosis (DATA). *Cardiovascular Drugs and Therapy*, *9*, 763–771.

Sissung, T. M., Price, D. K., Sparreboom, A., & Figg, W. D. (2006). Pharmacogenetics and regulation of human cytochrome *P*450 1B1: Implications in hormone-mediated tumor metabolism and a novel target for therapeutic intervention. *Molecular Cancer Research*, *4*, 135–150.

Sono, M., Roach, M. P., Coulter, E. D., & Dawson, J. H. (1996). Heme-containing oxygenases. *Chemical Reviews*, *96*, 2841–2847.

Spivack, S. D., Hurteau, G. J., Reilly, A. A., Aldous, K. M., Ding, X., & Kaminsky, L. S. (2001). CYP1B1 expression in human lung. *Drug Metabolism and Disposition*, *29*, 916–922.
Srivastava, S. K., & Singh, S. V. (2004). Cell cycle arrest, apoptosis induction and inhibition of nuclear factor kappa B activation in anti-proliferative activity of benzyl isothiocyanate against human pancreatic cancer cells. *Carcinogenesis*, *25*, 1701–1709.
Stahl, W., & Sies, H. (1996). Perspectives in biochemistry and biophysics: Lycopene: A biologically important carotenoid for humans? *Archives of Biochemistry and Biophysics*, *336*, 1–9.
Steinbrenner, H., & Sies, H. (2009). Protection against reactive oxygen species by selenoproteins. *Biochimica et Biophysica Acta*, *1790*, 1478–1485.
Stiborová, M., Frei, E., & Schmeiser, H. H. (2008). Biotransformation enzymes in development of renal injury and urothelial cancer caused by aristolochic acid. *Kidney International*, *73*, 1209–1211.
St-Pierre, J., Buckingham, J. A., Roebuck, S. J., & Brand, M. D. (2002). Topology of superoxide production from different sites in the mitochondrial electron transport chain. *The Journal of Biological Chemistry*, *277*, 44784–44790.
Su, T., Bao, Z., Zhang, Q. Y., Smith, T. J., Hong, J. Y., & Ding, X. (2000). Human cytochrome P450 CYP2A13: Predominant expression in the respiratory tract and its high efficiency metabolic activation of a tobacco-specific carcinogen, 4-(methylnitrosoamino)-1-(3-pyridyl)-1-butanone. *Cancer Research*, *60*, 5074–5079.
Sundaram, K., Sambandam, Y., Tsuruga, E., Wagner, C. L., & Reddy, S. V. (2014). 1α,25-Dihydroxyvitamin D_3 modulates CYP2R1 gene expression in human oral squamous cell carcinoma tumor cells. *Hormones and Cancer*, *5*, 90–97.
Szent-Györgyi, A. (1982). Biological oxidation and cancer. *International Journal of Quantum Chemistry: Quantum Biology Symposium*, *9*, 27–30.
Trachootham, D., Zhou, Y., Zhang, H., Demizu, Y., Chen, Z., Pelicano, H., et al. (2006). Selective killing of oncogenically transformed cells through a ROS-mediated mechanism by β-phenylethylisothiocyanate. *Cancer Cell*, *10*, 241–252.
Tsuji, K., Kwon, A. H., Yoshida, H., Qui, Z., Kaibori, M., Okumura, T., et al. (2005). Free radical scavenger (edaravone) prevents endotoxin-induced liver injury after partial hepatectomy in rats. *Journal of Hepatology*, *42*, 94–101.
Uzma, N., Kumar, B. S., & Anees, S. (2011). Red wine ameliorates CCl4-induced acute liver injury in rats. *Australian Journal of Biomedical Sciences*, *1*, 1–7.
Vagula, M. C., & Konieczko, E. M. (2012). Lipid peroxidation and polybrominated diphenyl ethers—A toxicological perspective. In A. Catala (Ed.), *Lipid peroxidation* (pp. 515–536). Rijeka, Croatia: InTech. http://dx.doi.org/10.5772/50345.
Valavanidis, A., Vlachogianne, T., Fiotakis, K., & Loridas, S. (2013). Pulmonary oxidative stress, inflammation and cancer: Respirable particulate matter, fibrous dusts and ozone as major causes of lung carcinogenesis through reactive oxygen species mechanisms. *International Journal of Environmental Research and Public Health*, *10*, 3886–3907.
Valko, M., Leibfritz, D., Moncol, J., Cronin, M. T. D., Mazur, M., & Telser, J. (2007). Free radicals and oxidants in normal physiological functions and human disease. *International Journal of Biochemistry and Cell Biology*, *39*, 44–84.
Venugopal, R., & Jaiswal, A. K. (1996). Nrf1 and Nrf2 positively and c-Fos and Fra 1 negatively regulate the human antioxidant response element-mediated expression of NAD(P)H:Quinone oxidoreductase 1 gene. *Proceedings of the National Academy of Sciences of the United States of America*, *93*, 14960–14965.
Veskoukis, A. S., Tsatsakis, A. M., & Kouretas, D. (2012). Dietary oxidative stress and antioxidant defense with an emphasis on plant extract administration. *Cell Stress and Chaperones*, *17*, 11–21.
Wächtershäuser, G. (1988). Before enzymes and templates: Theory of surface metabolism. *Microbiological Reviews*, *52*, 452–484.

Wahid, M., Mahjabeen, I., Baig, R. M., & Kayani, M. A. (2013). Expression of CYP1A1 and GSTP1 in human brain tumor tissues in Pakistan. *Asian Pacific Journal of Cancer Prevention, 14*, 7187–7191.

Wallace, D. C. (2005). A mitochondrial paradigm of metabolic and degenerative diseases, aging, and cancer: A dawn for evolutionary medicine. *Annual Review of Genetics, 39*, 359–407.

Waris, G., & Ahsan, H. (2006). Reactive oxygen species: Role in the development of cancer and various chronic conditions. *Journal of Carcinogenesis, 5*, 14–21.

Wickramasinghe, R. H., & Villee, C. A. (1975). Early role during chemical evolution for cytochrome P450 in oxygen detoxification. *Nature, 256*, 509–511.

Winterbourne, C. C., Hampton, M. B., Livesey, J. H., & Kettle, A. J. (2006). Modeling the reactions of superoxide and myeloperoxidase in the neutrophil phagosome: Implications for microbial killing. *The Journal of Biological Chemistry, 281*, 39860–39869.

Wiseman, H., & Halliwell, B. (1996). Damage to DNA by reactive oxygen and nitrogen species: Role in inflammatory diseases and progression to cancer. *Biochemical Journal, 313*, 17–29.

Wood, L. G., Gibson, P. G., & Garg, M. L. (2003). Biomarkers of lipid peroxidation, airway inflammation and asthma. *European Respiratory Journal, 21*, 177–186.

Woodruff, T., Blake, D. R., Freeman, J., Andrews, F. J., Salt, P., & Lunec, J. (1986). Is chronic synovitis an example of reperfusion injury? *Annals of the Rheumatic Diseases, 45*, 608–611.

Xia, Z., Dickens, M., Raingeaud, J., Davis, R. J., & Greenberg, M. E. (1995). Opposing effects of ERK and JNK-p38 MAP kinases on apoptosis. *Science, 270*, 1326–1331.

Yamamoto, T., Suzuki, T., Kobayashi, M., Wakabayashi, J., Maher, J., Motohashi, H., et al. (2008). Physiological significance of reactive cysteine residues of Keap 1 in determining Nrf2 activity. *Molecular and Cellular Biology, 28*, 2758–2770.

Yasui, H., Deo, K., Ogura, Y., Yoshida, H., Shiraga, T., Kagayama, A., et al. (2002). Evidence for singlet oxygen involvement in the rat and human cytochrome P450-dependent substrate oxidations. *Drug Metabolism and Pharmacokinetics, 17*, 416–426.

Yasui, H., Hayashi, S., & Sakurai, H. (2005). Possible involvement of singlet oxygen species as multiple oxidants in P450 catalytic reactions. *Drug Metabolism and Pharmacokinetics, 20*, 1–13.

Young, I. S., & Woodside, J. V. (2001). Antioxidants in health and disease. *Journal of Clinical Pathology, 54*, 176–186.

Yu, W., Chai, H., Zhao, H., Xie, X., Zheng, H., Wang, C., et al. (2012). Increased expression of CYP4Z1 promotes tumor angiogenesis and growth in human breast cancer. *Toxicology and Applied Pharmacology, 264*, 73–83.

Zangar, R. C., Davydov, D. R., & Verma, S. (2004). Mechanisms that regulate production of reactive oxygen species by cytochrome P450. *Toxicology and Applied Pharmacology, 199*, 316–331.

Zhang, R., Humphreys, I., Sahu, R. P., Shi, Y., & Srivastava, S. K. (2008). In vitro and in vivo induction of apoptosis by capsaicin in pancreatic cancer cells is mediated through ROS generation and mitochondrial death pathway. *Apoptosis, 13*, 1465–1478.

Zhou, L., Jing, Y., Styblo, M., Chen, Z., & Waxman, S. (2005). Glutathione-*S*-transferase π inhibits As_2O_3-induced apoptosis in lymphoma cells: Involvement of hydrogen peroxide catabolism. *Blood, 105*, 1198–1203.

CHAPTER THREE

Polymorphic Variants of Cytochrome P450: Relevance to Cancer and Other Diseases

Ann K. Daly[1]
Institute of Cellular Medicine, Newcastle University, Medical School, Newcastle upon Tyne, United Kingdom
[1]Corresponding author: e-mail address: a.k.daly@ncl.ac.uk

Contents

Abstract

Associations of cytochrome P450 (CYP) polymorphisms with risk of disease development have been reported widely. For lung cancer, a large number of studies on *CYP1A1*, *CYP2D6*, and *CYP2A6* polymorphisms have been performed. However, recent studies, including meta-analyses and genome-wide association studies, suggest that only the *CYP2A6* association, where genotypes associated with low activity decrease susceptibility possibly due to slower nicotine metabolism, appears significant. Associations with lung cancer susceptibility have also been reported for *CYP1A2*, *CYP1B1*, and *CYP2E1* polymorphisms but these, though biologically plausible, have not been well replicated. For cancers where exposure to xenobiotics other than tobacco smoke affects risk, CYP polymorphisms may also be relevant. Examples include *CYP3A* for hepatocellular carcinoma due to aflatoxin exposure, *CYP1A2* for colon cancer associated with heterocyclic arylamine exposure and *CYP2E1* for nitrosamine-related nasopharyngeal cancer. For other diseases, a well-established example relates to *CYP1B1* where homozygosity for rare mutations occurs in primary congenital glaucoma. Rare *CYP1B1* mutations and possibly polymorphisms may also contribute to risk for more common forms of glaucoma. CYP2C isoforms and CYP2J2 contribute to extrahepatic metabolism of arachidonic acid to epoxyeicosanoic acids which have effects in the cardiovascular system. Genotype for these isoforms may be relevant to risk of cardiovascular disease but evidence is still lacking. *CYP2C19* poor metabolizers may be at increased risk of endometriosis, and

Advances in Pharmacology, Volume 74
ISSN 1054-3589
http://dx.doi.org/10.1016/bs.apha.2015.03.001

CYP2E1 genotype may modulate risk of development of alcoholic liver disease. In conclusion, CYP polymorphisms are relevant to risk for some diseases but this may have been overstated in earlier studies.

ABBREVIATIONS

3-MC 3-methylcholanthrene
AhR aromatic hydrocarbon receptor
CYP cytochrome P450
EET epoxyeicosatrienoic acid
GWA genome-wide association
NNK 4-(methylnitrosamino)-1-(3-pyridyl)-1-butanone
PhIP 2-amino-3-methylimidazo[4,5-b]pyridine
POAG primary open-angle glaucoma

1. INTRODUCTION

For many years, cytochrome P450 (CYP) polymorphisms have been proposed to modulate susceptibility to cancer and other complex diseases. The early studies on this area were mainly on cancer and phenotype not genotype based, though once the genetic basis for common P450 polymorphisms became clear and genotyping assays were developed, associations with cancer continued to be reported. There is no evidence that CYP genotype in the tumor is relevant to cancer susceptibility and progression so the emphasis in this chapter is on germline DNA. Increasingly, it appears that CYP genotype may be more relevant to drug response in certain cancers than as a susceptibility factor since many previously reported associations with cancer are not standing up well to replication in the large genome-wide association (GWA) case-control studies now being reported. In addition, a relatively recent survey of pooled and meta-analyses involving at least 500 cancer cases for any gene as a risk factor for any cancer failed to show any associations with CYP genes when a false positive probability approach was used to assess significance (Dong et al., 2008). In spite of these concerns, it is worthwhile to consider the relevance of selected CYP polymorphisms to selected cancers, especially those associated with tobacco exposure. Lung cancer has a strong association with tobacco smoking (though certain other cancers also have weaker relationships with smoking) and will therefore be considered in detail in relation to a range of P450 isoforms that may contribute to metabolism of the wide range of chemicals found in tobacco smoke. The relevance of CYP polymorphisms to other cancers linked to exposure to xenobiotics and endogenous molecules subject to P450 metabolism will also be considered. The emphasis will be on xenobiotic metabolizing CYPs from families 1, 2, and 3. Other CYP families, contributing mainly to metabolism of endogenous compounds, may also be relevant to cancer susceptibility but are outside the scope of this chapter.

In addition to cancer, it has been suggested that CYP genotype may affect susceptibility to a range of other complex diseases. A large number of reports have appeared over the past 20–30 years but the majority of these have not been replicated subsequently. However, there are some exceptions to this and several selected examples, which relate to either xenobiotic or endogenous compound metabolizing activities of CYPs from families 1 to 3 where the association appears more convincing will be considered.

The relevance of CYP genes to risk for a variety of cancers has been considered controversial for some time with concerns about consistency and heterogeneity in proposed associations expressed in both meta-analyses and review articles (Agundez, 2004; Ingelman-Sundberg, 2002; Rodriguez-Antona, Gomez, Karlgren, Sim, & Ingelman-Sundberg, 2010; Rostami-Hodjegan, Lennard, Woods, & Tucker, 1998). The general shift in studies on disease susceptibility from candidate gene case-control studies to GWA studies has accentuated such concerns, with several apparently well-established disease associations previously reported for CYP genes not being confirmed using this approach. As discussed in detail elsewhere (Hardy & Singleton, 2009), GWA studies have a number of advantages over candidate-gene studies including their open nature in terms of genes being considered, the consistent standard of genotyping on a chip-based platform, the increasingly large number of cases being included, and an improved ability to correct for population stratification where cases and controls were imperfectly matched and which may have been a problem in early candidate-gene studies. However, it remains possible that small effects reported in candidate-gene studies but not confirmed in GWA studies could be real because of ethnic or phenotype differences between the studies, that the GWA did not include sufficient cases for the polymorphism to reach genome-wide significance or that coverage of polymorphisms in the gene of interest was inadequate on the GWA study chip. This general issue is considered in more detail below.

2. CYP GENOTYPE AND SUSCEPTIBILITY TO INDIVIDUAL CANCERS

2.1 Lung Cancer

Table 1 summarizes the most up-to-date position on possible associations of polymorphisms in a number of different isoforms with lung cancer susceptibility. Individual isoforms including their biological plausibility as risk factors are considered in detail below.

Table 1 Association of Cytochrome P450 Polymorphisms with Lung Cancer Susceptibility

Cytochrome P450 Isoform	Comments	References
CYP1A1	Pooled analyses suggested associations for rs4646903 and rs1048943 in Europeans but this has not been confirmed by more recent GWA studies	Vineis et al. (2003), Le Marchand et al. (2003), and Raimondi et al. (2005)
CYP1A2	Possible association for rs762551 but needs confirmation	Sachse, Brockmoller, Bauer, and Roots (1999) and Ma et al. (2014)
CYP1B1	Possible association for rs1056836 but needs confirmation	Xu, Zhou, Hang, and Shen (2012)
CYP2A6	Likely associations for several polymorphisms but not showing genome-wide significance in GWA studies	Fujieda et al. (2004) and Wassenaar et al. (2011)
CYP2C9	Unlikely	London, Sullivan-Klose, Daly, and Idle (1997) and Garcia-Martin et al. (2002)
CYP2C19	Still unclear	Wang, Song, et al. (2013)
CYP2D6	Unlikely	Rostami-Hodjegan et al. (1998)
CYP2E1	Possible association for rs2031920 but needs confirmation	Ye et al. (2014)

2.1.1 CYP1A1

The earliest studies suggesting an association between a genetic polymorphism involving a CYP gene and lung cancer susceptibility in humans appeared during the 1970s and were concerned with CYP1A1 which at that time was known as arylhydrocarbon hydroxylase. A phenotypic polymorphism affecting human CYP1A1 activity was reported in 1973 (Kellermann, Luyten-Kellermann, & Shaw, 1973). The extent of 3-methylcholanthrene (3-MC) induction of CYP1A1 activity in mitogen-stimulated lymphocytes showed considerable interindividual variability, with approximately 20% of subjects showing a "high-inducibility" phenotype. These findings were quite similar to those already reported for mice, where CYP1A1 is not inducible in some strains due to a genetic polymorphism in the aromatic hydrocarbon receptor (AhR) (Nebert & Gielen, 1972). It was also suggested that the high-inducibility phenotype was more common among lung cancer patients compared with controls (Kellermann, Shaw, & Luyten-Kellerman, 1973), but further studies failed to confirm this (Paigen et al., 1977; Prasad et al., 1979). Using a modified well-controlled protocol where only induced CYP1A1 activity was measured (due to uninduced levels being undetectable), the existence of wide interindividual variation in levels of induced enzyme activity and an apparent increased frequency of high levels among lung cancer patients was confirmed (Kouri et al., 1982).

When these original observations were reported, knowledge of *CYP1A1* genetics was relatively limited. The human gene (then called P1) was cloned and sequenced in the mid-1980s and several polymorphisms detectable with restriction enzymes described subsequently (Hayashi, Watanabe, Nakachi, & Kawajiri, 1991; Jaiswal, Gonzalez, & Nebert, 1985; Spurr, Gough, Stevenson, & Wolf, 1987). Two of these polymorphisms, a nonsynonymous polymorphism (rs1048943 which results in Ile462Val) (also known as m2) and another located in the 3′-noncoding region (rs4646903) (also known as m1), are in linkage disequilibrium (*CYP1A1*2B* allele) and have been well studied as putative genetic risk factors for lung cancer. Early studies showed increased frequencies of the variant allele among lung cancer patients, especially those from East Asia (for review see, Ikawa et al., 1995). A large pooled analysis involving over 2000 cases and 3000 controls suggested an association for rs4646903 in Europeans, particularly for squamous cell carcinoma and adenocarcinoma, but no association in East Asians (Vineis et al., 2003). Further pooled analyses on rs1048943 in similar populations also reported a significant association with this second polymorphism in Europeans (Le Marchand et al., 2003; Raimondi et al., 2005). However, despite the pooled analyses reporting odds ratios of approximately 2 for both rs4646903 and rs1048943, no evidence that any polymorphism in *CYP1A1* is a significant predictor of lung cancer risk has emerged more recently from five separate European and three East Asian GWA studies on lung cancer (Amos et al., 2008; Hu et al., 2011; Hung et al., 2008; Lan et al., 2012; McKay et al., 2008; Miki et al., 2010; Thorgeirsson et al., 2008; Wang et al., 2008). Though *CYP1A1* is located on chromosome 15q24 close to the location of a gene encoding the nicotinic acetylcholine receptor *CHRNA5*, the strongest risk factor for lung cancer which is located on chromosome 15q25 (Hung et al., 2008), there is no evidence from fine mapping studies that the signal reflects a contribution from *CYP1A1*. It remains possible that *CYP1A1* genotype could be relevant to a subset of lung cancers with a very specific phenotype, and the gene remains a biologically plausible candidate since it codes for a protein with a role in benzo[*a*]pyrene activation which is expressed in lung tissue from smokers (McLemore et al., 1990; Schmalix et al., 1993).

2.1.2 CYP2D6

The debrisoquine polymorphism was first reported in 1977 with a report that approximately 10% of Europeans could not metabolize this antihypertensive drug (Mahgoub, Idle, Dring, Lancaster, & Smith, 1977). The term "poor metabolizer" was used to describe this phenotype. The deficiency in metabolism of this drug, together with the antiarrhythmic agent sparteine (Eichelbaum, Spannbrucker, Steincke, & Dengler, 1979), was found to cosegregate in families with the trait inherited recessively. In 1984, it was suggested that poor metabolizers were less likely to develop lung cancer than those showing normal metabolism of debrisoquine (extensive metabolizers) (Ayesh, Idle, Ritchie, Crothers, & Hetzel, 1984). The basis for this association was suggested to be a role for the debrisoquine-metabolizing enzyme in the activation of chemical carcinogens, which was not implausible at that time due to the limited knowledge about the enzyme and its substrate specificity.

Soon after the study on debrisoquine metabolism in lung cancer patients, the CYP enzyme responsible for metabolism of debrisoquine and sparteine, now referred to as CYP2D6, was purified from human liver (Distlerath et al., 1985), followed by isolation of cDNA clones and mapping to chromosome 22 (Gonzalez et al., 1988). Subsequent genome mapping and cDNA sequencing enabled the identification of the main polymorphisms associated with absence of activity and development of genotyping assays to detect poor metabolizers (Daly, Armstrong, Monkman, Idle, & Idle, 1991; Gaedigk, Blum, Gaedigk, Eichelbaum, & Meyer, 1991; Gough et al., 1990; Hanioka, Kimura, Meyer, & Gonzalez, 1990; Heim & Meyer, 1990; Kagimoto, Heim, Kagimoto, Zeugin, & Meyer, 1990). These developments enabled the findings on lung cancer susceptibility obtained using debrisoquine phenotyping to be followed up in larger populations.

The initial follow-up studies on lung cancer risk used debrisoquine phenotyping but, following the identification of the molecular basis for the debrisoquine polymorphism, a number of independent studies investigated whether *CYP2D6* genotypes predicting the poor metabolizer phenotype were less common in lung cancer cases compared with controls. A meta-analysis prepared in 1998 remains a useful summary of the findings from the majority of studies on this issue (Rostami-Hodjegan et al., 1998). The overall meta-analysis found a small but significant decrease in lung cancer risk among poor metabolizers but this effect related only to studies where phenotyping was used. No significance was seen when genotyping studies only, which should be a more accurate predictor of phenotype and not subject to interference from other drugs, were considered. In particular, the two largest genotyping studies which each included more than 300 cases and 600 controls ailed to find a significant difference for the entire group of cases against controls, though some subgroup analyses were consistent with a lower poor metabolizer frequency in cases (London, Daly, et al., 1997; Wolf et al., 1992). There have been only a few additional studies since the meta-analysis appeared but these have also proved negative. In particular, a nested case-control study of individuals within the Physicians' Health Study did not provide any evidence that *CYP2D6* alleles associated with poor metabolism were predictors of lung cancer susceptibility (Perera et al., 2006). There is also no indication from any of the published GWA studies on lung cancer (see Section 2.1.1) of a signal in the region of *CYP2D6* on chromosome 22.

The basis for undertaking the initial studies on the relationship between debrisoquine and lung cancer susceptibility was the possibility that the enzyme responsible for debrisoquine metabolism might also have a role in carcinogen activation. Soon after initial cloning and expression of CYP2D6, some evidence for a role in both activation of 4-(methylnitrosamino)-1-(3-pyridyl)-1-butanone (NNK) and other nitrosamines and in nicotine metabolism was reported (Cholerton et al., 1994;

Crespi, Penman, Gelboin, & Gonzalez, 1991) and this, together with data suggesting that CYP2D6 was expressed in lung tissue (Ding & Kaminsky, 2003), provided a further basis for molecular epidemiology studies in this area. However, more detailed studies subsequently predicted that the contribution of CYP2D6 to either nitrosamine activation or nicotine metabolism was minimal and that other CYPs, including CYP2A6, were the main enzymes contributing to these processes (Benowitz, Jacob, & Perez-Stable, 1996; Islam, Wolf, Lennard, & Sternberg, 1991; Messina, Tyndale, & Sellers, 1997). For nicotine, there is some more recent evidence that in those with higher than normal CYP2D6 activity, who are termed ultrarapid metabolizers, and have at least one additional copy of *CYP2D6* (see Ingelman-Sundberg, 1999), there may be some contribution to nicotine metabolism by CYP2D6 (Caporaso et al., 2001; Saarikoski et al., 2000). There is no evidence currently that this is relevant to lung cancer susceptibility.

2.1.3 CYP2A6

CYP2A6 was originally purified and cloned as a coumarin hydroxylase, but was subsequently demonstrated to have major roles in both nicotine detoxication and nitrosamine activation (Messina et al., 1997; Tiano et al., 1993; Yamano, Tatsuno, & Gonzalez, 1990). These dual roles make it a compelling candidate gene for a role in lung cancer susceptibility since low CYP2A6 activity is predicted to result in a lower level of smoking due to slower clearance of nicotine together with a decreased level of activation of nitrosamines to carcinogenic species. The existence of individuals with CYP2A6 activity completely absent is well established. The first "absence of activity" allele to be identified was *CYP2A6*2* where a nonsynoymous polymorphism (L160H) (rs1801272) results in no enzyme activity (Fernandez-Salguero et al., 1995). This allele is found mainly in Europeans, though at a low population frequency. It was subsequently demonstrated that a second allele, *CYP2A6*4*, where most of the coding region is deleted, was common in East Asians with a frequency of approximately 2% for *CYP2A6*4* homozygotes (Oscarson et al., 1999). The relatively high frequency of the absence of activity phenotype in East Asians has led to a number of reports of association with both level of smoking and risk of lung cancer development in these ethnic groups, though there are also studies of Europeans and African-Americans reporting broadly similar conclusions.

In relation to smoking and *CYP2A6* genotype, early studies showed that those with *CYP2A6* variant alleles showed a lower level of smoking in terms of number of cigarettes consumed per day (Rao et al., 2000) and also that those with variant alleles found it easier to quit smoking (Gu, Hinks, Morton, & Day, 2000). A subsequent large study in Japanese found a significantly lower level of cigarettes smoked per day among *CYP2A6**4* homozygotes with significant decreases also in those homozygous for other low activity variant alleles or heterozygous for any low activity variant (Fujieda et al., 2004). GWA studies showed the strongest genetic association with level of smoking involved nicotinic acetylcholine receptor genes on chromosome 15 but a large meta-analysis of GWA studies on smoking behavior involving Europeans and white Americans confirmed a role for polymorphisms in the *CYP2A6* region on chromosome 19 in determining number of cigarettes smoked per day (Thorgeirsson et al., 2010). That study also investigated whether *CYP2A6* genotype was a significant risk factor for lung cancer and found evidence for a nominally significant but not genome-wide significant effect. While *CYP2A6* has not been found to be a genome-wide significant risk factor for lung cancer in any GWA study to date, several candidate-gene studies report highly significant associations. In particular, in the same study investigating *CYP2A6* genotype in relation to level of smoking in Japanese, Fujeida and colleagues found a significantly decreased risk for those homozygous for *CYP2A6**4* (odds ratio 0.3), though this appeared to be mainly relevant to heavy smokers (Fujieda et al., 2004). On the other hand, in a study involving white Americans, the risk of lung cancer in those wild type for *CYP2A6* was 1.6-fold higher than in those carrying low or absent activity alleles, but this appeared to be only the case in light smokers who consumed less than 20 cigarettes per day (Wassenaar et al., 2011). Recently, several additional *CYP2A6* alleles that are more common among African-Americans have been detected and characterized (Mwenifumbo, Zhou, Benowitz, Sellers, & Tyndale, 2010). In studies on lung cancer susceptibility involving two independent African-American populations, *CYP2A6*-decreased activity alleles were associated with a significantly decreased risk of lung cancer though, slightly unexpectedly, subgroup analysis suggested the decreased risk related only to males in both community groups with no effect for females (Wassenaar et al., 2015). The numbers of female cases were lower than males but this seems unlikely to explain the relatively large difference seen between the genders.

Currently, *CYP2A6* is the only CYP to show a consistent association with lung cancer in candidate-gene studies. The failure to see a genome-wide significant association in any GWA study up to now may be due to the overall effect size being small. Candidate-gene studies and one GWA suggest a consistently significant association between *CYP2A6* genotype and level of smoking. Several recent meta-analyses of candidate-gene studies, mainly involving East Asians populations, confirm the decreased lung cancer risk in those positive for variant alleles such as *CYP2A6*4* (Liu, Shu, Wang, Jin, & Lou, 2013; Wang, Zang, et al., 2013).

2.1.4 Other CYP Polymorphisms

In addition to the well-studied polymorphisms discussed above, the relationship between polymorphisms in *CYP1A2*, *CYP1B1*, *CYP2B6*, *CYP2C9*, *CYP2C19*, and *CYP2E1* and lung cancer has all been investigated, though in fewer studies. Each of these CYPs may contribute to metabolism of carcinogens found in tobacco smoke or to nicotine metabolism and are therefore plausible candidates for affecting susceptibility. There is no evidence for involvement of any of these genes in lung cancer from GWA studies, except that some evidence for an association with levels of smoking was reported for *CYP2B6* (Thorgeirsson et al., 2008). A number of candidate gene case-control studies, generally involving small numbers of subjects, concerning associations of various polymorphisms in the genes listed above have been reported, together with some recent meta-analyses. The recent meta-analyses need to be treated with some caution because of their use of small datasets without access to the original data and also a likely overemphasis on studies showing positive associations (since negative findings are less likely to be published) but are a useful summary of the available data for *CYP1A2*, *CYP1B1*, and *CYP2E1*.

For *CYP1A2*, the best studied polymorphism is rs762551 (−163C > A), which is also denoted *CYP1A2*1F* and may be associated with higher enzyme activity in smokers where expression has been induced by exposure to polycyclic aromatic hydrocarbons in tobacco smoke (Sachse et al., 1999). A meta-analysis of seven published studies relating to this polymorphism as a risk factor for lung cancer reported significant odds ratios of 1.2 approximately depending on the type of genetic model used (Ma et al., 2014).

CYP1B1 contributes to benzo[*a*]pyrene activation (Shimada et al., 1997) and the gene has several apparently functionally significant nonsynonymous polymorphisms (Hanna, Dawling, Roodi, Guengerich, & Parl, 2000) so is another plausible candidate gene. There are at least 10 published studies on the relationship of selected polymorphisms in this gene with lung cancer relating to four separate nonsynonymous polymorphisms. The best studied polymorphism is rs1056836 (L432V) with nine different reports representing 2500 cases and 3200 controls. Meta-analyses have examined several different genetic models but for a dominant model where carriers of the rare allele are compared with a group homozygous for the reference allele, a significant ($p=0.027$) with an odds ratio of 1.26 was reported (Xu et al., 2012). Data on the relevance of other variants are sparse though a role for some of these in susceptibility cannot be ruled out.

CYP2B6 is located adjacent to *CYP2A6* on chromosome 19 and due to linkage disequilibrium, it is difficult to separate effects from the two genes. However, the efficiency of CYP2B6 as an activator of nitrosamines, especially NNK, is higher than its efficiency as an oxidiser of nicotine, which is the reverse of CYP2A6 (Wassenaar, Dong, Amos, Spitz, & Tyndale, 2013). Despite the suggestion from the GWA study discussed above that *CYP2B6* genotype could be relevant to nicotine intake, the overall contribution of this isoform to nicotine metabolism is low (Lee et al., 2007). In a recent pilot study, some evidence suggesting an additional contribution by *CYP2B6* genotype to that of *CYP2A6* to risk of lung cancer susceptibility has been obtained, though additional studies involving larger populations or ethnic groups where low activity *CYP2B6* alleles are common are needed to confirm this (Wassenaar et al., 2013).

CYP2C9 contributes to benzo[*a*]pyrene metabolism (Shou, Korzekwa, Crespi, Gonzalez, & Gelboin, 1994) and this prompted some investigation of its relevance to lung cancer risk. However, two separate studies, investigating the relevance of the functionally significant *CYP2C9*2* and **3* alleles, found no evidence that either allele affected risk (Garcia-Martin et al., 2002; London, Sullivan-Klose, et al., 1997).

CYP2C19 may also contribute to benzo[*a*]pyrene and nitrosamine activation (Fujita & Kamataki, 2001; Yamazaki et al., 2004), though in view of low expression in most tissues, any contribution to cancer susceptibility is likely to be limited. A few studies in relation to lung cancer susceptibility involving this isoform have been performed. The first was based in France and involved CYP2C19 phenotyping using the well-established probe drug mephenytoin (Benhamou, Bouchardy, & Dayer, 1997). This reported no significant difference in the proportion of poor metabolizers between cases and controls. Similarly, in a study concerned with lung cancer in Europeans who developed the disease before 50, there was no significant association with the main European poor metabolizer allele, *CYP2C19*2* (Gemignani et al., 2007). A significantly higher genotypic CYP2C19 poor metabolizer frequency was reported in a group of 212 Chinese lung cancer patients compared with local controls (Shi & Chen, 2004). This is interesting in view of the higher frequency of CYP2C19 poor metabolizers in East Asia compared with Europe and there is also some further support for this in a recent meta-analysis (Wang, Song, Chen, & Yu, 2013).

CYP2E1 is well established to contribute to nitrosamine activation (Crespi et al., 1991) and is expressed in human lung tissue (Wheeler, Wrighton, & Guenthner, 1992), making it a plausible candidate gene for studies on lung cancer susceptibility. Two polymorphisms in noncoding regions of this gene which are in linkage disequilibrium have been studied widely as risk factors for lung cancer though whether either is functionally significant remains unclear. Both polymorphisms are relatively rare in Europeans, making well-powered studies difficult because of the large number of cases needed to show statistical significance unless the effect size was very large, but are common in East Asians. A recent meta-analysis is a useful summary of studies on *CYP2E1* as a risk factor for lung cancer (Ye et al., 2014). The best studied of the *CYP2E1* polymorphisms is located in a promoter region (rs2031920-often referred to as the *Rsa*I polymorphism with the variant allele also referred to as c2 and *CYP2E1*5*); it was originally suggested that the variant allele was associated with higher levels of transcription compared with the wild-type allele (Hayashi, Watanabe, & Kawajiri, 1991). This would predict that possession of the variant might be a risk factor for lung cancer. However, in the meta-analysis, carriage of the rs2031920 was found to be associated with a significantly decreased risk of lung cancer (odds ratio ~0.8) in 23 studies of East Asians, though no effect was seen in three studies involving Europeans. This is an interesting finding though the issue of whether this or any other polymorphism in *CYP2E1* affects function still needs more investigation. The absence of any *CYP2E1* signal in several GWA studies on lung cancer in East Asians is also a limitation.

2.2 Other Cancers

The association of selected CYP polymorphisms with a range of other cancers has also been quite widely investigated. Only selected examples involving both biologically plausible CYP candidate genes and several separate studies will be considered further here. A summary of the findings relating to these examples is provided in Table 2.

The *CYP3A* gene cluster is not generally considered a candidate for lung cancer susceptibility but its relevance to the risk of several other cancers where tobacco smoking is not an important risk factor has been investigated in a few studies. A possible association between *CYP3A* genotype and aflatoxin-related hepatocellular carcinoma is interesting, though difficult

Table 2 Association of Cytochrome P450 Polymorphisms with Cancers Other than Lung Cancer

Cytochrome P450 Isoform	Comments	References
CYP1A2	Possible association with colon cancer in phenotyping studies	Lang et al. (1994) and Le Marchand et al. (2001)
CYP2E1	Possible association of rs2031920 with nasopharyngeal cancer	Hildesheim et al. (1997)
CYP3A4/5	Associations with aflatoxin-related hepatocellular carcinoma need further investigation	Wojnowski et al. (2004)

to study. As reviewed recently, aflatoxins, which frequently contaminate dietary components such as maize and groundnuts in certain countries, are hepatocarcinogenic with hepatitis B infection also a contributor to the process (Wild & Gong, 2010). It is well established that CYP3A4 is the major enzyme involved in activation of aflatoxin B1 to a genotoxic 2,3-epoxide metabolite (Shimada & Guengerich, 1989) but there is also a contribution from the polymorphic *CYP3A5* (Kamdem, Meineke, Godtel-Armbrust, Brockmoller, & Wojnowski, 2006). CYP3A4 and related enzymes, particularly CYP3A5, show considerable interindividual variability in activity though currently only a small amount of this variability can be explained by known genetic polymorphisms (for review see, Perera, 2010). Up to the present, the main polymorphisms affecting the *CYP3A* locus that has been investigated in relation to risk of hepatocellular cancer are an upstream polymorphism in *CYP3A4* (rs2740574 (*CYP3A4**1B*)) which appears to be associated with a small increase in CYP3A4 expression (Rebbeck, Spitz, & Wu, 2004) and rs776746 which affects CYP3A5 RNA splicing and is the major cause of absence of CYP3A5 expression worldwide (Daly, 2006). In addition to genotyping, the ratio of urinary 6-beta-hydroxycortisol to cortisol is considered to be a useful biomarker of overall CYP3A4 activity. Studies in the Gambia where aflatoxin exposure and hepatitis B infection are both high have investigated the relevance of variability in CYP3A4 and CYP3A5 to levels of aflatoxin B1-epoxide adducts in serum. When both phenotypic CYP3A4 levels and expression of CYP3A5 based on genotype are combined, those with low CYP3A4 activity who also express CYP3A5 show a higher level of adduct formation. This suggests both *CYP3A4* and *CYP3A5* genotype may modulate risk of hepatocellular carcinoma (Wojnowski et al., 2004). Molecular epidemiology studies on patients with hepatocellular carcinoma induced by aflatoxin and hepatitis B exposure are needed to confirm this. Several Chinese-based GWA studies on hepatocellular carcinoma related to hepatitis B infection have not provided any evidence that the *CYP3A* locus contributes to risk, but it is uncertain whether these cases had also been exposed to aflatoxin (Jiang et al., 2013) and studies based in Africa where aflatoxin exposure is more widespread would likely be more informative.

Some of CYP polymorphisms mentioned as risk factors for lung cancer in Section 2.1.1 are also of possible relevance to certain other cancers where smoking is a less important risk factor. Exposure to activated forms of heterocyclic arylamines such as 2-amino-3-methylimidazo[4,5-b]pyridine (PhIP) is believed to be important in the development of colon cancer. The initial oxidation reaction in PhIP activation is performed by CYP1A2. The relationship between CYP1A2 and colon cancer has been mainly investigated by phenotyping studies with caffeine. Several studies have reported that caffeine rapid metabolizers (indicating high CYP1A2 activity) are more common among colon cancer cases than controls (Lang et al., 1994; Le Marchand et al., 2001). These findings would benefit from follow-up by *CYP1A2* genotyping but such studies do not appear to have been reported. Three independent GWA studies show a significant association between an upstream polymorphism in *CYP1A2* and coffee consumption, which makes the phenotypic studies performed with caffeine more credible as a predictor of CYP1A2 variability (Amin et al., 2012; Cornelis et al., 2011; Sulem et al., 2011). GWA studies on colon cancer susceptibility have not found any evidence for a contribution by *CYP1A2* (Whiffin et al., 2014), but it remains possible from the findings from the earlier candidate-gene studies that meat consumption also needs to be considered when assessing CYP1A2 variation as a risk factor.

Nasopharyngeal cancer is rare in most countries but relatively common in Southern China where a higher incidence is associated with infection with Epstein–Barr virus, together with environmental factors such as salted fish consumption and wood exposure. The environmental risk factors point to a role for nitrosamine exposure and because of the established role of CYP2E1 in activation of nitrosamines (see Section 2.1.4), the relevance of *CYP2E1* genotype to risk for development of this cancer has been investigated. As with the lung cancer studies, the main polymorphism investigated was rs2031920. In a study of approximately 350 cases of Chinese nasopharyngeal cancer cases and a similar number of controls, homozygosity for rs2031920 was associated with significantly increased risk of cancer and an odds ratio of 2.6 (Hildesheim et al., 1997). A subsequent study on Chinese families with more than one case of the disease also showed an association with this genotype (Yang et al., 2005). An independent study involving both families and unrelated cases and controls also reported a significantly increased frequency of several *CYP2E1* polymorphisms among cases, though not specifically with rs2031920 (Jia et al., 2009). The only study of non-Chinese reported an association with rs2031920 which was not significant, possibly due to small numbers of cases, but in the same direction as the studies based in China (Kongruttanachok et al., 2001). More recently, a GWA study on nasopharyngeal cancer involving approximately 1000 Chinese cases did not show a genome-wide significant signal in the *CYP2E1* region (Tang et al., 2012). The previous candidate gene findings for this disease therefore need to be treated with caution, though a study of 1000 cases might not be sufficient to show genome-wide significance if the contribution from *CYP2E1* is small.

3. OTHER CYP POLYMORPHISM-DISEASE ASSOCIATIONS

CYP polymorphisms have been suggested to be risk factors for a variety of diseases ranging from primary congenital glaucoma to endometriosis. In some cases, there is good biological plausibility for the association or the association has been discovered directly by genetic analysis but for some reports the biological plausibility is limited and the reasons for undertaking the study less clear. As discussed in Section 1, GWA studies have failed to confirm some of the previously reported associations even where several independent reports of associations found in candidate-gene studies may be available. Some examples where an association still seems possible or likely are discussed including disease associations involving *CYP1B1*, *CYP2C*, *CYP2E1*, and *CYP2J2*. A summary of the associations considered here is provided in Table 3.

There is a well-validated association involving *CYP1B1* with a rare disease termed primary congenital glaucoma (Stoilov et al., 1997). This is a rare disease among Europeans with an incidence of 1 in 10,000, but is more common in the Middle East at 1 in 2000. Unlike the possible associations of P450 genes with risk of developing certain cancers, this form of glaucoma is a single gene disorder with homozygosity for rare mutations in *CYP1B1* resulting in the development of the disease in early childhood. These rare

Table 3 Association of Cytochrome P450 Polymorphisms with Diseases Other than Cancer

Cytochrome P450 Isoform	Disease	Effect and Comments	References
CYP1B1	Primary congenital glaucoma	Rare mutations result in the disease	Stoilov, Akarsu, and Sarfarazi (1997) and Bhattacharjee et al. (2008)
CYP2C8/9	Cardiovascular disease	Suggested that impaired production of EETs by those positive for a range of variant alleles may increase risk of various cardiovascular diseases. Confirmation needed.	Shahabi, Siest, Meyer, and Visvikis-Siest (2014)
CYP2C19	Endometriosis	Several studies suggest CYP2C19 poor metabolizers at increased risk	Painter et al. (2014)
CYP2E1	Alcoholic liver disease	Association suggested with rs2031920	Zeng et al. (2013)
CYP2J2	Cardiovascular disease	Suggested that impaired production of EETs by this isoform in those positive for rs890293 may increase risk of various cardiovascular diseases. Confirmation needed.	Berlin, Sangkuhl, Klein, and Altman (2011)

mutations are generally deletions or insertions causing frameshifts that result in a truncated protein or nonsynonymous (missense) mutations resulting in an amino acid change (Vasiliou & Gonzalez, 2008). In addition to primary congenital glaucoma, *CYP1B1* mutations may contribute to other types of glaucoma affecting adults such as primary open-angle glaucoma (POAG) where other genes will also contribute to risk. There is some evidence that common *CYP1B1* polymorphisms including rs1056836 (L432V) may be associated with a higher risk of POAG development (Bhattacharjee et al., 2008). Mice lacking *cyp1b1* show abnormalities in ocular drainage. CYP1B1 is expressed in ocular tissue but the biological basis for why its absence due to rare mutations results in glaucoma is still unclear. It has been suggested that it may relate to roles for the enzyme in retinoid, steroid, or melatonin metabolism (Vasiliou & Gonzalez, 2008).

CYP2C isoforms, especially CYP2C8 and CYP2C9, contribute to arachidonic acid metabolism and convert this endogenous compound to epoxyeicosatrienoic acids (EETs) (reviewed by Fleming, 2014). These compounds have vasodilatory and antiinflammatory properties and may be relevant to a number of physiological processes including the regulation of vascular tone and angiogenesis. In addition to the CYP2C isoforms, another family 2 member, CYP2J2, also contributes to EET production from arachidonic acid. This isoform is expressed at low levels in the liver but relatively high levels in a range of extrahepatic tissues (Berlin et al., 2011). Both CYP2C8 and CYP2J2 are considered the most relevant isoforms in relation to the role of EET in the cardiovascular system (Edin et al., 2011). Polymorphisms in these genes have therefore received most attention as possible risk factors for diseases affecting the cardiovascular system but *CYP2C9* has also been considered. Certain *CYP2C8* and *CYP2C9* variant alleles such as *CYP2C8*3* and *CYP2C9*2* show linkage disequilibrium, so it is difficult to separate contributions to susceptibility from the individual isoforms. CYP2C8 generally shows higher levels of extrahepatic expression, including in endothelial cells, compared with CYP2C9. An upstream polymorphism in *CYP2J2*, (−76G > T; *CYP2J2*7*; rs890293), which has been reported to be associated with lower transcription (Spiecker et al., 2004), has been quite widely studied, together with *CYP2C8*2*, **3*, and **4*, as a risk factor for cardiovascular diseases. As described elsewhere (Berlin et al., 2011; Shahabi et al., 2014), a number of case-control candidate-gene studies on associations between *CYP2J2* and *CYP2C8/9* genotypes and susceptibility to coronary heart disease, coronary artery disease, myocardial infarction, and hypertension have been performed. Up to the present, these studies have yielded rather contradictory results with some reporting positive associations between the disease studied and possession of certain variant alleles but others reporting no effect. GWA studies in relation to cardiovascular diseases have so far failed to detect significant signals in the areas of these genes; however, so despite the biological plausibility of variation in EET production being relevant to risk of these diseases and a relatively large number of studies, evidence to support any effect is poor.

Endometriosis is a debilitating disease affecting 6–10% of women of reproductive age. A number of studies on genetic factors affecting susceptibility to this disease have been reported using GWA approaches. An early genome-wide linkage study in families with affected individuals found signals on several different chromosomes including chromosome 10 in the region of the *CYP2C* locus (Treloar et al., 2005). Subsequent GWA studies involving case-control designs also detected signals on several chromosomes but did not detect a genome-wide significant signal on chromosome 10 (Nyholt et al., 2012). Despite this, fine mapping of the chromosomal 10 region from the original family study has been performed. This showed that the association mapped to *CYP2C19* (Painter et al., 2011). A large case-control study also confirmed the *CYP2C19* association, especially with the *CYP2C19*2*-specific polymorphism rs4244285 (Painter et al., 2014). It is well established that CYP2C19 contributes to estrogen metabolism. Estrogen levels are considered relevant to risk of endometriosis development (Vercellini, Vigano, Somigliana, & Fedele, 2014) and a higher level (which might occur in a CYP2C19 poor metabolizer) could therefore affect susceptibility. The mapping study also indicated that the CYP2C19 ultrarapid metabolizer phenotype positive for rs12248560 is associated with a decreased risk of endometriosis development (Painter et al., 2011), although the effect in the subsequent case-control study was less significant (Painter et al., 2014). Independently of the genetic linkage study, the *CYP2C19*-endometriosis association had been reported in an earlier small candidate-gene study (Cayan, Ayaz, Aban, Dilek, & Gumus, 2009). A new study based in Brazil has also confirmed the association (Christofolini et al., 2015).

The relevance of CYP2E1 to ethanol metabolism is well established and a possible role in alcoholic liver disease has been studied for the past 40 years approximately (Lieber & DeCarli, 1970). In particular, rs2031920 (*CYP2E1*5*) has been well studied as a risk factor for development of alcoholic liver disease with a number of reports of positive associations together with a recent meta-analysis (Zeng et al., 2013). The basis for the association may involve either higher levels of acetaldehyde being produced from ethanol by CYP2E1 in those positive for rs2031920 if, as discussed in Section 2.1.4, the variant is associated with higher levels of transcription. It is also possible that higher levels of reactive oxygen species occur in those positive for the variant since CYP2E1-mediated reactions generate these species (Koop, 1992). The meta-analysis is a useful summary of all studies on the association of rs2031920 with alcoholic liver disease (Zeng et al., 2013). The overall analysis failed to show an overall significant association for carriage of the variant but the study highlights the small size of most published studies with 21 of the 27 considered involving fewer than 100 alcoholic liver disease cases. When the combined cases of European ethnicity compared with alcoholic controls only were considered, possession of the variant allele was associated with a significantly increased risk of alcoholic liver disease. A significant association in East Asians was also seen when homozygous variant cases only were compared with homozygous wild types. In view of these continuing slightly contradictory findings, the possible association of alcoholic liver disease with *CYP2E1* genotype needs further investigation in larger well-controlled studies to make a definitive conclusion on its significance. A report suggesting that *CYP2E1* genotype also contributes to interindividual variability in alcohol response (Webb et al., 2011) and might therefore have a role in alcoholism is an added complication to designing well-controlled studies. In addition, the relatively limited information on both functional significance of rs2031920 and other polymorphisms with which it is in linkage disequilibrium and also on possible variability in induction of gene expression by ethanol also makes clear conclusions difficult.

4. CONCLUSION

There is a large literature on CYP polymorphisms and disease susceptibility. However, there are problems and limitations with many reports in the area and now that the era of GWA studies and genome sequencing has arrived, the number of convincing associations involving P450 genes from families 1 to 3 is smaller than previously. The rigorous statistical analysis used in genome-wide studies means small associations may be missed unless the studies are sufficiently large and while this may be an explanation for failure to confirm some associations, it is also likely that there may have been problems with some of the previous reports such as genotyping errors, small numbers, and inadequate phenotyping. Associations that have stood up well over time include those between *CYP2A6* and levels of smoking, which may also affect risk of lung cancer and the association of rare *CYP1B1* variants with congenital primary glaucoma. There is an intermediate group where further investigation is needed, including *CYP3A5* with aflatoxin-related hepatocellular carcinoma; *CYP2E1* with lung cancer, nasopharyngeal cancer, and alcoholic liver disease; *CYP1A2* with lung cancer and colon cancer; *CYP2C* and *CYP2J2* with cardiovascular disease; and *CYP2C19* with endometriosis. Associations which can be considered very unlikely include *CYP2D6* with lung cancer. Though disease associations involving CYPs from families 1 to 3 have generally not emerged from GWA studies, other CYP families with roles in endogenous compound metabolism may contribute to risk of several diseases including hypertension (for example, polymorphisms in the region of *CYP17A1* and *CYP21A2* have been shown to contribute to blood pressure; Lu et al., 2015; Newton-Cheh et al., 2009) and risk of multiple sclerosis (signals close to *CYP24A1* and *CYP27B1*; Bahlo et al., 2009; Sawcer et al., 2011).

Further larger GWA studies combined with genome-wide sequencing may still identify new associations and confirm some suggested ones involving CYP families 1–3. However, any effects not yet discovered in GWA studies are likely to be small and therefore of limited value in predicting individual risk of disease.

CONFLICT OF INTEREST

The author has no conflicts of interest to declare.

REFERENCES

Agundez, J. A. (2004). Cytochrome P450 gene polymorphism and cancer. *Current Drug Metabolism*, *5*, 211–224.

Amin, N., Byrne, E., Johnson, J., Chenevix-Trench, G., Walter, S., Nolte, I. M., et al. (2012). Genome-wide association analysis of coffee drinking suggests association with CYP1A1/CYP1A2 and NRCAM. *Molecular Psychiatry*, *17*, 1116–1129.

Amos, C. I., Wu, X., Broderick, P., Gorlov, I. P., Gu, J., Eisen, T., et al. (2008). Genome-wide association scan of tag SNPs identifies a susceptibility locus for lung cancer at 15q25.1. *Nature Genetics*, *40*, 616–622.

Ayesh, R., Idle, J. R., Ritchie, J. C., Crothers, M. J., & Hetzel, M. R. (1984). Metabolic oxidation phenotypes as markers for susceptibility to lung cancer. *Nature*, *312*, 169–170.

Bahlo, M., Booth, D. R., Broadley, S. A., Brown, M. A., Foote, S. J., Griffiths, L. R., et al. (2009). Genome-wide association study identifies new multiple sclerosis susceptibility loci on chromosomes 12 and 20. *Nature Genetics*, *41*, 824–828.

Benhamou, S., Bouchardy, C., & Dayer, P. (1997). Lung cancer risk in relation to mephenytoin hydroxylation activity. *Pharmacogenetics*, *7*, 157–159.

Benowitz, N. L., Jacob, P., 3rd, & Perez-Stable, E. (1996). CYP2D6 phenotype and the metabolism of nicotine and cotinine. *Pharmacogenetics*, *6*, 239–242.

Berlin, D. S., Sangkuhl, K., Klein, T. E., & Altman, R. B. (2011). PharmGKB summary: Cytochrome P450, family 2, subfamily J, polypeptide 2: CYP2J2. *Pharmacogenetics and Genomics*, *21*, 308–311.

Bhattacharjee, A., Banerjee, D., Mookherjee, S., Acharya, M., Banerjee, A., Ray, A., et al. (2008). Leu432Val polymorphism in CYP1B1 as a susceptible factor towards predisposition to primary open-angle glaucoma. *Molecular Vision*, *14*, 841–850.

Caporaso, N. E., Lerman, C., Audrain, J., Boyd, N. R., Main, D., Issaq, H. J., et al. (2001). Nicotine metabolism and CYP2D6 phenotype in smokers. *Cancer Epidemiology, Biomarkers & Prevention*, *10*, 261–263.

Cayan, F., Ayaz, L., Aban, M., Dilek, S., & Gumus, L. T. (2009). Role of CYP2C19 polymorphisms in patients with endometriosis. *Gynecological Endocrinology*, *25*, 530–535.

Cholerton, S., Arpanahi, A., Mccracken, N., Boustead, C., Taber, H., Johnstone, E., et al. (1994). Poor metabolizers of nicotine and CYP2D6 polymorphism. *Lancet*, *343*, 62–63.

Christofolini, D. M., Amaro, A., Mafra, F., Sonnewend, A., Bianco, B., & Barbosa, C. P. (2015). CYP2C19 polymorphism increases the risk of endometriosis. *Journal of Assisted Reproduction and Genetics*, *32*, 91–94.

Cornelis, M. C., Monda, K. L., Yu, K., Paynter, N., Azzato, E. M., Bennett, S. N., et al. (2011). Genome-wide meta-analysis identifies regions on 7p21 (AHR) and 15q24 (CYP1A2) as determinants of habitual caffeine consumption. *PLoS Genetics*, *7*, e1002033.

Crespi, C. L., Penman, B. W., Gelboin, H. V., & Gonzalez, F. J. (1991). A tobacco smoke-derived nitrosamine, 4-(methylnitrosamino)-1-(3-pyridyl)-1-butanone, is activated by multiple human cytochrome P450s including the polymorphic human cytochrome P4502D6. *Carcinogenesis*, *12*, 1197–1201.

Daly, A. K. (2006). Significance of the minor cytochrome P450 3A isoforms. *Clinical Pharmacokinetics*, *45*, 13–31.

Daly, A. K., Armstrong, M., Monkman, S. C., Idle, M. E., & Idle, J. R. (1991). Genetic and metabolic criteria for the assignment of debrisoquine 4-hydroxylation (cytochrome P4502D6) phenotypes. *Pharmacogenetics*, *1*, 33–41.

Ding, X., & Kaminsky, L. S. (2003). Human extrahepatic cytochromes P450: Function in xenobiotic metabolism and tissue-selective chemical toxicity in the respiratory and gastrointestinal tracts. *Annual Review of Pharmacology and Toxicology*, *43*, 149–173.

Distlerath, L. M., Reilly, P. E., Martin, M. V., Davis, G. G., Wilkinson, G. R., & Guengerich, F. P. (1985). Purification and characterization of the human liver cytochromes P-450 involved in debrisoquine 4-hydroxylation and phenacetin O-deethylation, two prototypes for genetic polymorphism in oxidative drug metabolism. *The Journal of Biological Chemistry*, *260*, 9057–9067.

Dong, L. M., Potter, J. D., White, E., Ulrich, C. M., Cardon, L. R., & Peters, U. (2008). Genetic susceptibility to cancer: The role of polymorphisms in candidate genes. *JAMA*, *299*, 2423–2436.

Edin, M. L., Wang, Z., Bradbury, J. A., Graves, J. P., Lih, F. B., DeGraff, L. M., et al. (2011). Endothelial expression of human cytochrome P450 epoxygenase CYP2C8 increases susceptibility to ischemia-reperfusion injury in isolated mouse heart. *FASEB Journal*, *25*, 3436–3447.

Eichelbaum, M., Spannbrucker, N., Steincke, B., & Dengler, H. J. (1979). Defective N-oxidation of sparteine in man: A new pharmacogenetic defect. *European Journal of Clinical Pharmacology*, *16*, 183–187.

Fernandez-Salguero, P., Hoffman, S. M., Cholerton, S., Mohrenweiser, H., Raunio, H., Rautio, A., et al. (1995). A genetic polymorphism in coumarin 7-hydroxylation: Sequence of the human *CYP2A* genes and identification of variant *CYP2A6* alleles. *The American Journal of Human Genetics*, *57*, 651–660.

Fleming, I. (2014). The pharmacology of the cytochrome P450 epoxygenase/soluble epoxide hydrolase axis in the vasculature and cardiovascular disease. *Pharmacological Reviews*, *66*, 1106–1140.

Fujieda, M., Yamazaki, H., Saito, T., Kiyotani, K., Gyamfi, M. A., Sakurai, M., et al. (2004). Evaluation of CYP2A6 genetic polymorphisms as determinants of smoking behavior and tobacco-related lung cancer risk in male Japanese smokers. *Carcinogenesis*, *25*, 2451–2458.

Fujita, K., & Kamataki, T. (2001). Predicting the mutagenicity of tobacco-related N-nitrosamines in humans using 11 strains of Salmonella typhimurium YG7108, each coexpressing a form of human cytochrome P450 along with NADPH-cytochrome P450 reductase. *Environmental and Molecular Mutagenesis*, *38*, 339–346.

Gaedigk, A., Blum, M., Gaedigk, R., Eichelbaum, M., & Meyer, U. A. (1991). Deletion of the entire cytochrome-P450 Cyp2d6 gene as a cause of impaired drug-metabolism in poor metabolizers of the debrisoquine sparteine polymorphism. *The American Journal of Human Genetics, 48*, 943–950.

Garcia-Martin, E., Martinez, C., Ladero, J. M., Gamito, F. J., Rodriguez-Lescure, A., & Agundez, J. A. (2002). Influence of cytochrome P450 CYP2C9 genotypes in lung cancer risk. *Cancer Letters, 180*, 41–46.

Gemignani, F., Landi, S., Szeszenia-Dabrowska, N., Zaridze, D., Lissowska, J., Rudnai, P., et al. (2007). Development of lung cancer before the age of 50: The role of xenobiotic metabolizing genes. *Carcinogenesis, 28*, 1287–1293.

Gonzalez, F. J., Vilbois, F., Hardwick, J. P., McBride, O. W., Nebert, D. W., Gelboin, H. V., et al. (1988). Human debrisoquine 4-hydroxylase (P450IID1): cDNA and deduced amino acid sequence and assignment of the CYP2D locus to chromosome 22. *Genomics, 2*, 174–179.

Gough, A. C., Miles, J. S., Spurr, N. K., Moss, J. E., Gaedigk, A., Eichelbaum, M., et al. (1990). Identification of the primary gene defect at the cytochrome-P450 Cyp2d locus. *Nature, 347*, 773–776.

Gu, D. F., Hinks, L. J., Morton, N. E., & Day, I. N. (2000). The use of long PCR to confirm three common alleles at the CYP2A6 locus and the relationship between genotype and smoking habit. *Annals of Human Genetics, 64*, 383–390.

Hanioka, N., Kimura, S., Meyer, U. A., & Gonzalez, F. J. (1990). The human CYP2D locus associated with a common genetic defect in drug oxidation: A G1934—A base change in intron 3 of a mutant CYP2D6 allele results in an aberrant 3′ splice recognition site. *The American Journal of Human Genetics, 47*, 994–1001.

Hanna, I. H., Dawling, S., Roodi, N., Guengerich, F. P., & Parl, F. F. (2000). Cytochrome P450 1B1 (CYP1B1) pharmacogenetics: Association of polymorphisms with functional differences in estrogen hydroxylation activity. *Cancer Research, 60*, 3440–3444.

Hardy, J., & Singleton, A. (2009). Genomewide association studies and human disease. *The New England Journal of Medicine, 360*, 1759–1768.

Hayashi, S., Watanabe, J., & Kawajiri, K. (1991). Genetic polymorphisms in the 5′-flanking region change transcriptional regulation of the human cytochrome P450IIE1 gene. *Journal of Biochemistry, 110*, 559–565.

Hayashi, S., Watanabe, J., Nakachi, K., & Kawajiri, K. (1991). Genetic linkage of lung cancer-associated *Msp*I polymorphisms with amino acid replacement in the heme binding region of the human cytochrome P450IA1 gene. *Journal of Biochemistry, 110*, 407–411.

Heim, M., & Meyer, U. A. (1990). Genotyping of poor metabolisers of debrisoquine by allele-specific PCR amplification. *Lancet, 336*, 529–532.

Hildesheim, A., Anderson, L. M., Chen, C. J., Cheng, Y. J., Brinton, L. A., Daly, A. K., et al. (1997). CYP2E1 genetic polymorphisms and risk of nasopharyngeal carcinoma in Taiwan. *Journal of the National Cancer Institute, 89*, 1207–1212.

Hu, Z., Wu, C., Shi, Y., Guo, H., Zhao, X., Yin, Z., et al. (2011). A genome-wide association study identifies two new lung cancer susceptibility loci at 13q12.12 and 22q12.2 in Han Chinese. *Nature Genetics, 43*, 792–796.

Hung, R. J., McKay, J. D., Gaborieau, V., Boffetta, P., Hashibe, M., Zaridze, D., et al. (2008). A susceptibility locus for lung cancer maps to nicotinic acetylcholine receptor subunit genes on 15q25. *Nature, 452*, 633–637.

Ikawa, S., Uematsu, F., Watanabe, K., Kimpara, T., Osada, M., Hossain, A., et al. (1995). Assessment of cancer susceptibility in humans by use of genetic polymorphisms in carcinogen metabolism. *Pharmacogenetics, 5*(Spec No), S154–S160.

Ingelman-Sundberg, M. (1999). Duplication, multiduplication, and amplification of genes encoding drug-metabolizing enzymes: Evolutionary, toxicological, and clinical pharmacological aspects. *Drug Metabolism Reviews, 31*, 449–459.

Ingelman-Sundberg, M. (2002). Polymorphism of cytochrome P450 and xenobiotic toxicity. *Toxicology, 181–182*, 447–452.

Islam, S. A., Wolf, C. R., Lennard, M. S., & Sternberg, M. J. (1991). A three-dimensional molecular template for substrates of human cytochrome P450 involved in debrisoquine 4-hydroxylation. *Carcinogenesis*, *12*, 2211–2219.

Jaiswal, A. K., Gonzalez, F. J., & Nebert, D. W. (1985). Human P1-450 gene sequence and correlation of mRNA with genetic differences in benzo[a]pyrene metabolism. *Nucleic Acids Research*, *13*, 4503–4520.

Jia, W. H., Pan, Q. H., Qin, H. D., Xu, Y. F., Shen, G. P., Chen, L., et al. (2009). A case-control and a family-based association study revealing an association between CYP2E1 polymorphisms and nasopharyngeal carcinoma risk in Cantonese. *Carcinogenesis*, *30*, 2031–2036.

Jiang, D. K., Sun, J., Cao, G., Liu, Y., Lin, D., Gao, Y. Z., et al. (2013). Genetic variants in STAT4 and HLA-DQ genes confer risk of hepatitis B virus-related hepatocellular carcinoma. *Nature Genetics*, *45*, 72–75.

Kagimoto, M., Heim, M., Kagimoto, K., Zeugin, T., & Meyer, U. A. (1990). Multiple mutations of the human cytochrome P450IID6 gene (CYP2D6) in poor metabolizers of debrisoquine. Study of the functional significance of individual mutations by expression of chimeric genes. *The Journal of Biological Chemistry*, *265*, 17209–17214.

Kamdem, L. K., Meineke, I., Godtel-Armbrust, U., Brockmoller, J., & Wojnowski, L. (2006). Dominant contribution of P450 3A4 to the hepatic carcinogenic activation of aflatoxin B1. *Chemical Research in Toxicology*, *19*, 577–586.

Kellermann, G., Luyten-Kellermann, M., & Shaw, C. R. (1973). Genetic variation of aryl hydrocarbon hydroxylase in human lymphocytes. *The American Journal of Human Genetics*, *25*, 327–331.

Kellermann, G., Shaw, C. R., & Luyten-Kellerman, M. (1973). Aryl hydrocarbon hydroxylase inducibility and bronchogenic carcinoma. *The New England Journal of Medicine*, *289*, 934–937.

Kongruttanachok, N., Sukdikul, S., Setavarin, S., Kerekhjanarong, V., Supiyaphun, P., Voravud, N., et al. (2001). Cytochrome P450 2E1 polymorphism and nasopharyngeal carcinoma development in Thailand: A correlative study. *BMC Cancer*, *1*, 4.

Koop, D. R. (1992). Oxidative and reductive metabolism by cytochrome P450 2E1. *FASEB Journal*, *6*, 724–730.

Kouri, R. E., McKinney, C. E., Slomiany, D. J., Snodgrass, D. R., Wray, N. P., & McLemore, T. L. (1982). Positive correlation between high aryl-hydrocarbon hydroxylase-activity and primary lung-cancer as analyzed in cryopreserved lymphocytes. *Cancer Research*, *42*, 5030–5037.

Lan, Q., Hsiung, C. A., Matsuo, K., Hong, Y. C., Seow, A., Wang, Z., et al. (2012). Genome-wide association analysis identifies new lung cancer susceptibility loci in never-smoking women in Asia. *Nature Genetics*, *44*, 1330–1335.

Lang, N. P., Butler, M. A., Massengill, J., Lawson, M., Stotts, R. C., Hauer-Jensen, M., et al. (1994). Rapid metabolic phenotypes for acetyltransferase and cytochrome P4501A2 and putative exposure to food-borne heterocyclic amines increase the risk for colorectal cancer or polyps. *Cancer Epidemiology, Biomarkers & Prevention*, *3*, 675–682.

Lee, A. M., Jepson, C., Shields, P. G., Benowitz, N., Lerman, C., & Tyndale, R. F. (2007). CYP2B6 genotype does not alter nicotine metabolism, plasma levels, or abstinence with nicotine replacement therapy. *Cancer Epidemiology, Biomarkers & Prevention*, *16*, 1312–1314.

Le Marchand, L., Guo, C., Benhamou, S., Bouchardy, C., Cascorbi, I., Clapper, M. L., et al. (2003). Pooled analysis of the CYP1A1 exon 7 polymorphism and lung cancer (United States). *Cancer Causes & Control*, *14*, 339–346.

Le Marchand, L., Hankin, J. H., Wilkens, L. R., Pierce, L. M., Franke, A., Kolonel, L. N., et al. (2001). Combined effects of well-done red meat, smoking, and rapid N-acetyltransferase 2 and CYP1A2 phenotypes in increasing colorectal cancer risk. *Cancer Epidemiology, Biomarkers & Prevention*, *10*, 1259–1266.

Lieber, C. S., & DeCarli, L. M. (1970). Hepatic microsomal ethanol-oxidizing system. In vitro characteristics and adaptive properties in vivo. *The Journal of Biological Chemistry, 245*, 2505–2512.

Liu, Z. B., Shu, J., Wang, L. P., Jin, C., & Lou, Z. X. (2013). Cytochrome P450 2A6 deletion polymorphism and risk of lung cancer: A meta-analysis. *Molecular Biology Reports, 40*, 5255–5259.

London, S. J., Daly, A. K., Leathart, J. B., Navidi, W. C., Carpenter, C. C., & Idle, J. R. (1997). Genetic polymorphism of CYP2D6 and lung cancer risk in African-Americans and Caucasians in Los Angeles County. *Carcinogenesis, 18*, 1203–1214.

London, S. J., Sullivan-Klose, T., Daly, A. K., & Idle, J. R. (1997). Lung cancer risk in relation to the CYP2C9 genetic polymorphism among Caucasians in Los Angeles County. *Pharmacogenetics*, 7, 401–404.

Lu, X., Wang, L., Lin, X., Huang, J., Charles Gu, C., He, M., et al. (2015). Genome-wide association study in Chinese identifies novel loci for blood pressure and hypertension. *Human Molecular Genetics, 24*, 865–874.

Ma, Z., Guo, W., Gong, T., Niu, H. J., Wang, R. W., & Jiang, Y. G. (2014). CYP1A2 rs762551 polymorphism contributes to risk of lung cancer: A meta-analysis. *Tumour Biology, 35*, 2253–2257.

Mahgoub, A., Idle, J. R., Dring, L. G., Lancaster, R., & Smith, R. L. (1977). Polymorphic hydroxylation of debrisoquine in man. *Lancet, 2*, 584–586.

McKay, J. D., Hung, R. J., Gaborieau, V., Boffetta, P., Chabrier, A., Byrnes, G., et al. (2008). Lung cancer susceptibility locus at 5p15.33. *Nature Genetics, 40*, 1404–1406.

McLemore, T. L., Adelberg, S., Liu, M. C., McMahon, N. A., Yu, S. J., Hubbard, W. C., et al. (1990). Expression of CYP1A1 gene in patients with lung cancer: Evidence for cigarette smoke-induced gene expression in normal lung tissue and for altered gene regulation in primary pulmonary carcinomas. *Journal of the National Cancer Institute, 82*, 1333–1339.

Messina, E. S., Tyndale, R. F., & Sellers, E. M. (1997). A major role for CYP2A6 in nicotine C-oxidation by human liver microsomes. *The Journal of Pharmacology and Experimental Therapeutics, 282*, 1608–1614.

Miki, D., Kubo, M., Takahashi, A., Yoon, K. A., Kim, J., Lee, G. K., et al. (2010). Variation in TP63 is associated with lung adenocarcinoma susceptibility in Japanese and Korean populations. *Nature Genetics, 42*, 893–896.

Mwenifumbo, J. C., Zhou, Q., Benowitz, N. L., Sellers, E. M., & Tyndale, R. F. (2010). New CYP2A6 gene deletion and conversion variants in a population of Black African descent. *Pharmacogenomics, 11*, 189–198.

Nebert, D. W., & Gielen, J. E. (1972). Genetic regulation of aryl hydrocarbon hydroxylase induction in the mouse. *Federation Proceedings, 31*, 1315–1325.

Newton-Cheh, C., Johnson, T., Gateva, V., Tobin, M. D., Bochud, M., Coin, L., et al. (2009). Genome-wide association study identifies eight loci associated with blood pressure. *Nature Genetics, 41*, 666–676.

Nyholt, D. R., Low, S. K., Anderson, C. A., Painter, J. N., Uno, S., Morris, A. P., et al. (2012). Genome-wide association meta-analysis identifies new endometriosis risk loci. *Nature Genetics, 44*, 1355–1359.

Oscarson, M., McLellan, R. A., Gullsten, H., Yue, Q. Y., Lang, M. A., Bernal, M. L., et al. (1999). Characterisation and PCR-based detection of a CYP2A6 gene deletion found at a high frequency in a Chinese population. *FEBS Letters, 448*, 105–110.

Paigen, B., Gurtoo, H. L., Minowada, J., Houten, L., Vincent, R., Paigen, K., et al. (1977). Questionable relation of aryl hydrocarbon hydroxylase to lung-cancer risk. *The New England Journal of Medicine, 297*, 346–350.

Painter, J. N., Nyholt, D. R., Krause, L., Zhao, Z. Z., Chapman, B., Zhang, C., et al. (2014). Common variants in the CYP2C19 gene are associated with susceptibility to endometriosis. *Fertility and Sterility, 102*, 496–502, e495.

Painter, J. N., Nyholt, D. R., Morris, A., Zhao, Z. Z., Henders, A. K., Lambert, A., et al. (2011). High-density fine-mapping of a chromosome 10q26 linkage peak suggests association between endometriosis and variants close to CYP2C19. *Fertility and Sterility*, *95*, 2236–2240.

Perera, M. A. (2010). The missing linkage: What pharmacogenetic associations are left to find in CYP3A? *Expert Opinion on Drug Metabolism & Toxicology*, *6*, 17–28.

Perera, F. P., Tang, D., Brandt-Rauf, P., Santella, R. M., Mooney, L. V., Tu, Y. H., et al. (2006). Lack of associations among cancer and albumin adducts, ras p21 oncoprotein levels, and CYP1A1, CYP2D6, NAT1, and NAT2 in a nested case-control study of lung cancer within the physicians' health study. *Cancer Epidemiology, Biomarkers & Prevention*, *15*, 1417–1419.

Prasad, R., Prasad, N., Harrell, J. E., Thornby, J., Liem, J. H., Hudgins, P. T., et al. (1979). Aryl hydrocarbon hydroxylase inducibility and lymphoblast formation in lung cancer patients. *International Journal of Cancer*, *23*, 316–320.

Raimondi, S., Boffetta, P., Anttila, S., Brockmoller, J., Butkiewicz, D., Cascorbi, I., et al. (2005). Metabolic gene polymorphisms and lung cancer risk in non-smokers. An update of the GSEC study. *Mutation Research*, *592*, 45–57.

Rao, Y., Hoffmann, E., Zia, M., Bodin, L., Zeman, M., Sellers, E. M., et al. (2000). Duplications and defects in the CYP2A6 gene: Identification, genotyping, and in vivo effects on smoking. *Molecular Pharmacology*, *58*, 747–755.

Rebbeck, T. R., Spitz, M., & Wu, X. (2004). Assessing the function of genetic variants in candidate gene association studies. *Nature Reviews Genetics*, *5*, 589–597.

Rodriguez-Antona, C., Gomez, A., Karlgren, M., Sim, S. C., & Ingelman-Sundberg, M. (2010). Molecular genetics and epigenetics of the cytochrome P450 gene family and its relevance for cancer risk and treatment. *Human Genetics*, *127*, 1–17.

Rostami-Hodjegan, A., Lennard, M. S., Woods, H. F., & Tucker, G. T. (1998). Meta-analysis of studies of the CYP2D6 polymorphism in relation to lung cancer and Parkinson's disease. *Pharmacogenetics*, *8*, 227–238.

Saarikoski, S. T., Sata, F., Husgafvel-Pursiainen, K., Rautalahti, M., Haukka, J., Impivaara, O., et al. (2000). CYP2D6 ultrarapid metabolizer genotype as a potential modifier of smoking behaviour. *Pharmacogenetics*, *10*, 5–10.

Sachse, C., Brockmoller, J., Bauer, S., & Roots, I. (1999). Functional significance of a C→A polymorphism in intron 1 of the cytochrome P450 CYP1A2 gene tested with caffeine. *British Journal of Clinical Pharmacology*, *47*, 445–449.

Sawcer, S., Hellenthal, G., Pirinen, M., Spencer, C. C., Patsopoulos, N. A., Moutsianas, L., et al. (2011). Genetic risk and a primary role for cell-mediated immune mechanisms in multiple sclerosis. *Nature*, *476*, 214–219.

Schmalix, W. A., Maser, H., Kiefer, F., Reen, R., Wiebel, F. J., Gonzalez, F., et al. (1993). Stable expression of human cytochrome P450 1A1 cDNA in V79 Chinese hamster cells and metabolic activation of benzo[a]pyrene. *European Journal of Pharmacology*, *248*, 251–261.

Shahabi, P., Siest, G., Meyer, U. A., & Visvikis-Siest, S. (2014). Human cytochrome P450 epoxygenases: Variability in expression and role in inflammation-related disorders. *Pharmacology & Therapeutics*, *144*, 134–161.

Shi, W. X., & Chen, S. Q. (2004). Frequencies of poor metabolizers of cytochrome P450 2C19 in esophagus cancer, stomach cancer, lung cancer and bladder cancer in Chinese population. *World Journal of Gastroenterology*, *10*, 1961–1963.

Shimada, T., Gillam, E. M., Sutter, T. R., Strickland, P. T., Guengerich, F. P., & Yamazaki, H. (1997). Oxidation of xenobiotics by recombinant human cytochrome P450 1B1. *Drug Metabolism and Disposition*, *25*, 617–622.

Shimada, T., & Guengerich, F. P. (1989). Evidence for cytochrome P-450NF, the nifedipine oxidase, being the principal enzyme involved in the bioactivation of aflatoxins in human liver. *Proceedings of the National Academy of Sciences of the United States of America*, *86*, 462–465.

Shou, M., Korzekwa, K. R., Crespi, C. L., Gonzalez, F. J., & Gelboin, H. V. (1994). The role of 12 cDNA-expressed human, rodent and rabbit cytochromes P450 in the metabolism of benzo[*a*]pyrene and benzo[*a*]pyrene *trans*-7,8-diol. *Molecular Carcinogenesis*, *10*, 159–168.

Spiecker, M., Darius, H., Hankeln, T., Soufi, M., Sattler, A. M., Schaefer, J. R., et al. (2004). Risk of coronary artery disease associated with polymorphism of the cytochrome P450 epoxygenase CYP2J2. *Circulation*, *110*, 2132–2136.

Spurr, N. K., Gough, A. C., Stevenson, K., & Wolf, C. R. (1987). Msp-1 polymorphism detected with a cDNA probe for the P-450 I family on chromosome 15. *Nucleic Acids Research*, *15*, 5901.

Stoilov, I., Akarsu, A. N., & Sarfarazi, M. (1997). Identification of three different truncating mutations in cytochrome P4501B1 (CYP1B1) as the principal cause of primary congenital glaucoma (Buphthalmos) in families linked to the GLC3A locus on chromosome 2p21. *Human Molecular Genetics*, *6*, 641–647.

Sulem, P., Gudbjartsson, D. F., Geller, F., Prokopenko, I., Feenstra, B., Aben, K. K., et al. (2011). Sequence variants at CYP1A1-CYP1A2 and AHR associate with coffee consumption. *Human Molecular Genetics*, *20*, 2071–2077.

Tang, M., Lautenberger, J. A., Gao, X., Sezgin, E., Hendrickson, S. L., Troyer, J. L., et al. (2012). The principal genetic determinants for nasopharyngeal carcinoma in China involve the HLA class I antigen recognition groove. *PLoS Genetics*, *8*, e1003103.

Thorgeirsson, T. E., Geller, F., Sulem, P., Rafnar, T., Wiste, A., Magnusson, K. P., et al. (2008). A variant associated with nicotine dependence, lung cancer and peripheral arterial disease. *Nature*, *452*, 638–642.

Thorgeirsson, T. E., Gudbjartsson, D. F., Surakka, I., Vink, J. M., Amin, N., Geller, F., et al. (2010). Sequence variants at CHRNB3-CHRNA6 and CYP2A6 affect smoking behavior. *Nature Genetics*, *42*, 448–453.

Tiano, H. F., Hosokawa, M., Chulada, P. C., Smith, P. B., Wang, R. L., Gonzalez, F. J., et al. (1993). Retroviral mediated expression of human cytochrome P450 2A6 in C3H/10T1/2 cells confers transformability by 4-(methylnitrosamino)-1-(3-pyridyl)-1-butanone (NNK). *Carcinogenesis*, *14*, 1421–1427.

Treloar, S. A., Wicks, J., Nyholt, D. R., Montgomery, G. W., Bahlo, M., Smith, V., et al. (2005). Genomewide linkage study in 1,176 affected sister pair families identifies a significant susceptibility locus for endometriosis on chromosome 10q26. *The American Journal of Human Genetics*, 77, 365–376.

Vasiliou, V., & Gonzalez, F. J. (2008). Role of CYP1B1 in glaucoma. *Annual Review of Pharmacology and Toxicology*, *48*, 333–358.

Vercellini, P., Vigano, P., Somigliana, E., & Fedele, L. (2014). Endometriosis: Pathogenesis and treatment. *Nature Reviews. Endocrinology*, *10*, 261–275.

Vineis, P., Veglia, F., Benhamou, S., Butkiewicz, D., Cascorbi, I., Clapper, M. L., et al. (2003). CYP1A1 T3801 C polymorphism and lung cancer: A pooled analysis of 2451 cases and 3358 controls. *International Journal of Cancer*, *104*, 650–657.

Wang, Y., Broderick, P., Webb, E., Wu, X., Vijayakrishnan, J., Matakidou, A., et al. (2008). Common 5p15.33 and 6p21.33 variants influence lung cancer risk. *Nature Genetics*, *40*, 1407–1409.

Wang, H., Song, K., Chen, Z., & Yu, Y. (2013). Poor metabolizers at the cytochrome P450 2C19 loci is at increased risk of developing cancer in Asian populations. *PLoS One*, *8*, e73126.

Wang, L., Zang, W., Liu, J., Xie, D., Ji, W., Pan, Y., et al. (2013). Association of CYP2A6*4 with susceptibility of lung cancer: A meta-analysis. *PLoS One*, *8*, e59556.

Wassenaar, C. A., Dong, Q., Amos, C. I., Spitz, M. R., & Tyndale, R. F. (2013). Pilot study of CYP2B6 genetic variation to explore the contribution of nitrosamine activation to lung carcinogenesis. *International Journal of Molecular Sciences*, *14*, 8381–8392.

Wassenaar, C. A., Dong, Q., Wei, Q., Amos, C. I., Spitz, M. R., & Tyndale, R. F. (2011). Relationship between CYP2A6 and CHRNA5-CHRNA3-CHRNB4 variation and smoking behaviors and lung cancer risk. *Journal of the National Cancer Institute, 103*, 1342–1346.

Wassenaar, C. A., Ye, Y., Cai, Q., Aldrich, M. C., Knight, J., Spitz, M. R., et al. (2015). CYP2A6 reduced activity gene variants confer reduction in lung cancer risk in African American smokers-findings from two independent populations. *Carcinogenesis, 36*, 99–103.

Webb, A., Lind, P. A., Kalmijn, J., Feiler, H. S., Smith, T. L., Schuckit, M. A., et al. (2011). The investigation into CYP2E1 in relation to the level of response to alcohol through a combination of linkage and association analysis. *Alcoholism, Clinical and Experimental Research, 35*, 10–18.

Wheeler, C. W., Wrighton, S. A., & Guenthner, T. M. (1992). Detection of human lung cytochromes P450 that are immunochemically related to cytochrome P450IIE1 and cytochrome P450IIIA. *Biochemical Pharmacology, 44*, 183–187.

Whiffin, N., Hosking, F. J., Farrington, S. M., Palles, C., Dobbins, S. E., Zgaga, L., et al. (2014). Identification of susceptibility loci for colorectal cancer in a genome-wide meta-analysis. *Human Molecular Genetics, 23*, 4729–4737.

Wild, C. P., & Gong, Y. Y. (2010). Mycotoxins and human disease: A largely ignored global health issue. *Carcinogenesis, 31*, 71–82.

Wojnowski, L., Turner, P. C., Pedersen, B., Hustert, E., Brockmoller, J., Mendy, M., et al. (2004). Increased levels of aflatoxin-albumin adducts are associated with CYP3A5 polymorphisms in The Gambia, West Africa. *Pharmacogenetics, 14*, 691–700.

Wolf, C. R., Smith, C. A., Gough, A. C., Moss, J. E., Vallis, K. A., Howard, G., et al. (1992). Relationship between the debrisoquine hydroxylase polymorphism and cancer susceptibility. *Carcinogenesis, 13*, 1035–1038.

Xu, W., Zhou, Y., Hang, X., & Shen, D. (2012). Current evidence on the relationship between CYP1B1 polymorphisms and lung cancer risk: A meta-analysis. *Molecular Biology Reports, 39*, 2821–2829.

Yamano, S., Tatsuno, J., & Gonzalez, F. J. (1990). The CYP2A3-gene product catalyzes coumarin 7-hydroxylation in human liver-microsomes. *Biochemistry, 29*, 1322–1329.

Yamazaki, Y., Fujita, K., Nakayama, K., Suzuki, A., Nakamura, K., Yamazaki, H., et al. (2004). Establishment of ten strains of genetically engineered Salmonella typhimurium TA1538 each co-expressing a form of human cytochrome P450 with NADPH-cytochrome P450 reductase sensitive to various promutagens. *Mutation Research, 562*, 151–162.

Yang, X. R., Diehl, S., Pfeiffer, R., Chen, C. J., Hsu, W. L., Dosemeci, M., et al. (2005). Evaluation of risk factors for nasopharyngeal carcinoma in high-risk nasopharyngeal carcinoma families in Taiwan. *Cancer Epidemiology, Biomarkers & Prevention, 14*, 900–905.

Ye, X. H., Song, L., Peng, L., Bu, Z., Yan, S. X., Feng, J., et al. (2014). Association between the CYP2E1 polymorphisms and lung cancer risk: A meta-analysis. *Molecular Genetics and Genomics, 290*, 545–558.

Zeng, T., Guo, F. F., Zhang, C. L., Song, F. Y., Zhao, X. L., & Xie, K. Q. (2013). Roles of cytochrome P4502E1 gene polymorphisms and the risks of alcoholic liver disease: A meta-analysis. *PLoS One, 8*, e54188.

CHAPTER FOUR

Personalized Cancer Therapy Considering Cytochrome P450 Variability

Saskia Preissner*[,1], Maurizio Simmaco[†,‡], Giovanna Gentile[†,‡], Robert Preissner[§]

*Dental, Oral and Maxillary Medicine, Charité—Universitätsmedizin Berlin, Berlin, Germany
[†]NESMOS Department, Faculty of Medicine and Psychology, Sapienza University of Rome, Rome, Italy
[‡]Advanced Molecular Diagnostics Unit, Sant'Andrea Hospital, Rome, Italy
[§]Structural Bioinformatics Group, Charité—Universitätsmedizin Berlin, Berlin, Germany
[1]Corresponding author: e-mail address: saskia.preissner@charite.de

Contents

Abstract

The individual variability of pharmacokinetics is underestimated and few systematic studies exist in this field. In most cases, this leads to unwanted side effects or toxicity. In polychemotherapy, prodrugs (like ifosfamide), which have to be activated by cytochrome P450 enzymes (CYPs), play an important role. If patients are poor metabolizers for these drugs, the therapy will be ineffective. Furthermore, CYPs and transporters can be (over)expressed in target tissues, which is also not examined and considered in clinical routine. Here, we present a body map showing relevant enzymes in some organs and tissues. Finally, a typical case of a Caucasian chemotherapy patient with breast cancer is presented and discussed regarding a personalized cancer therapy considering the single nucleotide polymorphisms found via genotyping.

Advances in Pharmacology, Volume 74
ISSN 1054-3589
http://dx.doi.org/10.1016/bs.apha.2015.03.004

1. INTRODUCTION

The cytochrome P450 enzymes (CYPs) are major drug metabolizers (activation as well as inactivation of drugs). About 60 human CYPs have been identified (Ingelman-Sundberg, 2005), but most of the drugs are metabolized by a set of CYPs (Fig. 1). Regarding the entire number of about 2000 drug-metabolizing reactions, the CYPs 3A4, 2D6, 1A2, 2C9, and 2C19 are responsible for 60.0% of drug metabolism (Preissner et al., 2010). Twelve CYPs account for over 90.0% of drug metabolism (Table 1). Certain single nucleotide polymorphisms (SNPs) have a strong impact on CYP activity (Daly, 2003), as the major drug-metabolizing enzymes are polymorphic (Fig. 2). Especially in chemotherapy, these SNPs play a crucial role, as they can lead to adverse side effects due to malfunctioning *in vivo* activation of prodrugs and/or toxic drug levels (Morrow et al., 2011; Patterson & Murray, 2002).

This chapter summarizes the effect of common SNPs in cancer therapy and explains why it is necessary to establish personalized medicine in the therapy of cancer.

2. POLYMORPHIC CYPs

The family of CYPs is responsible for the metabolism of most of the xenobiotics, including drugs. As these CYPs are highly polymorphic, drug therapies are often ineffective. The effectiveness of administered drugs in cancer therapy was reported to be only about 25% (Spear, Heath-Chiozzi, & Huff, 2001). SNPs are one of the reasons for ineffective therapy. These SNPs

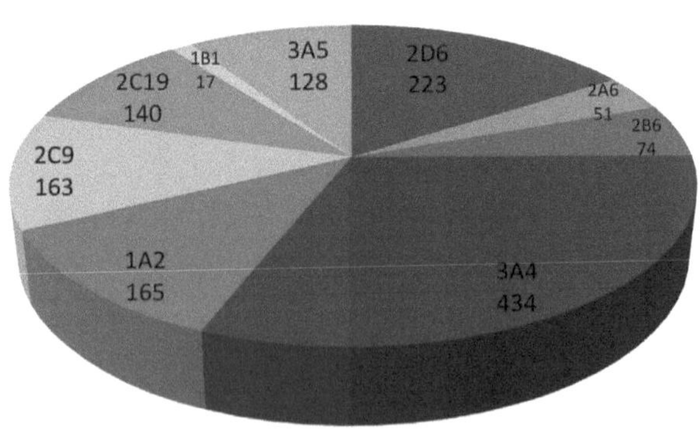

Figure 1 Major drug-metabolizing enzymes. The pie chart illustrates the number of drugs that can be metabolized by a specific CYP.

Table 1 Major Drug-Metabolizing Enzymes

CYP	Number of Drugs Metabolized	Number of Nucleotide Changes
3A4	434	46
2D6	223	114
1A2	165	41
2C9	163	39
2C19	140	29
3A5	128	26
2C8	112	17
3A7	98	5
2E1	81	18
2B6	74	57
2A6	51	68
1A1	50	21

The listed 12 CYPs are responsible for the metabolism of over 90% of the drugs. Additionally, these CYPs show a large number of relevant nucleotide changes.

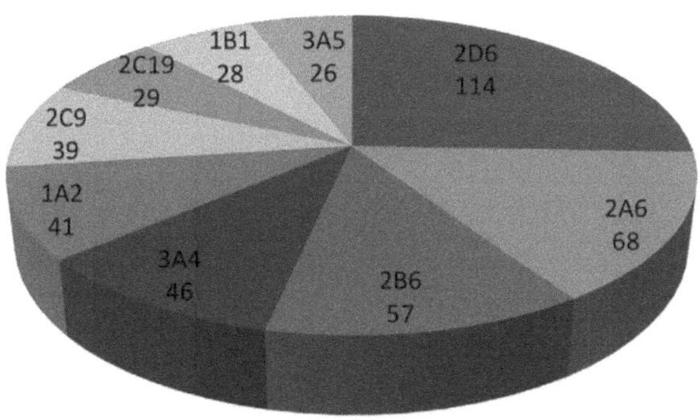

Figure 2 Polymorphic CYPs. The pie chart illustrates the number of known SNPs in different CYPs.

can lead to altered enzyme activity. Phenotypes compared to a normal (wild-type) metabolizer are poor, intermediate, extensive/rapid, or ultrarapid metabolizers (van der Weide & Steijns, 1999). In consequence, adaptations in drug dosages are necessary, but most of the patients are not screened for SNPs. The SuperCYP database shelters about 2000 CYP variants (Preissner et al., 2010). Additionally, the frequency of allele variants varies among different ethnicities (Rodriguez-Antona, Jande, Rane, & Ingelman-Sundberg, 2005).

In the Caucasian population, 34 different CYP alleles have been found to occur with a frequency >1% (Table 2; Preissner et al., 2013). CYP 2D6 and 2B6 had the largest number of alleles and both have impact on the drug metabolism. In CYP 2D6 maximum allele frequencies varied from 20.7% to 32.4%. The most frequent alleles are CYP 3A5*3C at 81.3% (decreased enzyme activity) followed by CYP 1A2*1F at 33.3% (higher inducibility).

Table 2 CYP SNP Frequencies in Caucasians

CYP	Allele	Caucasian (%)	Enzyme Activity	References (PMID)
1A1	*2A	19.0	Increase	19514967
1A2	*1F	33.3	Higher inducibility	12534642/22299824
	*1D	4.82	Decrease	12534642/20797314
2C9	*2	19.0% *1/*2 1.6% *2/*2 1.8% *2/*3	Decrease	15284536
	*3	9.0	Decrease	11678789
2C19	*2	16.0	Decrease	10460072
	*17	18.0	Increase	21247447
2D6	*3	2.04	Decrease	9012401
	*4	20.7	Decrease	9012401
	*4D	3.4	Decrease	11266079
	*4L	4.5	Decrease	11266079
	*5	4.1	No enzyme	9511177
	*6	1.3	Nonfunctional	9511177
	*7	1.0	Decrease	9089660
	*9	2.0	Decrease	9511177
	*10	8.0	Decrease	9511177/11505219
	*41	8.0	Decrease (expression)	15289790
2A6	*12	2.9	Decreased (expression)	16041240
	*1B	32.6	Increase	22850738
	*2	2.3	Decrease	11259354
	*4	1.0	No enzyme	11259354
	*9	7.1	Decreased	15475735

Table 2 CYP SNP Frequencies in Caucasians—cont'd

CYP	Allele	Caucasian (%)	Enzyme Activity	References (PMID)
3A4	*17	2.0	Decrease	11714865
	*1B	17.0	Increase (transcription)	12692107
	*2	2.7	Decrease	10668853
3A5	*3C	81.3	Decrease	17162466
	*3k/*10	2.0	Decrease	12893984
3A7	*2	8.0	Increase	15903124
2B6	*2	5.3	Increase	21746968/12242601
	*5	14.0	Decrease	11470993
	*4	5.0	Increase	14515060
	*6	25.2	Decrease (expression)	14515060
	*7	3.0	Decrease	12242601/14551287
	*22	3.0	Increase (transcription)	15722458

Polymorphisms that are relevant for Caucasians are shown here with CYP, allele, frequencies, and their effect on enzyme activity. PubMed IDs complete the table.

Alleles leading to increased metabolism are CYP 2A6*1B (30.0%), 3A4*1B (17.0%), 1A1*2A (19.0%), and 2C19*17 (18.0%). Decreased metabolism is attributed to 2D6*2A (32.4%), 2D6*4 (20.7%), and 2C9*2 (16.0%). Carrying the 2A6*4 allele (1.0%) leads to an inactive enzyme with no detectable substrate metabolism.

In Asian and African populations, CYP2A6*2 possessed a frequency of 28.0% and 62.0%, while a frequency of 8.0% is observed in Caucasians. This finding shows that it could be useful to take ethnicity information into account when genotyping patients.

3. PRODRUGS

Prodrugs are pharmacologically inactive medications that have to be converted to an active form through chemical reactions, such as hydrolysis or phosphorylation (Ortiz de Montellano, 2013). The idea of prodrugs in cancer therapy is to reduce unintended side effects by designing compounds that interact with specific targets. The majority of antineoplastic drugs are metabolized by CYPs and a large number of the prodrugs need activation through CYPs (Table 3). It is crucial to know if metabolizing or activating enzymes are altered in their activity to select drugs and/or adapt drug dosages.

Table 3 Prodrugs in Cancer Therapy

ATC	Drug	Reaction	Activation	Interactions	Structure
L01AA01	Cyclophosphamide	Hydroxylation	2B6, 3A4	2A6, 2B6, 2C18, 2C19, 2C8, 2C9, 3A4, 3A7	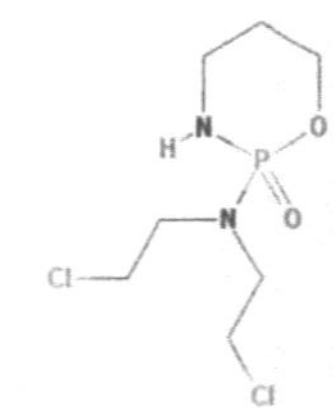
L01AA06	Ifosfamide	Hydroxylation	2B6, 3A4	2A6, 2B6, 2C18, 2C19, 2C8, 2C9, 3A4, 3A7	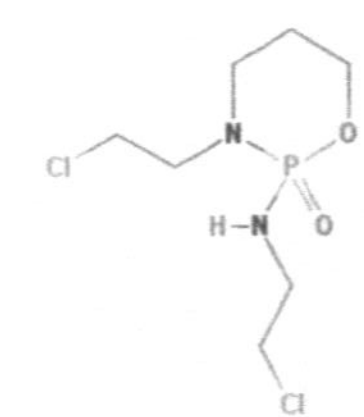

L01AA08	Trofosfamide	Hydroxylation	Alkylating agents	2B6, 3A4	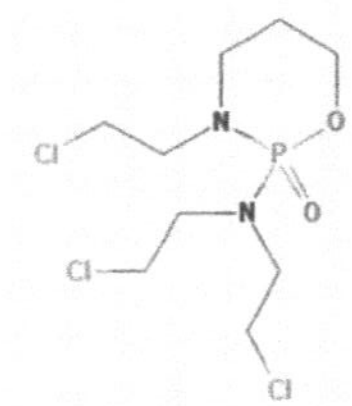
L01AB02	Treosulfan	Nonenzymatically	No enzyme	1A2, 3A4	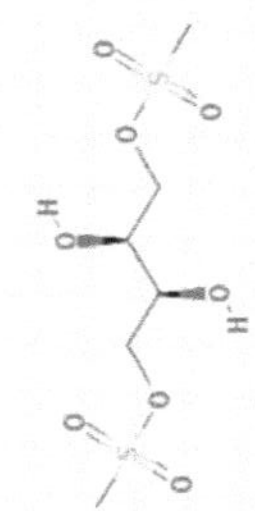

Continued

Table 3 Prodrugs in Cancer Therapy—cont'd

ATC	Drug	Reaction	Activation	Interactions	Structure
L01AX03	Temozolomide	Hydrolysis	Activated water molecule	3A4	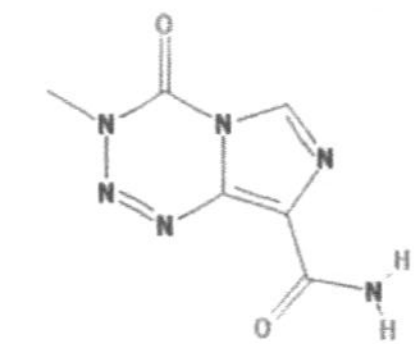
L01AX04	Dacarbazine	Demethylation	CYP (1A2 in mice)	1A1, 1A2, 2E1	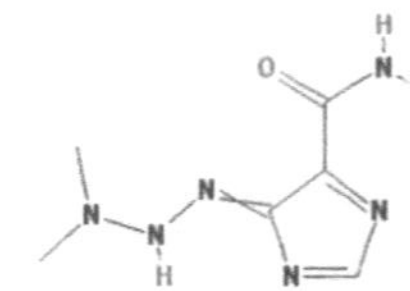
L01BC03	Tegafur → 5-fluorouracil	Hydroxylation	2A6	1A2, 2A6, 2C8, 2E1, 3A5	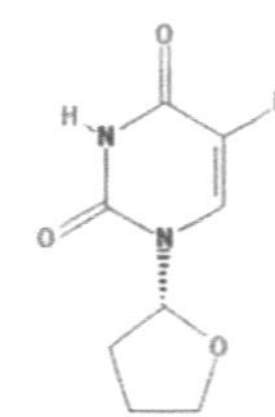

L01BC05	Gemcitabine	Phosphorylation	Deoxycytidine kinase	MRP5	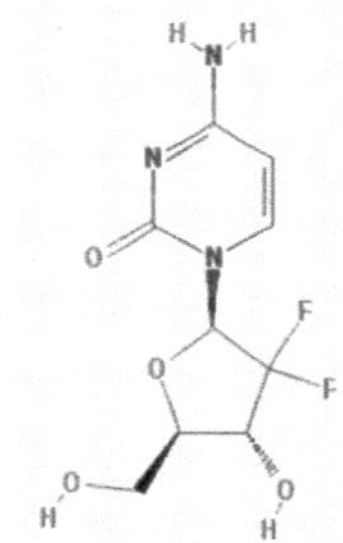
L01BC06	Capecitabine	Phosphorylation	Dihydropyrimidine dehydrogenase and thymidine phosphorylase	2C9	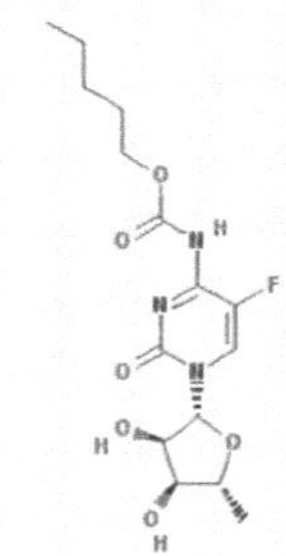

Continued

Table 3 Prodrugs in Cancer Therapy—cont'd

ATC	Drug	Reaction	Activation	Interactions	Structure
L01DC03	Mitomycin	Reduction	Bi- and trifunctional alkylating agents	1A2, MRP1	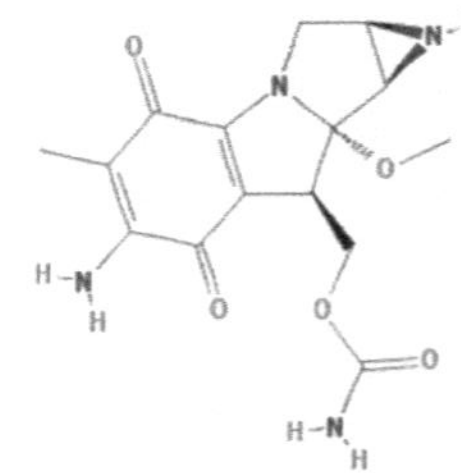
L01XA03	Oxaliplatin	Hydrolysis	No enzyme	1A1, 1A2, 1B1, 2E1, GST	

L01XB01	Procarbazine → azoprocarbazine	Oxidation	Mitochondrial monoamine oxidase, CYPs	1A1, 1A2, 1B1	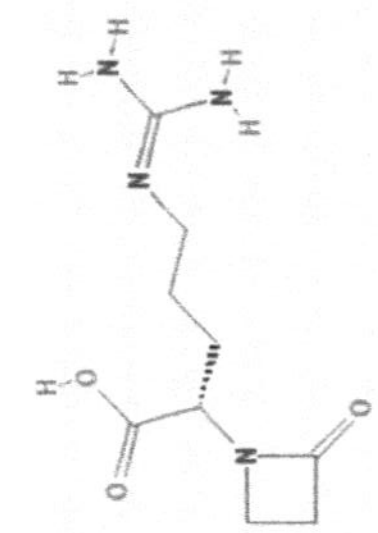
L01XX19	Irinotecan	Hydrolysis	3A4, 3A5	2B6, 3A4, 3A5, 3A7, UGT, MRP4	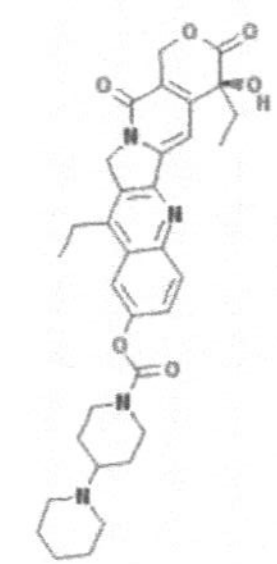

Antineoplastic agents are ordered by ATC-Code. The table lists activation reaction, activating enzyme, known CYP- or other enzymatic interactions, and the 2D structure of the prodrugs.

4. EXPRESSION OF CYPs IN DIFFERENT TISSUES

Mainly, CYPs are expressed in the liver, but interestingly high expression levels of some CYPs can be observed in particular tissues, which is illustrated in Fig. 3. The expression of 40 CYP isoforms in 41 tissues was investigated. Thirty-nine of these CYPs showed higher mRNA expression in one or more tissues. In the kidneys, a fourfold increase of CYP 2B6 was noted.

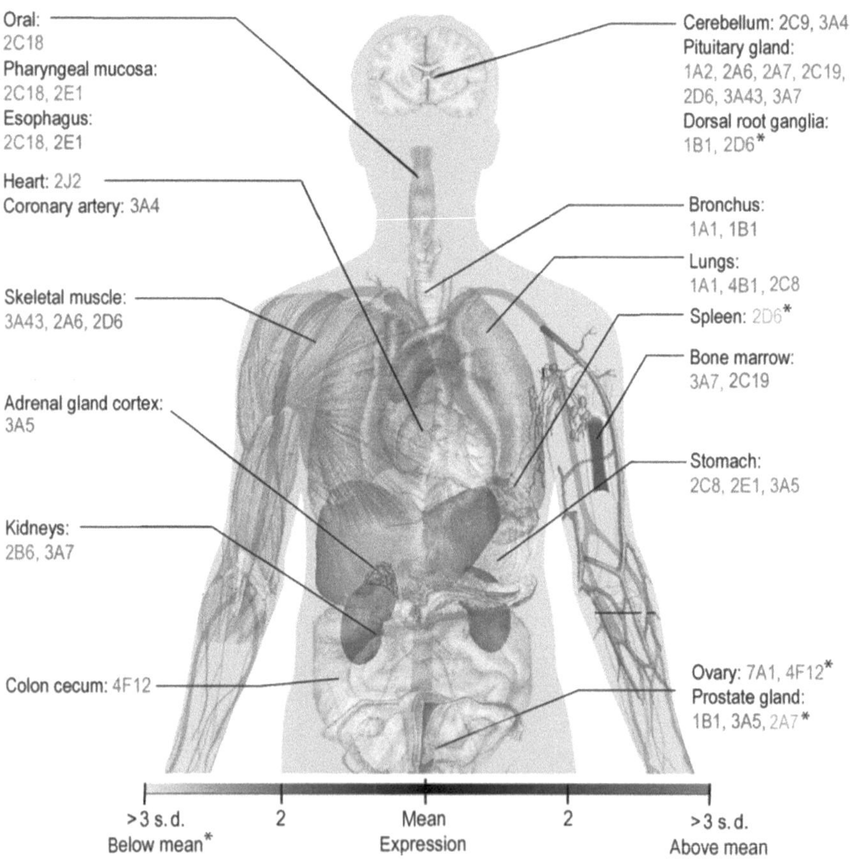

Figure 3 Body map of CYP expression. A map of the expression of selected CYPs in human organs excluding the liver. Expression values are relative to the mean expression in all organs. Red (gray in the print version) text indicates at least threefold higher expression of a CYP, while green (light gray in the print version) text indicates at least threefold lower expression. Average expression values were not included.

A fivefold higher expression of CYP 2C8 was found in the lungs and stomach. CYP 2C18 showed a high-level distribution in gastroesophageal areas, such as oral cavity, pharynx, and esophagus. Twofold lower expression was detected for CYP 2A1 in the esophagus, 2A7 in the prostate, as well as 2C9 and 2D6 in the spleen.

In the pituitary gland, CYPs 1A2, 2C19, and 2D6 are highly expressed, while the highest expression of CYP 2C9 was found in the cerebellum. CYP 2D6 and 2B6 had their greatest expression in skeletal muscle and kidneys.

The relevance of the differential distribution of the CYPs in various tissues has to be further investigated, but may have an influence on specific side effects of drugs. For example, cyclophosphamide therapy can lead to hyponatremia. As a prodrug, cyclophosphamide has to be converted by CYP 2B6 into its active form (Jounaidi, Chen, Veal, & Waxman, 2006). Cyclophosphamide induces an expression of aquaporins 1 and 7, which ultimately results in hyponatremia (Kim, Jo, Park, Han, & Kim, 2011). As CYP 2B6 shows high expression in kidneys, indicating that a higher level of active cyclophosphamide is likely to occur and lead to this undesirable side effect.

Some CYPs are even not expressed in the liver, but only in some tissues (Nishimura, Yaguti, Yoshitsugu, Naito, & Satoh, 2003).

These findings can be used to design target tissue-specific prodrugs, leading to an increased bioavailability at the target site and to a reduction of side effects. The variation in CYP expression among different tissues can also lead to problems in the efficacy of drugs, which are inactivated through higher expressed CYPs in the desired target tissues.

5. DRUG COCKTAIL OPTIMIZATION IN CHEMOTHERAPY

The clinical relevance of CYP SNPs has been shown in many studies, especially to prevent drug-induced deaths (Lund et al., 2011; Mehta & Davies, 2004; Morrow et al., 2012; Patterson & Murray, 2002; Potter et al., 2002). These cases mostly occur in chemotherapy, as drugs are administered in high doses.

More than 200,000 children under the age of 15 years are diagnosed with malignant tumors every year (Terracini, 2011). Acute lymphoblastic leukemia (ALL) is the most common cancer in children and has its incidence peak between 2 and 5 years. Although the overall survival rate of ALL has increased to almost 90% due to better treatment options

(Inaba, Greaves & Mullighan, 2013), 20% of the children suffer from a relapse and in these cases the long-term survival rate is only 45% (Ko et al., 2010).

Ifosfamide and cyclophosphamide are widely used drugs in the treatment of ALL relapses. Both are prodrugs and require activation mainly by the CYPs 2A6, 3A4, and 2C9 (Preiss et al., 2002) via a 4-hydroxylation. Since CYP 2A6 and 3A4 are the second and fourth most polymorphic enzymes (Preissner et al., 2013), it is likely that some patients will be poor metabolizers for one or both of these two CYPs.

The following case of a female patient diagnosed with a local advanced ductal breast cancer (HER2+, ER/PGR+, and cT1 cN2 stage IIIA) exemplarily shows how drug cocktail optimization can reduce side effects.

The proposed treatment was a neoadjuvant therapy with cyclophosphamide and epirubicin for every 3 weeks for four cycles followed by docetaxel and trastuzumab for every 3 weeks for four cycles. Premedication included ondansetron, aprepitant, dexamethasone, and chlorphenamine. Concomitant medications included omeprazole for treatment of gastroesophageal reflux disease.

Unfortunately, the administration of cyclophosphamide was interrupted during the first cycle due to the onset of diffuse adverse effects including dizziness, confusion, and extreme fatigue.

The patient was thus genotyped for the main polymorphic variant of CYP450 genes (CYP1A2, CYP2C8, CYP2C9, CYP2C19, CYP2D6, CYP3A4, and CYP3A5), to detect alteration affecting the metabolism of both chemotherapy and other medications.

The patient was a fast metabolizer for CYP3A4 and CYP3A5 and an intermediate metabolizer for CYP2C9 (Table 1). Considering this genotype, the originally administrated drug cocktail was tested for drug–drug interaction using the SuperCYP database. As shown in Fig. 4A, four of the administrated drugs act as inducers of CYP3A4 (ondansetron, dexamethasone, aprepitant, and omeprazole), and two of them as inducers of CYP3A5 (dexamethasone and omeprazole).

Since cyclophosphamide is activated by CYPs, increased CYP enzymatic activity is supposed to produce an excess of cytotoxic metabolites, leading to increased side effects. In this case, the patient, who had an increased CYP3A activity, was also exposed to the effect of inducer molecules.

Hence, according to the individual genomic profile, the drugs involved in CYP3A4 and CYP3A5 metabolism were substituted with the alternative drugs rabeprazole and betamethasone (inhibitors of CYP3A4) rather than

Name	1A1	1A2	1B1	2A6	2B6	2C8	2C9	2C18	2C19	2D6	2E1	3A4	3A5	3A7	3A43	4A11	11A	11B1	11B2	17A	19A	27A	excr	nocyp
Aprepitant		S					Inh Ind		Inh S			Inh Ind S	S	S										
Omeprazole	Ind S Inh	Inh Ind S	Ind			S	Inh S	S	S Ind Inh	Inh		Inh S Ind	Inh Ind S				Inh							
Dexamethason	Ind	Ind			Ind	Ind	Ind	Ind	Ind	Ind		Ind Inh S	Ind S	Ind S	Ind	Ind	Inh	Inh		Inh		Ind		
Ondansetron		Inh S					S			S Inh	S	S Ind	S	S										
Chlorphenami										Inh S		S	S	S										
Cyclophospha				S	S Ind	Ind S	Ind S	S	S			Inh S Ind		S										

A

Name	1A1	1A2	1B1	2A6	2B6	2C8	2C9	2C18	2C19	2D6	2E1	3A4	3A5	3A7	3A43	4A11	11A	11B1	11B2	17A	19A	27A	excr	nocyp
Chlorphenami										Inh S		S	S	S										
Cyclophospha				S	S Ind	Ind S	Ind S	S	S			Inh S Ind		S										
Rabeprazole	Ind	Ind							S Inh			S Inh												
Betamethason												Inh									S Ind			
palonosetron										S														

B

Figure 4 Drug cocktail optimization. (A) Drug cocktail before genome-based therapy adaptation. (B) Drug cocktail genome-based therapy adaptation.

omeprazole and dexamethasone, and palonosetron (no effect on CYP3A) rather than aprepitant and ondansetron (Fig. 4B).

The patient tolerated the therapy and proceeded with the scheduled neoadjuvant therapy without dose delay or reduction.

6. CONCLUSION

There is a lot of knowledge about drug interactions and the effect of SNPs in the literature, but only very few of this information is used in clinical practice. Especially in long-time or high-dose drug therapy individual SNPs can result in severe side effects and/or ineffective therapy (Binkhorst, Mathijssen, Jager, & van Gelder, 2015; El-Serafi et al., 2015).

The Vanderbilt University in Nashville, USA, is genotyping all patients with a specific risk profile (cardiovascular risk or transplantation patients) on certain variants to prevent secondary complications, such as reinfarction, rhabdomyolysis, or transplant rejection (Pulley et al., 2012).

As genotyping patients on common and relevant SNPs has become affordable, it should become a standard before administrating antineoplastic drugs in order to avoid harmful side effects (Madadi et al., 2010; Manage et al., 2014).

The development of target-specific prodrugs might be another chance to optimize chemotherapy and at the same time to lower undesired side effects (Mura, Bui, Couvreur, & Nicolas, 2015).

These goals are also supported by the European Commission in the "Horizon 2020" programme and hopefully will be achieved in clinical practice soon, as the techniques have been already established (Golubnitschaja, Kinkorova, & Costigliola, 2014).

CONFLICT OF INTEREST

The authors declare no conflict of interest.

REFERENCES

Binkhorst, L., Mathijssen, R. H., Jager, A., & van Gelder, T. (2015). Individualization of tamoxifen therapy: Much more than just CYP2D6 genotyping. *Cancer Treatment Reviews*, *41*(3), 289–299. http://dx.doi.org/10.1016/j.ctrv.2015.01.002.

Daly, A. K. (2003). Pharmacogenetics of the major polymorphic metabolizing enzymes. *Fundamental & Clinical Pharmacology*, *17*(1), 27–41.

El-Serafi, I., Fares, M., Abedi-Valugerdi, M., Afsharian, P., Moshfegh, A., Terelius, Y., et al. (2015). Cytochrome P450 2J2, a new key enzyme in cyclophosphamide bioactivation and a potential biomarker for hematological malignancies. *The Pharmacogenomics Journal*, http://dx.doi.org/10.1038/tpj.2014.82, (Epub ahead of print).

Golubnitschaja, O., Kinkorova, J., & Costigliola, V. (2014). Predictive, preventive and personalised medicine as the hardcore of 'Horizon 2020': EPMA position paper. *The EPMA Journal*, *5*(1), 6. http://dx.doi.org/10.1186/1878-5085-5-6.

Inaba, H., Greaves, M., & Mullighan, C. G. (2013). Acute lymphoblastic leukaemia. *Lancet*, *381*, 1943–1955. http://dx.doi.org/10.1016/S0140-6736(12)62187-4.

Ingelman-Sundberg, M. (2005). The human genome project and novel aspects of cytochrome P450 research. *Toxicology and Applied Pharmacology*, *207*(2 Suppl.), 52–56. http://dx.doi.org/10.1016/j.taap.2005.01.030.

Jounaidi, Y., Chen, C. S., Veal, G. J., & Waxman, D. J. (2006). Enhanced antitumor activity of P450 prodrug-based gene therapy using the low Km cyclophosphamide 4-hydroxylase P450 2B11. *Molecular Cancer Therapeutics*, *5*(3), 541–555. http://dx.doi.org/10.1158/1535-7163.MCT-05-0321.

Kim, S., Jo, C. H., Park, J. S., Han, H. J., & Kim, G. H. (2011). The role of proximal nephron in cyclophosphamide-induced water retention: Preliminary data. *Electrolyte Blood Press: E & BP*, *9*(1), 7–15. http://dx.doi.org/10.5049/EBP.2011.9.1.7.

Ko, R. H., Ji, L., Barnette, P., Bostrom, B., Hutchinson, R., Raetz, E., et al. (2010). Outcome of patients treated for relapsed or refractory acute lymphoblastic leukemia: A Therapeutic Advances in Childhood Leukemia Consortium study. *Journal of Clinical Oncology: Official Journal of the American Society of Clinical Oncology*, *28*(4), 648–654. http://dx.doi.org/10.1200/JCO.2009.22.2950.

Lund, B., Asberg, A., Heyman, M., Kanerva, J., Harila-Saari, A., Hasle, H., et al. (2011). Risk factors for treatment related mortality in childhood acute lymphoblastic leukaemia. *Pediatric Blood & Cancer*, *56*(4), 551–559. http://dx.doi.org/10.1002/pbc.22719.

Madadi, P., Hildebrandt, D., Gong, I. Y., Schwarz, U. I., Ciszkowski, C., Ross, C. J., et al. (2010). Fatal hydrocodone overdose in a child: Pharmacogenetics and drug interactions. *Pediatrics*, *126*(4), e986–e989. http://dx.doi.org/10.1542/peds.2009-1907.

Manage, D. P., Ma, L., Lauzon, J., Howell, A., Belch, A. R., Mackey, J. R., et al. (2014). Genotyping single nucleotide polymorphisms in human genomic DNA with an automated and self-contained PCR cassette. *The Journal of Molecular Diagnostics*, *16*(5), 550–557. http://dx.doi.org/10.1016/j.jmoldx.2014.04.004.

Mehta, P. A., & Davies, S. M. (2004). Pharmacogenetics of acute lymphoblastic leukemia. *Current Opinion in Hematology*, *11*(6), 434–438.

Morrow, P. K., Serna, R., Broglio, K., Pusztai, L., Nikoloff, D. M., & Hillman, G. R. (2011). Effect of CYP2D6 polymorphisms on breast cancer recurrence. *Cancer*, *118*(5), 1221–1227. http://dx.doi.org/10.1002/cncr.26407.

Morrow, P. K., Serna, R., Broglio, K., Pusztai, L., Nikoloff, D. M., Hillman, G. R., et al. (2012). Effect of CYP2D6 polymorphisms on breast cancer recurrence. *Cancer*, *118*(5), 1221–1227. http://dx.doi.org/10.1002/cncr.26407.

Mura, S., Bui, D. T., Couvreur, P., & Nicolas, J. (2015). Lipid prodrug nanocarriers in cancer therapy. *Journal of Controlled Release*. http://dx.doi.org/10.1016/j.jconrel.2015.01.021, (Epub ahead of print).

Nishimura, M., Yaguti, H., Yoshitsugu, H., Naito, S., & Satoh, T. (2003). Tissue distribution of mRNA expression of human cytochrome P450 isoforms assessed by high-sensitivity real-time reverse transcription PCR. *Yakugaku zasshi: Journal of the Pharmaceutical Society of Japan*, *123*(5), 369–375.

Ortiz de Montellano, P. R. (2013). Cytochrome P450-activated prodrugs. *Future Medicinal Chemistry*, *5*(2), 213–228. http://dx.doi.org/10.4155/fmc.12.197.

Patterson, L. H., & Murray, G. I. (2002a). Tumour cytochrome P450 and drug activation. *Current Pharmaceutical Design*, *8*(15), 1335–1347.

Potter, G. A., Patterson, L. H., Wanogho, E., Perry, P. J., Butler, P. C., Ijaz, T., et al. (2002). The cancer preventative agent resveratrol is converted to the anticancer agent piceatannol by the cytochrome P450 enzyme CYP1B1. *British Journal of Cancer, 86*(5), 774–778. http://dx.doi.org/10.1038/sj.bjc.6600197.

Preiss, R., Schmidt, R., Baumann, F., Hanschmann, H., Hauss, J., Geissler, F., et al. (2002). Measurement of 4-hydroxylation of ifosfamide in human liver microsomes using the estimation of free and protein-bound acrolein and codetermination of keto- and carboxyifosfamide. *Journal of Cancer Research and Clinical Oncology, 128*(7), 385–392. http://dx.doi.org/10.1007/s00432-002-0335-4.

Preissner, S. C., Hoffmann, M. F., Preissner, R., Dunkel, M., Gewiess, A., & Preissner, S. (2013). Polymorphic cytochrome P450 enzymes (CYPs) and their role in personalized therapy. *PLoS one, 8*(12), e82562. http://dx.doi.org/10.1371/journal.pone.0082562.

Preissner, S., Kroll, K., Dunkel, M., Senger, C., Goldsobel, G., Kuzman, D., et al. (2010). SuperCYP: A comprehensive database on cytochrome P450 enzymes including a tool for analysis of CYP-drug interactions. *Nucleic Acids Research, 38*(Database issue), D237–D243. http://dx.doi.org/10.1093/nar/gkp970.

Pulley, J. M., Denny, J. C., Peterson, J. F., Bernard, G. R., Vnencak-Jones, C. L., Ramirez, A. H., et al. (2012). Operational implementation of prospective genotyping for personalized medicine: The design of the Vanderbilt PREDICT project. *Clinical Pharmacology and Therapeutics, 92*(1), 87–95. http://dx.doi.org/10.1038/clpt.2011.371.

Rodriguez-Antona, C., Jande, M., Rane, A., & Ingelman-Sundberg, M. (2005). Identification and phenotype characterization of two CYP3A haplotypes causing different enzymatic capacity in fetal livers. *Clinical Pharmacology and Therapeutics*, 77(4), 259–270.

Spear, B. B., Heath-Chiozzi, M., & Huff, J. (2001). Clinical application of pharmacogenetics. *Trends in Molecular Medicine*, 7(5), 201–204.

Terracini, B. (2011). Epidemiology of childhood cancer. *Environmental Health, 10*(Suppl. 1), S8. http://dx.doi.org/10.1186/1476-069X-10-S1-S8.

van der Weide, J., & Steijns, L. S. (1999). Cytochrome P450 enzyme system: Genetic polymorphisms and impact on clinical pharmacology. *Annals of Clinical Biochemistry, 36*(Pt. 6), 722–729.

CHAPTER FIVE

Xenobiotic and Endobiotic Mediated Interactions Between the Cytochrome P450 System and the Inflammatory Response in the Liver

Benjamin L. Woolbright, Hartmut Jaeschke[1]
Department of Pharmacology, Toxicology and Therapeutics, Kansas University Medical Center, Kansas City, Kansas, USA
[1]Corresponding author: e-mail address: hjaeschke@kumc.edu

Contents

Abstract

The liver is a unique organ in the body as it has significant roles in both metabolism and innate immune clearance. Hepatocytes in the liver carry a nearly complete complement of drug metabolizing enzymes, including numerous cytochrome P450s. While a majority

Advances in Pharmacology, Volume 74
ISSN 1054-3589
http://dx.doi.org/10.1016/bs.apha.2015.04.001

of these enzymes effectively detoxify xenobiotics, or metabolize endobiotics, a subportion of these reactions result in accumulation of metabolites that can cause either direct liver injury or indirect liver injury through activation of inflammation. The liver also contains multiple populations of innate immune cells including the resident macrophages (Kupffer cells), a relatively large number of natural killer cells, and blood-derived neutrophils. While these cells are primarily responsible for clearance of pathogens, activation of these immune cells can result in significant tissue injury during periods of inflammation. When activated chronically, these inflammatory bouts can lead to fibrosis, cirrhosis, cancer, or death. This chapter will focus on interactions between how the liver processes xenobiotic and endobiotic compounds through the cytochrome P450 system, and how these processes can result in a response from the innate immune cells of the liver. A number of different clinically relevant diseases, as well as experimental models, are currently available to study mechanisms related to the interplay of innate immunity and cytochrome P450-mediated metabolism. A major focus of the chapter will be to evaluate currently understood mechanisms in the context of these diseases, as a way of outlining mechanisms that dictate the interactions between the P450 system and innate immunity.

ABBREVIATIONS

APAP acetaminophen
CYP cytochrome P450
DAMP damage associated molecular pattern
HMGB1 high mobility group box-1
ICAM-1 intercellular adhesion molecule-1
NAPDH oxidase nicotinamide adenine dinucleotide phosphate oxidase
PAMP pathogen associated molecular pattern

1. INTRODUCTION

Numerous metabolic processes including metabolism of both endobiotics and xenobiotics take place in the liver. A large number of these reactions are carried out by the cytochrome P450 family of proteins (CYPs). While a majority of these processes result in largely nontoxic quantities of metabolites with favorable excretion profiles, a subportion of these reactions result in toxic compounds that can directly elicit liver damage. Other compounds generated from CYPs can result in activation of the inflammatory response that can either exacerbate, or help to rectify ongoing liver damage. Moreover, inflammation during liver injury can affect expression and activity of CYPs, potentially altering ongoing drug metabolism and affecting toxicity by reducing or enhancing metabolism. The goal of this chapter will be to discuss how CYP-mediated activation of a number of different compounds affects immune cells either natively present, or recruited to the liver, as well as how persistent inflammation and cytokine production can alter expression levels of CYPs. This chapter will focus on clinically relevant models of liver injury and inflammation as model systems by which we can discern specific mechanisms related to how the CYP system interacts with cytokines produced by inflammatory cells and the different immune cells. As the liver has a well-defined and critical role in innate immunity, the focus of the chapter will be on interactions with the innate immune system, although adaptive immunity will discussed in brief. As the CYP system is discussed in detail throughout this book, the innate immune system present in the liver will be discussed as a preface to the chapter.

2. INNATE IMMUNE CELLS IN THE LIVER

2.1 Kupffer Cells

While the liver's involvement in metabolism is well established, the liver also has a significant role in innate immunity via clearance of pathogens and pathogenic particles from sinusoidal blood. The primary mediator of this action is the endogenous macrophage present in liver sinusoids, the Kupffer cell (reviewed in Dixon, Barnes, Tang, Pritchard, & Nagy, 2013). Kupffer cells are the first line of defense against xenobiotics and immunoreactive material that enters the liver from the gut. Kupffer cells primarily respond to pathogenic material by identifying and phagocytizing the material, removing it from the blood. In addition to phagocytosis of pathogens, Kupffer cells can also remove dead and dying endogenous cellular material from the bloodstream, facilitating a final barrier against potentially hazardous material before it enters systemic circulation. While it has long been known that Kupffer cells recognize and phagocytose pathogens in the hepatic sinusoids, recent research has entailed new signaling pathways by which Kupffer cells can recognize and react to pathogen associated molecular patterns (PAMPs) such as lipopolysaccharide (LPS) (Pallson-McDermott and O'Neill, 2004). Kupffer cells express a number of toll-like receptor proteins (TLR1–9) that can bind bacterial or viral products such as LPS, CpG DNA, single stranded RNA, and more (reviewed in Akira & Takeda, 2004). The binding of LPS to TLR4 is a classic example of the initiation of this signaling pathway. LPS in serum is recognized by LPS binding protein and CD14, which then mediate the binding of this complex to TLR4 present on macrophages in the sinusoids (Chow, Young, Golenbock, Christ, & Gusovsky, 1999). This activates a signaling pathway that recruits the protein MyD88, an adaptor protein for TLRs (Kawai, Adachi, Ogawa, Takeda, & Akira, 1999); although MyD88 independent effects can also occur (Kawai et al., 2001). This signaling pathway results in the activation of NF-κB and the generation of numerous cytokines, chemokines, and other inflammatory mediators (Jiang, Akashi, Miyake, & Petty, 2000). This serves to both, activate other surrounding macrophages, hepatocytes, and innate immune cells, as well as recruit neutrophils and other peripheral mediators of innate immunity. In addition to PAMPs, Kupffer cells can also recognize damage associated molecular patterns (DAMPs) through these same TLRs. Release of DAMPs occurs both actively from activated macrophages (Tang et al., 2007) and passively from necrotic cells (Scaffidi, Misteli, & Bianchi, 2002), and largely activates the same signaling mechanisms via the same TLRs as the corresponding PAMP. Thus, activation of Kupffer cells can occur due to both increased exposure to pathogenic material and increased exposure to proinflammatory cellular debris during liver injury.

Kupffer cells produce significant amounts of the apoptotic cytokine tumor necrosis factor alpha (TNF-α). TNF-α-induced apoptosis occurs through activation of the death receptor TNF receptor 1 (TNFR1) on the surface of a diverse array of cells, including hepatocytes (Yoon & Gores, 2002). TNFR1 contains an intracellular death domain which can oligomerize and recruit multiple proteins such as Fas-associated death domain, TNF receptor associated death domain and procaspase 8 to form a death-induced cell signaling (DISC) complex (Kischkel et al., 1995, 2001). In type I cells, those cells that require no mitochondrial amplification, this initiates an apoptotic cascade featuring activation of subsequent caspase such as caspase-3, and eventual cleavage of DNA by caspase activated DNAses (reviewed in Chen & Goeddel, 2002). Although certain stimuli can trigger apoptosis through a type I mechanism (Schüngel et al., 2009), hepatocytes generally act as type II cells and require mitochondrial amplification in order to fully undergo apoptosis (Bajt, Lawson, Vonderfecht, Gujral, & Jaeschke, 2000; Scaffidi et al., 1998; Yin et al., 1999). Caspase-8 cleaves Bid protein, which then translocates to the mitochondria (Yin et al., 1999). This results in release of mitochondrial proteins such as smac/diablo and cytochrome c (Du, Fang, Li, Li, & Wang, 2000; Liu, Kim, Yang, Jemmerson, & Wang,

1996), which when combined with ATP form a complex called the apoptosome that can activate caspase 9 and initiate procaspase-3 cleavage and amplify the apoptotic signal (reviewed in Jaeschke & Lemasters, 2003). Thus, Kupffer cells are capable of releasing TNF-α to induce a potent apoptotic signal. TNF-α is the major cytokine triggering apoptosis during galactosamine/endotoxin-induced liver injury (Rothe et al., 1993) and may serve as a source of liver injury in other models as well.

Activation of Kupffer cells also results in the production of reactive oxygen species (ROS) through the production of superoxide by NADPH oxidase (Jaeschke, 2011; Jaeschke & Farhood, 1991). Kupffer cell-mediated oxidant stress occurs largely in the hepatic vasculature where the Kupffer cells are located; however, enough oxidant stress is generated to directly damage hepatocytes, as inactivation of Kupffer cells is protective against ischemia–reperfusion injury in the liver (Jaeschke, Bautista, Spolarics and Spitzer, 1991; Jaeschke & Farhood, 1991). Kupffer cells are also capable of generating peroxynitrite through the expression of inducible nitric oxide synthase (Billiar et al., 1990); although there is limited evidence for the role of reactive nitrogen compounds in the pathophysiology of hepatic ischemia–reperfusion injury (Jaeschke, Schini, & Farhood, 1992; Wang, Mathews, Guido, Farhood, & Jaeschke, 1995). As such, production of ROS from Kupffer cells by NADPH oxidase is presumed to be the primary cause of Kupffer cell-mediated liver injury in relevant models. Kupffer cells do contain small amounts of cytochrome P450 2E1 (Koivisto, Mishin, Mak, Cohen, & Lieber, 1996), which metabolizes a number of hepatotoxicants including acetaminophen (McGill & Jaeschke, 2013) and ethanol (Lieber, 1997), although most of the CYP2E1 activity is present in hepatocytes (Koivisto et al., 1996). Thus, the major contribution of Kupffer cells to these models of injury is likely a combination of their phagocytic ability, their capacity for ROS production, their ability to generate proapoptotic molecules such as TNF-α and their generation of other proinflammatory and antiinflammatory cytokines during liver injury.

2.2 Neutrophils

Polymorphonuclear leukocytes (PMNs), or neutrophils, are the major circulating population of leukocytes in the body. Neutrophils are generated in the bone marrow, but due to the large vascular volume in the sinusoids, and the liver being a low-flow organ, significant numbers of neutrophils are

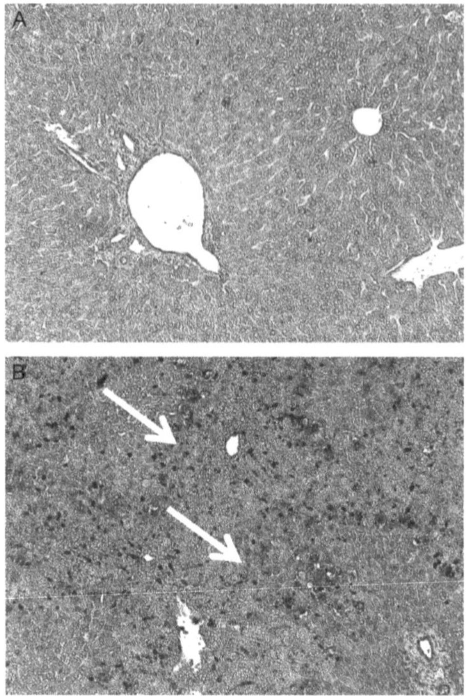

Figure 1 Ly6B-positive neutrophils present 24 h after control saline (A) or acetaminophen overdose in mice (B). Neutrophils localize to areas of injury in this model; however, neutrophils are constantly present in the sinusoids as in (A). White arrows points to areas of neutrophil accumulation.

present in the sinusoids at any given time (Fig. 1). Neutrophils generally do not induce liver injury while sitting in the hepatic vasculature, and can accumulate to significant degrees before any onset of injury (Chosay, Essani, Dunn, & Jaeschke, 1997). Neutrophil extravasation coincides with increases in liver injury parameters in a number of models, and as blockade of neutrophil extravasation or neutrophil recruitment is protective, neutrophils have been implicated in the injury process of a number of a different liver pathologies including obstructive cholestasis (Gujral, et al., 2003; Gujral, Liu, Farhood, Hinson, & Jaeschke, 2004; Yang et al., 2014), ischemia–reperfusion injury (Jaeschke, Farhood, & Smith, 1990; Jaeschke & Woolbright, 2012), nonalcoholic steatohepatitis (Harley et al., 2014), alcoholic hepatitis (Bertola, Park, & Gao, 2013), galactosamine/endotoxin-mediated liver injury (Jaeschke, Bautista, et al., 1991; Jaeschke, Farhood, et al., 1991; Jaeschke et al., 1998), and shock-induced liver trauma (Leung et al., 2014). Recruitment of neutrophils is thought to take place through multiple different mechanisms dependent on the etiology of the injury. Numerous pathologies feature release of cytokines and chemokines that can actively recruit neutrophils via their respective receptors (Bajt, Farhood, & Jaeschke, 2001; Lentsch, Yoshidome, Cheadle, Miller, & Edwards, 1998). DAMPs released from dying cells have been suggested to recruit neutrophils to areas of injury during liver intoxication as well (Imaeda et al., 2009). This process is thought to occur through DAMP mediated activation of a recently identified complex called the inflammasome, which helps mediate leukocyte recruitment in a number of models (Kubes & Mehal, 2012); although the degree to which these activities can contribute to liver injury remains controversial in certain models (Jaeschke, Williams, Ramachandran, & Bajt, 2012; Williams et al., 2011, Williams, Bajt, Farhood, & Jaeschke, 2010; Williams, Farhood, & Jaeschke, 2010). Activation of TLRs by DAMP products stimulates formation of an inflammasome complex. A classic example of this is the multiprotein complex containing apoptosis-associated speck-like protein containing a CARD, a NACHT, LRR, and PYD domains-containing protein 3 protein (NALP3), and caspase-1—or the NALP3 inflammasome (Schroder & Tschopp, 2010). The caspase-1 activity of this complex cleaves pro-IL-1ß to generate the active cytokine IL-1ß, which binds to IL-1 receptors and stimulates neutrophil recruitment. Once recruited to the hepatic sinusoids, neutrophils must extravasate into the hepatic parenchyma for neutrophil-mediated cell killing, which is typically mediated by interactions between adhesion molecules such as intercellular adhesion molecule-1 (ICAM-1) present on hepatocytes and sinusoidal endothelial cells, and $ß_2$ integrins present on neutrophils (Jaeschke & Smith, 1997). Extravasated neutrophils can exacerbate or initiate hepatic injury through release of reactive oxygen species generated by NADPH oxidase as well as degranulation and release of cytotoxic enzymes (Jaeschke, 2011). Neutrophils are known to express the enzyme myeloperoxidase, which can generate hypochlorous acid from hydrogen peroxide and chloride anions. HOCl modified proteins are detectable in a number of different liver pathologies with significant neutrophil involvement and are a direct marker of neutrophil-mediated oxidant stress and neutrophil-mediated liver injury (Gujral, Farhood, Bajt, Jaeschke, 2003; Gujral, Hinson, Farhood and Jaeschke, 2004;Gujral, Hinson, & Jaeschke, 2004; Hasegawa, Malle, Farhood, & Jaeschke, 2005; Üllen et al., 2013). Through release of ROS and degranulation, neutrophils create a potent localized oxidant stress that can kill hepatocytes (Jaeschke, 2011). The potential for a neutrophil-mediated hepatic cell death *in vivo* will be discussed in detail in future sections of this chapter, as well as, the biology and physiology behind neutrophil recruitment during multiple models of liver injury.

2.3 Natural Killer and Natural Killer T Cells

The liver also contains a significant number of natural kill cells (previously known as Pit cells) which act as an effector population in the liver (Godfrey, Hammond, Poulton, Smyth, & Baxter, 2000). This includes multiple subtypes such as NK, NKT, iNKT, and γδT (Lysakova-Devine & O'Farrelly, 2014). Natural killer cells are capable of producing multiple cytokines including interferon-gamma (IFN-γ), and TNF-α and additionally may be capable of inducing cell death in hepatic parenchymal cells via release of enzymes such as perforin and granzymes (Vermijlen et al., 1999). Release of a number of these cytokines can affect survival of other local effector populations, including neutrophils, by either inducing or inhibiting cell death in these populations via release of IL-4 or IFN-γ (Wang, Feng, Park, Yin, & Gao, 2013). The effect of NK and NKT cells on liver injury is an ongoing topic of research (Tian, Chen, & Gao, 2013.)

2.4 Other Immune Cells of the Liver

Dendritic cells are an antigen presenting class of innate immune cells and classically thought of as a link between innate immunity and adaptive immunity. While their role in liver toxicity has not been studied to the same extent as other innate immune populations, a few studies have been done that indicate dendritic cells may have a role in certain P450-mediated liver injuries, and subsequent inflammation. Depletion of dendritic cells can exacerbate acetaminophen toxicity (Connolly et al., 2011). This is likely due to increased formation of cytokines such as IL-6 shown to be proregenerative in human patients (Antoniades et al., 2006). It should be noted that dendritic cell depleted mice had higher levels of neutrophils in their nonparenchymal fraction after dendritic cell depletion; however, this was excluded a cause of increased injury (Connolly et al., 2011). Myeloid derived suppressor cells (MDSC) may also be involved in some forms of drug-induced immune-mediated liver injury as a recent study indicates that MDSCs may link innate and adaptive immunity and play a key role in immune tolerance to molecules such as halothane (Chakraborty et al., 2015). Further work is required in these cell populations before a full understanding of how they function and how they affect interactions between P450s and the innate immune system.

3. CROSS-TALK BETWEEN LIVER CYPs AND INFLAMMATION AFTER EXPOSURE TO TOXICANTS

3.1 Alcoholic Liver Injury

Excessive consumption of alcohol is a major source of morbidity and mortality, especially in the West (Gao & Bataller, 2011). Alcoholic liver injury is a progressive injury with a majority of patients experiencing simple steatosis, and no further liver dysfunction. A subpopulation of these patients undergo a profound change from largely benign steatosis to an inflammatory state called steatohepatitis, which can then progress to fibrosis, cirrhosis, and eventually cancer. At any point during this progression, patients can also enter a clinically distinct syndrome referred to as alcoholic hepatitis, wherein the liver goes into an acute inflammatory state with severe liver dysfunction and liver injury. As bouts of alcoholic hepatitis occur simultaneously with ongoing steatohepatitis, it is presumed that alcoholic hepatitis can help drive the injury process toward fibrosis and cirrhosis. A significant quantity of research has focused on how metabolism of alcohol by cytochrome P450s and other drug metabolizing enzymes stimulates the onset of alcoholic steatohepatitis and alcoholic hepatitis.

The role of cytochrome P450 enzymes in alcohol-induced liver injury has been somewhat controversial, especially that of CYP2E1 (Tsukamoto, 2000). Alcohol is metabolized primarily via alcohol- and aldehyde dehydrogenase to the relatively harmless chemical acetic acid (Lieber, 2005); although a portion of alcohol is metabolized by CYP2E1. Protein levels of CYP2E1 are also induced by alcohol both in the liver (Lu & Cederbaum, 2008) and in the gut (Hakkak, Korourian, Ronis, Ingelman-Sundberg, & Badger, 1996). CYP2E1 metabolizes millimolar quantities of ethanol into micromolar quantities of acetaldehyde, a more reactive and more toxic intermediate. During this process CYP2E1 also produces ROS that may contribute to the total ethanol-induced oxidative stress (Lu & Cederbaum, 2008). Autoantibodies against CYP2E1 have been found in alcoholic patients suggesting a potential proimmunogenic role (Clot et al., 1996). However, there was surprisingly little difference in liver histology and serum AST levels between CYP2E1$^{-/-}$ mice and their WT counterparts when ethanol was administered via intragastric feeding (Kono et al., 1999). Moreover, administration of a cytochrome P450 inhibitor did not affect oxidative stress, indicating the vast majority of oxidative stress occurs independently of cytochrome P450 function (Isayama et al., 2003). Instead, most of the oxidative stress is likely caused by other sources such as NADPH oxidase (Kono et al., 2001) and inducible nitric oxide synthase (McKim et al., 2003) expressed largely in inflammatory cells such as Kupffer cells and recruited neutrophils (Arteel, 2003). Thus, most likely, CYP2E1 contributes minimally to oxidative stress during chronic alcohol exposure.

Despite these findings, recent information suggests CYP2E1 might play a role during binge ethanol intake (Abdelmegeed et al., 2013; Yun et al., 2014). Binge ethanol intake is increasingly noted as a likely contributor to alcoholic liver injury both in experimental models (Bertola et al., 2013; Ding et al., 2010) and in clinical practice (Plunk, Syed-Mohammed, Cavazos-Rehg, Bierut, & Grucza, 2014). Inhibition of CYP2E1 limits accumulation of triglycerides and reduces steatosis during binge alcohol treatment (Wu, Wang, Zhou, Yang, & Cederbaum, 2012). CYP2E1 has also been presumed to enhance injury during acute binge after chronic exposure (Bertola et al., 2013). As neutrophil recruitment is limited without administration of an additional binge after chronic administration of alcohol, metabolism through CYP2E1 may be linked to neutrophil recruitment during alcoholic liver injury in murine models, although this hypothesis remains untested. As neutrophil accumulation is a well-known aspect of the human pathophysiology both in patients with alcoholic hepatitis and alcoholic cirrhosis rodent models that feature neutrophil accumulation may more accurately recapitulate the human pathophysiology (Mathews, Xu, Wang, Bertola, & Gao, 2014). Given that binge alcohol is likely to overwhelm normal enzymatic capacity and increase the contribution of alcohol inducible CYP2E1 to alcohol metabolism, future studies investigating the effect of CYP2E1 on binge and chronic + binge-induced alcoholic liver injury may be warranted, especially in the context of advanced disease states, such as alcoholic hepatitis. This section is summarized in Fig. 2.

3.2 Acetaminophen-Induced Liver Injury

In contrast to alcohol, acetaminophen (APAP) is a classic example of cytochrome P450-mediated formation of a reactive metabolite (McGill & Jaeschke, 2013; Potter et al., 1973). While the majority of acetaminophen is either glucuronidated or sulfated and subsequently excreted, a subportion is metabolized by cytochrome P450s, largely CYP2E1, CYP2A1, and CYP3A4, into the reactive metabolite *N*-acetyl-p-benzoquinone imine (NAPQI) (McGill & Jaeschke, 2013). As formation of the reactive metabolite NAPQI is the initiating step for liver injury, drug metabolism becomes

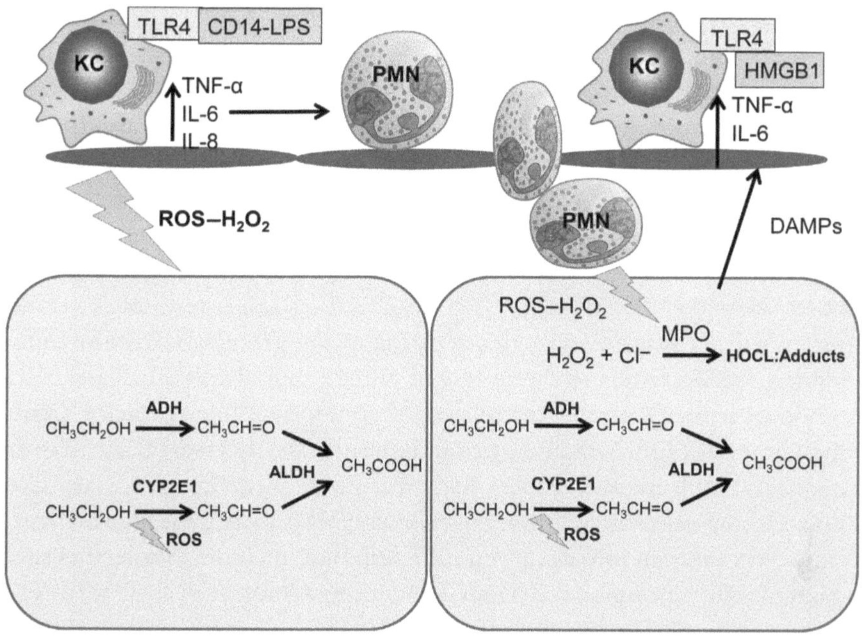

Figure 2 A simple model of alcohol-induced liver injury. A majority of alcohol is metabolized by alcohol dehydrogenase to acetaldehyde and then by aldehyde dehydrogenase into acetic acid. A subportion of alcohol may be metabolized by CYP2E1 during binge alcohol dosing. This causes intracellular stress due to excessive acetaldehyde production. Kupffer cell activation by LPS and other DAMPs results in additional oxidative stress, as well as release of cytokines that recruit cytotoxic neutrophils. Neutrophils use NADPH to generate superoxide radical and hydrogen peroxide. Hydrogen peroxide is converted to hypochlorous acid by myeloperoxidase, which can then adduct proteins and cause significant oxidative stress. The combined oxidant stress results in hepatocyte death. TNF-α, tumor necrosis factor-α; IL, interleukin; TLR, toll-like receptor; KC, Kupffer cell; LPS, lipopolysaccharide; ADH, alcohol dehydrogenase; ALDH, aldehyde dehydrogenase; DAMP, damage associated molecular pattern; H_2O_2, hydrogen peroxide; MPO, myeloperoxidase; HOCl, hypochlorous acid; PMN, polymorphonuclear leukocytes; HMGB1, high mobility group box-1; ROS, reactive oxygen species.

a critical point in the assessment of APAP-induced liver injury (Jaeschke, McGill, Williams, & Ramachandran, 2011). Inhibition of drug metabolism by potential therapeutics will prevent all future downstream sequelae, and mask actual drug effects, preventing a realistic assessment of the pharmacological action of the potential therapeutic; thus, studies where inhibition of drug metabolism occurs must be interpreted with caution. Production of NAPQI results in adduction of mitochondrial proteins by NAPQI and initiates mitochondrial oxidant stress that causes a significant portion of the injury (Dahlin, Miwa, Lu, & Nelson, 1984; Jaeschke, McGill, & Ramachandran, 2012). This initial oxidant stress leads to the activation (phosphorylation) of c-jun N terminal kinase (JNK) (Gunawan et al., 2006). The phosphorylated form of JNK translocates to mitochondria and substantially enhances the mitochondrial oxidant stress (Hanawa et al., 2008; Saito, Lemasters, & Jaeschke, 2010). The amplified mitochondrial oxidant stress results in opening of the mitochondrial membrane permeability transition pore and initiates cellular necrosis (Gujral, Knight, Farhood, Bajt, & Jaeschke, 2002; Kon, Kim, Jaeschke, & Lemasters, 2004). Liver injury is further magnified by translocation of the protein Bax to the mitochondria, which results in the release of mitochondrial endonucleases, such as endonuclease G, as well as release of apoptosis-inducing factor (AIF), which amplifies DNA fragmentation and enhances necrosis (Bajt, Cover, Lemasters, & Jaeschke, 2006; Bajt, Farhood, Lemasters, & Jaeschke, 2008). During necrosis, a number of cellular constituents are leaked out from hepatocytes that can initiate an immune response, including molecules such as high mobility group box-1 (HMGB1) protein (Antoine et al., 2012), ATP (Hoque et al., 2012), mitochondrial DNA (McGill et al., 2012), nuclear DNA fragments (McGill et al., 2012), and more. As these molecules exacerbate inflammation through binding of DAMP receptors such as TLR4, it has been proposed that APAP contains a second phase of injury that is independent of drug metabolism, and instead occurs through the activation of the innate immune system (Imaeda et al., 2009; Kubes & Mehal, 2012; Liu, Han, Gunawan, & Kaplowitz, 2006). While a majority of these mechanisms on intracellular dysfunction during APAP are widely accepted in the literature, there is considerable controversy over whether or not the inflammatory cascade that follows the initial cell death mediates any portion of APAP-induced liver injury (Imaeda et al., 2009; Jaeschke, 2008; Lawson, Farhood, Hopper, Bajt, & Jaeschke, 2000; Liu et al., 2006; Williams et al., 2014), even though it has been established for over 30 years that modulators of inflammation can have profound effects on CYP activity (Renton & Dickson, 1984). While it is understood that inflammation and inflammatory mediators can affect APAP metabolism through drug metabolizing enzymes such as CYP2E1, there remains substantial debate over whether or not these effects are critical to the actual pathogenesis or secondary effects (Feng et al., 2014; Jaeschke, Williams et al., 2012; Martin-Murphy, Kominsky, Orlicky, Donohue, & Ju, 2013).

Kupffer cells have been repeatedly suggested to be a major component of acetaminophen-induced liver injury over the last 20 years (Choi, Ban,

Kim, & Hong, 2015; Michael, Pumford, Mayeux, Niesman, & Hinson, 1999), although this topic remains controversial due to conflicts in the data (Ito et al., 2004; Ju et al., 2002; You et al., 2013). Pretreatment of mice with gadolinium chloride, which reduces the capacity of Kupffer cells to produce ROS (Liu et al., 1995), was shown to protect against APAP (Michael et al., 1999); however, pretreatment with clodronated liposomes, a more specific macrophage toxin, has opposite results (Ju et al., 2002). Kupffer cells were presumed to contribute to injury via production of ROS via NADPH oxidase (Michael et al., 1999). However, mice with deficiency of NADPH oxidase activity had similar levels of oxidative stress compared to WT mice after APAP, indicating Kupffer cell-mediated oxidative stress contributed minimally to APAP-induced liver (James, McCullough, Knight, Jaeschke, & Hinson, 2003). Some of this confusion may be due to interactions between treatments and drug metabolism. It was recently proposed that APAP-induced liver injury is aggravated by Kupffer cells-derived cytokines induced through the purinergic receptor P2X7, as blockade with the inhibitor A438079 was protective against APAP (Hoque et al., 2012); however, it was subsequently demonstrated that this inhibitor functions as a direct inhibitor of P450 activity, indicating the actual protective effect was mediated by inhibition of drug metabolism (Xie et al., 2013). Recent studies indicating a role for macrophages have also failed to fully examine drug metabolism effects related to treatment regimes, and thus, must be interpreted cautiously (Choi et al., 2015). Together, the preponderance of evidence is in favor of intracellular mechanisms of injury mediating a majority of APAP-induced liver injury (Jaeschke, McGill et al., 2012).

Recent studies in man have illustrated a separate role for macrophages during APAP-induced liver injury. Studies in human patients indicate the macrophage population is expanded in patients, especially at sites of necrosis, and is derived of both endogenous Kupffer cells that are initially depleted after APAP overdose, and recruited monocytes that differentiate into macrophages (Antoniades et al., 2012). These macrophages have a largely antiinflammatory, or Type-II, cytokine profile with high levels of expression of cytokines such as IL-10 and secretory leukocytes protease inhibitor (Antoniades et al., 2014). Moreover, these cells stimulate angiogenesis and while reducing the activation of status of neutrophils, which serves to regenerate new hepatic architecture and limit excess inflammatory damage (Zigmond et al., 2014). Thus, any potential Kupffer cell-mediated injury during APAP overdose likely plays a secondary role to regenerative effects seen as the injury resolves.

Another function of recruited monocytes and Kupffer cells during APAP overdose is the recruitment of other inflammatory cells. Both recruited monocytes and Kupffer cells are capable of producing multiple cytokines including TNF-α, IL-6, and more (Dixon et al., 2013). Neutrophils are known to accumulate at sites of injury after APAP in murine models of APAP overdose (Fig. 1). Initial reports indicated a lack of a role for neutrophils after APAP-induced liver injury (Lawson et al., 2000); however, this was challenged when it was shown that pretreatment with a neutrophil depleting antibody (anti-Gr-1) was protective against APAP (Liu et al., 2006). However, the interpretation of these data has been questioned, as pretreatment with the Gr-1 antibody can cause adaptive changes with induction of protective genes (Jaeschke & Liu, 2007). In addition, ICAM-$1^{-/-}$ mice were not protected (Cover et al., 2006). This has led to significant controversy in the field regarding the role of neutrophils (Jaeschke, Williams et al, 2012; Krenkel, Mossanen, & Tacke, 2014; Kubes & Mehal, 2012). While evidence in both human patients (Williams et al., 2014) and mice (Connolly et al., 2011; Cover et al., 2006; Lawson et al., 2000; Williams, Bajt et al., 2010) does not support the hypothesis of neutrophil-mediated injury, a number of papers have shown a link between metabolism of APAP and inflammation that might explain why some models show modulation of injury with therapeutic regimens designed at mitigating inflammation. Short-term treatment with IL-22 protected mice from APAP-induced liver injury through a signal transducer and activator of transcription 3-dependent mechanism, suggesting intracellular signaling mechanisms were present that attenuated the injury (Feng et al., 2014). On the other hand IL-22 transgenic mice that chronically overexpress IL-22 had increased expression of CY2E1, which lead to an exacerbation of the injury (Feng et al., 2014). Similarly, Jα-$18^{-/-}$ mice and $CD1d^{-/-}$ mice, which are largely depleted of iNKT and NKT cells, respectively, both had higher levels of injury after APAP overdose (Martin-Murphy et al., 2013). This was attributed to higher expression of CYP2E1, due to higher ketone body levels present in these mice (Martin-Murphy et al., 2013). This was in contrast to previous data using NK cell depleting antibodies that suggested NK cells might exacerbate APAP-induced liver injury (Liu, Govindarajan, & Kaplowitz, 2004); although the involvement of NK/NKT cells was triggered by the use of DMSO as vehicle for APAP (Masson, Carpenter, Graf, & Pohl, 2008). Even so, further studies showed protection against APAP-induced liver injury in Jα-$18^{-/-}$ mice due to higher glutathione levels (Downs, Aw, Liu, Adegboyega, & Ajuebor, 2012). The contrast in these studies points out two critical issues: first, assessments of metabolism are critical to the understanding of therapeutic potential, as even seemingly unrelated interventions can have profound effects on metabolism that might otherwise mask interpretation of data and second, studies performed in knockout mice must be carefully controlled and interpreted as contradictory data can occur in these models due to adaptive mechanisms. Moreover, chronic depletion of inflammatory cells can have completely different effects than immediate antibody-based depletion schemes. Investigations using multiple models of depletion, as well as overexpression systems, if relevant, are advised due to potentially unforeseen alterations in basic metabolism. This section is summarized in Fig. 3.

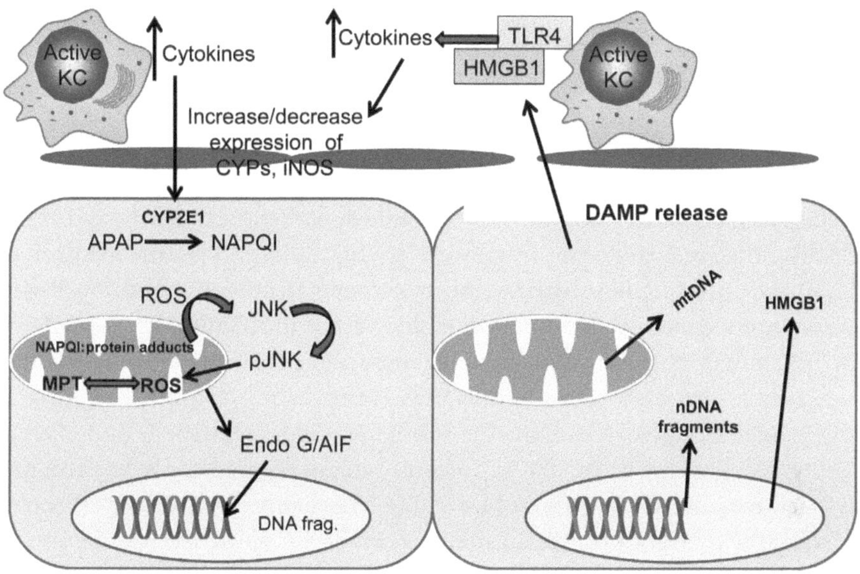

Figure 3 A simplified model of APAP toxicity. CYP2E1 converts APAP to NAPQI resulting in an intracellular oxidant stress. This results in phosphorylation of JNK and translocation of pJNK to the mitochondria. Translocation of pJNK amplifies the mitochondrial stress and causes the mitochondrial permeability transition pore. This results in the release of nucleases such as endonuclease G and apoptosis-inducing factor that cause DNA fragmentation. This results in cell death, causing release of DAMPs such as mitochondrial DNA, nuclear DNA fragments, and HMGB1. These DAMPs can then act on Kupffer cells, which secrete cytokines that can act on CYP2E1 by either reducing or increasing expression, resulting in magnification or reduction of the injury to subsequent hepatocytes. KC, Kupffer cell; APAP, acetaminophen; NAPQI, *N*-acetyl-*p*-benzoquinone imine; JNK, c-jun N terminal kinase; Endo G, endonuclease G; DAMPs, damage associated molecular patterns; CYPs, cytochrome P450; AIF, apoptosis-inducing factor; nDNA, nuclear DNA; mtDNA, mitochondrial DNA; HMGB1, high mobility group box-1; TLR, toll-like receptor; MPT, mitochondrial permeability transition pore; iNOS, inducible nitric oxide synthase; DNA frag., DNA fragments.

3.3 Halothane Hepatitis, *Isoniazid*, and Idiosyncratic Drug-Induced Liver Injury

Halothane is a volatile anesthetic currently used predominantly in developing countries, due to its low cost (Habibollahi et al., 2011). Unlike a number of the other volatile anesthetics, halothane induces drug-induced liver injury only in a subpopulation of patients (Uetrecht, 2009). Halothane-induced hepatitis is thought to be mediated by the conversion of halothane to trifluoroacetyl chloride (TFA) by CYPs, predominantly CYP2E1 (Bourdi, Amouzadeh, Rushmore, Martin, & Pohl, 2001). CYP2E1 has also been proposed as a cell surface autoantigen in murine models that exacerbates the inflammatory response (Eliasson & Kenna, 1996). Accumulation of TFA in tissue is thought to be partially responsible for the adaptive immune response that results in immune-mediated liver injury (Bourdi et al., 2001); thus, CYP activity is a major determinant of halothane-induced liver injury. This immune response is strongly associated with previous exposure to halothane, leading to the "hapten hypothesis," or the idea that haptens—protein-drug conjugates that are formed during halothane metabolism—can develop during repeated halothane exposure, and result in this immune response (reviewed in Ju, 2009). The mechanism of halothane-induced hepatitis has been somewhat controversial, both due to its unpredictability in animal models and the problems with identifying the inflammatory cells that are directly responsible for liver injury. Although some studies have proposed neutrophils as the main cytotoxic leukocyte in this model (Dugan, MacDonald, Roth, & Ganey, 2010; You, Cheng, Reilly, Wegmann, & Ju, 2006), recent evidence suggested a limited role for neutrophils, with eosinophils being the likely contributor (Proctor et al., 2013). Eosinophil recruitment is driven by epithelial production of thymic stromal lymphopoietin (Proctor et al., 2014). More studies are required in this area, especially in regards to metabolism of halothane to its metabolites, to fully understand what drives halothane-mediated hepatitis, especially in vulnerable subpopulations.

Halothane-induced hepatitis is a classic example of idiosyncratic drug-induced liver injury (IDILI), which is defined primarily by two different facets: one, the injury is very rare and unpredictable (idiosyncratic) and two, the injury usually only occurs on repeat exposure to the drug (Uetrecht, 2009). IDILI is especially troubling as a clinical manifestation of liver injury as the mechanisms are poorly defined, almost impossible to predict, and diagnosis can be difficult due to the lack of specific parameters. IDILI is largely considered to be an immune-mediated pathophysiology rather than an injury caused by the parent drug, although substantial debate still occurs as to whether adaptive immunity fully explains the mechanism (Uetrecht, 2009). Evidence is favor of an immune response that largely comes from clinical data. Even in patients who undergo mild increases in ALT after exposure to a drug, a majority of patients will undergo adaptation and ALT levels will return to normal (Watkins, 2005). What is unknown is why some patients undergo continued liver injury, with immune intolerance being the most common explanation (Eksteen, Afford, Wigmore, Holt, & Adams, 2007). Upon rechallenge with a drug, these patients undergo a rapid adaptive immune response that results in DILI. Despite the current lack of understanding of the immune element, what is largely agreed upon is that metabolic activation of the drug by CYPs and other drug metabolizing enzymes probably plays a key role, and that autoantibodies against drug metabolizing enzymes are present during DILI in some cases (Mizutani et al., 2005). Although there is only limited evidence for the pathogenic role for these autoantibodies in a majority of IDILI type injuries, their presence is an independent predictor of outcome in patients with numerous drug-mediated liver injuries (Sutti, Rigamonti, Vidali, & Albano, 2014). A recent study using a novel model of drug (halothane)-induced allergic hepatitis suggests that autoantibodies against protein adducts may be critical to liver injury in IDILI (Chakraborty et al., 2015). Further research in this area is needed, as both removal of pharmaceuticals from the market due to previously undescribed DILI, and clinical liver injury, can be attributed to IDILI and our current lack of understanding of this topic.

The other most commonly observed form of IDILI is due to isoniazid toxicity (Metushi, Cai, Zhu, Nakagawa, & Uetrecht, 2011). Isoniazid is a first line drug for treatment of tuberculosis, and is commonly administered in combination with rifampin to enhance efficacy. Isoniazid carries a number of commonalities with halothane, including delayed increases in serum transaminases, delayed liver injury, a small subportion (~20%) of susceptible individuals in the population, the idea that the major toxicity is due to a reactive metabolite (either acetylhydrazine or hydrazine in the case of isoniazid) generated by P450s, and the idea of liver "adapation" to the drug (Metushi et al., 2011). While the exact mechanism of toxicity of isoniazid is very poorly understood, it is thought that one of the reactive metabolites results in activation of the immune system and immune-mediated liver injury (Metushi et al., 2011). Isoniazed administration results in autoantibodies against both isoniazid and CYP2E1 in human patients and is associated with increases in Th17 and IL-10 cytokines (Metushi, Cai, Vega, Grant, &

Uetrecht, 2014; Metushi, Sanders, Acute Liver Study Group, Lee, & Uetrecht, 2014), all indicating a complex inflammatory environment after administration of isoniazid. It remains to be determined how metabolism of isoniazid by P450s results in formation of reactive metabolites that stimulate the immune system and block immune tolerance. Additionally, a greater understanding of which immune populations mediate the injury is required for further progress in the area.

4. CROSS-TALK BETWEEN INFLAMMATION AND LIVER CYPs BY ENDOGENOUS MACROMOLECULES

4.1 Cholestasis

While the cytochrome P450 family is well characterized in its role in drug metabolism, a substantial amount of P450 activity is involved in metabolism of endogenous macromolecules. With the recent discovery of the farnesoid X receptor (FXR) as a nuclear receptor for bile acids, increased focus has been placed on bile acids as signaling molecules (Makishima et al., 1999; reviewed in Li & Chiang, 2015). Generation of bile acids is dependent upon multiple CYP-mediated processes, with the rate limiting step being the conversion of cholesterol to 7α hydroxyl-cholesterol by CYP7A1 (reviewed in Li & Chiang, 2015). A number of different events can result in the regulation of CYP7A1, including activation of FXR by bile acids (Chiang, 2009), activation of FGFR4 by its ligand FGF15/FGF19 (Inagaki et al., 2005), and down-regulation by increased exposure of hepatocytes to cytokines (Feingold, Spady, Pollock, Moser, & Grunfeld, 1996). This interaction between cytokine levels and CYP7A1 was first noted in macrophages (Feingold et al., 1996), although it has subsequently been identified to occur in human hepatocytes as well (Li, Jahan, & Chiang, 2006). This might be species dependent though, as some cytokines such as TGF-ß increased CYP7A1 levels in rats (Li, Ma, & Chiang, 2008). As inflammation is a noted aspect of cholestasis (Gujral et al., 2003; Saito & Maher, 2000; Woolbright & Jaeschke, 2012; Woolbright et al., 2013, 2014), and *in vivo* models of cholestatic liver injury such as the bile duct ligation (BDL) model have increased expression of numerous cytokines during the initial injury period (Gujral, Liu et al., 2004), it is probable that cytokines help to protect the liver from excess accumulation bile acids in hepatocytes during cholestasis. Inflammation is a critical regulator of cholestatic liver injury in multiple murine models (Gujral et al., 2003; Kodali, Wu, Lahiji, Brown, & Maher, 2006) and is present even in models where it does not directly mediate the injury (Woolbright et al., 2014), where it is accompanied by significant cytokine production (Woolbright et al., 2014). BDL causes a dramatic increase in multiple serum cytokine levels between 3 and 72 h after BDL (Gujral, Liu et al., 2004; Saito & Maher, 2000). These increases occur concurrently with retention of biliary constituents including bile acids (Zhang et al., 2012). Thus, both the retention of bile acids during the actual cholestasis and the subsequent production of cytokines by local inflammatory cells downregulate CYP7A1 and other CYPs such as CYP27A1 in a redundant fashion. Recently, it was noted that administration of bile acids to hepatocytes can increase the expression of chemokines, including mouse keratinocyte factor and macrophage inhibitory protein-2 in an FXR independent fashion (Allen, Jaeschke, & Copple, 2011). This might serve as a further redundancy through paracrine actions of cytokines produced directly by hepatocytes. Thus, while established mechanisms of FXR signaling during cholestasis are likely critical to the compensatory responses associated with cholestatic hepatocytes, the induction of cytokines and their subsequent paracrine response on hepatocytes serve as an additional mechanism of regulation of bile acid levels through modulation of CYP expression.

In addition to the response from hepatocytes, Kupffer cells may help to control bile acid homeostasis during cholestasis as well. Kupffer cells express the bile acid receptor TGR5, which can activate anti-inflammatory cascades when activated by bile acids (Keitel, Donner, Winandy, Kubitz, & Häussinger, 2008; Wang, Chen, Yu, Forman, & Huang, 2011). The presence of these anti-inflammatory cytokines may help limit excessive inflammation, and normalize CYP7A1 levels. TGR5 may also help to directly limit increases in intrahepatic bile acid levels as whole body TGR5 knockout mice had increased liver and plasma bile acid levels after partial hepatectomy, although a direct interaction between TGR5 and CYP7A1 was not shown (Pean et al., 2013). The role of TGR5 on macrophages, hepatocytes, and sinusoidal endothelial cells in the liver and their regulation of inflammation and the CYP system is an ongoing area of research.

4.2 Lipopolysaccharide

LPS or endotoxin is a component of cell walls of gram-negative bacteria that increases in concentration in serum during alcoholic liver injury, and in septic shock. LPS activates TLR4 present on inflammatory cells to initiate a potent inflammatory response and induction of multiple cytokine expression cascades (Chow et al., 1999). LPS administration reduces both CYP7A1 and

CYP27A1 levels in syrian hamsters (Feingold et al., 1996; Memon, Moser, Shigenaga, Grunfeld, & Feingold, 2001). This may be through the previously mentioned effects of cytokines on CYP7A1 and CYP27A1 expression, or may be due to direct effects on FXR (Kim et al., 2003). LPS also reduces levels of hepatobiliary transporters such as the sodium taurocholate cotransporting polypeptide indicating the FXR response as a transcriptional regulator is likely a part of the effect (Green, Beier, & Gollan, 1996). LPS also affects hepatic levels of specific drug metabolizing enzymes. Administration of, even very low quantities, LPS to human patients results in decreases in hepatic CYP activity (Shedlofsky, Israel, McClain, Hill, & Blouin, 1994). This has recently been modeled in human hepatocyte cell culture, as Hep-aRG cells, a bipotential line of hepatoma cells that express drug metabolizing enzymes including CYPs, are responsive to LPS, and treatment with LPS results in downregulation of CYPs including CYP1A2, CYP2B6, and CYP3A4 (Rubin et al., 2015). Administration of IL-6 had similar effects, indicating cytokines are also likely capable of affecting drug metabolism (Rubin et al., 2015). LPS also protects against the aforementioned APAP-induced liver injury (Liu et al., 2000). Administration of LPS before APAP reduces CYP expression through cytokines such as IL-1α that in turn blocks drug metabolism and protects the mouse against injury (Liu, Sendelbach, Parkinson, & Klaassen, 2000). Thus, chronic inflammation related to septic infections can result in alterations in CYP levels that can affect normal metabolic processes and drug metabolism.

5. CYPs IN IMMUNE CELLS AND THEIR ROLE IN INFLAMMATION

While a majority of the CYP activity is present in epithelial tissue, with hepatocytes expressing and maintaining significant activity, small amounts of CYP activity is present in immune cells themselves. While this expression is quite modest in most cases, there are a few examples of CYPs in immune cells potentially affecting inflammation. Most notably, immune cells express the aryl hydrocarbon receptor (AhR) and respond to prototypical AhR ligands such as 2,3,7,8-tetrachlorodibenzo-*p*-dioxin (TCDD) (Germolec, Adams, & Luster, 1995) or polyaromatic hydrocarbons (Houser, Raha, & Vickers, 1992) with an increase in CYP1A1 expression (Germolec et al., 1995). This induction is selective for specific lymphoid tissues and only occurs in those that express high concentrations of AhR (Germolec et al., 1995). TCDD is a noted immunosuppressive agent with numerous toxic actions in mice (Kerkvliet et al., 2002). Recently it was shown that not all of the actions of AhR might be immunosuppressive though, as activation of AhR regulates T_{reg} and T_h17 cell formation in a ligand-dependent manner (Quintana et al., 2008; Veldhoen et al., 2008). While activation of CYP1A1 was observed in all mice with functional AhR in these experiments, the role of CYP1A1 in this process has not been directly investigated. Assessment of the role of CYPs in the induction of T cell differentiation caused by TCDD and other AhR ligands may provide fruitful information on the relatively untested idea that CYPs commonly recognized for drug metabolism might also have a role in endogenous metabolic activity.

6. MITOCHONDRIA SPECIFIC P450s AND THEIR ROLE IN XENOBIOTIC AND ENDOBIOTIC METABOLISM AND INFLAMMATION

Mitochondria in liver express a number of CYP families, including 24A and 27A involved in steroidal biogenesis (Omura, 2006). While a majority of xenobiotic metabolism occurs through microsomal P450s, mitochondria express small quantities of xenobiotic metabolizing P450s including CYP2E1, 2D6, 1A1, and more (Anandatheerthavarada et al., 1997; Avadhani, Sangar, Bansal, & Bajpai, 2011). Mitochondrial CYP2E1 is likely generated in the same fashion as microsomal CYP2E1 and instead targeted to the mitochondria posttranslationally and it appears that like microsomal CYP2E1, the mitochondrial form is inducible (Avadhani et al., 2011). Mitochondrial CYP2E1 may be important to both alcoholic liver injury and acetaminophen-induced liver injury, as a model where mitochondrial CYP2E1 is induced exclusively *in vitro* resulted in both APAP and ethanol-induced hepatotoxicity (Knockaert, Fromenty, & Robin, 2011). Given the importance of CYPs such as CYP2E1 to drug-induced toxicity and subsequent inflammation, a role for mitochondrial 2E1 may develop in both of these toxicities.

7. CONCLUSION

In summary, interactions occur between immune cells, cytokine production and the cytochrome P450 system that mediate a number of different aspects of liver disease and immunology. While a considerable amount of effort has gone into the delineation of these mechanisms, there is a great deal of work left to be done in this field. Experimental models of alcoholic liver injury and APAP hepatotoxicity provide excellent, clinically relevant models to better understand how specific CYPs such as CYP2E1 can interact with immunity. At the same time, cytokine production and bile acid retention during cholestasis has obvious implications for feedback-mediated effects of endogenous metabolism through CYP7A1. Continued research in these areas focusing on novel aspects of how CYPs interact with different inflammatory states will yield novel information that can improve patient safety and therapeutic efficacy of drugs.

CONFLICT OF INTEREST

The authors have no conflicts of interest to declare.

ACKNOWLEDGMENTS

Work in the authors' laboratory was supported in part by the National Institutes of Health grants R01 DK070195 and DK102142, and by grants from the National Center for Research Resources (5P20RR021940-07) and the National Institute of General Medical Sciences (8 P20 GM103549-07) from the National Institutes of Health. Dr. B.L.W. was supported by the "Training Program in Environmental Toxicology" T32 ES007079-26A2 from the National Institute of Environmental Health Sciences.

REFERENCES

Abdelmegeed, M. A., Banerjee, A., Jang, S., Yoo, S. H., Yun, J. W., Gonzalez, F. J., et al. (2013). CYP2E1 potentiates binge alcohol-induced gut leakiness, steatohepatitis, and apoptosis. *Free Radical Biology and Medicine*, *65*, 1238–1245.

Akira, S., & Takeda, K. (2004). Toll-like receptor signalling. *Nature Reviews Immunology*, *4*, 499–511.

Allen, K., Jaeschke, H., & Copple, B. L. (2011). Bile acids induce inflammatory genes in hepatocytes: A novel mechanism of inflammation during obstructive cholestasis. *American Journal of Pathology*, *178*, 175–186.

Anandatheerthavarada, H. K., Addya, S., Dwivedi, R. S., Biswas, G., Mullick, J., & Avadhani, N. G. (1997). Localization of multiple forms of inducible cytochromes P450 in rat liver mitochondria: Immunological characteristics and patterns of xenobiotic substrate metabolism. *Archives of Biochemistry and Biophysics*, *339*, 136–150.

Antoine, D. J., Jenkins, R. E., Dear, J. W., Williams, D. P., McGill, M. R., Sharpe, M. R., et al. (2012). Molecular forms of HMGB1 and keratin-18 as mechanistic biomarkers for mode of cell death and prognosis during clinical acetaminophen hepatotoxicity. *Journal of Hepatology*, *56*, 1070–1079.

Antoniades, C. G., Berry, P. A., Davies, E. T., Hussain, M., Bernal, W., Vergani, D., et al. (2006). Reduced monocyte HLA-DR expression: A novel biomarker of disease severity and outcome inacetaminophen-induced acute liver failure. *Hepatology*, *44*, 34–43.

Antoniades, C. G., Khamri, W., Abeles, R. D., Taams, L. S., Triantafyllou, E., Possamai, L. A., et al. (2014). Secretory leukocyte protease inhibitor: A pivotal mediator of anti-inflammatory responses in acetaminophen-induced acute liver failure. *Hepatology*, *59*, 1564–1576.

Antoniades, C. G., Quaglia, A., Taams, L. S., Mitry, R. R., Hussain, M., Abeles, R., et al. (2012). Source and characterization of hepatic macrophages in acetaminophen-induced acute liver failure in humans. *Hepatology*, *56*, 735–746.

Arteel, G. E. (2003). Oxidants and antioxidants in alcohol-induced liver disease. *Gastroenterology*, *124*, 778–790.

Avadhani, N. G., Sangar, M. C., Bansal, S., & Bajpai, P. (2011). Bimodal targeting of cytochrome P450s to endoplasmic reticulum and mitochondria: The concept of chimeric signals. *FEBS Journal*, *278*, 4218–4229.

Bajt, M. L., Cover, C., Lemasters, J. J., & Jaeschke, H. (2006). Nuclear translocation of endonuclease G and apoptosis-inducing factor during acetaminophen-induced liver cell injury. *Toxicological Sciences*, *94*, 217–225.

Bajt, M. L., Farhood, A., & Jaeschke, H. (2001). Effects of CXC chemokines on neutrophil activation and sequestration in hepatic vasculature. *American Journal of Physiology. Gastrointestinal and Liver Physiology*, *281*, G1188–G1195.

Bajt, M. L., Farhood, A., Lemasters, J. J., & Jaeschke, H. (2008). Mitochondrial bax translocation accelerates DNA fragmentation and cell necrosis in a murine model of acetaminophen hepatotoxicity. *Journal of Pharmacology and Experimental Therapeutics*, *324*, 8–14.

Bajt, M. L., Lawson, J. A., Vonderfecht, S. L., Gujral, J. S., & Jaeschke, H. (2000). Protection against Fas receptor-mediated apoptosis in hepatocytes and nonparenchymal cells by a caspase-8 inhibitor in vivo: Evidence for a postmitochondrial processing of caspase-8. *Toxicological Sciences*, *58*, 109–117.

Bertola, A., Park, O., & Gao, B. (2013). Chronic plus binge ethanol feeding synergistically induces neutrophil infiltration and liver injury in mice: A critical role for E-selectin. *Hepatology*, *58*, 1814–1823.

Billiar, T. R., Lysz, T. W., Curran, R. D., Bentz, B. G., Machiedo, G. W., & Simmons, R. L. (1990). Hepatocyte modulation of Kupffer cell prostaglandin E2 production in vitro. *Journal of Leukocyte Biology*, *47*, 305–311.

Bourdi, M., Amouzadeh, H. R., Rushmore, T. H., Martin, J. L., & Pohl, L. R. (2001). Halothane-induced liver injury in outbred guinea pigs: Role of trifluoroacetylated protein adducts in animal susceptibility. *Chemical Research in Toxicology*, *14*, 362–370.

Chakraborty, M., Fullerton, A. M., Semple, K., Chea, L. S., Proctor, W. R., Bourdi, M., et al. (2015). Drug-induced allergic hepatitis developed in mice when myeloid-derived suppressor cells were depleted prior to halothane treatment. *Hepatology*. http://dx.doi.org/10.1002/hep.27764.

Chen, G., & Goeddel, D. V. (2002). TNF-R1 signaling: A beautiful pathway. *Science*, *31*, 1634–1635.

Chiang, J. Y. (2009). Bile acids: Regulation of synthesis. *Journal of Lipid Research*, *50*, 1955–1966.

Choi, D. Y., Ban, J. O., Kim, S. C., & Hong, J. T. (2015). CCR5 knockout mice with C57BL6 background are resistant to acetaminophen-mediated hepatotoxicity due to decreased macrophages migration into the liver. *Archives of Toxicology*, *89*, 211–220.

Chosay, J. G., Essani, N. A., Dunn, C. J., & Jaeschke, H. (1997). Neutrophil margination and extravasation in sinusoids and venules of liver during endotoxin-induced injury. *American Journal of Physiology*, *272*, G1195–G1200.

Chow, J. C., Young, D. W., Golenbock, D. T., Christ, W. J., & Gusovsky, F. (1999). Toll-like receptor-4 mediates lipopolysaccharide-induced signal transduction. *Journal of Biological Chemistry*, *274*, 10689–10692.

Clot, P., Albano, E., Eliasson, E., Tabone, M., Aricò, S., Israel, Y., et al. (1996). Cytochrome P4502E1 hydroxyethyl radical adducts as the major antigen in autoantibody formation among alcoholics. *Gastroenterology*, *111*, 206–216.

Connolly, M. K., Ayo, D., Malhotra, A., Hackman, M., Bedrosian, A. S., Ibrahim, J., et al. (2011). Dendritic cell depletion exacerbates acetaminophen hepatotoxicity. *Hepatology*, *54*, 959–968.

Cover, C., Liu, J., Farhood, A., Malle, E., Waalkes, M. P., Bajt, M. L., et al. (2006). Pathophysiological role of the acute inflammatory response during acetaminophen hepatotoxicity. *Toxicology and Applied Pharmacology*, *216*, 98–107.

Dahlin, D. C., Miwa, G. T., Lu, A. Y., & Nelson, S. D. (1984). N-acetyl-p-benzoquinone imine: A cytochrome P-450-mediated oxidation product of acetaminophen. *Proceedings of the National Academy of Sciences of the United States of America*, *81*, 1327–1331.

Ding, W. X., Li, M., Chen, X., Ni, H. M., Lin, C. W., Gao, W., et al. (2010). Autophagy reduces acute ethanol-induced hepatotoxicity and steatosis in mice. *Gastroenterology*, *139*, 1740–1752.

Dixon, L. J., Barnes, M., Tang, H., Pritchard, M. T., & Nagy, L. E. (2013). Kupffer cells in the liver. *Comprehensive Physiology*, *3*, 785–797.

Downs, I., Aw, T. Y., Liu, J., Adegboyega, P., & Ajuebor, M. N. (2012). Vα14iNKT cell deficiency prevents acetaminophen-induced acute liver failure by enhancing hepatic glutathione and altering APAP metabolism. *Biochemical and Biophysical Research Communications*, *428*, 245–251.

Du, C., Fang, M., Li, Y., Li, L., & Wang, X. (2000). Smac, a mitochondrial protein that promotes cytochrome c-dependent caspase activation by eliminating IAP inhibition. *Cell*, *102*, 33–42.

Dugan, C. M., MacDonald, A. E., Roth, R. A., & Ganey, P. E. (2010). A mouse model of severe halothane hepatitis based on human risk factors. *Journal of Pharmacology and Experimental Therapeutics*, *333*, 364–372.

Eksteen, B., Afford, S. C., Wigmore, S. J., Holt, A. P., & Adams, D. H. (2007). Immune-mediated liver injury. *Seminars in Liver Disease*, *27*, 351–366.

Eliasson, E., & Kenna, J. G. (1996). Cytochrome P450 2E1 is a cell surface autoantigen in halothane hepatitis. *Molecular Pharmacology*, *50*, 573–582.

Feingold, K. R., Spady, D. K., Pollock, A. S., Moser, A. H., & Grunfeld, C. (1996). Endotoxin, TNF, and IL-1 decrease cholesterol 7 alpha-hydroxylase mRNA levels and activity. *Journal of Lipid Research*, *37*, 223–228.

Feng, D., Wang, Y., Wang, H., Weng, H., Kong, X., Martin-Murphy, B. V., et al. (2014). Acute and chronic effects of IL-22 on acetaminophen-induced liver injury. *Journal of Immunology*, *193*, 2512–2518.

Gao, B., & Bataller, R. (2011). Alcoholic liver disease: Pathogenesis and new therapeutic targets. *Gastroenterology*, *141*, 1572–1585.

Germolec, D. R., Adams, N. H., & Luster, M. I. (1995). Comparative assessment of metabolic enzyme levels in macrophage populations of the F344 rat. *Biochemical Pharmacology*, *50*, 1495–1504.

Godfrey, D. I., Hammond, K. J., Poulton, L. D., Smyth, M. J., & Baxter, A. G. (2000). NKT cells: Facts, functions and fallacies. *Immunology Today*, *21*, 573–583.

Green, R. M., Beier, D., & Gollan, J. L. (1996). Regulation of hepatocyte bile salt transporters by endotoxin and inflammatory cytokines in rodents. *Gastroenterology*, *111*, 193–198.

Gujral, J. S., Farhood, A., Bajt, M. L., & Jaeschke, H. (2003). Neutrophils aggravate acute liver injury during obstructive cholestasis in bile duct-ligated mice. *Hepatology*, *38*, 355–363.

Gujral, J. S., Hinson, J. A., Farhood, A., & Jaeschke, H. (2004). NADPH oxidase-derived oxidant stress is critical for neutrophil cytotoxicity during endotoxemia. *American Journal of Physiology. Gastrointestinal and Liver Physiology*, *287*, G243–G252.

Gujral, J. S., Hinson, J. A., & Jaeschke, H. (2004). Chlorotyrosine protein adducts are reliable biomarkers of neutrophil-induced cytotoxicity in vivo. *Comparative Hepatology*, *14*(Suppl 1), S48.

Gujral, J. S., Knight, T. R., Farhood, A., Bajt, M. L., & Jaeschke, H. (2002). Mode of cell death after acetaminophen overdose in mice: Apoptosis or oncotic necrosis? *Toxicological Sciences*, *67*, 322–328.

Gujral, J. S., Liu, J., Farhood, A., Hinson, J. A., & Jaeschke, H. (2004). Functional importance of ICAM-1 in the mechanism of neutrophil-induced liver injury in bile duct-ligated mice. *American Journal of Physiology. Gastrointestinal and Liver Physiology*, *286*, G499–G507.

Gunawan, B. K., Liu, Z. X., Han, D., Hanawa, N., Gaarde, W. A., & Kaplowitz, N. (2006). c-Jun N-terminal kinase plays a major role in murine acetaminophen hepatotoxicity. *Gastroenterology*, *131*, 165–178.

Habibollahi, P., Mahboobi, N., Esmaeili, S., Safari, S., Dabbagh, A., & Alavian, S. M. (2011). Halothane-induced hepatitis: A forgotten issue in developing countries: Halothane-induced hepatitis. *Hepatitis Monthly*, *11*, 3–6.

Hakkak, R., Korourian, S., Ronis, M. J., Ingelman-Sundberg, M., & Badger, T. M. (1996). Effects of diet and ethanol on the expression and localization of cytochromes P450 2E1 and P450 2C7 in the colon of male rats. *Biochemical Pharmacology*, *51*, 61–69.

Hanawa, N., Shinohara, M., Saberi, B., Gaarde, W. A., Han, D., & Kaplowitz, N. (2008). Role of JNK translocation to mitochondria leading to inhibition of mitochondria bioenergetics in acetaminophen-induced liver injury. *Journal of Biological Chemistry*, *283*, 13565–13577.

Harley, I. T., Stankiewicz, T. E., Giles, D. A., Softic, S., Flick, L. M., Cappelletti, M., et al. (2014). IL-17 signaling accelerates the progression of nonalcoholic fatty liver disease in mice. *Hepatology*, *59*, 1830–1839.

Hasegawa, T., Malle, E., Farhood, A., & Jaeschke, H. (2005). Generation of hypochlorite-modified proteins by neutrophils during ischemia-reperfusion injury in rat liver: Attenuation by ischemic preconditioning. *American Journal of Physiology. Gastrointestinal and Liver Physiology*, *289*, G760–G767.

Hoque, R., Sohail, M. A., Salhanick, S., Malik, A. F., Ghani, A., Robson, S. C., et al. (2012). P2X7 receptor-mediated purinergic signaling promotes liver injury in acetaminophen hepatotoxicity in mice. *American Journal of Physiology. Gastrointestinal and Liver Physiology*, *302*, G1171–G1179.

Houser, W. H., Raha, A., & Vickers, M. (1992). Induction of CYP1A1 gene expression in H4-II-E rat hepatoma cells by benzo[e]pyrene. *Molecular Carcinogenesis*, *5*, 232–237.

Imaeda, A. B., Watanabe, A., Sohail, M. A., Mahmood, S., Mohamadnejad, M., Sutterwala, F. S., et al. (2009). Acetaminophen-induced hepatotoxicity in mice is dependent on Tlr9 and the Nalp3 inflammasome. *Journal of Clinical Investigation*, *119*, 305–314.

Inagaki, T., Choi, M., Moschetta, A., Peng, L., Cummins, C. L., & McDonald, J. G. (2005). Fibroblast growth factor 15 functions as an enterohepatic signal to regulate bile acid homeostasis. *Cell Metabolism*, *2*, 217–225.

Isayama, F., Froh, M., Bradford, B. U., McKim, S. E., Kadiiska, M. B., Connor, H. D., et al. (2003). The CYP inhibitor 1-aminobenzotriazole does not prevent oxidative stress associated with alcohol-induced liver injury in rats and mice. *Free Radical Biology and Medicine*, *35*, 1568–1581.

Ito, Y., Abril, E. R., Bethea, N. W., & McCuskey, R. S. (2004). Role of nitric oxide in hepatic microvascular injury elicited by acetaminophen in mice. *American Journal of Physiology. Gastrointestinal and Liver Physiology*, *286*, G60–G67.

Jaeschke, H. (2008). Innate immunity and acetaminophen-induced liver injury: Why so many controversies? *Hepatology*, *48*, 699–701.

Jaeschke, H. (2011). Reactive oxygen and mechanisms of inflammatory liver injury: Present concepts. *Journal of Gastroenterology and Hepatology*, *26*(Suppl 1), 173–179.

Jaeschke, H., Bautista, A. P., Spolarics, Z., & Spitzer, J. J. (1991). Superoxide generation by Kupffer cells and priming of neutrophils during reperfusion after hepatic ischemia. *Free Radical Research Communications*, *15*, 277–284.

Jaeschke, H., & Farhood, A. (1991). Neutrophil and Kupffer cell-induced oxidant stress and ischemia-reperfusion injury in rat liver. *American Journal of Physiology*, *260*, G355–G362.

Jaeschke, H., Farhood, A., & Smith, C. W. (1990). Neutrophils contribute to ischemia/reperfusion injury in rat liver in vivo. *FASEB Journal*, *4*, 3355–3359.

Jaeschke, H., Farhood, A., & Smith, C. W. (1991). Neutrophil-induced liver cell injury in endotoxin shock is a CD11b/CD18-dependent mechanism. *American Journal of Physiology*, *261*, G1051–G1056.

Jaeschke, H., Fisher, M. A., Lawson, J. A., Simmons, C. A., Farhood, A., & Jones, D. A. (1998). Activation of caspase 3 (CPP32)-like proteases is essential for TNF-alpha-induced hepatic parenchymal cell apoptosis and neutrophil-mediated necrosis in a murine endotoxin shock model. *Journal of Immunology*, *160*, 3480–3486.

Jaeschke, H., & Lemasters, J. J. (2003). Apoptosis versus oncotic necrosis in hepatic ischemia/reperfusion injury. *Gastroenterology*, *125*, 1246–1257.

Jaeschke, H., & Liu, J. (2007). Neutrophil depletion protects against murine acetaminophen hepatotoxicity: Another perspective. *Hepatology*, *45*, 1588–1589.

Jaeschke, H., McGill, M. R., & Ramachandran, A. (2012). Oxidant stress, mitochondria, and cell death mechanisms in drug-induced liver injury: Lessons learned from acetaminophen hepatotoxicity. *Drug Metabolism Reviews*, *44*, 88–106.

Jaeschke, H., McGill, M. R., Williams, C. D., & Ramachandran, A. (2011). Current issues with acetaminophen hepatotoxicity—A clinically relevant model to test the efficacy of natural products. *Life Sciences*, *25*, 737–745.

Jaeschke, H., Schini, V. B., & Farhood, A. (1992). Role of nitric oxide in the oxidant stress during ischemia/reperfusion injury of the liver. *Life Sciences*, *50*, 1797–1804.

Jaeschke, H., & Smith, C. W. (1997). Mechanisms of neutrophil-induced parenchymal cell injury. *Journal of Leukocyte Biology*, *61*, 647–653.

Jaeschke, H., Williams, C. D., Ramachandran, A., & Bajt, M. L. (2012). Acetaminophen hepatotoxicity and repair: The role of sterile inflammation and innate immunity. *Liver International*, *32*, 8–20.

Jaeschke, H., & Woolbright, B. L. (2012). Current strategies to minimize hepatic ischemia-reperfusion injury by targeting reactive oxygen species. *Transplantation Reviews*, *26*, 103–114.

James, L. P., McCullough, S. S., Knight, T. R., Jaeschke, H., & Hinson, J. A. (2003). Acetaminophen toxicity in mice lacking NADPH oxidase activity: Role of peroxynitrite formation and mitochondrial oxidant stress. *Free Radical Research*, *37*, 1289–1297.

Jiang, Q., Akashi, S., Miyake, K., & Petty, H. R. (2000). Lipopolysaccharide induces physical proximity between CD14 and toll-like receptor 4 (TLR4) prior to nuclear translocation of NF-kappa B. *Journal of Immunology*, *165*, 3541–3554.

Ju, C. (2009). The role of haptic macrophages in regulation of idiosyncratic drug reactions. *Toxicologic Pathology*, *37*, 12–17.

Ju, C., Reilly, T. P., Bourdi, M., Radonovich, M. F., Brady, J. N., George, J. W., et al. (2002). Protective role of Kupffer cells in acetaminophen-induced hepatic injury in mice. *Chemical Research in Toxicology*, *15*, 1504–1513.

Kawai, T., Adachi, O., Ogawa, T., Takeda, K., & Akira, S. (1999). Unresponsiveness of MyD88-deficient mice to endotoxin. *Immunity*, *11*, 115–122.

Kawai, T., Takeuchi, O., Fujita, T., Inoue, J., Mühlradt, P. F., Sato, S., et al. (2001). Lipopolysaccharide stimulates the MyD88-independent pathway and results in activation of IFN-regulatory factor 3 and the expression of a subset of lipopolysaccharide-inducible genes. *Journal of Immunology*, *167*, 5887–5894.

Keitel, V., Donner, M., Winandy, S., Kubitz, R., & Häussinger, D. (2008). Expression and function of the bile acid receptor TGR5 in Kupffer cells. *Biochemical and Biophysical Research Communications*, *372*, 78–84.

Kerkvliet, N. I., Shepherd, D. M., & Baecher-Steppan, L. (2002). T lymphocytes are direct, aryl hydrocarbon receptor (AhR)-dependent targets of 2,3,7,8-tetrachlorodibenzo-p-dioxin (TCDD): AhR expression in both CD4+ and CD8+ T cells is necessary for full suppression of a cytotoxic T lymphocyte response by TCDD. *Toxicology and Applied Pharmacology*, *185*, 146–152.

Kim, M. S., Shigenaga, J., Moser, A., Feingold, K., & Grunfeld, C. (2003). Repression of farnesoid X receptor during the acute phase response. *Journal of Biological Chemistry*, *278*, 8988–8995.

Kischkel, F. C., Hellbardt, S., Behrmann, I., Germer, M., Pawlita, M., Krammer, P. H., et al. (1995). Cytotoxicity-dependent APO-1 (Fas/CD95)-associated proteins form a death-inducing signaling complex (DISC) with the receptor. *EMBO Journal*, *14*, 5579–5588.

Kischkel, F. C., Lawrence, D. A., Tinel, A., LeBlanc, H., Virmani, A., Schow, P., et al. (2001). Death receptor recruitment of endogenous caspase-10 and apoptosis initiation in the absence of caspase-8. *Journal of Biological Chemistry*, *276*, 46639–46646.

Knockaert, L., Fromenty, B., & Robin, M. A. (2011). Mechanisms of mitochondrial targeting of cytochrome P450 2E1: Physiopathological role in liver injury and obesity. *FEBS Journal*, *278*, 4252–4260.

Kodali, P., Wu, P., Lahiji, P. A., Brown, E. J., & Maher, J. J. (2006). ANIT toxicity toward mouse hepatocytes in vivo is mediated primarily by neutrophils via CD18. *American Journal of Physiology. Gastrointestinal and Liver Physiology*, *291*, G355–G363.

Koivisto, T., Mishin, V. M., Mak, K. M., Cohen, P. A., & Lieber, C. S. (1996). Induction of cytochrome P-4502E1 by ethanol in rat Kupffer cells. *Alcoholism, Clinical and Experimental Research*, *20*, 207–212.

Kon, K., Kim, J. S., Jaeschke, H., & Lemasters, J. J. (2004). Mitochondrial permeability transition in acetaminophen-induced necrosis and apoptosis of cultured mouse hepatocytes. *Hepatology*, *40*, 1170–1179.

Kono, H., Bradford, B. U., Yin, M., Sulik, K. K., Koop, D. R., Peters, J. M., et al. (1999). CYP2E1 is not involved in early alcohol-induced liver injury. *American Journal of Physiology*, *277*, G1259–G1267.

Kono, H., Rusyn, I., Uesugi, T., Yamashina, S., Connor, H. D., Dikalova, A., et al. (2001). Diphenyleneiodonium sulfate, an NADPH oxidase inhibitor, prevents early alcohol-induced liver injury in the rat. *American Journal of Physiology. Gastrointestinal and Liver Physiology*, *280*, G1005–G1012.

Krenkel, O., Mossanen, J. C., & Tacke, F. (2014). Immune mechanisms in acetaminophen-induced acute liver failure. *Hepatobiliary Surgery and Nutrition*, *3*, 331–343.

Kubes, P., & Mehal, W. Z. (2012). Sterile inflammation in the liver. *Gastroenterology*, *143*, 1158–1172.

Lawson, J. A., Farhood, A., Hopper, R. D., Bajt, M. L., & Jaeschke, H. (2000). The hepatic inflammatory response after acetaminophen overdose: Role of neutrophils. *Toxicological Sciences*, *54*, 509–516.

Lentsch, A. B., Yoshidome, H., Cheadle, W. G., Miller, F. N., & Edwards, M. J. (1998). Chemokine involvement in hepatic ischemia/reperfusion injury in mice: Roles for macrophage inflammatory protein-2 and KC. *Hepatology*, *27*, 1172–1177.

Leung, C. H., Caldarone, C. A., Wang, F., Venkateswaran, S., Ailenberg, M., Vadasz, B., et al. (2014). Remote ischemic conditioning prevents lung and liver injury after hemorrhagic shock/resuscitation: Potential role of a humoral plasma factor. *Annals of Surgery*, Sept 1. (Epub ahead of print).

Li, T., & Chiang, J. Y. (2015). Bile acids as metabolic regulators. *Current Opinion in Gastroenterology*, *31*, 159–165.

Li, T., Jahan, A., & Chiang, J. Y. (2006). Bile acids and cytokines inhibit the human cholesterol 7 alpha-hydroxylase gene via the JNK/c-jun pathway in human liver cells. *Hepatology*, *43*, 1202–1210.

Li, T., Ma, H., & Chiang, J. Y. (2008). TGFbeta1, TNFalpha, and insulin signaling crosstalk in regulation of the rat cholesterol 7alpha-hydroxylase gene expression. *Journal of Lipid Research*, *49*, 1981–1989.

Lieber, C. S. (1997). Cytochrome P-4502E1: Its physiological and pathological role. *Physiological Reviews*, *77*, 517–544.

Lieber, C. S. (2005). Metabolism of alcohol. *Clinics in Liver Disease, 9*, 1–35.
Liu, Z. X., Govindarajan, S., & Kaplowitz, N. (2004). Innate immune system plays a critical role in determining the progression and severity of acetaminophen hepatotoxicity. *Gastroenterology, 127*, 1760–1774.
Liu, Z. X., Han, D., Gunawan, B., & Kaplowitz, N. (2006). Neutrophil depletion protects against murine acetaminophen hepatotoxicity. *Hepatology, 43*, 1220–1230.
Liu, X., Kim, C. N., Yang, J., Jemmerson, R., & Wang, X. (1996). Induction of apoptotic program in cell-free extracts: Requirement for dATP andcytochrome c. *Cell, 86*, 147–157.
Liu, P., McGuire, G. M., Fisher, M. A., Farhood, A., Smith, C. W., & Jaeschke, H. (1995). Activation of Kupffer cells and neutrophils for reactive oxygen formation is responsible for endotoxin-enhanced liver injury after hepatic ischemia. *Shock, 3*, 56–62.
Liu, J., Sendelbach, L. E., Parkinson, A., & Klaassen, C. D. (2000). Endotoxin pretreatment protects against the hepatotoxicity of acetaminophen and carbon tetrachloride: Role of cytochrome P450 suppression. *Toxicology, 147*, 167–176.
Lu, Y., & Cederbaum, A. I. (2008). CYP2E1 and oxidative liver injury by alcohol. *Free Radical Biology and Medicine, 44*, 723–738.
Lysakova-Devine, T., & O'Farrelly, C. (2014). Tissue-specific NK cell populations and their origin. *Journal of Leukocyte Biology, 96*, 981–990.
Makishima, M., Okamoto, A. Y., Repa, J. J., Tu, H., Learned, R. M., Luk, A., et al. (1999). Identification of a nuclear receptor for bile acids. *Science, 284*, 1362–1365.
Martin-Murphy, B. V., Kominsky, D. J., Orlicky, D. J., Donohue, T. M., Jr., & Ju, C. (2013). Increased susceptibility of natural killer T-cell-deficient mice to acetaminophen-induced liver injury. *Hepatology, 57*, 1575–1584.
Masson, M. J., Carpenter, L. D., Graf, M. L., & Pohl, L. R. (2008). Pathogenic role of natural killer T and natural killer cells in acetaminophen-induced liver injury in mice is dependent on the presence of dimethyl sulfoxide. *Hepatology, 48*, 889–897.
Mathews, S., Xu, M., Wang, H., Bertola, A., & Gao, B. (2014). Animals models of gastrointestinal and liver diseases. Animal models of alcohol-induced liver disease: Pathophysiology, translational relevance, and challenges. *American Journal of Physiology. Gastrointestinal and Liver Physiology, 306*, G819–G823.
McGill, M. R., & Jaeschke, H. (2013). Metabolism and disposition of acetaminophen: Recent advances in relation to hepatotoxicity and diagnosis. *Pharmaceutical Research, 30*, 2174–2187.
McGill, M. R., Sharpe, M. R., Williams, C. D., Taha, M., Curry, S. C., & Jaeschke, H. (2012). The mechanism underlying acetaminophen-induced hepatotoxicity in humans and mice involves mitochondrial damage and nuclear DNA fragmentation. *Journal of Clinical Investigation, 122*, 1574–1583.
McKim, S. E., Gäbele, E., Isayama, F., Lambert, J. C., Tucker, L. M., Wheeler, M. D., et al. (2003). Inducible nitric oxide synthase is required in alcohol-induced liver injury: Studies with knockout mice. *Gastroenterology, 125*, 1834–1844.
Memon, R. A., Moser, A. H., Shigenaga, J. K., Grunfeld, C., & Feingold, K. R. (2001). In vivo and in vitro regulation of sterol 27-hydroxylase in the liver during the acute phase response. Potential role of hepatocyte nuclear factor-1. *Journal of Biological Chemistry, 276*, 30118–30126.
Metushi, I. G., Cai, P., Vega, L., Grant, D. M., & Uetrecht, J. (2014). Paradoxical attenuation of autoimmune hepatitis by oral isoniazid in wild-type and N-acetyltransferase-deficient mice. *Drug Metabolism and Disposition, 42*, 963–973.
Metushi, I. G., Cai, P., Zhu, X., Nakagawa, T., & Uetrecht, J. P. (2011). A fresh look at the mechanism of isoniazid-induced hepatotoxicity. *Clinical Pharmacology and Therapeutics, 89*, 911–914.

Metushi, I. G., Sanders, C., Acute Liver Study Group, Lee, W. M., & Uetrecht, J. (2014). Detection of anti-isoniazid and anti-cytochrome P450 antibodies in patients with isoniazid-induced liver failure. *Hepatology*, *59*, 1084–1093.

Michael, S. L., Pumford, N. R., Mayeux, P. R., Niesman, M. R., & Hinson, J. A. (1999). Pretreatment of mice with macrophage inactivators decreases acetaminophen hepatotoxicity and the formation of reactive oxygen and nitrogen species. *Hepatology*, *30*, 186–195.

Mizutani, T., Shinoda, M., Tanaka, Y., Kuno, T., Hattori, A., Usui, T., et al. (2005). Autoantibodies against CYP2D6 and other drug-metabolizing enzymes in autoimmune hepatitis type 2. *Drug Metabolism Reviews*, *37*, 235–252.

Omura, T. (2006). Mitochondrial P450s. *Chemico-Biological Interactions*, *163*, 86–93.

Pålsson-McDermott, E. M., & O'Neill, L. A. (2004). Signal transduction by the lipopolysaccharide receptor, Toll-like receptor-4. *Immunology*, *113*, 153–162.

Péan, N., Doignon, I., Garcin, I., Besnard, A., Julien, B., & Liu, B. (2013). The receptor TGR5 protects the liver from bile acid overload during liver regeneration in mice. *Hepatology*, *58*, 1451–1460.

Plunk, A. D., Syed-Mohammed, H., Cavazos-Rehg, P., Bierut, L. J., & Grucza, R. A. (2014). Alcohol consumption, heavy drinking, and mortality: Rethinking the j-shaped curve. *Alcoholism, Clinical and Experimental Research*, *38*, 471–478.

Potter, W. Z., Davis, D. C., Mitchell, J. R., Jollow, D. J., Gillette, J. R., & Brodie, B. B. (1973). Acetaminophen-induced hepatic necrosis. 3. Cytochrome P-450-mediated covalent binding in vitro. *Journal of Pharmacology and Experimental Therapeutics*, *187*, 203–210.

Proctor, W. R., Chakraborty, M., Chea, L. S., Morrison, J. C., Berkson, J. D., Semple, K., et al. (2013). Eosinophils mediate the pathogenesis of halothane-induced liver injury in mice. *Hepatology*, *57*, 2026–2036.

Proctor, W. R., Chakraborty, M., Fullerton, A. M., Korrapati, M. C., Ryan, P. M., Semple, K., et al. (2014). Thymic stromal lymphopoietin and interleukin-4 mediate the pathogenesis of halothane-induced liver injury in mice. *Hepatology*, *60*, 1741–1752.

Quintana, F. J., Basso, A. S., Iglesias, A. H., Korn, T., Farez, M. F., & Bettelli, E. (2008). Control of T(reg) and T(H)17 cell differentiation by the aryl hydrocarbon receptor. *Nature*, *453*, 65–71.

Renton, K. W., & Dickson, G. (1984). The prevention of acetaminophen-induced hepatotoxicity by the interferon inducer poly(rI. rC). *Toxicology and Applied Pharmacology*, *72*, 40–45.

Rothe, J., Lesslauer, W., Lötscher, H., Lang, Y., Koebel, P., Köntgen, F., et al. (1993). Mice lacking the tumour necrosis factor receptor 1 are resistant to TNF-mediated toxicity but highly susceptible to infection by Listeria monocytogenes. *Nature*, *364*, 798–802.

Rubin, K., Janefeldt, A., Andersson, L., Berke, Z., Grime, K., & Andersson, T. B. (2015). HepaRG cells as human-relevant in vitro model to study the effects of inflammatory stimuli on cytochrome P450 isoenzymes. *Drug Metabolism and Disposition*, *43*, 119–125.

Saito, C., Lemasters, J. J., & Jaeschke, H. (2010). c-Jun N-terminal kinase modulates oxidant stress and peroxynitrite formation independent of inducible nitric oxide synthase in acetaminophen hepatotoxicity. *Toxicology and Applied Pharmacology*, *246*, 8–17.

Saito, J. M., & Maher, J. J. (2000). Bile duct ligation in rats induces biliary expression of cytokine-induced neutrophil chemoattractant. *Gastroenterology*, *118*, 1157–1168.

Scaffidi, C., Fulda, S., Srinivasan, A., Friesen, C., Li, F., Tomaselli, K. J., et al. (1998). Two CD95 (APO-1/Fas) signaling pathways. *EMBO Journal*, *17*, 1675–1687.

Scaffidi, P., Misteli, T., & Bianchi, M. E. (2002). Release of chromatin protein HMGB1 by necrotic cells triggers inflammation. *Nature*, *418*, 191–195.

Schroder, K., & Tschopp, J. (2010). The inflammasomes. *Cell*, *140*, 821–832.

Schüngel, S., Buitrago-Molina, L. E., Nalapareddy, P. D., Lebofsky, M., Manns, M. P., Jaeschke, H., et al. (2009). The strength of the Fas ligand signal determines whether hepatocytes act as type 1 or type 2 cells in murine livers. *Hepatology*, *50*, 1558–1566.
Shedlofsky, S. I., Israel, B. C., McClain, C. J., Hill, D. B., & Blouin, R. A. (1994). Endotoxin administration to humans inhibits hepatic cytochrome P450-mediated drug metabolism. *Journal of Clinical Investigation*, *94*, 2209–2214.
Sutti, S., Rigamonti, C., Vidali, M., & Albano, E. (2014). CYP2E1 autoantibodies in liver diseases. *Redox Biology*, *3*, 72–78.
Tang, D., Shi, Y., Kang, R., Li, T., Xiao, W., Wang, H., et al. (2007). Hydrogen peroxide stimulates macrophages and monocytes to actively release HMGB1. *Journal of Leukocyte Biology*, *81*, 741–747.
Tian, Z., Chen, Y., & Gao, B. (2013). Natural killer cells in liver disease. *Hepatology*, *57*, 1654–1662.
Tsukamoto, H. (2000). Cyp2e1 and ALD. *Hepatology*, *32*, 154–156.
Uetrecht, J. (2009). Immunoallergic drug-induced liver injury in humans. *Seminars in Liver Disease*, *29*, 383–392.
Üllen, A., Singewald, E., Konya, V., Fauler, G., Reicher, H., Nusshold, C., et al. (2013). Myeloperoxidase-derived oxidants induce blood-brain barrier dysfunction in vitro and in vivo. *PloS One*, *14*, e64034.
Veldhoen, M., Hirota, K., Westendorf, A. M., Buer, J., Dumoutier, L., & Renauld, J. C. (2008). The aryl hydrocarbon receptor links TH17-cell-mediated autoimmunity to environmental toxins. *Nature*, *453*, 106–109.
Vermijlen, D., Luo, D., Robaye, B., Seynaeve, C., Baekeland, M., & Wisse, E. (1999). Pit cells (Hepatic natural killer cells) of the rat induce apoptosis in colon carcinoma cells by the perforin/granzyme pathway. *Hepatology*, *29*, 51–56.
Wang, Y. D., Chen, W. D., Yu, D., Forman, B. M., & Huang, W. (2011). The G-protein-coupled bile acid receptor, Gpbar1 (TGR5), negatively regulates hepatic inflammatory response through antagonizing nuclear factor κ light-chain enhancer of activated B cells (NF-κB) in mice. *Hepatology*, *54*, 1421–1432.
Wang, H., Feng, D., Park, O., Yin, S., & Gao, B. (2013). Invariant NKT cell activation induces neutrophil accumulation and hepatitis: Opposite regulation by IL-4 and IFN-γ. *Hepatology*, *58*, 1474–1485.
Wang, Y., Mathews, W. R., Guido, D. M., Farhood, A., & Jaeschke, H. (1995). Inhibition of nitric oxide synthesis aggravates reperfusion injury after hepatic ischemia and endotoxemia. *Shock*, *4*, 282–288.
Watkins, P. B. (2005). Idiosyncratic liver injury: Challenges and approaches. *Toxicologic Pathology*, *33*, 1–5.
Williams, C. D., Antoine, D. J., Shaw, P. J., Benson, C., Farhood, A., Williams, D. P., et al. (2011). Role of the Nalp3 inflammasome in acetaminophen-induced sterile inflammation and liver injury. *Toxicology and Applied Pharmacology*, *252*, 289–297.
Williams, C. D., Bajt, M. L., Farhood, A., & Jaeschke, H. (2010). Acetaminophen-induced hepatic neutrophil accumulation and inflammatory liver injury in CD18-deficient mice. *Liver International*, *30*, 1280–1292.
Williams, C. D., Bajt, M. L., Sharpe, M. R., McGill, M. R., Farhood, A., & Jaeschke, H. (2014). Neutrophil activation during acetaminophen hepatotoxicity and repair in mice and humans. *Toxicology and Applied Pharmacology*, *275*, 122–133.
Williams, C. D., Farhood, A., & Jaeschke, H. (2010). Role of caspase-1 and interleukin-1beta in acetaminophen-induced hepatic inflammation and liver injury. *Toxicology and Applied Pharmacology*, *247*, 169–178.

Woolbright, B. L., Antoine, D. J., Jenkins, R. E., Bajt, M. L., Park, B. K., & Jaeschke, H. (2013). Plasma biomarkers of liver injury and inflammation demonstrate a lack of apoptosis during obstructive cholestasis in mice. *Toxicology and Applied Pharmacology*, *273*, 524–531.

Woolbright, B. L., & Jaeschke, H. (2012). Novel insight into mechanisms of cholestatic liver injury. *World Journal of Gastroenterology*, *18*, 4985–4993.

Woolbright, B. L., Li, F., Xie, Y., Farhood, A., Fickert, P., Trauner, M., et al. (2014). Lithocholic acid feeding results in direct hepato-toxicity independent of neutrophil function in mice. *Toxicology Letters*, *228*, 56–66.

Wu, D., Wang, X., Zhou, R., Yang, L., & Cederbaum, A. I. (2012). Alcohol steatosis and cytotoxicity: The role of cytochrome P4502E1 and autophagy. *Free Radical Biology and Medicine*, *53*, 1346–1357.

Xie, Y., Williams, C. D., McGill, M. R., Lebofsky, M., Ramachandran, A., & Jaeschke, H. (2013). Purinergic receptor antagonist A438079 protects against acetaminophen-induced liver injury by inhibiting p450 isoenzymes, not by inflammasome activation. *Toxicological Sciences*, *131*, 325–335.

Yang, M., Ramachandran, A., Yan, H. M., Woolbright, B. L., Copple, B. L., Fickert, P., et al. (2014). Osteopontin is an initial mediator of inflammation and liver injury during obstructive cholestasis after bile duct ligation in mice. *Toxicology Letters*, *224*, 186–195.

Yin, X. M., Wang, K., Gross, A., Zhao, Y., Zinkel, S., Klocke, B., et al. (1999). Bid-deficient mice are resistant to Fas-induced hepatocellular apoptosis. *Nature*, *400*, 886–891.

Yoon, J. H., & Gores, G. J. (2002). Death receptor-mediated apoptosis and the liver. *Journal of Hepatology*, *37*, 400–410.

You, Q., Cheng, L., Reilly, T. P., Wegmann, D., & Ju, C. (2006). Role of neutrophils in a mouse model of halothane-induced liver injury. *Hepatology*, *44*, 1421–1431.

You, Q., Holt, M., Yin, H., Li, G., Hu, C. J., & Ju, C. (2013). Role of hepatic resident and infiltrating macrophages in liver repair after acute injury. *Biochemical Pharmacology*, *86*, 836–843.

Yun, J. W., Son, M. J., Abdelmegeed, M. A., Banerjee, A., Morgan, T. R., Yoo, S. H., et al. (2014). Binge alcohol promotes hypoxic liver injury through a CYP2E1-HIF-1α-dependent apoptosis pathway in mice and humans. *Free Radical Biology and Medicine*, *77*, 183–194.

Zhang, Y., Hong, J. Y., Rockwell, C. E., Copple, B. L., Jaeschke, H., & Klaassen, C. D. (2012). Effect of bile duct ligation on bile acid composition in mouse serum and liver. *Liver International*, *32*, 58–69.

Zigmond, E., Samia-Grinberg, S., Pasmanik-Chor, M., Brazowski, E., Shibolet, O., Halpern, Z., et al. (2014). Infiltrating monocyte-derived macrophages and resident Kupffer cells display different ontogeny and functions in acute liver injury. *Journal of Immunology*, *193*, 344–353.

CHAPTER SIX

Role of Cytochrome P450s in Inflammation

Peter Christmas[1]
Department of Biology, Radford University, Radford, Virginia, USA
[1]Corresponding author: e-mail address: pchristma2@radford.edu

Contents

Abstract

Cytochrome P450 epoxygenases and hydroxylases play a regulatory role in the activation and suppression of inflammation by generating or metabolizing bioactive mediators. CYP2C and CYP2J epoxygenases convert arachidonic acid to anti-inflammatory epoxyeicosatrienoic acids, which have protective effects in a variety of disorders including cardiovascular disease and metabolic syndrome. CYP4A and CYP4F hydroxylases have the ability to metabolize multiple substrates related to the regulation of inflammation and lipid homeostasis, and it is a challenge to determine which substrates are physiologically relevant for each enzyme; the best-characterized activities include generation of 20-hydroxyeicosatetraenoic acid and inactivation of leukotriene B_4. The expression of hepatic drug-metabolizing cytochrome P450s is modulated by cytokines during inflammation, resulting in changes to the pharmacokinetics of prescribed medications. Cytochrome P450s are therefore the focus of intersecting challenges in the pharmacology of inflammation: not only do they represent targets for development of new anti-inflammatory drugs but they also contribute to variability in drug efficacy or toxicity in inflammatory disease. Animal models and primary hepatocytes have been used extensively to study the effects of cytokines on cytochrome P450 expression and

Advances in Pharmacology, Volume 74
ISSN 1054-3589
http://dx.doi.org/10.1016/bs.apha.2015.03.005

activity. However, it is difficult to predict changes in drug exposure in patients because the response to inflammation varies depending on the disease state, its time course, and the cytochrome P450 involved. In these circumstances, the development of endogenous markers of cytochrome P450 metabolism might provide a useful tool to reevaluate drug dosage and choice of therapy.

ABBREVIATIONS

CAM cell adhesion molecule
COX cyclooxygenase
CYP cytochrome P450
DHA docosahexaenoic acid
DHET dihydroxyeicosatrienoic acid
EET epoxyeicosatrienoic acid
EPA eicosapentaenoic acid
HETE hydroxyeicosatetraenoic acid
HO heme oxygenase
IBD inflammatory bowel disease
IL interleukin
LOX lipoxygenase
LPS lipopolysaccharide
LT leukotriene
NFκB nuclear factor kappa B
PG prostaglandin
PPAR peroxisome proliferator-activated receptor
PUFA polyunsaturated fatty acid
sEH soluble epoxide hydrolase
TLR Toll-like receptor
TNF tumor necrosis factor
VEGF vascular endothelial growth factor

1. INTRODUCTION

Some cytochrome P450s (CYPs) play an active role in inflammation by converting fatty acids to pro- or anti-inflammatory mediators. Most focus has been on the products of arachidonic acid (eicosatetraenoic acid), a 20-carbon ω-6 polyunsaturated fatty acid (PUFA), but there is growing interest in the metabolism of other PUFAs, including ω-3 fatty acids such as docosahexaenoic acid (DHA) and eicosapentaenoic acid (EPA). The role of CYPs in arachidonic acid metabolism is summarized in Fig. 1. CYP epoxygenases convert arachidonic acid to anti-inflammatory epoxyeicosatrienoic acids (EETs) and include enzymes in the CYP2C and CYP2J families. CYP hydroxylases convert arachidonic acid to

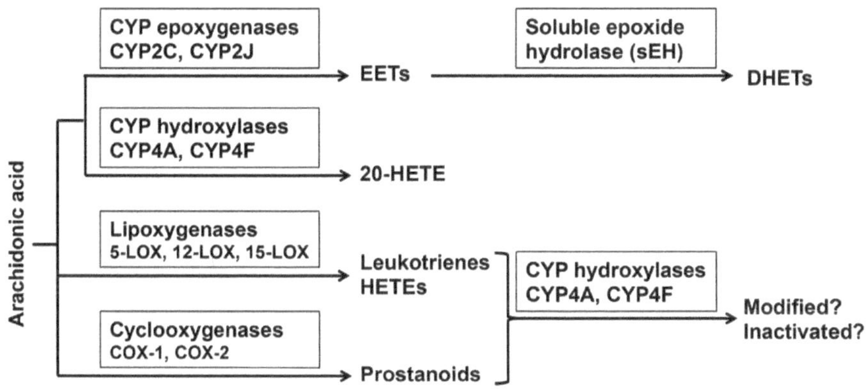

Figure 1 Role of cytochrome P450s in arachidonic acid metabolism.

proinflammatory 20-hydroxyeicosatetraenoic acid (20-HETE) and include enzymes in the CYP4A and CYP4F families. Therefore, CYPs can participate in the suppression, activation, and resolution of inflammation. CYPs are often described as the third arm of arachidonic acid metabolism; the other two arms are the cyclooxygenase (COX) pathway that produces prostanoids such as prostaglandins (PGs), and the lipoxygenase (LOX) pathway that produces leukotrienes (LTs) and a variety of HETEs. CYPs can inactivate or modify products of these other pathways, and this adds to the complexity of regulatory networks. CYP activity is determined mainly by the availability of substrate: inflammatory cytokines or other signals stimulate release of arachidonic acid from cell membranes by phospholipases, and the products of arachidonic acid (eicosanoids) then depend on which CYP enzymes are expressed in a particular cell type. However, CYP expression levels can change during inflammation and disease progression, and this adds another dynamic to the regulation of inflammation that is poorly understood. In addition, the expression of hepatic drug-metabolizing CYPs is modified during inflammation, resulting in changes in the clearance of drugs. This was originally considered to be a nonspecific downregulation of CYP expression, as the liver diverts its resources to the production of acute phase proteins. However, detailed analysis of animal models and primary hepatocytes has shown that individual CYPs are regulated by inflammatory cytokines in diverse ways. The consequences for drug pharmacokinetics are difficult to predict and depend on the disease context and the particular CYP involved. These issues are relevant to classic inflammatory responses to injury and infection and also to chronic inflammatory diseases such as arthritis, cardiovascular disease, and cancer.

2. IMPACT OF INFLAMMATION ON EXPRESSION AND ACTIVITY OF DRUG-METABOLIZING CYTOCHROME P450s

2.1 Animal Models

In the last few decades, a variety of animal models have been used to study changes in expression and activity of hepatic CYPs during inflammation. These include live pathogen infections, treatment with chemicals to induce injury, and injection of lipopolysaccharide (LPS; endotoxin) and other agents that mimic bacterial infection (Aitken, Richardson, & Morgan, 2006; Morgan, 1997; Renton, 2004). The best-characterized model is administration of LPS, a major constituent of the outer membrane of Gram-negative bacteria. LPS initiates inflammatory responses by binding to Toll-like receptors (TLRs) such as TLR4 in macrophages and other cells. TLR activation stimulates the release of tumor necrosis factor α (TNFα) and interleukin-1β (IL-1β), which in turn stimulate release of other cytokines such as IL-6. Multiple studies have shown a rapid downregulation of hepatic CYPs following administration of LPS *in vivo* and following direct addition of cytokines (including TNFα, IL-1β, and IL-6) to primary hepatocyte cultures *in vitro*. Downregulation was measured using assays for CYP mRNA, CYP protein, and clearance of probe substrates. A survey of rat LPS studies reported that the magnitude of CYP suppression was generally about one- to twofold (Yang & Lee, 2008). This is consistent with values reported for other disease models (Dickmann, McBride, et al., 2012).

A mouse model of live bacterial infection with *Citrobacter rodentium*, a Gram-negative pathogen, resulted in a different pattern of hepatic CYP mRNA expression compared to sterile inflammation induced by LPS: there was a more graded downregulation of different CYPs, and some CYPs were unaffected or induced (Richardson et al., 2006). *C. rodentium* causes an intestinal infection in mice which is equivalent to human enteropathogenic *Escherichia coli* infection, and the resulting colitis can be used as a model for inflammatory bowel disease (IBD). Different patterns of CYP expression were observed when comparing intestinal and systemic infections of *C. rodentium* (Chaluvadi, Kinloch, et al., 2009), and when comparing these models to dextran sulfate sodium colitis, a form of chemically induced IBD (Chaluvadi, Nyagode, Kinloch, & Morgan, 2009). These and other studies reveal a selective modulation of CYPs depending on disease state, rather than a nonspecific downregulation of CYPs as a uniform response to inflammation. This is probably the result of differential regulation of CYPs by cytokines and varying profiles of cytokines in different diseases.

Live *C. rodentium* infections were studied in knockout mice lacking TLR4, IL-6, interferon-γ (IFNγ), TNFα receptor-1 (TNFR1), and IL-1 receptor, using assays for a large set of CYP mRNAs, and some cytokine-specific effects were observed (Kinloch, Lee, van Rooijen, & Morgan, 2011; Nyagode, Lee, & Morgan, 2010; Richardson et al., 2006). For example, downregulation of hepatic *Cyp3a11* and *Cyp3a25* was abolished in mice lacking TNFR1 but unchanged in mice lacking the IL-I receptor (Kinloch et al., 2011). A TNFα antagonist selectively blocked the downregulation of *Cyp3a11* and *Cyp3a25* in wild-type mice, demonstrating that downregulation can be reversed (desuppressed) by drug treatment (Nyagode, Jahangardi, Merrell, Tansey, & Morgan, 2014). The expression of most CYPs was unchanged in knockout mice compared to wild type, despite the ability of individual cytokines to downregulate CYP expression in primary hepatocytes. This suggests that there is some redundancy in the actions of cytokines *in vivo*, which might complicate the use of cytokines as biomarkers to predict the response to inflammation.

A growing number of model systems are being used to explore the diverse disease states that modify CYP expression and to demonstrate that the pattern of change varies according to disease. Studies with rodent models of arthritis established a correlation between severity of inflammation, suppression of hepatic CYP expression, and the clearance of CYP substrates and demonstrated that antibodies to TNFα and IL-6 desuppressed CYP expression (Ashino et al., 2007; Ling & Jamali, 2009; Piquette-Miller & Jamali, 1995). Serum cytokines were measured in a mouse model of collagen antibody-induced arthritis, and similar combinations and concentrations of cytokines were then used to stimulate mouse hepatocytes *in vitro* (Dickmann, McBride, et al., 2012). A comparison of *in vivo* and *in vitro* data showed some discordance in the patterns of downregulation of CYPs, demonstrating the difficulties of *in vitro–in vivo* extrapolation. Transplantation of the Engelbreth-Holm-Swarm (EHS) mouse sarcoma, a solid tumor model, into IL-6 knockout mice resulted in a reduction of hepatic *Cyp3a* expression that could be reversed with an anti-IL-6 antibody (Kacevska et al., 2013). These studies showed that hepatic CYP downregulation in cancer was mediated by IL-6 derived from the tumor, not the host.

2.2 Primary Human Hepatocytes

Primary hepatocyte culture is the standard model for investigating human CYP expression and activity. Early studies investigated cytokine-dependent downregulation of induced CYP expression because the higher CYP levels were easier to measure (Aitken et al., 2006). The nuclear receptors pregnane X receptor, constitutive androstane receptor, and aryl hydrocarbon receptor regulate induction of human CYPs following addition of inducing agents to cells. More recent studies have demonstrated cytokine-dependent downregulation of constitutive CYP expression and show that this occurs at physiologically relevant concentrations of cytokines (Aitken & Morgan, 2007; Dallas et al., 2012; Dickmann, Patel, Rock, Wienkers, & Slatter, 2011, Dickmann, Patel, Wienkers, & Slatter, 2012; Evers et al., 2013). Variations on the hepatocyte model include the use of HepaRG cells, a human hepatoma cell line that can be terminally differentiated (Rubin et al., 2015). These reports emphasize that hepatocyte monocultures do not reflect the complex interactions between multiple cytokines and cell types *in vivo* and cannot accurately predict responses to cytokines and therapeutic proteins in humans (Dickmann et al., 2011, Dickmann, Patel, et al., 2012; Evers et al., 2013). Cocultures of primary hepatocytes with Kupffer cells and other nonparenchymal cells have been used in attempts to more closely approximate *in vivo* conditions (Chen, Han, Yang, & Rodrigues, 2011; Sunman, Hawke, LeCluyse, & Kashuba, 2004). The feasibility of this approach is supported by the recent development of a long-term human hepatocyte-Kupffer cell coculture model that is more stable and sensitive than hepatocyte monocultures (Nguyen et al., 2015).

A summary of cytokine effects on CYP expression in primary human hepatocytes is shown in Table 1. CYP3A4 has been extensively studied because it is responsible for the metabolism of a large number of prescribed drugs. It was downregulated by most cytokines tested in primary human hepatocytes, but the dose-dependent inhibition of CYP3A4 expression and activity by IL-6 was one of the most sensitive with IC_{50} values in the pg/ml range (Dickmann et al., 2011; Evers et al., 2013). An anti-IL-6 monoclonal antibody partially reversed the inhibition (Dickmann et al., 2011). This is significant because elevated IL-6 levels are associated with multiple disease states, and a monoclonal antibody to the IL-6 receptor (tocilizumab) is an approved drug (Tanaka & Kishimoto, 2014). IL-6 circulates in picomolar concentrations under normal conditions in humans and may increase by as much as 1000-fold during infection and trauma. Interestingly, inhibition of CYP3A4 expression and activity by IL-1β showed

Table 1 Summary of Major Drug-Metabolizing CYPs and Effect of Cytokines

CYP Enzyme	%[a]	Substrates with Narrow Therapeutic Window[b]	Effect of Cytokines on CYP Expression in Primary Human Hepatocytes[c]
CYP3A4[d]	30.2	Alfentanil, cisapride, cyclosporine, ergotamine, fentanyl, pimozide, quinidine, sirolimus, tacrolimus, terfenadine Anticancer drugs such as docetaxel, etoposide, vinblastine, vincristine, vinorelbine	⇓ IL-6, TNFα, IL-1β, IFNγ, TGFβ
CYP2D6	20	Pimozide Thioridazine	⇓ TNFα ⇓⇔ IL-6, IL-1β
CYP2C9	12.8	Phenytoin Warfarin	⇓ IL-6, TGFβ ⇓⇔ IL-1β ⇔ TNFα, IFNγ
CYP1A2	8.9	Theophylline Tizanidine	⇓ IL-6, TNFα, IL-1β
CYP2B6	7.2		⇓ IL-6, IFNγ ⇔ TNFα, IL-1β, TGFβ
CYP2C19	6.8	S-Mephenytoin	⇓ TGFβ ⇓⇔ IL-6, IL-1β ⇔ TNFα, IFNγ
CYP2C8	4.7	Paclitaxel	⇓ IL-6, TNFα, IL-1β, IFNγ, TGFβ

[a]Value indicates % of human drug metabolism pathways for which the CYP is the major contributing isoform, as estimated by Zanger and Schwab (2013). CYP2A6 (3.4%), 2E1 (3%), and 2J2 (3%) are not listed.

[b]A more comprehensive list of anticancer drugs that are substrates for CYPs is given by Rodriguez-Antona and Ingelman-Sundberg (2006).

[c]Cytokine data are compiled from Aitken and Morgan (2007), Dickmann et al. (2011), Dickmann, Patel, et al. (2012), Dallas et al. (2012), Evers et al. (2013), Nguyen et al. (2015). ⇓, downregulation; ⇔, no effect; ⇓⇔, divergent results.

[d]CYP3A5 is expressed at about 10% of the level of CYP3A4 in liver and may contribute to metabolism of some of the drugs listed (% value is for CYP3A4 and CYP3A5). Expression data are specific for CYP3A4.

low IC_{50} values comparable to IL-6 in the hepatocyte coculture model, indicating that Kupffer cells enhance the effects of IL-1β (Nguyen et al., 2015). CYP2C8, 2C9, 2C19, and 2J2 are CYP epoxygenases that generate bioactive eicosanoids in liver and extrahepatic tissues, so the enzymes have dual functions in drug metabolism and inflammation regulation. They are differentially regulated by cytokines in primary hepatocytes (Table 1), suggesting that the patterns of expression in inflammatory disease will be complex. The signaling pathways that link cytokines to changes in CYP gene expression include both transcriptional and posttranscriptional regulatory mechanisms and are summarized in recent reviews (Morgan, 2009; Zordoky & El-Kadi, 2009).

2.3 Clinical Studies

An influenza epidemic in 1982 triggered severe drug toxicity in children who were receiving theophylline to treat asthma, and this is one of the first recorded examples of human disease impacting CYP metabolism (Kraemer et al., 1982). The inflammatory response to the virus probably caused down-regulation of hepatic CYP1A2 and led to an increased plasma concentration and half-life of the drug. Infections with several viruses, including herpes simplex, adenovirus, hepatitis C, and human immunodeficiency virus, have since been reported to suppress CYP metabolism and reduce drug clearance (Christensen & Hermann, 2012). Administration of low doses of LPS to healthy volunteers caused reduced clearance of several CYP-metabolized drugs (Shedlofsky, Israel, McClain, Hill, & Blouin, 1994). A separate study concluded that CYP-mediated drug metabolism was decreased in children with sepsis and was related to the degree of inflammation and organ failure (Carcillo et al., 2003). Altered drug disposition has also been reported in patients with chronic inflammatory diseases such as rheumatoid arthritis and cancer (Christensen & Hermann, 2012; Gandhi, Moorthy, & Ghose, 2012; Harvey & Morgan, 2014). Increases in drug exposure up to threefold have been observed in these studies. In general, the effects are small, and adverse consequences may be limited to drugs with a narrow therapeutic window (Table 1). For example, small changes in the pharmacokinetics of many cancer chemotherapy drugs can lead to either treatment failure (reduced efficacy) or life-threatening reactions (increased toxicity). There is wide interindividual variability in the response of cancer patients to these drugs, and different levels of inflammation may contribute to this variability by altering hepatic CYP expression and drug clearance (Slaviero, Clarke, & Rivory, 2003).

A number of clinical studies have reported an inverse relationship between plasma cytokine levels and CYP-metabolized drug clearance in patients. Increased IL-6 and TNFα plasma concentrations were correlated with suppression of CYP1A2 and CYP2C19 activity in patients with congestive heart failure, using caffeine and mephenytoin as probe substrates (Frye, Schneider, Frye, & Feldman, 2002). Similarly, IL-6 plasma levels negatively correlated with CYP3A4-dependent metabolism of cyclosporine in bone marrow-transplanted patients (Chen et al., 1994). Studies on cancer patients established a relationship between the presence of extrahepatic tumors, IL-6 plasma concentration, and reduced hepatic CYP3A4 activity measured by the erythromycin breath test (Rivory, Slaviero, & Clarke, 2002). This was associated with increased toxicity from docetaxel and vinorelbine (Slaviero et al., 2003). A recent proof of principle clinical study demonstrated that an anti-IL-6 receptor monoclonal antibody (tocilizumab) could desuppress CYP3A4 activity and increase drug clearance in rheumatoid arthritis patients (Schmitt, Kuhn, Zhang, Kivitz, & Grange, 2011). Twelve patients received simvastatin, a CYP3A4 substrate, on days 1, 15, and 43, and a single dose of tocilizumab on day 8. On day 15, there was a doubling of simvastatin clearance and a decrease in area under the curve (AUC) of simvastatin exposure to 43%, compared to day 1. On day 43, the AUC was 61% and trending toward baseline values. There were smaller effects of tocilizumab on other CYP-metabolized drugs (Kim, Ostor, & Nisar, 2012).

Therapeutic proteins, such as cytokines and monoclonal antibodies, are not metabolized by CYPs and are not usually considered relevant to small molecule drug–drug interactions involving CYP enzymes. However, therapeutic proteins can interact with CYPs by changing their expression, and this is another potential source of adverse drug reactions. There have been many reports of altered CYP expression and drug metabolism in patients receiving therapeutic cytokines such as interferons and interleukins (Christensen & Hermann, 2012; Harvey & Morgan, 2014; Morgan, 2009). Treatments with monoclonal antibodies to the IL-6 receptor (tocilizumab) and TNFα (adalimumab, etanercept, golimumab, infliximab, certolizumab) are approved for inflammatory diseases such as rheumatoid arthritis. These antibodies are cytokine modulators because they inhibit the actions of cytokines; therefore, they have the potential to desuppress CYP activity and increase drug clearance. The consequences could be beneficial (reduced drug toxicity) or harmful (reduced drug efficacy and treatment failure) depending on the circumstance.

It is difficult to predict the effects of inflammation and therapeutic proteins on drug disposition in patients. The responses probably vary depending on the type of disease, its time course, and the CYP involved. In addition, there is considerable variation in CYP expression between individuals based on age, sex, genetics, and other factors (Shahabi, Siest, Meyer, & Visvikis-Siest, 2014). A personalized approach would facilitate decisions about adjustment of drug doses and choice of therapy, but this depends on the development of suitable biomarkers. Measurement of plasma levels of inflammatory cytokines, such as IL-6, might be useful in some cases but is complicated by the pleiotropic nature of cytokine action. A more direct measurement of CYP activity is preferred, and there is progress in the development of endogenous markers of CYP metabolism (Lutz, Bittner, Ufer, & Lutz, 2010; Shin et al., 2013). This strategy avoids the need to administer probe substrates that might have unpredictable effects on physiology. It is reported that blood leukocytes express the most relevant drug-metabolizing CYPs, and that expression levels correlate with hepatic CYP expression (Siest et al., 2009), so isolation and analysis of blood cells might provide an alternative strategy for monitoring CYP activity.

3. PRO- AND ANTI-INFLAMMATORY ROLES OF CYTOCHROME P450s

3.1 CYP Epoxygenases

CYP epoxygenases convert arachidonic acid to four bioactive EET regioisomers: 5,6-, 8,9-, 11,12-, and 14,15-EET (Spector & Kim, 2015). The main human epoxygenases have been identified as CYP2C8, CYP2C9, and CYP2J2, although a number of other CYPs have the capacity to generate EETs. These enzymes are also involved in drug metabolism in the liver, so they have dual functions. The main EET inactivation pathway involves soluble epoxide hydrolase (sEH), which converts EETs to less active dihydroxyeicosatrienoic acids (DHETs). EET levels can be increased in model systems (*in vitro*, *ex vivo*, *in vivo*) using selective inhibitors of sEH, synthetic EET analogs that have a longer half-life, sEH knockout mice, and transgenic mice with endothelial expression of human CYP2J2 and CYP2C8 (Deng et al., 2011; Pillarisetti & Khanna, 2012; Sudhahar, Shaw, & Imig, 2010). Numerous studies have established that EETs generated in tissues such as the endothelium, myocardium, and kidney exhibit a range of anti-inflammatory and cardiovascular protective effects (Deng, Theken, & Lee, 2010; Fleming, 2014; Imig, 2012; Shahabi, Siest, & Visvikis-Siest, 2014). CYP epoxygenases are not expressed in all endothelia, but are localized to certain arterial regions including the coronary, renal, and cerebral vasculatures. However, CYP-derived EETs are incorporated into the membranes of red blood cells, and these EET reservoirs can be released at locations that are distant from their site of synthesis, so the range of EET action might be more extensive than expected from the expression pattern of CYP epoxygenases (Jiang, Anderson, & McGiff, 2010).

3.1.1 Anti-Inflammatory Function of EETs

In the inflammatory response to infection, cytokines such as TNFα and IL-1 activate NFκB (nuclear factor kappa B) signaling in endothelial cells and promote surface expression of cell adhesion molecules (CAMs). Circulating neutrophils and monocytes attach to the CAMs, infiltrate into tissues at the site of infection, and attack invading pathogens. Leukocyte defense mechanisms are relatively nonspecific and can damage healthy tissue, so the response must be controlled. Anti-inflammatory signals have important roles as controls and are not restricted to turning inflammation off during the resolution phase. These signals also suppress inappropriate activation during the noninflammatory state and prevent excessive amplification of the inflammatory response during the activation phase (Nathan, 2002). The endothelium plays a central role in maintaining the correct pro-/anti-inflammatory balance (Pober & Sessa, 2007). The first evidence that CYP epoxygenases have anti-inflammatory functions came from studies using cultured human umbilical vein endothelial cells and isolated murine carotid arteries. Cytokine-induced NFκB signaling, expression of CAMs, and leukocyte adhesion were attenuated by exogenous administration of EETs at low physiological concentrations, or overexpression of CYP2J2 (Node et al., 1999). This has been confirmed by a large number of studies. Inhibition of NFκB signaling is the main anti-inflammatory mechanism of EET action, but there is complex interplay with the peroxisome proliferator-activated receptors (PPARs), heme oxygenase (HO) enzymes, and inducible COX-2 and LOX-5 enzymes (Deng et al., 2010; Shahabi, Siest, & Visvikis-Siest, 2014).

EETs were originally identified as endothelium-derived relaxing factors that regulate vascular tone in certain arteries. They are produced by CYP epoxygenases in response to stimuli such as acetylcholine, bradykinin, and mechanical stress, and they promote vasodilation and blood flow by hyperpolarizing vascular smooth muscle cells (Campbell & Fleming, 2010). These EETs might also contribute to maintenance of the noninflammatory state in the endothelium, although the beneficial effects of EETs are difficult to observe in experimental systems in the absence of a pathologic stimulus. There is good experimental evidence that the systemic inflammatory response to LPS administration is reduced in transgenic mice with endothelial expression of human CYP2J2 and CYP2C8, sEH knockout mice, and mice treated with sEH inhibitors (Deng et al., 2011; Schmeizer et al., 2005). CYP epoxygenases are downregulated in tissues such as liver, heart, and kidney, following LPS administration in rodents, with a corresponding decrease in EET levels (Anwar-Mohamed, Zordoky, Aboutabl, & El-Kadi, 2010; Theken et al., 2011). It is speculated that loss of anti-inflammatory EETs would drive more inflammation and less EET production in a vicious cycle, and this might be a mechanism for the development of chronic inflammation following an initial insult in cardiovascular tissues (Shahabi, Siest, & Visvikis-Siest, 2014). The process probably varies depending on the disease and the tissue location. Indeed, recent studies observed the opposite effect: LPS induced an increase in CYP2J2 expression in endothelial cells and monocytes (Askari et al., 2014; Bystrom et al., 2013). In the case of monocytes, CYP2J2 induction was required for acquisition of a phagocytic phenotype as the cells differentiated into macrophages, and deficient CYP2J2 induction and bacterial clearance were observed in macrophages from patients with Crohn's disease (Bystrom et al., 2013).

3.1.2 Chronic Inflammation and Cardiovascular Disease

Inflammation is a unifying mechanism underlying many chronic diseases, including cardiovascular disease. In chronic inflammatory diseases, the sites of epithelial cell activation differ (arterial endothelial cells in atherosclerosis, tubular epithelial cells in renal disease, and enterocytes in IBD), and the profile of leukocyte infiltration can vary (often includes T lymphocytes), but the underlying process is similar (Libby, 2007). Inhibition or knockout of sEH reduced leukocyte infiltration and tissue damage in several rodent models of chronic inflammation, including tobacco smoke-induced lung inflammation (Smith et al., 2005), hypertension-induced renal injury (Manhiani et al., 2009), and IBD (Zhang et al., 2012). In atherosclerosis, monocytes and T lymphocytes leave the bloodstream and accumulate in arterial walls, forming the atherosclerotic plaque (Libby, 2006). These plaques often form at sites of nonlaminar flow, a condition that is known to reduce EET levels in cultured endothelial cells (Liu et al., 2005). Some animal models of accelerated atherosclerosis show attenuated plaque formation in response to CYP2J2 overexpression or inhibition of sEH (Fleming, 2014; Wang, Ulu, Zhang, & Hammock, 2010). Associations between the risk of developing coronary artery disease and polymorphisms in genes coding for CYP2J2, CYP2C8/9, and sEH have been discovered (Oni-Orisan, Alsaleh, Lee, & Seubert, 2014). There is also a significant negative correlation between EET levels and vascular inflammation in patients with coronary artery disease (Schuck et al., 2013).

The anti-inflammatory and vasodilatory properties of EETs are components of a broad spectrum of activities that provide protective effects against cardiovascular disease. EETs modulate processes involved in wound healing that are intimately related to inflammation: clotting, pain, extracellular matrix remodeling, angiogenesis, cell proliferation, and cell survival. They are profibrinolytic, antiplatelet, proangiogenic, and analgesic. They are also antihypertensive mediators in the renal feedback mechanisms that regulate blood pressure. The potential therapeutic value of increasing EET activity is relevant to multiple stages of cardiovascular disease including hypertension, atherosclerosis, thrombosis, fibrosis, and ischemia–reperfusion injury. EETs produce their effects by activating a variety of signaling pathways and ion channels. The functions of EETs in cardiovascular disease, and their signaling mechanisms, are described in detail in a number of recent reviews (Deng et al., 2010; Fan, Muroya, & Roman, 2015; Fleming, 2014; Imig, 2012; Imig, Simpkins, Renic, & Harder, 2011; Oni-Orisan et al., 2014; Shahabi, Siest, & Visvikis-Siest, 2014; Zhang, Kodani, & Hammock, 2014). CYP epoxygenases are expressed in endothelium, kidney tubules, cardiomyocytes, certain neurons, and glia. They are upregulated in hypoxia, and elevated EET production stimulates cell proliferation and angiogenesis that assists tissue recovery after ischemic injury to the heart and brain. EETs promote survival and function of cardiomyocytes, neurons, and glial cells, which further limits the extent and consequences of tissue injury following myocardial infarction and stroke.

3.1.3 Inflammation and Metabolic Syndrome

Cardiovascular disease is part of the collection of disorders referred to as metabolic syndrome. These disorders include diabetes, obesity, fatty liver disease, and dyslipidemia, and they are assumed to have common underlying causes because of their tendency to cluster in individuals. There is growing evidence for the theory that inflammation is a driving force in initiation and progression of these diseases. The classic inflammatory process is a response to injury or infection, and it produces high levels of cytokines. Metabolic syndrome is associated with chronic "low-grade inflammation" where low levels of cytokines can have deleterious effects on insulin signaling and other metabolic functions, even in the absence of leukocyte infiltration into tissues (Chassaing & Gewirtz, 2014; Straub, 2014). The inflammation can occur in the absence of injury or infection and may be related to other environmental factors such as diet. Theories for causative agents that trigger inflammation include diet-induced changes to gut microbiota (Caesar, Fak, & Backhed, 2010; Chassaing & Gewirtz, 2014), overconsumption that leads to endoplasmic reticulum (ER) stress (Chaudhari, Talwar, Parimisetty, Lefebvre d'Hellencourt, & Ravanan, 2014; Hotamisligil, 2010), excess fructose metabolism in the liver (Lustig, 2013), imbalance in the dietary ω-6/ω-3 PUFA ratio (Simopoulos, 2008), and the formation of inflammatory adipocytes in obesity (Greenberg & Obin, 2006).

The links between eicosanoids and metabolic syndrome have been intensely investigated (Hardwick et al., 2013), and recently, there has been increased interest in the role of EETs. Obesity was significantly correlated with low plasma EET levels in a clinical study (Theken et al., 2012). Multiple studies have associated increased EET levels with a reduction in inflammation and other symptoms of disease in rodent models of obesity and diabetes (Fleming, 2014), and fatty liver disease (Chen et al., 2015; Schuck et al., 2014). CYP epoxygenase expression in locations such as liver, endothelial cells, pancreas, and mesenchymal stem cells is relevant to these activities. Reported mechanisms of action of EETs include inhibition of NFκB signaling and interaction with PPARs (Li et al., 2015), interplay with the HO system and inhibition of adipogenesis (Abraham et al., 2014; Kim et al., 2010; Sodhi et al., 2009), and attenuation of ER stress in liver and other tissues (Bettaieb et al., 2013; Lopez-Vicario et al., 2015; Xu et al., 2013). In addition, the ω-3 PUFAs EPA and DHA are efficiently metabolized by CYP epoxygenases and converted to fatty acid epoxides with beneficial activities that can be more potent than EETs (Fischer et al., 2014; Zhang et al., 2014). Transgenic mice that express the *Caenorhabditis elegans fat-1* gene can covert ω-6 to ω-3 fatty acids and have tissues that are enriched in endogenous ω-3 PUFAs (Kang, 2007). Compared to wild type, *fat-1* mice are more resistant to obesity and inflammation induced by a high-fat diet, and treatment with an sEH inhibitor increased fatty acid epoxide levels and further enhanced the beneficial effects (Lopez-Vicario et al., 2015). Many interrelated factors are involved in the development of metabolic syndrome, and it is difficult to disentangle cause and effect, but this is an important problem because of the need to treat causes rather than symptoms.

3.1.4 Inflammation and Cancer

Inflammation is one of the hallmarks of cancer, and tumors resemble "wounds that do not heal" (Dvorak, 1986; Hanahan & Weinberg, 2011).

Local production of inflammatory mediators in the tumor microenvironment supports growth of the tumor, and angiogenesis is essential for the tumor to grow beyond a diameter of about 2 mm. Hypoxia and vascular endothelial growth factor (VEGF) induce expression of various CYP epoxygenases, and increased EET production stimulates cell proliferation and angiogenesis in normal wound healing and cancer (Fleming, 2014; Panigrahy, Greene, Pozzi, Wang, & Zeldin, 2011; Shahabi, Siest, & Visvikis-Siest, 2014). Endothelial expression of human CYP2C8 and CYP2J2, and application of EETs, not only accelerated wound healing and vascularization in mice (Panigrahy et al., 2013) but also stimulated tumor growth and metastasis in several mouse models of cancer (Panigrahy et al., 2012). A potential role for CYP epoxygenases in human cancer was first described in studies of CYP2J2 (Jiang et al., 2005, 2007). PPARα ligands have antitumorigenic properties that are mediated in part by down-regulation of CYP epoxygenases and reduced EET production, and CYP2C9 was found to be selectively expressed in the vasculature of human renal tumors (Pozzi et al., 2010). Some studies show CYP overexpression in cancer matched with elevated levels of EETs in blood, urine, or biopsy samples (Chen, Wei, et al., 2011; Wei et al., 2014). However, there is considerable variability in the expression of CYP epoxygenases in human cancers, with over- and underexpression reported (Panigrahy et al., 2011), and the significance of these differences is unclear.

The CYP epoxygenase pathway is an attractive target for pharmaceutical intervention. There is potential therapeutic value of increasing EET activity for inflammatory and cardiovascular disease, metabolic syndrome, tissue regeneration, and pain management. sEH inhibitors have already been tested in human subjects and are undergoing continued development (Chen, Whitcomb, et al., 2012; Imig & Hammock, 2009). It is unlikely that the global reduction of EET metabolism by sEH inhibition will be without side effects, and there is concern that increasing EET levels will promote cancer (Panigrahy et al., 2011). CYP epoxygenases are expressed differentially in tissues, they produce different profiles of EET regioisomers, these EETs have variable activities and potencies, and there is evidence for the existence of EET receptors in some signaling pathways (Imig, 2012). This knowledge might be exploited for more specific pharmaceutical intervention. For example, it might be possible to achieve a desired therapeutic effect without promoting tumor angiogenesis. In addition, the expression of CYP epoxygenases in cancer provides an opportunity for using selective CYP inhibitors in cancer therapy (Chen & Wang, 2013). It is reported that ω-3 PUFA epoxides inhibit angiogenesis and tumor metastasis, the opposite effect of EETs, which provides more potential options for therapy (Zhang et al., 2014).

3.2 CYP Hydroxylases

3.2.1 20-HETE Generation

CYP4 enzymes catalyze end-chain hydroxylation of fatty acids, at the terminal (ω) carbon or other positions (ω-1, ω-2, ω-3). ω-Hydroxylation of arachidonic acid generates 20-HETE, a bioactive mediator that has been extensively studied in animal models. The development of selective CYP hydroxylase inhibitors and 20-HETE antagonists has made it possible to distinguish the contributions of 20-HETE and EETs, which often have opposite functions (Imig, 2013). 20-HETE is proinflammatory and stimulates NFκB activation and expression of CAMs and cytokines when overexpressed in endothelial cells (Ishizuka et al., 2008). The 20-HETE/EET + DHET formation ratio is a potential biomarker of inflammatory status (Theken et al., 2011). 20-HETE regulates vascular tone and is a potent vasoconstrictor in arterioles in certain locations such as the kidney, heart, and brain (Gebremedhin et al., 2000; Nithipatikom et al., 2004; Sun, Falck, Harder, & Roman, 1999). It is a prohypertensive signal in the renal microvasculature but has antihypertensive activity in the kidney tubules where it inhibits Na^+ transport and retention (Wu, Gupta, Garcia, Ding, & Schwartzman, 2014). 20-HETE has similar proangiogenic properties to EETs, and it supports cell proliferation and vascularization in cancer (Alexanian & Sorokin, 2013; Chen, Ackerman, & Guo, 2012; Moreno, 2009). Targeting 20-HETE is a potential therapeutic strategy for cancer and other diseases such as hypertension and stroke (Edson & Rettie, 2013; Elshenawy, Anwar-Mohamed, & El-Kadi, 2013).

A number of human CYP4A and 4F hydroxylases are potential sources of bioactive 20-HETE. Enzymes in the human CYP4 family are summarized in Table 2. CYP4A11 is efficient at ω-hydroxylation of medium-chain length (C10–C16) fatty acids, but has low activity for arachidonic acid. However, CYP4A11 is expressed in endothelial progenitor cells and is reported to be the main source of 20-HETE that promotes angiogenesis following induction by hypoxia and VEGF (Chen et al., 2014). There are six genes in the human CYP4F family: *CYP4F2*, *4F3*, *4F8*, *4F11*, *4F12*, and *4F22* (Kalsotra & Strobel, 2006). The *CYP4F3* gene generates two distinct enzymes by alternative splicing: CYP4F3A and 4F3B (Christmas et al., 2001). Enzymes with capability of converting arachidonic acid to

Table 2 The Human CYP4 Family

CYP Enzyme[a]	Putative Substrates[b]	Possible Disease Associations[c]
CYP4A11	Medium-chain fatty acids (C10–C16) Arachidonic acid	Hypertension
CYP4F2	LTB_4 Arachidonic acid Very long-chain fatty acids (C22–C26) Vitamin E, vitamin K_1 Drugs: fingolimod	Hypertension Celiac disease Crohn's disease
CYP4F3A	LTB_4 EETs and other fatty acid epoxides	Celiac disease
CYP4F3B	LTB_4 Arachidonic acid ω-3 PUFAs: EPA, DHA Very long-chain fatty acids (C22–C26) EETs and other fatty acid epoxides	
CYP4F8	Prostaglandin endoperoxides: PGH_1 and PGH_2	Prostate cancer
CYP4F11	Drugs: benzphetamine, chlorpromazine, erythromycin, ethylmorphine, theophylline	
CYP4F12	Prostaglandin endoperoxides: PGH_1 and PGH_2 Drugs: ebastine	
CYP4F22	?	Lamellar ichthyosis
CYP4V2	Medium-chain fatty acids	Bietti's crystalline dystrophy
CYP4X1	?	
CYP4Z1	?	

[a]CYP4A22 and CYP4B1 are not listed; they are expressed at the mRNA level, but the enzymes may be nonfunctional. CYP4F3A and CYP4F3B are alternative splice forms of the *CYP4F3* gene.
[b]See Kalsotra and Strobel (2006) and Corcos et al. (2012) for references.
[c]Association between disease risk and genetic polymorphisms is strongest for *CYP4F22* and *CYP4V2* genes (Kelly, Nakano, Rohatgi, Yarov-Yarovoy, & Rettie, 2011), but more controversial for *CYP4A11*, *CYP4F2*, and *CYP4F3* genes (references in text). CYP4F8 is associated with prostate cancer based on expression studies (Vainio et al., 2011).

20-HETE *in vitro* include CYP4F2 and CYP4F3B (Harmon et al., 2006; Lasker et al., 2000). Transgenic mouse models that express human CYP4F2 increase bioactive 20-HETE production *in vivo* (Cheng et al., 2014; Liu et al., 2009), and sequence variants in CYP4F2 and CYP4A11 are associated with hypertension and changes in urinary 20-HETE levels in a number of patient populations (Wu et al., 2014). Induction of CYP4F3B expression results in higher levels of 20-HETE synthesis in cultured human liver cells (Antoun et al., 2008; Plee-Gautier et al., 2012). The ω-3 PUFAs EPA and DHA are also substrates for ω-hydroxylation by CYP4F3B *in vitro*, and they competitively inhibit conversion of arachidonic acid to 20-HETE (Harmon et al., 2006). This is another possible mechanism for the anti-inflammatory effects of ω-3 PUFAs.

CYP4 enzymes have catalytic activity for multiple substrates *in vitro*, and a particular enzyme might have the ability to generate 20-HETE, inactivate proinflammatory mediators such as leukotriene B_4 (LTB_4), metabolize drugs, and participate in lipid homeostasis by catabolizing fatty acids (Table 2). It is possible that the enzymes have different functions in different tissue locations, and it is a challenge to determine which substrates are physiologically relevant. A growing number of eicosanoids (including PG analogues, fatty acid epoxides, HETEs, and lipoxins), drugs, and lipids have been identified as potential substrates for CYP4F enzymes (Corcos et al., 2012; Kalsotra & Strobel, 2006), but the significance of these diverse reactions is not yet established. Some enzymes are still classified as orphan CYPs (such as CYP4X1 and CYP4Z1), because their substrates are poorly defined, so the list of biological functions attributed to the CYP4 family will continue to expand.

3.2.2 LTB_4 Inactivation

CYP4F family enzymes were originally identified as LTB_4 ω-hydroxylases. LTB_4 is a potent proinflammatory mediator generated by the 5-lipoxygenase pathway of arachidonic acid metabolism, and ω-hydroxylation initiates one of the pathways for LTB_4 catabolism and inactivation (Murphy & Gijon, 2007). CYP4F2, 4F3A, and 4F3B can convert LTB_4 to 20-hydroxy LTB_4 (Kalsotra & Strobel, 2006). The major pathway for the inactivation of LTB_4 in human neutrophils involves ω-hydroxylation by CYP4F3A (Kikuta et al., 1993; Shak & Goldstein, 1984). The specific expression of CYP4F3A in neutrophils, and its high activity for LTB_4, suggests that this is a specialized function of the enzyme. CYP-dependent ω-hydroxylation and inactivation of LTB_4 is induced by retinoids in skin cells and might contribute to the beneficial effects of retinoids in the treatment of inflammatory skin diseases (Du, Yin, Morrow, Strobel, & Keeney, 2009; Kalsotra et al., 2008). CYP4F enzymes are also expressed in neurons and glia, and LTB_4 hydroxylase activity limits neuroinflammation in mouse models (Sehgal et al., 2011). Genetic association studies link variants of the *CYP4F3* and *CYP4F2* genes with celiac disease and Crohn's disease, inflammatory disorders of the gastrointestinal tract (Costea et al., 2010; Curley, Monsuur, Wapenaar, Rioux, & Wijmenga, 2006). The relevance of LTB_4 inactivation is inferred from studies that showed reduced LTB_4 ω-hydroxylase activity in neutrophils and colonic mucosa from patients with IBD (Ikehata et al., 1995).

Changes in expression of CYP4A and 4F enzymes have been measured in various models of inflammation (Anwar-Mohamed et al., 2010; Kalsotra et al., 2003; Richardson et al., 2006; Stoilov et al., 2006; Theken et al., 2011). The direction of change (induction or repression) does not necessarily distinguish pro- or anti-inflammatory function. An increase in CYP4F expression in the liver is thought to control excessive inflammation by removing LTB_4, therefore serving an anti-inflammatory function. However, there are numerous reports of CYP epoxygenase downregulation in inflammation, despite a known anti-inflammatory function in the generation of EETs. Careful analysis of the time course of CYP4F expression in a rat model of traumatic brain injury demonstrated changes in CYP4F mRNA and protein that inversely correlated with LTB_4 levels in the brain, consistent with a role for CYP4Fs in the resolution of inflammation (Wang et al., 2008). These studies are consistent with a temporal switch in LTB_4 hydroxylase expression with initial downregulation during the activation phase of inflammation and then increased expression over an extended period during the resolution phase. A different pattern of expression is observed for human CYP4F3A in neutrophils, a short-lived cell associated with the early stages of inflammation. CYP4F3A is expressed at a high constitutive level in neutrophils (Christmas et al., 2003), both before and after stimulating the cells with LTB_4. Pro- and anti-inflammatory controls probably operate simultaneously during the activation stage of inflammation to control the magnitude of the response (Nathan, 2002).

Differences in CYP isoforms and physiology between species often make it difficult to extrapolate data from animal models to humans, and vice versa. The human *CYP4F3* gene is alternatively spliced in a tissue-specific manner and generates two distinct enzymes defined by mutually exclusive use of exons 3 and 4 (Christmas et al., 2001). CYP4F3A contains exon 4 and is the LTB_4 ω-hydroxylase in human neutrophils. CYP4F3B contains exon 3 and is expressed in liver, kidney, and other tissues. It has lower activity for LTB_4 and higher activity for arachidonic acid, compared to CYP4F3A. The mouse homologue of CYP4F3A was identified as *Cyp4f18* (Christmas et al., 2006), and knockout of *Cyp4f18* abolishes LTB_4 hydroxylase activity in mouse neutrophils (Winslow et al., 2014). However, there are significant differences between mice and humans. The *Cyp4f18* gene is not alternatively spliced: it generates a single enzyme that resembles CYP4F3A in sequence, neutrophil localization, and high activity for LTB_4 but not arachidonic acid. The products of LTB_4 hydroxylation in mice are 19-hydroxy LTB_4, and to a lesser extent 18-hydroxy LTB_4, whereas human CYP4F3A generates 20-hydroxy LTB_4. It is not known whether ω-1 and ω-2 hydroxylation of LTB_4 represents an efficient inactivation pathway, so this might be a significant difference. In addition, mouse neutrophils have an alternative pathway of LTB_4 metabolism that involves a 12-hydroxydehydrogenase. Knockout of *Cyp4f18* does not impact inflammation in a mouse model of renal ischemia–reperfusion (Winslow et al., 2014), although knockout of the genes for LTB_4 synthesis does have observable effects in this model (Patel et al., 2004). It would appear that mice have redundant pathways of LTB_4 inactivation in neutrophils, and it is possible that *Cyp4f18* has an alternative function in these cells.

4. CONCLUSION

CYPs are difficult to study because of a number of issues. CYPs have broad and overlapping substrate specificity, and it is a challenge to identify physiologically relevant substrates among multiple possibilities. There are differences in physiology and CYP isoforms expressed in different species, which make it difficult to compare humans and animal models. It is sometimes difficult to distinguish products, such as the ω, ω-1, and ω-2 metabolites of CYP hydroxylases, which is pharmacologically important. For example, 19-HETE is produced by a number of CYPs and is an antagonist of 20-HETE (Alonso-Galicia, Falck, Reddy, & Roman, 1999). Despite these problems, new innovations have led to significant advances in knowledge. The development of specific CYP inhibitors, and reagents that antagonize or stabilize CYP products, was necessary to determine the biological actions of CYP enzymes. Transgenic mouse models have been created that overexpress human CYPs in specific tissues. Genetic and biomarker-based studies are being used to associate disease risk with CYP variants or alterations in CYP activity in human subjects (Oni-Orisan et al., 2014), and there is growing understanding of the factors that account for variability in CYP expression and activity among individuals (Shahabi, Siest, Meyer, et al., 2014). Future progress will benefit from the continued application of these strategies and other emerging technologies. Proteomic and lipidomic approaches make it possible to more accurately identify CYPs and multiple products (Astarita, Kendall, Dennis, & Nicolaou, 2015; Kawakami et al., 2011) and have considerable potential for analyzing complex patterns of eicosanoid production in physiological settings. Improvements in hepatocyte culture systems might lead to greater predictive value of *in vitro* data on the effects of cytokines on CYP activity. The development of endogenous markers of CYP metabolism will enable CYP activity to be measured during disease progression and may provide new opportunities for personalized medicine. A greater understanding of the role of diet in inflammation and disease, such as changes in ω-6/ω-3 PUFA ratio (Fischer et al., 2014), could have considerable impact on disease prevention. The mechanisms that link CYP activity to inflammatory diseases, including the broad spectrum of disorders associated with metabolic syndrome, are being defined in greater detail. This increases the chances of developing specific pharmaceutical interventions with clinical benefit.

CONFLICT OF INTEREST

The author declares that there are no conflicts of interest.

ACKNOWLEDGMENTS

The author acknowledges the support of the Department of Biology and College of Science and Technology at Radford University.

REFERENCES

Abraham, N. G., Sodhi, K., Silvis, A. M., Vanella, L., Favero, G., Rezzani, R., et al. (2014). CYP2J2 targeting to endothelial cells attenuates adiposity and vascular dysfunction in mice fed a high fat diet by reprogramming adipocyte differentiation. *Hypertension*, *64*(6), 1352–1361.

Aitken, A. E., & Morgan, E. T. (2007). Gene-specific effects of inflammatory cytokines on cytochrome P4502C, 2B6 and 3A4 mRNA levels in human hepatocytes. *Drug Metabolism and Disposition*, *35*(9), 1687–1693.

Aitken, A. E., Richardson, T. A., & Morgan, E. T. (2006). Regulation of drug-metabolizing enzymes and transporters in inflammation. *Annual Review of Pharmacology and Toxicology*, *46*, 123–149.

Alexanian, A., & Sorokin, A. (2013). Targeting 20-HETE producing enzymes in cancer—Rationale, pharmacology, and clinical potential. *OncoTargets and Therapy*, *6*, 243–255.

Alonso-Galicia, M., Falck, J. R., Reddy, K. M., & Roman, R. J. (1999). 20-HETE agonists and antagonists in the renal circulation. *American Journal of Physiology. Renal Physiology*, *277*(5), F790–F796.

Antoun, J., Goulitquer, S., Amet, Y., Dreano, Y., Salaun, J. P., Corcos, L., et al. (2008). CYP4F3B is induced by PGA_1 in human liver cells: A regulation of the 20-HETE synthesis. *The Journal of Lipid Research*, *49*(10), 2135–2141.

Anwar-Mohamed, A., Zordoky, B. N., Aboutabl, M. E., & El-Kadi, A. O. (2010). Alteration of cardiac cytochrome P450-mediated arachidonic acid metabolism in response

to lipopolysaccharide-induced acute systemic inflammation. *Pharmacological Research, 61*(5), 410–418.
Ashino, T., Arima, Y., Shioda, S., Iwakura, Y., Numazawa, S., & Yoshida, T. (2007). Effect of interleukin-6 neutralization on CYP3A11 and metallothionein-1/2 expressions in arthritic mouse liver. *European Journal of Pharmacology, 558*(1–3), 199–207.
Askari, A. A., Thomson, S., Edin, M. L., Lih, F. B., Zeldin, D. C., & Bishop-Bailey, D. (2014). Basal and inducible anti-inflammatory epoxygenase activity in endothelial cells. *Biochemical and Biophysical Research Communications, 446*(2), 633–637.
Astarita, G., Kendall, A. C., Dennis, E. A., & Nicolaou, A. (2015). Targeted lipidomic strategies for oxygenated metabolites of polyunsaturated fatty acids. *Biochimica et Biophysica Acta, 1851*(4), 456–468.
Bettaieb, A., Nagata, N., AbouBechara, D., Chahed, S., Morisseau, C., Hammock, B. D., et al. (2013). Soluble epoxide hydrolase deficiency or inhibition attenuates diet-induced endoplasmic reticulum stress in liver and adipose tissue. *The Journal of Biological Chemistry, 288*(20), 14189–14199.
Bystrom, J., Thomson, S. J., Johansson, J., Edin, M. L., Zeldin, D. C., Gilroy, D. W., et al. (2013). Inducible CYP2J2 and its product 11,12-EET promotes bacterial phagocytosis: A role for CYP2J2 deficiency in the pathogenesis of Crohn's disease? *PLoS One, 8*(9), e75107.
Caesar, R., Fak, F., & Backhed, F. (2010). Effects of gut microbiota on obesity and atherosclerosis via modulation of inflammation and lipid metabolism. *Journal of Internal Medicine, 268*(4), 320–328.
Campbell, W. B., & Fleming, I. (2010). Epoxyeicosatrienoic acids and endothelium-dependent responses. *Pflügers Archiv, 459*(6), 881–895.
Carcillo, J. A., Doughty, L., Kofos, D., Frye, R. F., Kaplan, S. S., Sasser, H., et al. (2003). Cytochrome P450 mediated-drug metabolism is reduced in children with sepsis-induced multiple organ failure. *Intensive Care Medicine, 29*(6), 980–984.
Chaluvadi, M. R., Kinloch, R. D., Nyagode, B. A., Richardson, T. A., Raynor, M. J., Sherman, M., et al. (2009). Regulation of hepatic cytochrome P450 expression in mice with intestinal or systemic infections of *Citrobacter rodentium. Drug Metabolism and Disposition, 37*(2), 366–374.
Chaluvadi, M. R., Nyagode, B. A., Kinloch, R. D., & Morgan, E. T. (2009). TLR4-dependent and -independent regulation of hepatic cytochrome P450 in mice with chemically-induced inflammatory bowel disease. *Biochemical Pharmacology,* 77(3), 464–471.
Chassaing, B., & Gewirtz, A. T. (2014). Gut microbiota, low-grade inflammation, and metabolic syndrome. *Toxicologic Pathology, 42*(1), 49–53.
Chaudhari, N., Talwar, P., Parimisetty, A., Lefebvre d'Hellencourt, C., & Ravanan, P. (2014). A molecular web: Endoplasmic reticulum stress, inflammation, and oxidative stress. *Frontiers in Cellular Neuroscience, 8*, 213.
Chen, L., Ackerman, R., & Guo, A. M. (2012). 20-HETE in neovascularization. *Prostaglandins & Other Lipid Mediators, 98*(3-4), 63–68.
Chen, L., Ackerman, R., Saleh, M., Gotlinger, K. H., Kessler, M., Mendelowitz, L. G., et al. (2014). 20-HETE regulates the angiogenic functions of human endothelial progenitor cells and contributes to angiogenesis in vivo. *Journal of Pharmacology and Experimental Therapeutics, 348*(3), 442–451.
Chen, C., Han, Y.-H., Yang, Z., & Rodrigues, A. D. (2011). Effect of interferon-α2b on the expression of various drug-metabolizing enzymes and transporters in co-cultures of freshly prepared human primary hepatocytes. *Xenobiotica, 41*(6), 476–485.
Chen, Y. L., Le Vraux, V., Leneveu, A., Dreyfus, F., Stheneur, A., Florentin, I., et al. (1994). Acute-phase response, interleukin-6, and alteration of cyclosporine pharmacokinetics. *Clinical Pharmacology & Therapeutics, 55*(6), 649–660.

Chen, C., & Wang, D. W. (2013). CYP epoxygenase derived EETs: From cardiovascular protection to human cancer therapy. *Current Topics in Medicinal Chemistry, 13*(12), 1454–1469.

Chen, C., Wei, X., Rao, X., Wu, J., Yang, S., Chen, F., et al. (2011). Cytochrome P450 2J2 is highly expressed in hematologic malignant disease and promotes tumor cell growth. *The Journal of Pharmacology and Experimental Therapeutics, 336*(2), 344–355.

Chen, D., Whitcomb, R., MacIntyre, E., Tran, V., Do, Z. N., Sabry, J., et al. (2012). Pharmacokinetics and pharmacodynamics of AR9281, an inhibitor of soluble epoxide hydrolase, in single- and multiple-doses in healthy human subjects. *The Journal of Clinical Pharmacology, 52*(3), 319–328.

Chen, G., Xu, R., Zhang, S., Wang, Y., Wang, P., Edin, M. L., et al. (2015). CYP2J2 overexpression attenuates nonalcoholic fatty liver disease induced by high-fat diet in mice. *American Journal of Physiology. Endocrinology and Metabolism, 308*(2), E97–E110.

Cheng, J., Edin, M. L., Hoopes, S. L., Li, H., Bradbury, J. A., Graves, J. P., et al. (2014). Vascular characterization of mice with endothelial expression of cytochrome P450 4F2. *The FASEB Journal, 28*(7), 2915–2931.

Christensen, H., & Hermann, M. (2012). Immunological response as a source to variability in drug metabolism and transport. *Frontiers in Pharmacology, 3*, 8.

Christmas, P., Carlesso, N., Shang, H., Cheng, S.-M., Weber, B. M., Preffer, F. I., et al. (2003). Myeloid expression of cytochrome P450 4F3 is determined by a lineage-specific alternative promoter. *The Journal of Biological Chemistry, 278*(27), 25133–25142.

Christmas, P., Jones, J. P., Patten, C. J., Rock, D. A., Zheng, Y., Cheng, S.-M., et al. (2001). Alternative splicing determines the function of CYP4F3 by switching substrate specificity. *The Journal of Biological Chemistry, 276*(41), 38166–38172.

Christmas, P., Tolentino, K., Primo, V., Zemski Berry, K., Murphy, R. C., Chen, M., et al. (2006). Cytochrome P-450 4F18 is the leukotriene B_4 ω-1/ω-2 hydroxylase in mouse polymorphonuclear leukocytes. *The Journal of Biological Chemistry, 281*(11), 7189–7196.

Corcos, L., Lucas, D., Le Jossic-Corcos, C., Dreano, Y., Simon, B., Plee-Gautier, E., et al. (2012). Human cytochrome P450 4F3: Structure, functions, and prospects. *Drug Metabolism and Drug Interactions, 27*(2), 63–71.

Costea, I., Mack, D. R., Israel, D., Morgan, K., Krupoves, A., Seidman, E., et al. (2010). Genes involved in the metabolism of poly-unsaturated fatty-acids (PUFA) and risk for Crohn's disease in children & young adults. *PLoS One, 5*(12), e15672.

Curley, C. R., Monsuur, A. J., Wapenaar, M. C., Rioux, J. D., & Wijmenga, C. (2006). A functional candidate screen for coeliac disease genes. *European Journal of Human Genetics, 14*(11), 1215–1222.

Dallas, S., Sensenhauser, C., Batheja, A., Singer, M., Markowska, M., Zakszewski, C., et al. (2012). De-risking bio-therapeutics for possible drug interactions using cryopreserved human hepatocytes. *Current Drug Metabolism, 13*(7), 923–929.

Deng, Y., Edin, M. L., Theken, K. N., Schuck, R. N., Flake, G. P., Kannon, M. A., et al. (2011). Endothelial CYP epoxygenase overexpression and soluble epoxide hydrolase disruption attenuate acute vascular inflammatory responses in mice. *The FASEB Journal, 25*(2), 703–713.

Deng, Y., Theken, K. N., & Lee, C. R. (2010). Cytochrome P450 epoxygenases, soluble epoxide hydrolase and the regulation of cardiovascular inflammation. *Journal of Molecular and Cellular Cardiology, 48*(2), 331–341.

Dickmann, L. J., McBride, H. J., Patel, S. K., Miner, K., Wienkers, L. C., & Slatter, J. G. (2012). Murine collagen antibody induced arthritis (CAIA) and primary mouse hepatocyte culture as models to study cytochrome P450 suppression. *Biochemical Pharmacology, 83*(12), 1682–1689.

Dickmann, L. J., Patel, S. K., Rock, D. A., Wienkers, L. C., & Slatter, J. G. (2011). Effects of interleukin-6 (IL-6) and an anti-IL-6 monoclonal antibody on drug-metabolizing enzymes in human hepatocyte culture. *Drug Metabolism and Disposition, 39*(8), 1415–1422.

Dickmann, L. J., Patel, S. K., Wienkers, L. C., & Slatter, J. G. (2012). Effects of interleukin 1β (IL-1β) and IL-1β/interleukin 6 (IL-6) combinations on drug metabolizing enzymes in human hepatocyte culture. *Current Drug Metabolism, 13*(7), 930–937.

Du, L., Yin, H., Morrow, J. D., Strobel, H. W., & Keeney, D. S. (2009). 20-Hydroxylation is the CYP-dependent and retinoid-inducible leukotriene B_4 inactivation pathway in human and mouse skin cells. *Archives of Biochemistry and Biophysics, 484*(1), 80–86.

Dvorak, H. F. (1986). Tumors: wounds that do not heal. Similarities between tumor stroma generation and wound healing. *The New England Journal of Medicine, 315*(26), 1650–1659.

Edson, K. Z., & Rettie, A. E. (2013). CYP4 enzymes as potential drug targets: Focus on enzyme multiplicity, inducers and inhibitors, and therapeutic modulation of 20-hydroxyeicosatetraenoic acid (20-HETE) synthase and fatty acid ω-hydroxylase activities. *Current Topics in Medicinal Chemistry, 13*(12), 1429–1440.

Elshenawy, O. H., Anwar-Mohamed, A., & El-Kadi, A. O. (2013). 20-Hydroxyeicosatetraenoic acid is a potential therapeutic target in cardiovascular diseases. *Current Drug Metabolism, 14*(6), 706–719.

Evers, R., Dallas, S., Dickmann, L. J., Fahmi, O. A., Kenny, J. R., Kraynov, E., et al. (2013). Critical review of preclinical approaches to investigate cytochrome P450-mediated therapeutic protein drug-drug interactions and recommendations for best practices: A white paper. *Drug Metabolism and Disposition, 41*(9), 1598–1609.

Fan, F., Muroya, Y., & Roman, R. J. (2015). Cytochrome P450 eicosanoids in hypertension and renal disease. *Current Opinion in Nephrology and Hypertension, 24*(1), 37–46.

Fischer, R., Konkel, A., Mehling, H., Blossey, K., Gapelyuk, A., Wessel, N., et al. (2014). Dietary omega-3 fatty acids modulate the eicosanoid profile in man primarily via the CYP-epoxygenase pathway. *The Journal of Lipid Research, 55*(6), 1150–1164.

Fleming, I. (2014). The pharmacology of the cytochrome P450 epoxygenase/soluble epoxide hydrolase axis in the vasculature and cardiovascular disease. *Pharmacological Reviews, 66*(4), 1106–1140.

Frye, R. F., Schneider, V. M., Frye, C. S., & Feldman, A. M. (2002). Plasma levels of TNF-alpha and IL-6 are inversely related to cytochrome P450-dependent drug metabolism in patients with congestive heart failure. *Journal of Cardiac Failure, 8*(5), 315–319.

Gandhi, A., Moorthy, B., & Ghose, R. (2012). Drug disposition in pathophysiological conditions. *Current Drug Metabolism, 13*(9), 1327–1344.

Gebremedhin, D., Lange, A. R., Lowry, T. F., Taheri, M. R., Birks, E. K., Hudetz, A. G., et al. (2000). Production of 20-HETE and its role in autoregulation of cerebral blood flow. *Circulation Research, 87*(1), 60–65.

Greenberg, A. S., & Obin, M. S. (2006). Obesity and the role of adipose tissue in inflammation and metabolism. *The American Journal of Clinical Nutrition, 83*(2), 461S–465S.

Hanahan, D., & Weinberg, R. A. (2011). Hallmarks of cancer: The next generation. *Cell, 144*(5), 646–674.

Hardwick, J. P., Eckman, K., Kwang Lee, Y., Abdelmegeed, M. A., Esterle, A., Chilian, W. M., et al. (2013). Eicosanoids in metabolic syndrome. *Advances in Pharmacology, 66*, 157–266.

Harmon, S. D., Fang, X., Kaduce, T. L., Hu, S., Raj Gopal, V., Falck, J. R., et al. (2006). Oxygenation of omega-3 fatty acids by human cytochrome P450 4F3B: Effect on 20-hydroxyeicosatetraenoic acid production. *Prostaglandins, Leukotrienes, and Essential Fatty Acids, 75*(3), 169–177.

Harvey, R. D., & Morgan, E. T. (2014). Cancer, inflammation, and therapy: Effects on cytochrome P450-mediated drug metabolism and implications for novel immunotherapeutic agents. *Clinical Pharmacology & Therapeutics, 96*(4), 449–457.

Hotamisligil, G. S. (2010). Endoplasmic reticulum stress and the inflammatory basis of metabolic disease. *Cell*, *140*(6), 900–917.
Ikehata, A., Hiwatashi, N., Kinouchi, Y., Yamazaki, H., Ito, K., & Toyota, T. (1995). Altered leukotriene B_4 metabolism in colonic mucosa with inflammatory bowel disease. *Scandinavian Journal of Gastroenterology*, *30*(1), 44–49.
Imig, J. D. (2012). Epoxides and soluble epoxide hydrolase in cardiovascular physiology. *Physiological Reviews*, *92*(1), 101–130.
Imig, J. D. (2013). Epoxyeicosatrienoic acids, 20-hydroxyeicosatetraenoic acid, and renal microvascular function. *Prostaglandins & Other Lipid Mediators*, *104–105*, 2–7.
Imig, J. D., & Hammock, B. D. (2009). Soluble epoxide hydrolase as a therapeutic target for cardiovascular disease. *Nature Reviews. Drug Discovery*, *8*(10), 794–805.
Imig, J. D., Simpkins, A. N., Renic, M., & Harder, D. R. (2011). Cytochrome P450 eicosanoids and cerebral vascular function. *Expert Reviews in Molecular Medicine*, *13*, e7.
Ishizuka, T., Cheng, J., Singh, H., Vitto, M. D., Manthati, V. L., Falck, J. R., et al. (2008). 20-Hydroxyeicosatetraenoic acid stimulates nuclear factor-κB activation and the production of inflammatory cytokines in human endothelial cells. *The Journal of Pharmacology and Experimental Therapeutics*, *324*(1), 103–110.
Jiang, H., Anderson, G. D., & McGiff, J. C. (2010). Red blood cells (RBCs), epoxyeicosatrienoic acids (EETs) and adenosine triphosphate (ATP). *Pharmacological Reports*, *62*(3), 468–474.
Jiang, J. G., Chen, C. L., Card, J. W., Yang, S., Chen, J. X., & Fu, X. N. (2005). Cytochrome P450 2J2 promotes the neoplastic phenotype of carcinoma cells and is up-regulated in human tumors. *Cancer Research*, *65*(11), 4707–4715.
Jiang, J. G., Ning, Y. G., Chen, C., Ma, D., Liu, Z. J., Yang, S., et al. (2007). Cytochrome P450 epoxygenase promotes human cancer metastasis. *Cancer Research*, *67*(14), 6665–6674.
Kacevska, M., Mahns, A., Sharma, R., Clarke, S. J., Robertson, G. R., & Liddle, C. (2013). Extra-hepatic cancer represses hepatic drug metabolism via interleukin (IL)-6 signalling. *Pharmaceutical Research*, *30*(9), 2270–2278.
Kalsotra, A., Cui, X., Antonovic, L., Robida, A. M., Morgan, E. T., & Strobel, H. W. (2003). Inflammatory prompts produce isoform-specific changes in the expression of leukotriene B_4 ω-hydroxylases in rat liver and kidney. *FEBS Letters*, *555*(2), 236–242.
Kalsotra, A., Du, L., Wang, Y., Ladd, P. A., Kikuta, Y., Duvic, M., et al. (2008). Inflammation resolved by retinoid X receptor-mediated inactivation of leukotriene signaling pathways. *The FASEB Journal*, *22*(2), 538–547.
Kalsotra, A., & Strobel, H. W. (2006). Cytochrome P450 4F subfamily: At the crossroads of eicosanoid and drug metabolism. *Pharmacology & Therapeutics*, *112*(3), 589–611.
Kang, J. X. (2007). Fat-1 transgenic mice: A new model for omega-3 research. *Prostaglandins, Leukotrienes, and Essential Fatty Acids*, 77(5-6), 263–267.
Kawakami, H., Ohtsuki, S., Kamiie, J., Suzuki, T., Abe, T., & Terasaki, T. (2011). Simultaneous absolute quantification of 11 cytochrome P450 isoforms in human liver microsomes by liquid chromatography tandem mass spectrometry with in silico target peptide selection. *Journal of Pharmaceutical Sciences*, *100*(1), 341–352.
Kelly, E. J., Nakano, M., Rohatgi, P., Yarov-Yarovoy, V., & Rettie, A. E. (2011). Finding homes for orphan cytochrome P450s: CYP4V2 and CYP4F22 in disease states. *Molecular Interventions*, *11*(2), 124–132.
Kikuta, Y., Kusunose, E., Endo, K., Yamamoto, S., Sogawa, K., Fujii-Kuriyama, Y., et al. (1993). A novel form of cytochrome P-450 family 4 in human polymorphonuclear leukocytes. cDNA cloning and expression of leukotriene B_4 omega-hydroxylase. *The Journal of Biological Chemistry*, *268*(13), 9376–9380.
Kim, S., Ostor, A. J., & Nisar, M. K. (2012). Interleukin-6 and cytochrome-P450, reason for concern? *Rheumatology International*, *32*(9), 2601–2604.

Kim, D. H., Vanella, L., Inoue, K., Burgess, A., Gotlinger, K., Manthati, V. L., et al. (2010). Epoxyeicosatrienoic acid agonist regulates human mesenchymal stem cell-derived adipocytes through activation of HO-1-pAKT signaling and a decrease in PPARγ. *Stem Cells and Development, 19*(12), 1863–1873.

Kinloch, R. D., Lee, C.-M., van Rooijen, N., & Morgan, E. T. (2011). Selective role for tumor necrosis factor-α, but not interleukin-1 or Kupffer cells, in down-regulation of CYP3A11 and CYP3A25 in livers of mice infected with a noninvasive intestinal pathogen. *Biochemical Pharmacology, 82*(3), 312–321.

Kraemer, M. J., Furukawa, C. T., Koup, J. R., Shapiro, G. G., Pierson, W. E., & Bierman, C. W. (1982). Altered theophylline clearance during an influenza B outbreak. *Pediatrics, 69*(4), 476–480.

Lasker, J. M., Chen, W. B., Wolf, I., Bloswick, B. P., Wilson, P. D., & Powell, P. K. (2000). Formation of 20-hydroxyeicosatetraenoic acid, a vasoactive and natriuretic eicosanoid, in human kidney. Role of CYP4F2 and CYP4A11. *The Journal of Biological Chemistry, 275*(6), 4118–4126.

Li, R., Xu, X., Chen, C., Wang, Y., Gruzdev, A., Zeldin, D. C., et al. (2015). CYP2J2 attenuates metabolic dysfunction in diabetic mice by reducing hepatic inflammation via the PPARγ. *American Journal of Physiology. Endocrinology and Metabolism, 308*(4), E270–E282.

Libby, P. (2006). Inflammation and cardiovascular disease mechanisms. *The American Journal of Clinical Nutrition, 83*(2), 456S–460S.

Libby, P. (2007). Inflammatory mechanisms: the molecular basis of inflammation and disease. *Nutrition Reviews, 65*(12), S140–S146.

Ling, S., & Jamali, F. (2009). The effect of infliximab on hepatic cytochrome P450 and pharmacokinetics of verapamil in rats with pre-adjuvant arthritis: A drug-disease and drug-drug interaction. *Basic & Clinical Pharmacology & Toxicology, 105*(1), 24–29.

Liu, Y., Zhang, Y., Schmeizer, K., Lee, T.-S., Fang, X., Zhu, Y., et al. (2005). The anti-inflammatory effect of laminar flow: The role of PPARγ, epoxyeicosatrienoic acids, and soluble epoxide hydrolase. *Proceedings of the National Academy of Sciences of the United States of America, 102*(46), 16747–16752.

Liu, X., Zhao, Y., Wang, L., Yang, X., Zheng, Z., Zhang, Y., et al. (2009). Overexpression of cytochrome P450 4F2 in mice increases 20-hydroxyeicosatetraenoic acid production and arterial blood pressure. *Kidney International, 75*(12), 1288–1296.

Lopez-Vicario, C., Alcaraz-Quiles, J., Garcia-Alonso, V., Rius, B., Hwang, S. H., Titos, E., et al. (2015). Inhibition of soluble epoxide hydrolase modulates inflammation and autophagy in obese adipose tissue and liver: Role for omega-3 epoxides. *Proceedings of the National Academy of Sciences of the United States of America, 112*(2), 536–541.

Lustig, R. H. (2013). Fructose: It's "alcohol without the buzz" *Advances in Nutrition, 4*(2), 226–235.

Lutz, U., Bittner, N., Ufer, M., & Lutz, W. K. (2010). Quantification of cortisol and 6 beta-hydroxycortisol in human urine by LC-MS/MS, and gender-specific evaluation of the metabolic ratio as biomarker of CYP3A activity. *Journal of Chromatography. B, Analytical Technologies in the Biomedical and Life Sciences, 878*(1), 97–101.

Manhiani, M., Quigley, J. E., Knight, S. F., Tasoobshirazi, S., Moore, T., Brands, M. W., et al. (2009). Soluble epoxide hydrolase gene deletion attenuates renal injury and inflammation with DOCA-salt hypertension. *American Journal of Physiology. Renal Physiology, 297*(3), F740–F748.

Moreno, J. J. (2009). New aspects of the role of hydroxyeicosatetraenoic acids in cell growth and cancer development. *Biochemical Pharmacology, 77*(1), 1–10.

Morgan, E. T. (1997). Regulation of cytochromes P450 during inflammation and infection. *Drug Metabolism Reviews, 29*(4), 1129–1188.

Morgan, E. T. (2009). Impact of infectious and inflammatory disease on cytochrome P450-mediated drug metabolism and pharmacokinetics. *Clinical Pharmacology & Therapeutics*, *85*(4), 434–438.
Murphy, R. C., & Gijon, M. A. (2007). Biosynthesis and metabolism of leukotrienes. *Biochemical Journal*, *405*(3), 379–395.
Nathan, C. (2002). Points of control in inflammation. *Nature*, *420*(6917), 846–852.
Nguyen, T. V., Ukairo, O., Khetani, S. R., McVay, M., Kanchagar, C., Seghezzi, W., et al. (2015). Establishment of a hepatocyte-Kupffer cell co-culture model for assessment of proinflammatory cytokine effects on metabolizing enzymes and drug transporters. *Drug Metabolism and Disposition*, *43*(5), 774–785.
Nithipatikom, K., Gross, E. R., Endsley, M. P., Moore, J. M., Isbell, M. A., Falck, J. R., et al. (2004). Inhibition of cytochrome P450ω-hydroxylase. A novel endogenous cardioprotective pathway. *Circulation Research*, *95*(8), e65–e71.
Node, K., Huo, Y., Ruan, X., Yang, B., Spiecker, M., Ley, K., et al. (1999). Anti-inflammatory properties of cytochrome P450 epoxygenase-derived eicosanoids. *Science*, *285*(5431), 1276–1279.
Nyagode, B. A., Jahangardi, R., Merrell, M. D., Tansey, M. G., & Morgan, E. T. (2014). Selective effects of a therapeutic protein targeting tumor necrosis factor-alpha on cytochrome P450 regulation during infectious colitis: Implications for disease-dependent drug-drug interactions. *Pharmacology Research & Perspectives*, *2*(1), e00027.
Nyagode, B. A., Lee, C.-M., & Morgan, E. T. (2010). Modulation of hepatic cytochrome P450s by *Citrobacter rodentium* infection in interleukin-6- and interferon-γ-null mice. *The Journal of Pharmacology and Experimental Therapeutics*, *335*(2), 480–488.
Oni-Orisan, A., Alsaleh, N., Lee, C. R., & Seubert, J. M. (2014). Epoxyeicosatrienoic acids and cardioprotection: The road to translation. *Journal of Molecular and Cellular Cardiology*, *74*, 199–208.
Panigrahy, D., Edin, M. L., Lee, C. R., Huang, S., Bielenberg, D. R., Butterfield, C. E., et al. (2012). Epoxyeicosanoids stimulate multiorgan metastasis and tumor dormancy escape in mice. *The Journal of Clinical Investigation*, *122*(1), 178–191.
Panigrahy, D., Greene, E. R., Pozzi, A., Wang, D. W., & Zeldin, D. C. (2011). EET signaling in cancer. *Cancer and Metastasis Reviews*, *30*(3-4), 525–540.
Panigrahy, D., Kalish, B. T., Huang, S., Bielenberg, D. R., Le, H. D., Yang, J., et al. (2013). Epoxyeicosanoids promote organ and tissue regeneration. *Proceedings of the National Academy of Sciences of the United States of America*, *110*(33), 13528–13533.
Patel, N. S., Cuzzocrea, S., Chatterjee, P. K., Di Paola, R., Sautebin, L., Britti, D., et al. (2004). Reduction of renal ischemia-reperfusion injury in 5-lipoxygenase knockout mice and by the 5-lipoxygenase inhibitor zileuton. *Molecular Pharmacology*, *66*(2), 220–227.
Pillarisetti, S., & Khanna, I. (2012). Targeting soluble epoxide hydrolase for inflammation and pain—An overview of pharmacology and the inhibitors. *Inflammation & Allergy Drug Targets*, *11*(2), 143–158.
Piquette-Miller, M., & Jamali, F. (1995). Influence of severity of inflammation on the disposition kinetics of propranolol enantiomers in ketoprofen-treated and untreated adjuvant arthritis. *Drug Metabolism and Disposition*, *23*(2), 240–245.
Plee-Gautier, E., Antoun, J., Goulitquer, S., Le Jossic-Corcos, C., Simon, B., Amet, Y., et al. (2012). Statins increase cytochrome P450 4F3-mediated eicosanoids production in human liver cells: A PXR dependent mechanism. *Biochemical Pharmacology*, *84*(4), 571–579.
Pober, J. S., & Sessa, W. C. (2007). Evolving functions of endothelial cells in inflammation. *Nature Reviews. Immunology*, 7(10), 803–815.
Pozzi, A., Popescu, V., Yang, S., Mei, S., Shi, M., Puolitaival, S. M., et al. (2010). The anti-tumorigenic properties of peroxisomal proliferator-activated receptor alpha are

arachidonic acid epoxygenase-mediated. *The Journal of Biological Chemistry, 285*(17), 12840–12850.

Renton, K. W. (2004). Cytochrome P450 regulation and drug biotransformation during inflammation and infection. *Current Drug Metabolism, 5*(3), 235–243.

Richardson, T. A., Sherman, M., Antonovic, L., Kardar, S. S., Strobel, H. W., Kalman, D., et al. (2006). Hepatic and renal cytochrome P450 gene regulation during *Citrobacter rodentium* infection in wildtype and Toll-like receptor 4 mutant mice. *Drug Metabolism and Disposition, 34*(3), 354–360.

Rivory, L. P., Slaviero, K. A., & Clarke, S. J. (2002). Hepatic cytochrome P450 3A drug metabolism is reduced in cancer patients who have an acute-phase response. *British Journal of Cancer, 87*(3), 277–280.

Rodriguez-Antona, C., & Ingelman-Sundberg, M. (2006). Cytochrome P450 pharmacogenetics and cancer. *Oncogene, 25*(11), 1679–1691.

Rubin, K., Janefeldt, A., Andersson, L., Berke, Z., Grime, K., & Andersson, T. B. (2015). HepaRG cells as human-relevant in vitro model to study the effects of inflammatory stimuli on cytochrome P450 isoenzymes. *Drug Metabolism and Disposition, 43*(1), 119–125.

Schmeizer, K. R., Kubala, L., Newman, J. W., Kim, I. H., Eiserich, J. P., & Hammock, B. D. (2005). Soluble epoxide hydrolase is a therapeutic target for acute inflammation. *Proceedings of the National Academy of Sciences of the United States of America, 102*(28), 9772–9777.

Schmitt, C., Kuhn, B., Zhang, X., Kivitz, A. J., & Grange, S. (2011). Disease-drug-drug interaction involving tocilizumab and simvastatin in patients with rheumatoid arthritis. *Clinical Pharmacology & Therapeutics, 89*(5), 735–740.

Schuck, R. N., Theken, K. N., Edin, M. L., Caughey, M., Bass, A., Ellis, K., et al. (2013). Cytochrome P450-derived eicosanoids and vascular dysfunction in coronary artery disease patients. *Atherosclerosis, 227*(2), 442–448.

Schuck, R. N., Zha, W., Edin, M. L., Gruzdev, A., Vendrov, K. C., Miller, T. M., et al. (2014). The cytochrome P450 epoxygenase pathway regulates the hepatic inflammatory response in fatty liver disease. *PLoS One, 9*(10), e110162.

Sehgal, N., Agarwal, V., Valli, R. K., Joshi, S. D., Antonovic, L., Strobel, H. W., et al. (2011). Cytochrome P4504f, a potential therapeutic target limiting neuroinflammation. *Biochemical Pharmacology, 82*(1), 53–64.

Shahabi, P., Siest, G., Meyer, U. A., & Visvikis-Siest, S. (2014). Human cytochrome P450 epoxygenases: Variability in expression and role in inflammation-related disorders. *Pharmacology & Therapeutics, 144*(2), 134–161.

Shahabi, P., Siest, G., & Visvikis-Siest, S. (2014). Influence of inflammation on cardiovascular protective effects of cytochrome P450 epoxygenase-derived epoxyeicosatrienoic acids. *Drug Metabolism Reviews, 46*(1), 33–56.

Shak, S., & Goldstein, I. M. (1984). Omega-oxidation is the major pathway for the catabolism of leukotriene B_4 in human polymorphonuclear leukocytes. *The Journal of Biological Chemistry, 259*(16), 10181–10187.

Shedlofsky, S. I., Israel, B. C., McClain, C. J., Hill, D. B., & Blouin, R. A. (1994). Endotoxin administration to humans inhibits hepatic cytochrome P450-mediated drug metabolism. *The Journal of Clinical Investigation, 94*(6), 2209–2214.

Shin, K. H., Choi, M. H., Lim, K. S., Yu, K. S., Jang, I. J., & Cho, J. Y. (2013). Evaluation of endogenous metabolic markers of hepatic CYP3A activity using metabolic profiling and midazolam clearance. *Clinical Pharmacology & Therapeutics, 94*(5), 601–609.

Siest, G., Jeannesson, E., Marteau, J. B., Samara, A., Pfister, M., & Visvikis-Siest, S. (2009). Drug metabolizing enzymes and transporters mRNA in peripheral blood mononuclear cells of healthy subjects: Biological variations and importance of pre-analytical steps. *Current Drug Metabolism, 10*(4), 410–419.

Simopoulos, A. P. (2008). The importance of the omega-6/omega-3 fatty acid ratio in cardiovascular disease and other chronic diseases. *Experimental Biology and Medicine*, *233*(6), 674–688.

Slaviero, K. A., Clarke, S. J., & Rivory, L. P. (2003). Inflammatory response: An unrecognized source of variability in the pharmacokinetics and pharmacodynamics of cancer chemotherapy. *Lancet Oncology*, *4*(4), 224–232.

Smith, K. R., Pinkerton, K. E., Watanabe, T., Pedersen, T. L., Ma, S. J., & Hammock, B. D. (2005). Attenuation of tobacco smoke-induced lung inflammation by treatment with a soluble epoxide hydrolase inhibitor. *Proceedings of the National Academy of Sciences of the United States of America*, *102*(6), 2186–2191.

Sodhi, K., Inoue, K., Gotlinger, K. H., Canestraro, M., Vanella, L., Kim, D. H., et al. (2009). Epoxyeicosatrienoic acid agonist rescues the metabolic syndrome phenotype of HO-2-null mice. *Journal of Pharmacology and Experimental Therapeutics*, *331*(3), 906–916.

Spector, A. A., & Kim, H. Y. (2015). Cytochrome P450 epoxygenase pathway of polyunsaturated fatty acid metabolism. *Biochimica et Biophysica Acta*, *1851*(4), 356–365.

Stoilov, I., Krueger, W., Mankowski, D., Guernsey, L., Kaur, A., Glynn, J., et al. (2006). The cytochromes P450 (CYP) response to allergic inflammation of the lung. *Archives of Biochemistry and Biophysics*, *456*(1), 30–38.

Straub, R. H. (2014). Insulin resistance, selfish brain, and selfish immune system: An evolutionarily positively selected program used in chronic inflammatory diseases. *Arthritis Research & Therapy*, *16*(Suppl. 2), S4.

Sudhahar, V., Shaw, S., & Imig, J. D. (2010). Epoxyeicosatrienoic acid analogs and vascular function. *Current Medicinal Chemistry*, *17*(12), 1181–1190.

Sun, C. W., Falck, J. R., Harder, D. R., & Roman, R. J. (1999). Role of tyrosine kinase and PKC in the vasoconstrictor response to 20-HETE in renal arterioles. *Hypertension*, *33*(1), 414–418.

Sunman, J. A., Hawke, R. L., LeCluyse, E. L., & Kashuba, A. D. M. (2004). Kupffer cell-mediated IL-2 suppression of CYP3A activity in human hepatocytes. *Drug Metabolism and Disposition*, *32*(3), 359–363.

Tanaka, T., & Kishimoto, T. (2014). The biology and medical implications of interleukin-6. *Cancer Immunology Research*, *2*(4), 288–294.

Theken, K. N., Deng, Y., Kannon, M. A., Miller, T. M., Poloyac, S. M., & Lee, C. R. (2011). Activation of the acute inflammatory response alters cytochrome P450 expression and eicosanoid metabolism. *Drug Metabolism and Disposition*, *39*(1), 22–29.

Theken, K. N., Schuck, R. N., Edin, M. L., Tran, B., Ellis, K., Bass, A., et al. (2012). Evaluation of cytochrome P450-derived eicosanoids in humans with stable atherosclerotic cardiovascular disease. *Atherosclerosis*, *222*(2), 530–536.

Vainio, P., Gupta, S., Ketola, K., Mirtti, T., Mpindi, J.-P., Kohonen, P., et al. (2011). Arachidonic acid pathway members PLA2G7, HPGD, EPHX2, and CYP4F8 identified as putative novel therapeutic targets in prostate cancer. *The American Journal of Pathology*, *178*(2), 525–536.

Wang, Y. X., Ulu, A., Zhang, L. N., & Hammock, B. (2010). Soluble epoxide hydrolase in atherosclerosis. *Current Atherosclerosis Reports*, *12*(3), 174–183.

Wang, Y., Zhao, J., Kalsotra, A., Turman, C. M., Grill, R. J., Dash, P. K., et al. (2008). CYP4Fs expression in rat brain correlates with changes in LTB_4 levels after traumatic brain injury. *Journal of Neurotrauma*, *25*(10), 1187–1194.

Wei, X., Zhang, D., Dou, X., Niu, N., Huang, W., Bai, J., et al. (2014). Elevated 14,15-epoxyeicosatrienoic acid by increasing of cytochrome P450 2C8, 2C9 and 2J2 and decreasing of soluble epoxide hydrolase associated with aggressiveness of human breast cancer. *BMC Cancer*, *14*, 841.

Winslow, V., Vaivoda, R., Vasilyev, A., Dombkowski, D., Douaidy, K., Stark, C., et al. (2014). Altered leukotriene B_4 metabolism in CYP4F18-deficient mice does not impact inflammation following renal ischemia. *Biochimica et Biophysica Acta*, *1841*(6), 868–879.
Wu, C. C., Gupta, T., Garcia, V., Ding, Y., & Schwartzman, M. L. (2014). 20-HETE and blood pressure regulation: Clinical implications. *Cardiology in Review*, *22*(1), 1–12.
Xu, X., Tu, L., Feng, W., Ma, B., Li, R., Zheng, C., et al. (2013). CYP2J3 gene delivery up-regulated adiponectin expression via reduced endoplasmic reticulum stress in adipocytes. *Endocrinology*, *154*(5), 1743–1753.
Yang, K. H., & Lee, M. G. (2008). Effects of endotoxin derived from Escherichia coli lipopolysaccharide on the pharmacokinetics of drugs. *Archives of Pharmacal Research*, *31*(9), 1073–1086.
Zanger, U. M., & Schwab, M. (2013). Cytochrome P450 enzymes in drug metabolism: Regulation of gene expression, enzyme activities, and impact of genetic variation. *Pharmacology & Therapeutics*, *138*(1), 103–141.
Zhang, G., Kodani, S., & Hammock, B. D. (2014). Stabilized epoxygenated fatty acids regulate inflammation, pain, angiogenesis and cancer. *Progress in Lipid Research*, *53*, 108–123.
Zhang, W., Yang, A. L., Liao, J., Li, H., Dong, H., Chung, Y. T., et al. (2012). Soluble epoxide hydrolase gene deficiency or inhibition attenuates chronic active inflammatory bowel disease in IL-10(−/−) mice. *Digestive Diseases and Sciences*, *57*(10), 2580–2591.
Zordoky, B. N., & El-Kadi, A. O. (2009). Role of NF-kappaB in the regulation of cytochrome P450 enzymes. *Current Drug Metabolism*, *10*(2), 164–178.

CHAPTER SEVEN

Cytochrome P450-CYP2 Family-Epoxygenase Role in Inflammation and Cancer

Chen Chen, Dao Wen Wang[1]
Department of Internal Medicine and Gene Therapy Center, Tongji Hospital, Tongji Medical College, Huazhong University of Science and Technology, Wuhan, PR China
[1]Corresponding author: e-mail address: dwwang@tjh.tjmu.edu.cn

Contents

Abstract

Arachidonic acids are converted to eicosanoid mediators by different enzyme systems: cyclooxygenase, lipoxygenase, and cytochrome P450 (CYP) monooxygenase pathways (ω/ω-1-hydroxylases and epoxygenases). Of 57 putatively functional human CYPs, only about a dozen enzymes are responsible for xenobiotic metabolism. CYP2 family is the predominant epoxygenase isoform abundantly expressed in the endothelium,

Advances in Pharmacology, Volume 74
ISSN 1054-3589
http://dx.doi.org/10.1016/bs.apha.2015.04.005

myocardium, and kidney in humans. Numerous studies have demonstrated the cardiovascular protective effects of CYP epoxygenases and eicosatrienoic acids ranging from vasodilation, antihypertensive, proangiogenesis, antiatherosclerosis, and cardiac protection. The roles of CYP2 family and their metabolites in inflammation and cancer biology have recently attracted great attention. Here, we review the recent progress on polymorphisms, distribution and function of CYP2 family, and their roles in inflammation and cancer.

ABBREVIATIONS

AAA abdominal aortic aneurysm
AAs arachidonic acids
Ang II angiotensin II
AT2R angiotensin II receptor type 2
ATO arsenic trioxide
CYP cytochrome P450
DHA docosahexaenoic acid
EETs eicosatrienoic acids
eNOS endothelial nitric oxide synthase
EPA eicosapentaenoic acid
EpDPE epoxydocosapentaenoic acids
EpETE epoxyeicosatetraenoic acids
HETEs hydroxyeicosatetraenoic acids
ISO isoproterenol
MMP matrix metalloprotein
NF-κB nuclear factor-kappaB
NNK 4-(methylnitrosamino)-1-(3-pyridyl)-butanone
PPARs peroxisome proliferator-activated receptors
PUFAs polyunsaturated fatty acids
rAAV recombinant adeno-associated virus
ROS reactive oxygen species
sEH soluble epoxide hydrolase
SERCA2a sarcoplasmic/endoplasmic reticulum calcium ATPase
TNF-α tumor necrosis factor-α
VSMCs vascular smooth muscle cells

1. INTRODUCTION

Inflammation is the immune system's response to infection and injury and has been implicated in the pathogeneses of multiple diseases, including cancer. Inflammation is an intrinsically beneficial event that leads to removal of offending factors and restoration of tissue structure and physiological function. The acute phase of inflammation is characterized by the rapid influx of blood granulocytes, typically neutrophils, followed by monocytes that mature into inflammatory macrophages that subsequently proliferate and thereby affect the functions of resident tissue macrophages. This process causes the cardinal signs of acute inflammation: rubor (redness), calor (heat), tumor (swelling), and dolor (pain). Once the initiating noxious stimulus is removed via phagocytosis, the inflammatory reaction can decrease and resolve. During the resolution of inflammation, granulocytes are eliminated, and macrophages and lymphocytes return to normal preinflammatory numbers and phenotypes. The usual outcome of the acute inflammatory program is successful resolution and repair of tissue damage, rather than persistence and dysfunction of the inflammatory response, which can lead to scarring and loss of organ function (Nathan, 2002; Ricciotti & FitzGerald, 2011).

The hallmarks of cancer comprise six biological capabilities acquired during the multistep development of human tumors, including sustaining proliferative signaling, evading growth suppressors, resisting cell death, enabling replicative immortality, inducing angiogenesis, and activating tissue invasion and metastasis. Underlying these hallmarks is genome instability, which generates the genetic diversity that expedites their acquisition, and inflammation, which fosters multiple hallmark functions. Conceptual progress in the last decade has added two emerging hallmarks of potential generality to this list-reprogramming of energy metabolism and evading immune destruction (Hanahan & Weinberg, 2000, 2011).

Chronic inflammation has long been associated with the development of cancer (Hudson et al., 1999). Over the past decades, manipulation of pharmacological inhibitors or genes involved in the determination or functions of various immune cell types showed that inflammation plays diverse and critical roles in fostering tumorigenesis (Hanahan & Weinberg, 2011). Recent studies have identified tumor-promoting functions of inflammatory cells including macrophage subtypes, mast cells, and neutrophils, as well as T and B lymphocytes (DeNardo, Andreu, & Coussens, 2010; Murdoch, Muthana, Coffelt, & Lewis, 2008). Moreover, a growing list of signaling molecules released by inflammatory cells has been identified as effectors of tumor-promoting actions. These include the tumor growth factor (epidermal growth factor, EGF), the angiogenic growth factor (vascular endothelial growth factor, VEGF), other proangiogenic factors such as fibroblast growth factor 2, chemokines, and cytokines that amplify the inflammatory state; in addition, the inflammatory cells may produce proangiogenic and/or proinvasive matrix-degrading enzymes, including matrix metalloprotein-9 (MMP-9) and other MMPs, cysteine cathepsin proteases, and Heparanse (Qian & Pollard, 2010). Consistent with their expression of these diverse effectors, tumor-infiltrating inflammatory cells have been shown to induce and enhance sustained tumor angiogenesis, stimulate cancer cell proliferation, and facilitate invasion and migration (Joyce & Pollard, 2009; Mantovani, 2010; Mantovani, Allavena, Sica, & Balkwill, 2008).

Accumulating data suggest that metabolism mediated by cytochrome P450 (CYP) enzymes play an important role in the cardiovascular system. Human (CYP) epoxygenase is dominantly expressed in the heart and exhibits cardioprotective effects on cardiomyocyte apoptosis, ischemia reperfusion injury, hypertrophy, and heart failure (Cai et al., 2013; Monti et al., 2008; Seubert et al., 2004; Wang et al., 2014; Zhao et al., 2012b). Moreover, CYP2 isoforms (e.g., CYP2C and CYP2J) convert AA into eicosatrienoic acids (EETs), which have important roles in other body systems, such as protection of renal functions (Chen, Wang, et al., 2011; Chen, Xu, et al., 2012b; Ma, Lin, et al., 2013; Makita et al., 1994; Wang et al., 2003; Xiao et al., 2010; Zhao et al., 2012a), amelioration of pulmonary arterial hypertension and pulmonary ischemia/reperfusion injury (Chen et al., 2014; Feng et al., 2013; Zheng et al., 2010), attenuation of insulin resistance, and the regulation of lipid/glucose metabolism (Chen et al., 2014; Li et al., 2015; Ma, Xiong, et al., 2013; Xu, Tu, et al., 2013; Xu, Tu, Wang, Fang, & Wang, 2011). However, the role of CYP epoxygenase in inflammation and cancer did not attract attention until recently. Recently, several laboratories have shown that AA and its CYP epoxygenase metabolites play significant roles in inflammation and cancer biology (Chen & Wang, 2013; Panigrahy, Greene, Pozzi, Wang, & Zeldin, 2011; Xu, Zhang, & Wang, 2011). The association between CYP epoxygenases and cancer phenotype was detected (Jiang et al., 2005; Jiang, Ning, et al., 2007; Panigrahy et al., 2012). Subsequent work with synthetic inhibitors of CYP epoxygenases indicated that CYP epoxygenases, especially CYP2J2, may be a promising therapeutic target for cancer therapy (Chen et al., 2009, 2011; Pozzi et al., 2007, 2010).

2. OVERVIEW OF CYTOCHROME P450-CYP2 FAMILY-EPOXYGENASE

Now, it is well known that AA is converted to eicosanoid mediators by three distinct enzyme systems: cyclooxygenase (COX-1 and COX-2), lipoxygenase (5-LOX, 12-LOX, 15-LOXa, 15-LOXb), and CYP monooxygenase pathways (ω-1-hydroxylases and epoxygenases) (Brash, 2001). The CYP pathway produces two types of eicosanoid products: atypical hydroxyeicosatetraenoic acids (HETEs), formed by CYP oxidases dependent on NADPH, and EETs, formed by CYP epoxygenases (Capdevila, Falck, & Harris, 2000). Epoxidation can occur at any of the four double bonds of AA, resulting in four regioisomers, 5,6-, 8,9-, 11,12-, and 14,15-EETs. In most tissues, EETs are quickly metabolized into their less active dihydroxy derivatives mainly by soluble epoxide hydrolase (sEH) (Campbell & Falck, 2007).

CYP enzymes are membrane bound, comprised of a superfamily of heme-containing terminal oxidases, and contribute to the oxidative metabolism of endogenous molecules as well as xenobiotics. The number of CYP gene products now surpasses 2000 among various species, representing the largest family of proteins currently listed in gene data bank (Chen & Wang, 2013). Up to now, 57 different genes, arranged in 18 families and 42 subfamilies, and 46 pseudogenes have been found in human CYP superfamily (Table 1). Among them, only about a dozen enzymes, belonging to the CYP1, -2, and -3 families, are responsible for the xenobiotic metabolism.

The CYP2 family contains 16 full-length genes, which all have 9 exons and 8 introns. The genes are spread over different chromosomes and organized in multigene clusters containing one or several subfamilies (Hoffman, Nelson, & Keeney, 2001). The three largest gene clusters are the CYP2ABFGST cluster on chromosome 19q13.2, which contains the CYP2A6 and CYP2B6 genes, the CYP2C cluster on chromosome 10q23.33 with the CYP2C8, CYP2C9, and CYP2C19 genes, and the CYP2D cluster on chromosome 22q13.1–2 with the only functional gene CYP2D6(Zanger & Schwab, 2013). In the evolution of rodents, many of the CYP2 subfamilies expanded tremendously, making the identification of true orthologues between mouse and human CYPs especially challenging (Nelson et al., 2004). Most pharmacologically important CYP2 genes are highly polymorphic, in particular CYP2A6, CYP2B6, CYP2C9, CYP2C19, and CYP2D6.

3. SUBSTRATES AND GENETIC POLYMORPHISMS OF CYP2 FAMILY

3.1 CYP2A Subfamily

There are three full-length human CYP2A genes 2A6, 2A7, and 2A13 and a split pseudogene CYP2A18P on a 370 kb gene cluster on chromosome

Table 1 The 57 Human Cytochrome P450s

CYP family	Isoforms
1	1A1, 1A2, 1B1
2	2A6, 2A7, 2A13, 2B6, 2C8, 2C9, 2C18, 2C19, 2D6, 2E1, 2F1, 2J2, 2R1, 2S1, 2U1, 2W1
3	3A4, 3A5, 3A7, 3A43
4	4A11, 4A22, 4B1, 4F2, 4F3, 4F8, 4F11, 4F12, 4F22, 4V2, 4X1, 4Z1
5	5A1
7	7A1, 7B1
8	8A1, 8B1
11	11A1, 11B1, 11B2
17	17A1
19	19A1
20	20A1
21	21A1
24	24A1
26	26A1, 26B1, 26C1
27	27A1, 27B1, 27C1
39	39A1
46	46A1
51	51A1

19q13.2 which also contains genes and pseudogenes of other subfamilies (Hoffman et al., 2001). However, only CYP2A6 and CYP2A13 are functional, whereas CYP2A7 apparently encodes a nonfunctional gene (Ding, Lake, Friedberg, & Wolf, 1995).

CYP2A subfamily displays restricted substrate specificity. There are only a few drugs that are cleared solely or mainly by this subfamily, except coumarin and nicotine. Some of the other drugs that are partially metabolized by this subfamily include the anesthetics halothane and methoxyflurane; the antiepileptics losigamone and valproic acid; anticancer drugs tegafur and doxifluridine; and disulfiram, used for alcohol withdrawal (Krau, 2013).

Recently, CYP2A6 and CYP2A13 enzymes were shown to be high-affinity metabolizers of nicotine and several other carcinogens.

The CYP2A6 gene is highly polymorphic with approximately 40 annotated allelic variants, such as deleted or duplicated genes, gene conversions, nucleotide deletions and insertions, as well as coding and noncoding SNPs. Both increased and decreased gene expression and enzyme activity have been associated with genetic variation in CYP2A6 gene (Zanger & Schwab, 2013). Tobacco consumption is the main identifiable risk to cancer, contributing to the majority of tumors in upper aerodigestive tissues. There are approximately eight different nitrosamines present in mainstream cigarette smoke, but the main ones are the tobacco-specific nitrosamines, 4-(methylnitrosamino)-1-(3-pyridyl)-butanone (NNK) and *N*-nitrosonornicotine. CYP2A6 and CYP2A13 are responsible for NNK activation (Rossini, de Almeida Simao, Albano, & Pinto, 2008). The polymorphisms which confer decreased CYP2A6 enzymatic activity, including the *4 allele, are related to deficient nicotine metabolism, and thus lead to lower cigarette consumption (Rao et al., 2000). Also, these polymorphisms were reported to be associated with reduced susceptibility to lung, oral, head, and neck cancers, particularly in smokers (Canova et al., 2009; Fujieda et al., 2004; Wassenaar et al., 2014). It should be pointed out that association studies of these alleles carried out in oriental individuals are likely to have more statistical power, because of the higher allele frequencies. Negative or inconclusive findings for the association were reported in Caucasian subjects (London, Idle, Daly, & Coetzee, 1999). Recently, a CYP2A13 haplotype carrying variant alleles of T478C/T494C was found to be associated with reduced risk of head and neck cancer (Sharma, Ahuja, Panda, & Khullar, 2010).

3.2 CYP2B Subfamily

The functional CYP2B6 gene and the nonfunctional pseudogene CYP2B7P are located in a tandem head-to-tail arrangement within the large CYP2ABFGST gene cluster on chromosome 19 (Nelson et al., 2004).

The CYP2B6 substrate selectivity accepts diverse chemicals, including not only clinically used drugs but also a large number of environmental chemicals. Therapeutically important drugs metabolized mainly by CYP2B6 include the prodrug cyclophosphamide, which is converted to the direct precursor of the cytotoxic metabolites, phosphoramide mustard and acrolein, by 4-hydroxylation; the nonnucleoside reverse transcriptase inhibitor, efavirenz; the atypical antidepressant and smoking cessation agent bupropion; the antimalarial artemisinin; the anesthetics propofol and ketamine; the μ-opioid agonist methadone; and others (Turpeinen & Zanger, 2012).

The CYP2B6 gene is extensively polymorphic with numerous variants in the coding and noncoding regions, including 29 distinct star alleles with variant amino acid sequence or with proven functional effects. Recent studies have suggested that the genotyping for CYP2B6 alleles (CYP2B6*4 (rs2279343), CYP2B6*5 (rs3211371), and CYP2B6*9 (rs3745274)) does not help in personalizing cancer therapy with tamoxifen or cyclophosphamide (Fleeman et al., 2011; Haroun et al., 2015).

3.3 CYP2C Subfamily

The human CYP2C subfamily consists of four highly homologous genes CYP2C18–CYP2C19–CYP2C9–CYP2C8 which are localized in a ~390 kb gene cluster on chromosome 10q23.3.

The major enzyme CYP2C9 not only metabolizes weakly acidic substances including the anticoagulant warfarin, the anticonvulsants phenytoin and valproic acid, the angiotensin receptor blockers candesartan and losartan, oral antidiabetics like glibenclamide and tolbutamide, and most nonsteroidal anti-inflammatory drugs but also metabolizes endogenous substances, in particular arachidonic acid (AA) and some steroids (Lee, Goldstein, & Pieper, 2002).

CYP2C8 is mainly responsible for the metabolism of the antidiabetics (rosiglitazone and pioglitazone), the antiarrhythmic (amiodarone), the natural anticancer drug (paclitaxel), and the antimalarial (amodiaquine, chloroquine, and dapsone) (Kerb et al., 2009).

The CYP2C19 enzyme was shown to be the major enzyme for the inactivating metabolism of proton pump inhibitors including omeprazole and pantoprazole and for the metabolic activation of the anticoagulant clopidogrel to the active 2-oxo metabolite (Boulenc et al., 2012).

Moreover, the CYP2C family converts AAs into four regioisomers of EETs: 5,6-, 8,9-, 11,12-, and 14,15-EETs. Accumulating data suggest that metabolism mediated by CYP enzymes play an important role in the cardiovascular system and tumorigenesis.

Apart from rare variants with no (*5, *7), reduced (*8) or unknown activity, three alleles with amino acid changes, CYP2C8*2 and *3, and *4 are more common and have potential clinical relevance

(Zanger & Schwab, 2013). CYP2C8 variants were reported to moderately alter pharmacokinetics and effects of anticancer drug paclitaxel frequently used in treatment of women with ovarian or breast cancer (Leskela et al., 2011). It was found that CYP2C9*3, CYP2C19*2, and CYP2C19*3 polymorphisms resulted in a reduced elimination rate of anticancer agent indisulam (Zandvliet et al., 2007). Because CYP2C9 catalyzes AA epoxygenation, it was suggested that the polymorphism of CYP2C9 may be associated with cancer. Recent studies have showed that individuals lacking CYP2C9*2 or CYP2C9*3 alleles are at increased risk of colorectal cancer (Martinez et al., 2001).

3.4 CYP2D Subfamily

CYP2D6 is the only protein-coding gene of the CYP2D subfamily, which harbors two more pseudogenes, CYP2D7 and CYP2D8P, located on chromosome 22q13.1.

The number of drugs metabolized primarily by CYP2D6 is very large, including antiarrhythmics (e.g., propafenone, mexiletine, flecainide), tricyclic and second-generation antidepressants (e.g., amitriptyline, paroxetine, venlafaxin), antipsychotics (aripiprazole, risperidone), β-blockers (bufuralol, metroprolol), as well as anticancer drugs, in particular the selective estrogen receptor modifier tamoxifen, several opioid analgesics including codeine and tramadol, and many others.

A large number of structural variations exist on the CYP2D locus. Although the first association between CYP2D6 enzyme activity and lung cancer risk was described in 1984, the role of CYP2D6 in lung cancer risk is still a controversial topic (Ayesh, Idle, Ritchie, Crothers, & Hetzel, 1984). Although most studies failed to identify a significant association, some studies reported a minor positive association (Agundez et al., 2001; Legrand et al., 1996; London et al., 1997). The overall results indicated a modest association of CYP2D6 polymorphism with lung cancer risk. However, more detailed studies on the role of the CYP2D6*9 allele and CYP2D6 gene duplications are required to fully elucidate the proposed association (Agundez, 2004). Other types of malignant diseases were also studied, and the results showed that individuals carrying two or more functional genes are at increased risk of liver cancer (Agundez et al., 1995) and melanoma (Dolzan, Rudolf, & Breskvar, 1995), while absence of relevant associations was found in acute leukemia (Sinnett, Krajinovic, & Labuda, 2000), as well as prostate (Agundez et al., 1998), bladder (Anwar, Abdel-Rahman,

El-Zein, Mostafa, & Au, 1996), brain (Elexpuru-Camiruaga et al., 1995), renal (Gallou et al., 2001), and colorectal cancers (Sachse et al., 2002).

3.5 CYP2E Subfamily

CYP2E1 is the only gene of the CYP2E subfamily located on chromosome 10q26.3. CYP2E1 displays a substrate preference for low-molecular-weight molecules, including ethanol, acetone, and other organic solvents, narcotics like halothane, and some drugs including chlorzoxazone and paracetamol (Zanger & Schwab, 2013). Only 13 variant alleles of CYP2E1 have been detected so far, and studies have failed to provide conclusive evidence for cancer risk likely because these variants are extremely rare (Fairbrother et al., 1998; Hu et al., 1997).

3.6 CYP2J Subfamily

CYP2J2 was originally cloned from a human liver cDNA library and found to be highly expressed in the human heart (Wu, Moomaw, Tomer, Falck, & Zeldin, 1996). The human CYP2J subfamily has only a single gene, CYP2J2, which encodes a 502 amino acid microsomal P450 protein. In rabbits (CYP2J1), primates, and dogs (CYP2J2), like humans, only one CYP2J has been identified. Rats (CYP2J3, CYP2J4, CYP2J10, CYP2J13, CYP2J16) and mice (CYP2J5, CYP2J6, CYP2J8, CYP2J9, CYP2J11, and CYP2J12) have multiple CYP2J isoforms and pseudogenes (CYP2J7, CYP2J14, and CYP2J15) (Nelson, 2009).

The role of CYP2J2 in drug metabolism has not been fully evaluated. CYP2J2 is one of the major P450 enzymes to metabolize AA predominantly via NADPH-dependent olefin epoxidation to 20-HETE and all four regioisomeric EETs, i.e., the 5,6-, 8,9-, 11,12-, and 14,15-EETs, which play important roles in inflammation and cancer (Chen & Wang, 2013; Xu, Zhang, et al., 2011).

In addition to AA, CYP2J2 can metabolize polyunsaturated fatty acids (PUFAs) into PUFA epoxides (Zeldin, 2001). CYP epoxygenases preferentially oxidize linoleic acid, the 18-carbon ω-6 PUFA, to epoxyoctadecenoic acids, and the ω-3 PUFA eicosapentaenoic acid (EPA) and docosahexaenoic acid (DHA) to epoxyeicosatetraenoic acids (EpETE) and epoxydocosapentaenoic acids (EpDPE), respectively (Spector & Kim, 2015). By using a panel of 15 human recombinant CYPs, it was found that CYP2C9/2C19, 1A2, and 2J2 were the most efficient in EPA and DHA epoxidations (Fer et al., 2008). The most abundant DHA regioisomer is

19,20-EpDPE, while the most abundant EPA regioisomer is 17,18-EpETE (Oliw, 1994).

To date, at least nine variants of CYP2J2 have been identified and named CYP2J2*2 to *10 (Xu, Ju, Hao, Wang, & Li, 2013). The expression and function of CYP2J2 are speculated to be important in the cardiovascular homeostasis and pathogenesis of cardiovascular diseases, so studies were focused on the potential associations mainly between the specific polymorphisms in CYP2J2 and cardiovascular disease risk not cancer.

4. CYP2 FAMILY IN INFLAMMATION

Beyond their contribution to the metabolism of xenobiotics, CYP epoxygenases are actively involved in the metabolism of endogenous substances, which play important roles in inflammation and cancer. The main human CYP epoxygenases, i.e., CYP2C8, CYP2C9, CYP2C19, and CYP2J2, convert AA to four regioisomer EETs (Fig. 1). EETs possess a wide range of protective effects including anti-inflammatory effects (Shahabi, Siest, Meyer, & Visvikis-Siest, 2014).

Inflammation is a complex process with endothelial activation, leukocyte–endothelial adhesion, and subsequent leukocyte transmigration across

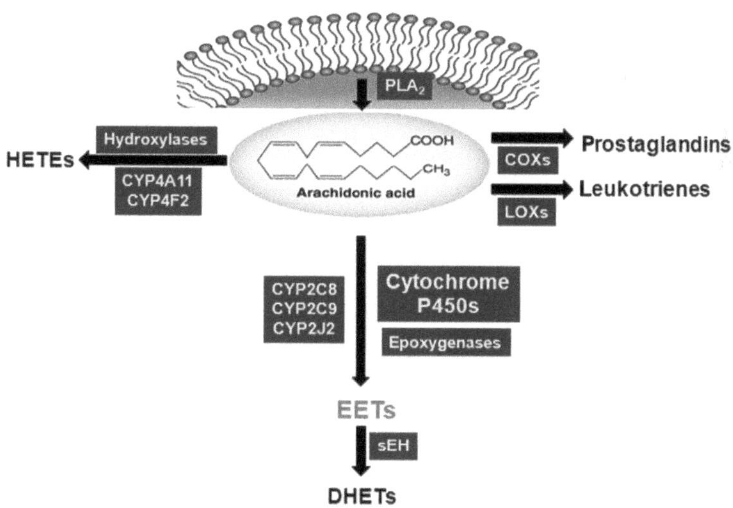

Figure 1 The metabolism of arachidonic acids. AA is converted to eicosanoid mediators by three distinct enzyme systems: cyclooxygenase (COX-1 and COX-2), lipoxygenase (5-LOX, 12-LOX, 15-LOXa, 15-LOXb), and cytochrome P450 (CYP) monooxygenase pathways (ω-hydroxylases and epoxygenases).

the endothelium being the primary events in the inflammatory response. As early as 1990, it was found that 14,15-EET significantly enhanced U937 cell attachment to endothelial cells with maximal cell attachment, which indicated that 14,15-EET may play a substantial role in inflammation and/or atherogenesis by inducing monocyte attachment to endothelial cells (Pritchard, Tota, Stemerman, & Wong, 1990). Recently, a growing body of *in vivo* and *in vitro* studies demonstrated that EETs, especially 11,12-EET, exhibited a broad spectrum of anti-inflammatory activity against acute and chronic inflammation.

4.1 Activation of Nuclear Factor-kappaB

At the beginning of inflammation, activation of nuclear factor-kappaB (NF-κB), which is a central mediator of this process, induces transcriptional upregulation of endothelial cytokine, chemokine, and cellular adhesion molecule expression and drives the subsequent adherence of leukocytes to the endothelium. In 1999, the vascular anti-inflammatory properties of EETs were originally described by Node et al. (1999). Physiological concentrations of EETs or overexpression of CYP2J2 decreased cytokine-induced endothelial cell adhesion molecule expression (VCAM-1, E-selectin, and ICAM-1), and EETs prevented leukocyte adhesion to the vascular wall by a mechanism involving inhibition of transcription factor NF-κB and IκB kinase. The inhibitory effects of EETs were independent of their membrane-hyperpolarizing effects, since BK_{Ca++} channel inhibition with charybdotoxin or iberiotoxin did not alter the impact of 11,12-EET, suggesting that these molecules play an important nonvasodilatory role in vascular inflammation. Interestingly, 14,15-EET administration yielded no anti-inflammatory effect.

Subsequently, a large number of human and animal studies have demonstrated that CYPs epoxygenases can target different steps of NF-κB pathway among different species, types of cells, types of cytokine stimulus, and the EET regioisomers (Falck et al., 2003; Fleming et al., 2001; Liu et al., 2005; Xu et al., 2006). Compared with wild-type controls, CYP2J2 and CYP2C8 transgenic mice each exhibited a significant attenuation of endotoxin-induced activation of NF-κB signaling, cellular adhesion molecule, chemokine and cytokine expression, and neutrophil infiltration in lung *in vivo* (Deng et al., 2011). Furthermore, attenuation of endotoxin-induced NF-κB activation and cellular adhesion molecule and chemokine expression was observed in primary pulmonary endothelial cells isolated from CYP2J2 and CYP2C8 transgenic mice (Deng et al., 2011). This attenuation was inhibited by a putative EET receptor antagonist and CYP epoxygenase inhibitor, directly implicating CYP epoxygenase-derived EETs with the observed anti-inflammatory phenotype (Deng et al., 2011). Recently, it was demonstrated that EETs exert beneficial effects on ox-LDL-induced inflammation primarily through the inhibition of LOX-1 receptor upregulation, MAPK phosphorylation, and NF-κB activation and through the upregulation of CYP2J4 expression (Jiang et al., 2014).

It was also found that 11,12-EET acted as a second messenger in angiotensin II receptor type 2 (AT2R)-mediated anti-inflammatory effects. Direct stimulation of AT2R increased CYP2C/2J-mediated production of 11,12-EET resulting in the modulation of NF-kB activity and reduction of IL-6 (Matavelli, Huang, & Siragy, 2011; Rompe et al., 2010). And EETs and CYP2J2 inhibited tumor necrosis factor-α (TNF-α)-induced apoptosis in pulmonary artery endothelial cells and transforming growth factor-β1-induced migration in pulmonary artery smooth muscle cells (Feng et al., 2013).

4.2 Activation of Peroxisome Proliferator-Activated Receptors

In addition to regulating glucose/lipid metabolism, peroxisome proliferator-activated receptors (PPARs) elicit various anti-inflammatory effects upon activation including inhibition of NF-κB activation (Plutzky, 2003). Recent studies have demonstrated that laminar flow activated endogenous PPARγ in endothelial cells, which is ligand dependent. The flow-induced production of PPARγ ligands was through the PLA2–CYP450 pathway, and the induced PPARγ ligands exerted anti-inflammatory effects in several types of cells (Liu et al., 2004). EETs or CYP2J2 overexpression can prevent TNF-α-induced cardiac cell injury and cardiac dysfunction by inhibiting apoptosis, reducing inflammation, and enhancing PPARγ expression (Zhao, Wang, et al., 2012b). Overexpression of human CYP2J2 in HEK293 cells resulted in a synergistic activation of PPARα, -β/δ, and -γ reporter gene activity, and EETs were able to induce PPARα reporter activity (Wray et al., 2009). In diabetic mice, CYP2J2 expression attenuated the diabetic phenotype and insulin resistance via inhibition of NF-κB and mitogen-activated protein kinase signaling pathways and activation of PPARγ (Li et al., 2015).

In angiotensin II (Ang II)-induced abdominal aortic aneurysm (AAA) mice, CYP2J2 overexpression attenuated MMP expression and activity, elastin degradation, and AAA formation, which was associated with reduced aortic inflammation and macrophage infiltration. In cultured vascular smooth muscle cells (VSMCs), recombinant adeno-associated virus (rAAV)-mediated CYP2J2 overexpression and EETs markedly suppressed Ang II-induced inflammatory cytokine expression. Moreover, overexpressed CYP2J2 and EETs inhibited Ang II-induced macrophage migration in a VSMC–macrophage coculture system. These protective effects were mediated by PPARγ activation (Cai et al., 2013). In human umbilical vein endothelial cells and macrophages, CYP2C8-derived EETs exerted antivascular inflammatory and antioxidative effects, at least in part, through the activation of PPARγ (Liu, Wang, Ding, Wang, & Zeng, 2014). Recently, it was demonstrated that Tie2-CYP2J2 gene targeting restored high-fat diet-mediated decreases in vascular heme oxygenase-1, CYP2C44, sEH, phosphorylated endothelial nitric oxide synthase (eNOS), phosphorylated PKB, and phosphorylated AMPK expression, thus improving vascular function. These changes translated into decreased inflammation and oxidative stress within adipose tissue and decreased PPARγ, CCAAT/enhancer-binding protein alpha, mesoderm-specific transcript, and adipocyte 2 expression and increased uncoupling protein 1 and uncoupling protein 2 expression, reflecting the effects of vascular EET overproduction on adipogenesis.

4.3 Expression of eNOS

The role of eNOS and NO in the regulation of vascular homeostasis is highly complicated. The impacts of CYP-derived EETs on eNOS expression/NO biosynthesis pathways were described by Wang et al. (2003). In cultured bovine aortic endothelial cells, incubation with EETs or transfection with CYP epoxygenase enzymes resulted in increased eNOS mRNA and protein expression and increased metabolic conversion of L-arginine to L-citrulline. These effects were mediated by activation of MAPK and PKC signaling. Similarly, both overexpression of CYP epoxygenases and exogenous addition of EETs have been associated with a robust upregulation of eNOS expression and activity in other types of cells via various pathways, including MAPK, PI3K/Akt, and PKC (Jiang, Chen, et al., 2007; Zhang, Cui, Geng, Zeng, & Tang, 2008).

Moreover, activation of eNOS and subsequent NO release is related with insulin resistance, which is also related with inflammation. In fructose-treated rats, CYP2J3 overexpression increased EET generation, reduced blood pressure, and reversed insulin resistance as determined by insulin resistance index *in vivo*. Furthermore, administration of eNOS inhibitor L-NMMA partially abolished the beneficial effects of CYP2J3 gene delivery on hypertension and insulin resistance induced by fructose intake, and the possible mechanisms include increased ET-1, ETA-receptor mRNA expression and reduced insulin sensitivity in peripheral tissues and organs characterized by reduced activity of IRS-1/PI3K/AKT and AMPK signaling pathways (Xu, Tu, et al., 2011).

4.4 Endoplasmic Reticulum Stress

Endoplasmic Reticulum (ER) is a central organelle of eukaryotic cells that participates in lipid synthesis, protein folding and maturation, and calcium storage. Various cellular stresses, such as inflammation, hypoxia, oxidative stress, Ca^{2+} depletion of ER stores, and excessive accumulation of unfolded protein can lead to impairment of ER function. Initially, CYP2A5 was suggested to be related to G6PD deficiency, in which ER stress may be involved (Nichols & Kirby, 2008).

Wang et al. first provided evidences for CYP2J2-derived EETs as important regulators of intracellular Ca^{2+} levels and ER homeostasis *in vivo* and *in vitro* (Wang et al., 2014). Transgenic mice with cardiomyocyte-specific CYP2J2 expression were protected against heart failure compared with wild-type mice. Isoproterenol (ISO) or Ang II administration induced ER stress and apoptosis, and increased levels of intracellular Ca^{2+}. These phenotypes were abolished by CYP2J2 overexpression *in vivo* or exogenous EETs treatment of cardiomyocytes *in vitro*. ISO or Ang II reduced sarcoplasmic/endoplasmic reticulum calcium ATPase (SERCA2a) expression in hearts or isolated cardiomyocytes; however, loss of SERCA2a expression was prevented in CYP2J2 transgenic hearts *in vivo* or in cardiomyocytes treated with EETs *in vitro*. The reduction of SERCA2a activity was concomitant with increased oxidation of SERCA2a. EETs reversed SERCA2a oxidation through increased expression of antioxidant enzymes and reduced reactive oxygen species (ROS) levels. Similarly, it was detected that CYP2J3 gene delivery upregulated adiponectin expression and excretion in adipose tissue of rats treated with high-fat diet through inhibition of ER stress, which can decrease adiponectin expression (Xu, Tu, et al., 2013).

5. CYP2 FAMILY IN CANCER

While the protective effects of CYP epoxygenase in cardiovascular system have been investigated extensively, the role of CYP epoxygenase in cancer remains poorly characterized (Chen & Wang, 2013). Accumulating data showed that the CYP epoxygenase genes may be not only associated with cancer risk as demonstrated by polymorphisms but also are important factors for tumorigenesis (Dai et al., 2001; Rahman, Korzekwa, Grogan, Gonzalez, & Harris, 1994; Rodriguez-Antona, Gomez, Karlgren, Sim, & Ingelman-Sundberg, 2010).

5.1 Expression of CYP2 Epoxygenase in Cancer

In the early 1990s, a study was conducted to compare the levels of expression between normal and pathological human liver tissues. A group of 100 human liver samples obtained from three different network sources was divided into groups of normal, cirrhotic, metastatic cancer, and other disease groups, and these samples were analyzed for amounts of cytochrome P450 IA2, IIC, IIE1, and IIIA (CYP1A2, CYP2C, CYP2E1, and CYP3A) and epoxide hydrolase per unit of microsomal protein using sodium dodecyl sulfate-polyacrylamide gel electrophoresis and immunochemical staining (Guengerich & Turvy, 1991). CYP2C expression was increased in metastatic liver tissues compared with normal liver tissues (Guengerich & Turvy, 1991). Later, the distribution of sEH, CYP2C8, 2C9, and 2J2 in human neoplastic tissues was detected using tissue microarrays (Enayetallah, French, & Grant, 2006). CYP2C9 was the most abundantly expressed epoxygenase. Both CYP2C8 and CYP2C9 were expressed in human malignant neoplasms such as lung adenocarcinoma, breast ductal carcinoma, colon adenocarcinoma, basal cell carcinoma, bladder transitional cell carcinoma, ovarian adenocarcinoma, endometrial carcinoma, and adenocarcinoma of the prostate (Enayetallah et al., 2006). Moreover, CYP2C44 epoxygenase was detected in the tumor vessels of a xenograft model of human nonsmall cell lung cancer in mice (Pozzi et al., 2005). CYP2C9 epoxygenase was found to be selectively expressed in the vasculature of human renal tumors (Pozzi et al., 2010). Recently, tissue distribution and role of CYP2C8, 2C9, 2J2, and sEH in human breast carcinogenesis were determined by quantitative RT-PCR and immunohistochemical staining (Wei et al., 2014). The results showed that elevated EET levels in breast cancer tissues were associated with upregulation of CYP2C8,

2C9, and 2J2, and downregulation of sEH, and were also associated with aggressive cell behavior in breast cancer patients.

5.2 Functions of CYP2 Epoxygenase in Cancer

Although little is known about the role of CYP2 epoxygenase signaling in cancer progression, emerging evidences indicate that CYP epoxygenases and the metabolites they generate are involved in tumor biology. The direct roles of CYP2 epoxygenase in cancer were first studied in 2005 (Jiang et al., 2005). A very strong and selective CYP2J2 expression was detected in human carcinoma tissues in 101 of 130 patients (77%) as well as 8 human carcinoma cell lines, but the levels were undetectable in adjacent normal tissues and nontumor human cell lines by Western blot, RT-PCR, and immunohistochemical staining. In addition, forced overexpression of CYP2J2, and CYP BM3F87V or addition of EETs in cultured carcinoma cell lines *in vitro* markedly accelerated proliferation, and protected carcinoma cells from apoptosis induced by TNF-α in cultures. In contrast, antisense 2J2 transfection or addition of epoxygenase inhibitors 17-ODYA inhibited proliferation and accelerated cell apoptosis induced by TNF-α. Examination of signaling pathways on the effects of CYP2J2 and EETs revealed activation of MAPK and PI3K/AKT systems and elevation of EGFR phosphorylation level. Moreover, MMPs activated by EETs in cancer cells can lead to EET-induced transactivation of EGFR, which subsequently releases the proangiogenic factor HB-EGF in cancer cell lines (Cheng et al., 2010). These results strongly suggest that CYP epoxygenase 2J2 plays a previously unknown role in promotion of the neoplastic cellular phenotype and in the pathogenesis of a variety of human cancers. Later, it was discovered that CYP epoxygenase overexpression or EETs treatment promoted tumor metastasis independent of effects on tumor growth (Jiang, Ning, et al., 2007). In four different human cancer cell lines, overexpression of CYP2J2 or CYP102 F87V with an associated increase in EETs production or addition of synthetic EETs significantly induced Transwell migration, invasion of cells, cell adhesion to fibronectin, and colony formation in soft agar. In contrast, the epoxygenase inhibitor 17-ODYA or infection with the antisense rAAV-CYP2J2 vector inhibited cell migration, invasion, and adhesion with an associated reduction in EET production. CYP overexpression also enhanced lung metastatic potential in athymic BALB/c mice and enhanced angiogenesis in and around primary tumors compared with control cells. Lung metastasis was abolished by infection with the antisense rAAV-CYP2J2 vector. Treatment of carcinoma cells *in vitro* with exogenous EETs or the overexpression of CYP2J2 resulted in an increase in MMP-9 and CD44 expression and decreased CD82 and nm23 expression.

Subsequently, several CYP epoxygenases have been detected in various tumor cells *in vitro* and *in vivo*. Selective CYP2J2 expression was found in 5 human-derived malignant hematological cell lines and in leukemia cells from peripheral blood and bone marrow in 36 of 42 patients (86%) with malignant hematologic diseases, and CYP2J2 promoted xenograft tumor growth in leukemia models (Chen, Wei, et al., 2011). Arsenic trioxide (As_2O_3; ATO) was first used clinically to induce differentiation and apoptosis of all-*trans*-retinoic acid-resistant acute promyelocytic leukemia, and its clinical effects were induced by elevated intracellular ROS, disrupted cellular redox equilibrium, and mitochondrial dysfunction. Superoxide dismutase and catalase are two enzymes that can detoxify ROS, and it was found that 11,12-EET pretreatment can increase the expression of these two antioxidant enzymes. The diminished levels of ROS inhibit ATO-induced apoptosis. Further, 11,12-EET prevented ATO-induced activation of p38 MAPK, JNK, caspase-3, and caspase-9. Therefore, 11,12-EET pretreatment can attenuate the formation of ROS, diminish mitochondrial dysfunction and caspase activation. Taken together, these data indicated that 11,12-EET inhibited apoptosis induced by ATO through a mechanism that involves induction of antioxidant proteins and attenuation of ROS-mediated mitochondrial dysfunction (Liu et al., 2011). Notably, using transgenic mice with high-endothelial CYP2 epoxygenases levels (Tie2-CYP2C8-Tr, Tie2-CYP2J2-Tr), it was demonstrated that CYP2 epoxygenases and their metabolites were critical for primary tumor growth and metastasis in a variety of mouse models of cancer (Panigrahy et al., 2012).

Moreover, in a mouse xenograft model of tumorigenesis, disruption of the host Cyp2c44 gene caused marked reductions in tumor volume, mass, and vascularization (Pozzi et al., 2010). Other studies showed that esophageal cancer cell proliferation was in part mediated by CYP2C9 in an 11,12-EET-dependent fashion (Schmelzle et al., 2011). CYP2C9 expression correlated with high Ki-67 labeling in esophageal adenocarcinoma and Ki-67-positive tumor cells in esophageal adenocarcinoma and esophageal squamous cell carcinoma. Selective inhibition of CYP2C9 decreased tumor cell proliferation *in vitro*, which was abolished by 11,12-EET. Cell-cycle analysis using FACS revealed that inhibition of CYP2C9 leads to a G0/G1 phase cell-cycle arrest.

Collectively, these data demonstrated that CYP2 epoxygenases have previously unrecognized role in promoting carcinoma metastasis and may therefore have important therapeutic implications.

5.3 Clinical Relevance

Overexpression of CYP2 epoxygenases and EETs contributes to the pathological processes in various kinds of cancer. It is possible that inhibitors of CYP may exhibit antitumorigenic properties.

The high-affinity, selective CYP2J2 inhibitors which were related to terfenadine were designed and improved since 2006 (Lafite et al., 2006; Lafite, Dijols, Zeldin, Dansette, & Mansuy, 2007). Then, four of these inhibitors (compounds 4, 5, 11, and 26) were tested for antitumor effectiveness *in vitro* and *in vivo* (Chen et al., 2009). In Tca-8113 cells, the CYP2J2 inhibitors decreased EET production by approximately 60%, whereas they had no effect on CYP2J2 mRNA or protein expression. Compound 26 (C26) inhibited the proliferation of human tumor cells, reduced their ability to adhere, invade, and migrate, and attenuated activation of EGFR and PI3K/Akt pathways. Inhibition of CYP2J2 also significantly potentiated human tumor cell apoptosis and caused a corresponding increase in caspase-3 activity and change in expression of apoptosis-related proteins Bax and Bcl-2. In murine xenograft models using MDA-MB-435 cells, treatment with C26 significantly repressed tumor growth, decreased lung metastasis, and was associated with increased expression of the anticancer genes CD82 and nm23, without causing toxicity. Moreover, hematologic malignant processes were inhibited by oral administration of C26 in Tie2-CYP2J2 transgenic mice and in severe combined immunodeficiency xenograft mice (Chen, Wei, et al., 2011). These data suggest that CYP2J2 inhibitors hold significant promise for use in the treatment of neoplastic diseases.

Besides C26, it was also found that the antitumorigenic effects of Wyeth, a PPARα-selective ligand, were PPARα species-independent and associated with reductions in the expression of an endothelial epoxygenase and that PPARα ligand activation could serve as a novel and potentially effective target for the development of low toxicity antiangiogenic cancer therapies.

It is worth noting that one important regulator of CYP epoxygenase function is the microRNAs, which represents a class of promising endogenous drugs. Recently, a new mechanism for controlling the tumor-promoting functions of CYP2J2 associated with microRNAs was demonstrated. Luciferase and Western blot assays revealed that CYP2J2 was regulated by let-7b. In addition, let-7b decreased the enzymatic activity of endogenous CYP2J2. Furthermore, let-7b diminished cell proliferation and promoted apoptosis of tumor cells via posttranscriptional repression of CYP2J2. Tumor xenografts were induced in nude mice by subcutaneous injection of MDA-MB-435 cells. Let-7b significantly inhibited the tumor phenotype by targeting CYP2J2. Moreover, quantitative RT-PCR and Western blotting were used to determine the expression levels of let-7b and CYP2J2 protein from 18-matched lung squamous cell cancer and adjacent normal lung tissues; the expression level of CYP2J2 was inversely proportional to that of let-7b (Chen, Chen, et al., 2012a).

6. CONCLUSION

There is considerable evidence that CYP2 family epoxygenases and their metabolites play important roles in the initiation and development of inflammation and cancer (Fig. 2). Although the specific mechanisms remain incompletely understood, accumulating evidences suggest that CYP2 epoxygenases are potential therapeutic targets for inflammation

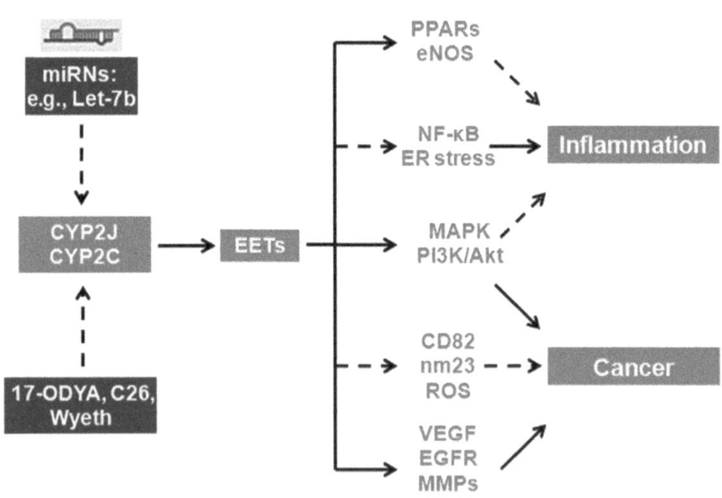

Figure 2 CYP2 family epoxygenases and their metabolites play important roles in the initiation and development of inflammation and cancer. Chemical inhibitors or miRNAs may serve as potential cancer therapy via the inhibition of CYP2C/2J-EET system.

and cancer. Meanwhile, it should be noted that enzymes from other CYP families, such as CYP1A, can also mediate the similar activity despite CYP2 family (Diani-Moore, Papachristou, Labitzke, & Rifkind, 2006).

CONFLICT OF INTEREST

The authors have no conflicts of interest to declare.

ACKNOWLEDGMENT

This work was supported by projects of 973 programs (2012CB518004) and NSFC (31130031 and 30930039).

REFERENCES

Agundez, J. A. (2004). Cytochrome P450 gene polymorphism and cancer. *Current Drug Metabolism*, *5*(3), 211–224.

Agundez, J. A., Gallardo, L., Ledesma, M. C., Lozano, L., Rodriguez-Lescure, A., Pontes, J. C., et al. (2001). Functionally active duplications of the CYP2D6 gene are more prevalent among larynx and lung cancer patients. *Oncology*, *61*(1), 59–63, http://dx.doi.org/10.1159/000055354.

Agundez, J. A., Ledesma, M. C., Benitez, J., Ladero, J. M., Rodriguez-Lescure, A., Diaz-Rubio, E., et al. (1995). CYP2D6 genes and risk of liver cancer. *Lancet*, *345*(8953), 830–831.

Agundez, J. A., Martinez, C., Olivera, M., Gallardo, L., Ladero, J. M., Rosado, C., et al. (1998). Expression in human prostate of drug- and carcinogen-metabolizing enzymes: Association with prostate cancer risk. *British Journal of Cancer*, *78*(10), 1361–1367.

Anwar, W. A., Abdel-Rahman, S. Z., El-Zein, R. A., Mostafa, H. M., & Au, W. W. (1996). Genetic polymorphism of GSTM1, CYP2E1 and CYP2D6 in Egyptian bladder cancer patients. *Carcinogenesis*, *17*(9), 1923–1929.

Ayesh, R., Idle, J. R., Ritchie, J. C., Crothers, M. J., & Hetzel, M. R. (1984). Metabolic oxidation phenotypes as markers for susceptibility to lung cancer. *Nature*, *312*(5990), 169–170.

Boulenc, X., Djebli, N., Shi, J., Perrin, L., Brian, W., Van Horn, R., et al. (2012). Effects of omeprazole and genetic polymorphism of CYP2C19 on the clopidogrel active metabolite. *Drug Metabolism and Disposition*, *40*(1), 187–197. http://dx.doi.org/10.1124/dmd.111.040394.

Brash, A. R. (2001). Arachidonic acid as a bioactive molecule. *The Journal of Clinical Investigation*, *107*(11), 1339–1345. http://dx.doi.org/10.1172/JCI13210.

Cai, Z., Zhao, G., Yan, J., Liu, W., Feng, W., Ma, B., et al. (2013). CYP2J2 overexpression increases EETs and protects against angiotensin II-induced abdominal aortic aneurysm in mice. *The Journal of Lipid Research*, *54*(5), 1448–1456. http://dx.doi.org/10.1194/jlr.M036533.

Campbell, W. B., & Falck, J. R. (2007). Arachidonic acid metabolites as endothelium-derived hyperpolarizing factors. *Hypertension*, *49*(3), 590–596. http://dx.doi.org/10.1161/01.HYP.0000255173.50317.fc.

Canova, C., Hashibe, M., Simonato, L., Nelis, M., Metspalu, A., Lagiou, P., et al. (2009). Genetic associations of 115 polymorphisms with cancers of the upper aerodigestive tract across 10 European countries: The ARCAGE project. *Cancer Research*, *69*(7), 2956–2965. http://dx.doi.org/10.1158/0008-5472.CAN-08-2604.

Capdevila, J. H., Falck, J. R., & Harris, R. C. (2000). Cytochrome P450 and arachidonic acid bioactivation. Molecular and functional properties of the arachidonate monooxygenase. *The Journal of Lipid Research*, *41*(2), 163–181.

Chen, F., Chen, C., Yang, S., Gong, W., Wang, Y., Cianflone, K., et al. (2012). Let-7b inhibits human cancer phenotype by targeting cytochrome P450 epoxygenase 2J2. *PLoS One*, 7(6), e39197. http://dx.doi.org/10.1371/journal.pone.0039197.

Chen, C., Li, G., Liao, W., Wu, J., Liu, L., Ma, D., et al. (2009). Selective inhibitors of CYP2J2 related to terfenadine exhibit strong activity against human cancers in vitro and in vivo. *The Journal of Pharmacology and Experimental Therapeutics*, *329*(3), 908–918. http://dx.doi.org/10.1124/jpet.109.152017.

Chen, C., & Wang, D. W. (2013). CYP epoxygenase derived EETs: From cardiovascular protection to human cancer therapy. *Current Topics in Medicinal Chemistry*, *13*(12), 1454–1469.

Chen, G., Wang, P., Zhao, G., Xu, G., Gruzdev, A., Zeldin, D. C., et al. (2011). Cytochrome P450 epoxygenase CYP2J2 attenuates nephropathy in streptozotocin-induced diabetic mice. *Prostaglandins & Other Lipid Mediators*, *96*(1–4), 63–71. http://dx.doi.org/10.1016/j.prostaglandins.2011.06.009.

Chen, C., Wei, X., Rao, X., Wu, J., Yang, S., Chen, F., et al. (2011). Cytochrome P450 2J2 is highly expressed in hematologic malignant diseases and promotes tumor cell growth. *The Journal of Pharmacology and Experimental Therapeutics*, *336*(2), 344–355. http://dx.doi.org/10.1124/jpet.110.174805.

Chen, G., Xu, R., Wang, Y., Wang, P., Zhao, G., Xu, X., et al. (2012). Genetic disruption of soluble epoxide hydrolase is protective against streptozotocin-induced diabetic nephropathy. *American Journal of Physiology, Endocrinology and Metabolism*, *303*(5), E563–E575. http://dx.doi.org/10.1152/ajpendo.00591.2011.

Chen, G., Xu, R., Zhang, S., Wang, Y., Wang, P., Edin, M. L., et al. (2014). CYP2J2 overexpression attenuates non-alcoholic fatty liver disease induced by high fat diet in mice. *American Journal of Physiology Endocrinology and Metabolism*. http://dx.doi.org/10.1152/ajpendo.00366.2014, ajpendo 00366 02014.

Chen, W., Zheng, G., Yang, S., Ping, W., Fu, X., Zhang, N., et al. (2014). CYP2J2 and EETs protect against oxidative stress and apoptosis in vivo and in vitro following lung ischemia/reperfusion. *Cellular Physiology and Biochemistry*, *33*(6), 1663–1680. http://dx.doi.org/10.1159/000362950.

Cheng, L. M., Jiang, J. G., Sun, Z. Y., Chen, C., Dackor, R. T., Zeldin, D. C., et al. (2010). The epoxyeicosatrienoic acid-stimulated phosphorylation of EGF-R involves the activation of metalloproteinases and the release of HB-EGF in cancer cells. *Acta Pharmacologica Sinica*, *31*(2), 211–218. http://dx.doi.org/10.1038/aps.2009.184.

Dai, D., Zeldin, D. C., Blaisdell, J. A., Chanas, B., Coulter, S. J., Ghanayem, B. I., et al. (2001). Polymorphisms in human CYP2C8 decrease metabolism of the anticancer drug paclitaxel and arachidonic acid. *Pharmacogenetics*, *11*(7), 597–607.

DeNardo, D. G., Andreu, P., & Coussens, L. M. (2010). Interactions between lymphocytes and myeloid cells regulate pro- versus anti-tumor immunity. *Cancer and Metastasis Reviews*, *29*(2), 309–316. http://dx.doi.org/10.1007/s10555-010-9223-6.

Deng, Y., Edin, M. L., Theken, K. N., Schuck, R. N., Flake, G. P., Kannon, M. A., et al. (2011). Endothelial CYP epoxygenase overexpression and soluble epoxide hydrolase disruption attenuate acute vascular inflammatory responses in mice. *FASEB Journal*, *25*(2), 703–713. http://dx.doi.org/10.1096/fj.10-171488.

Diani-Moore, S., Papachristou, F., Labitzke, E., & Rifkind, A. B. (2006). Induction of CYP1A and cyp2-mediated arachidonic acid epoxygenation and suppression of 20-hydroxyeicosatetraenoic acid by imidazole derivatives including the aromatase inhibitor vorozole. *Drug Metabolism and Disposition*, *34*(8), 1376–1385. http://dx.doi.org/10.1124/dmd.106.009498.

Ding, S., Lake, B. G., Friedberg, T., & Wolf, C. R. (1995). Expression and alternative splicing of the cytochrome P-450 CYP2A7. *Biochemical Journal*, *306*(Pt 1), 161–166.

Dolzan, V., Rudolf, Z., & Breskvar, K. (1995). Human CYP2D6 gene polymorphism in Slovene cancer patients and healthy controls. *Carcinogenesis*, *16*(11), 2675–2678.

Elexpuru-Camiruaga, J., Buxton, N., Kandula, V., Dias, P. S., Campbell, D., McIntosh, J., et al. (1995). Susceptibility to astrocytoma and meningioma: Influence of allelism at glutathione S-transferase (GSTT1 and GSTM1) and cytochrome P-450 (CYP2D6) loci. *Cancer Research*, *55*(19), 4237–4239.

Enayetallah, A. E., French, R. A., & Grant, D. F. (2006). Distribution of soluble epoxide hydrolase, cytochrome P450 2C8, 2C9 and 2J2 in human malignant neoplasms. *Journal of Molecular Histology*, *37*(3–4), 133–141. http://dx.doi.org/10.1007/s10735-006-9050-9.

Fairbrother, K. S., Grove, J., de Waziers, I., Steimel, D. T., Day, C. P., Crespi, C. L., et al. (1998). Detection and characterization of novel polymorphisms in the CYP2E1 gene. *Pharmacogenetics*, *8*(6), 543–552.

Falck, J. R., Reddy, L. M., Reddy, Y. K., Bondlela, M., Krishna, U. M., Ji, Y., et al. (2003). 11,12-Epoxyeicosatrienoic acid (11,12-EET): Structural determinants for inhibition of TNF-alpha-induced VCAM-1 expression. *Bioorganic & Medicinal Chemistry Letters*, *13*(22), 4011–4014.

Feng, W., Xu, X., Zhao, G., Li, G., Liu, T., Zhao, J., et al. (2013). EETs and CYP2J2 inhibit TNF-alpha-induced apoptosis in pulmonary artery endothelial cells and TGF-beta1-induced migration in pulmonary artery smooth muscle cells. *International Journal of Molecular Medicine*, *32*(3), 685–693. http://dx.doi.org/10.3892/ijmm.2013.1435.

Fer, M., Dreano, Y., Lucas, D., Corcos, L., Salaun, J. P., Berthou, F., et al. (2008). Metabolism of eicosapentaenoic and docosahexaenoic acids by recombinant human cytochromes P450. *Archives of Biochemistry and Biophysics*, *471*(2), 116–125. http://dx.doi.org/10.1016/j.abb.2008.01.002.

Fleeman, N., Martin Saborido, C., Payne, K., Boland, A., Dickson, R., Dundar, Y., et al. (2011). The clinical effectiveness and cost-effectiveness of genotyping for CYP2D6 for the management of women with breast cancer treated with tamoxifen: A systematic review. *Health Technology Assessment*, *15*(33), 1–102. http://dx.doi.org/10.3310/hta15330.

Fleming, I., Michaelis, U. R., Bredenkotter, D., Fisslthaler, B., Dehghani, F., Brandes, R. P., et al. (2001). Endothelium-derived hyperpolarizing factor synthase (Cytochrome P450 2C9) is a functionally significant source of reactive oxygen species in coronary arteries. *Circulation Research*, *88*(1), 44–51.

Fujieda, M., Yamazaki, H., Saito, T., Kiyotani, K., Gyamfi, M. A., Sakurai, M., et al. (2004). Evaluation of CYP2A6 genetic polymorphisms as determinants of smoking behavior and tobacco-related lung cancer risk in male Japanese smokers. *Carcinogenesis*, *25*(12), 2451–2458. http://dx.doi.org/10.1093/carcin/bgh258.

Gallou, C., Longuemaux, S., Delomenie, C., Mejean, A., Martin, N., Martinet, S., et al. (2001). Association of GSTT1 non-null and NAT1 slow/rapid genotypes with von Hippel-Lindau tumour suppressor gene transversions in sporadic renal cell carcinoma. *Pharmacogenetics*, *11*(6), 521–535.

Guengerich, F. P., & Turvy, C. G. (1991). Comparison of levels of several human microsomal cytochrome P-450 enzymes and epoxide hydrolase in normal and disease states using immunochemical analysis of surgical liver samples. *The Journal of Pharmacology and Experimental Therapeutics*, *256*(3), 1189–1194.

Hanahan, D., & Weinberg, R. A. (2000). The hallmarks of cancer. *Cell*, *100*(1), 57–70.

Hanahan, D., & Weinberg, R. A. (2011). Hallmarks of cancer: The next generation. *Cell*, *144*(5), 646–674. http://dx.doi.org/10.1016/j.cell.2011.02.013.

Haroun, F., Al-Shaar, L., Habib, R. H., El-Saghir, N., Tfayli, A., Bazarbachi, A., et al. (2015). Effects of CYP2B6 genetic polymorphisms in patients receiving cyclophosphamide combination chemotherapy for breast cancer. *Cancer Chemotherapy and Pharmacology*, *75*(1), 207–214. http://dx.doi.org/10.1007/s00280-014-2632-4.

Hoffman, S. M., Nelson, D. R., & Keeney, D. S. (2001). Organization, structure and evolution of the CYP2 gene cluster on human chromosome 19. *Pharmacogenetics*, *11*(8), 687–698.

Hu, Y., Oscarson, M., Johansson, I., Yue, Q. Y., Dahl, M. L., Tabone, M., et al. (1997). Genetic polymorphism of human CYP2E1: Characterization of two variant alleles. *Molecular Pharmacology*, *51*(3), 370–376.

Hudson, J. D., Shoaibi, M. A., Maestro, R., Carnero, A., Hannon, G. J., & Beach, D. H. (1999). A proinflammatory cytokine inhibits p53 tumor suppressor activity. *The Journal of Experimental Medicine*, *190*(10), 1375–1382.

Jiang, J. G., Chen, C. L., Card, J. W., Yang, S., Chen, J. X., Fu, X. N., et al. (2005). Cytochrome P450 2J2 promotes the neoplastic phenotype of carcinoma cells and is up-regulated in human tumors. *Cancer Research*, *65*(11), 4707–4715. http://dx.doi.org/10.1158/0008-5472.CAN-04-4173.

Jiang, J. G., Chen, R. J., Xiao, B., Yang, S., Wang, J. N., Wang, Y., et al. (2007). Regulation of endothelial nitric-oxide synthase activity through phosphorylation in response to epoxyeicosatrienoic acids. *Prostaglandins & Other Lipid Mediators*, *82*(1–4), 162–174. http://dx.doi.org/10.1016/j.prostaglandins.2006.08.005.

Jiang, J. G., Ning, Y. G., Chen, C., Ma, D., Liu, Z. J., Yang, S., et al. (2007). Cytochrome p450 epoxygenase promotes human cancer metastasis. *Cancer Research*, *67*(14), 6665–6674. http://dx.doi.org/10.1158/0008-5472.CAN-06-3643.

Jiang, J. X., Zhang, S. J., Liu, Y. N., Lin, X. X., Sun, Y. H., Shen, H. J., et al. (2014). EETs alleviate ox-LDL-induced inflammation by inhibiting LOX-1 receptor expression in rat pulmonary arterial endothelial cells. *European Journal of Pharmacology*, *727*, 43–51. http://dx.doi.org/10.1016/j.ejphar.2014.01.045.

Joyce, J. A., & Pollard, J. W. (2009). Microenvironmental regulation of metastasis. *Nature Reviews Cancer*, *9*(4), 239–252. http://dx.doi.org/10.1038/nrc2618.

Kerb, R., Fux, R., Morike, K., Kremsner, P. G., Gil, J. P., Gleiter, C. H., et al. (2009). Pharmacogenetics of antimalarial drugs: Effect on metabolism and transport. *The Lancet Infectious Diseases*, *9*(12), 760–774. http://dx.doi.org/10.1016/S1473-3099(09)70320-2.

Krau, S. D. (2013). Cytochrome p450 part 2: What nurses need to know about the cytochrome p450 family systems. *The Nursing Clinics of North America*, *48*(4), 681–696. http://dx.doi.org/10.1016/j.cnur.2013.09.003.

Lafite, P., Dijols, S., Buisson, D., Macherey, A. C., Zeldin, D. C., Dansette, P. M., et al. (2006). Design and synthesis of selective, high-affinity inhibitors of human cytochrome P450 2J2. *Bioorganic & Medicinal Chemistry Letters*, *16*(10), 2777–2780. http://dx.doi.org/10.1016/j.bmcl.2006.02.004.

Lafite, P., Dijols, S., Zeldin, D. C., Dansette, P. M., & Mansuy, D. (2007). Selective, competitive and mechanism-based inhibitors of human cytochrome P450 2J2. *Archives of Biochemistry and Biophysics*, *464*(2), 155–168. http://dx.doi.org/10.1016/j.abb.2007.03.028.

Lee, C. R., Goldstein, J. A., & Pieper, J. A. (2002). Cytochrome P450 2C9 polymorphisms: A comprehensive review of the in-vitro and human data. *Pharmacogenetics*, *12*(3), 251–263.

Legrand, M., Stucker, I., Marez, D., Sabbagh, N., Lo-Guidice, J. M., & Broly, F. (1996). Influence of a mutation reducing the catalytic activity of the cytochrome P450 CYP2D6 on lung cancer susceptibility. *Carcinogenesis*, *17*(10), 2267–2269.

Leskela, S., Jara, C., Leandro-Garcia, L. J., Martinez, A., Garcia-Donas, J., Hernando, S., et al. (2011). Polymorphisms in cytochromes P450 2C8 and 3A5 are associated with paclitaxel neurotoxicity. *The Pharmacogenomics Journal*, *11*(2), 121–129. http://dx.doi.org/10.1038/tpj.2010.13.

Li, R., Xu, X., Chen, C., Wang, Y., Gruzdev, A., Zeldin, D. C., et al. (2015). CYP2J2 attenuates metabolic dysfunction in diabetic mice by reducing hepatic inflammation via the PPARgamma. *American Journal of Physiology Endocrinology and Metabolism*, *308*(4), E270–E282. http://dx.doi.org/10.1152/ajpendo.00118.2014, ajpendo 00118 02014.

Liu, L., Chen, C., Gong, W., Li, Y., Edin, M. L., Zeldin, D. C., et al. (2011). Epoxyeicosatrienoic acids attenuate reactive oxygen species level, mitochondrial dysfunction, caspase activation, and apoptosis in carcinoma cells treated with arsenic trioxide. *The Journal of Pharmacology and Experimental Therapeutics*, *339*(2), 451–463. http://dx.doi.org/10.1124/jpet.111.180505.

Liu, W., Wang, B., Ding, H., Wang, D. W., & Zeng, H. (2014). A potential therapeutic effect of CYP2C8 overexpression on anti-TNF-alpha activity. *International Journal of Molecular Medicine*, *34*(3), 725–732. http://dx.doi.org/10.3892/ijmm.2014.1844.

Liu, Y., Zhang, Y., Schmelzer, K., Lee, T. S., Fang, X., Zhu, Y., et al. (2005). The antiinflammatory effect of laminar flow: The role of PPARgamma, epoxyeicosatrienoic acids, and soluble epoxide hydrolase. *Proceedings of the National Academy of Sciences of the United States of America*, *102*(46), 16747–16752. http://dx.doi.org/10.1073/pnas.0508081102.

Liu, Y., Zhu, Y., Rannou, F., Lee, T. S., Formentin, K., Zeng, L., et al. (2004). Laminar flow activates peroxisome proliferator-activated receptor-gamma in vascular endothelial cells. *Circulation*, *110*(9), 1128–1133. http://dx.doi.org/10.1161/01.CIR.0000139850.08365.EC.

London, S. J., Daly, A. K., Leathart, J. B., Navidi, W. C., Carpenter, C. C., & Idle, J. R. (1997). Genetic polymorphism of CYP2D6 and lung cancer risk in African-Americans and Caucasians in Los Angeles County. *Carcinogenesis*, *18*(6), 1203–1214.

London, S. J., Idle, J. R., Daly, A. K., & Coetzee, G. A. (1999). Genetic variation of CYP2A6, smoking, and risk of cancer. *Lancet*, *353*(9156), 898–899. http://dx.doi.org/10.1016/S0140-6736(98)04984-8.

Ma, F., Lin, F., Chen, C., Cheng, J., Zeldin, D. C., Wang, Y., et al. (2013). Indapamide lowers blood pressure by increasing production of epoxyeicosatrienoic acids in the kidney. *Molecular Pharmacology*, *84*(2), 286–295. http://dx.doi.org/10.1124/mol.113.085878.

Ma, B., Xiong, X., Chen, C., Li, H., Xu, X., Li, X., et al. (2013). Cardiac-specific overexpression of CYP2J2 attenuates diabetic cardiomyopathy in male streptozotocin-induced diabetic mice. *Endocrinology*, *154*(8), 2843–2856. http://dx.doi.org/10.1210/en.2012-2166.

Makita, K., Takahashi, K., Karara, A., Jacobson, H. R., Falck, J. R., & Capdevila, J. H. (1994). Experimental and/or genetically controlled alterations of the renal microsomal cytochrome P450 epoxygenase induce hypertension in rats fed a high salt diet. *The Journal of Clinical Investigation*, *94*(6), 2414–2420. http://dx.doi.org/10.1172/JCI117608.

Mantovani, A. (2010). Molecular pathways linking inflammation and cancer. *Current Molecular Medicine*, *10*(4), 369–373.

Mantovani, A., Allavena, P., Sica, A., & Balkwill, F. (2008). Cancer-related inflammation. *Nature*, *454*(7203), 436–444. http://dx.doi.org/10.1038/nature07205.

Martinez, C., Garcia-Martin, E., Ladero, J. M., Sastre, J., Garcia-Gamito, F., Diaz-Rubio, M., et al. (2001). Association of CYP2C9 genotypes leading to high enzyme activity and colorectal cancer risk. *Carcinogenesis*, *22*(8), 1323–1326.

Matavelli, L. C., Huang, J., & Siragy, H. M. (2011). Angiotensin AT(2) receptor stimulation inhibits early renal inflammation in renovascular hypertension. *Hypertension*, *57*(2), 308–313. http://dx.doi.org/10.1161/HYPERTENSIONAHA.110.164202.

Monti, J., Fischer, J., Paskas, S., Heinig, M., Schulz, H., Gosele, C., et al. (2008). Soluble epoxide hydrolase is a susceptibility factor for heart failure in a rat model of human disease. *Nature Genetics*, *40*(5), 529–537. http://dx.doi.org/10.1038/ng.129.

Murdoch, C., Muthana, M., Coffelt, S. B., & Lewis, C. E. (2008). The role of myeloid cells in the promotion of tumour angiogenesis. *Nature Reviews Cancer*, *8*(8), 618–631. http://dx.doi.org/10.1038/nrc2444.
Nathan, C. (2002). Points of control in inflammation. *Nature*, *420*(6917), 846–852. http://dx.doi.org/10.1038/nature01320.
Nelson, D. R. (2009). The cytochrome p450 homepage. *Human Genomics*, *4*(1), 59–65.
Nelson, D. R., Zeldin, D. C., Hoffman, S. M., Maltais, L. J., Wain, H. M., & Nebert, D. W. (2004). Comparison of cytochrome P450 (CYP) genes from the mouse and human genomes, including nomenclature recommendations for genes, pseudogenes and alternative-splice variants. *Pharmacogenetics*, *14*(1), 1–18.
Nichols, K. D., & Kirby, G. M. (2008). Expression of cytochrome P450 2A5 in a glucose-6-phosphate dehydrogenase-deficient mouse model of oxidative stress. *Biochemical Pharmacology*, *75*(5), 1230–1239. http://dx.doi.org/10.1016/j.bcp.2007.10.032.
Node, K., Huo, Y., Ruan, X., Yang, B., Spiecker, M., Ley, K., et al. (1999). Anti-inflammatory properties of cytochrome P450 epoxygenase-derived eicosanoids. *Science*, *285*(5431), 1276–1279.
Oliw, E. H. (1994). Oxygenation of polyunsaturated fatty acids by cytochrome P450 monooxygenases. *Progress in Lipid Research*, *33*(3), 329–354.
Panigrahy, D., Edin, M. L., Lee, C. R., Huang, S., Bielenberg, D. R., Butterfield, C. E., et al. (2012). Epoxyeicosanoids stimulate multiorgan metastasis and tumor dormancy escape in mice. *The Journal of Clinical Investigation*, *122*(1), 178–191. http://dx.doi.org/10.1172/JCI58128.
Panigrahy, D., Greene, E. R., Pozzi, A., Wang, D. W., & Zeldin, D. C. (2011). EET signaling in cancer. *Cancer and Metastasis Reviews*, *30*(3–4), 525–540. http://dx.doi.org/10.1007/s10555-011-9315-y.
Plutzky, J. (2003). Medicine. PPARs as therapeutic targets: reverse cardiology? *Science*, *302*(5644), 406–407. http://dx.doi.org/10.1126/science.1091172.
Pozzi, A., Ibanez, M. R., Gatica, A. E., Yang, S., Wei, S., Mei, S., et al. (2007). Peroxisomal proliferator-activated receptor-alpha-dependent inhibition of endothelial cell proliferation and tumorigenesis. *The Journal of Biological Chemistry*, *282*(24), 17685–17695. http://dx.doi.org/10.1074/jbc.M701429200.
Pozzi, A., Macias-Perez, I., Abair, T., Wei, S., Su, Y., Zent, R., et al. (2005). Characterization of 5,6- and 8,9-epoxyeicosatrienoic acids (5,6- and 8,9-EET) as potent in vivo angiogenic lipids. *The Journal of Biological Chemistry*, *280*(29), 27138–27146. http://dx.doi.org/10.1074/jbc.M501730200.
Pozzi, A., Popescu, V., Yang, S., Mei, S., Shi, M., Puolitaival, S. M., et al. (2010). The anti-tumorigenic properties of peroxisomal proliferator-activated receptor alpha are arachidonic acid epoxygenase-mediated. *The Journal of Biological Chemistry*, *285*(17), 12840–12850. http://dx.doi.org/10.1074/jbc.M109.081554.
Pritchard, K. A., Jr., Tota, R. R., Stemerman, M. B., & Wong, P. Y. (1990). 14,15-Epoxyeicosatrienoic acid promotes endothelial cell dependent adhesion of human monocytic tumor U937 cells. *Biochemical and Biophysical Research Communications*, *167*(1), 137–142.
Qian, B. Z., & Pollard, J. W. (2010). Macrophage diversity enhances tumor progression and metastasis. *Cell*, *141*(1), 39–51. http://dx.doi.org/10.1016/j.cell.2010.03.014.
Rahman, A., Korzekwa, K. R., Grogan, J., Gonzalez, F. J., & Harris, J. W. (1994). Selective biotransformation of taxol to 6 alpha-hydroxytaxol by human cytochrome P450 2C8. *Cancer Research*, *54*(21), 5543–5546.
Rao, Y., Hoffmann, E., Zia, M., Bodin, L., Zeman, M., Sellers, E. M., et al. (2000). Duplications and defects in the CYP2A6 gene: Identification, genotyping, and in vivo effects on smoking. *Molecular Pharmacology*, *58*(4), 747–755.

Ricciotti, E., & FitzGerald, G. A. (2011). Prostaglandins and inflammation. *Arteriosclerosis, Thrombosis, and Vascular Biology*, *31*(5), 986–1000. http://dx.doi.org/10.1161/ATVBAHA.110.207449.

Rodriguez-Antona, C., Gomez, A., Karlgren, M., Sim, S. C., & Ingelman-Sundberg, M. (2010). Molecular genetics and epigenetics of the cytochrome P450 gene family and its relevance for cancer risk and treatment. *Human Genetics*, *127*(1), 1–17. http://dx.doi.org/10.1007/s00439-009-0748-0.

Rompe, F., Artuc, M., Hallberg, A., Alterman, M., Stroder, K., Thone-Reineke, C., et al. (2010). Direct angiotensin II type 2 receptor stimulation acts anti-inflammatory through epoxyeicosatrienoic acid and inhibition of nuclear factor kappaB. *Hypertension*, *55*(4), 924–931. http://dx.doi.org/10.1161/HYPERTENSIONAHA.109.147843.

Rossini, A., de Almeida Simao, T., Albano, R. M., & Pinto, L. F. (2008). CYP2A6 polymorphisms and risk for tobacco-related cancers. *Pharmacogenomics*, *9*(11), 1737–1752. http://dx.doi.org/10.2217/14622416.9.11.1737.

Sachse, C., Smith, G., Wilkie, M. J., Barrett, J. H., Waxman, R., Sullivan, F., et al. (2002). A pharmacogenetic study to investigate the role of dietary carcinogens in the etiology of colorectal cancer. *Carcinogenesis*, *23*(11), 1839–1849.

Schmelzle, M., Dizdar, L., Matthaei, H., Baldus, S. E., Wolters, J., Lindenlauf, N., et al. (2011). Esophageal cancer proliferation is mediated by cytochrome P450 2C9 (CYP2C9). *Prostaglandins & Other Lipid Mediators*, *94*(1–2), 25–33. http://dx.doi.org/10.1016/j.prostaglandins.2010.12.001.

Seubert, J., Yang, B., Bradbury, J. A., Graves, J., Degraff, L. M., Gabel, S., et al. (2004). Enhanced postischemic functional recovery in CYP2J2 transgenic hearts involves mitochondrial ATP-sensitive K+ channels and p42/p44 MAPK pathway. *Circulation Research*, *95*(5), 506–514. http://dx.doi.org/10.1161/01.RES.0000139436.89654.c8.

Shahabi, P., Siest, G., Meyer, U. A., & Visvikis-Siest, S. (2014). Human cytochrome P450 epoxygenases: Variability in expression and role in inflammation-related disorders. *Pharmacology & Therapeutics*, *144*(2), 134–161. http://dx.doi.org/10.1016/j.pharmthera.2014.05.011.

Sharma, R., Ahuja, M., Panda, N., & Khullar, M. (2010). Polymorphisms in CYP2A13 and UGT1A7 genes and head and neck cancer susceptibility in North Indians. *Oral Diseases*, *16*(8), 760–768. http://dx.doi.org/10.1111/j.1601-0825.2010.01683.x.

Sinnett, D., Krajinovic, M., & Labuda, D. (2000). Genetic susceptibility to childhood acute lymphoblastic leukemia. *Leukemia & Lymphoma*, *38*(5–6), 447–462. http://dx.doi.org/10.3109/10428190009059264.

Spector, A. A., & Kim, H. Y. (2015). Cytochrome P epoxygenase pathway of polyunsaturated fatty acid metabolism. *Biochimica et Biophysica Acta*, *1851*(4), 356–365. http://dx.doi.org/10.1016/j.bbalip.2014.07.020.

Turpeinen, M., & Zanger, U. M. (2012). Cytochrome P450 2B6: Function, genetics, and clinical relevance. *Drug Metabolism and Drug Interactions*, *27*(4), 185–197. http://dx.doi.org/10.1515/dmdi-2012-0027.

Wang, H., Lin, L., Jiang, J., Wang, Y., Lu, Z. Y., Bradbury, J. A., et al. (2003). Up-regulation of endothelial nitric-oxide synthase by endothelium-derived hyperpolarizing factor involves mitogen-activated protein kinase and protein kinase C signaling pathways. *The Journal of Pharmacology and Experimental Therapeutics*, *307*(2), 753–764. http://dx.doi.org/10.1124/jpet.103.052787.

Wang, X., Ni, L., Yang, L., Duan, Q., Chen, C., Edin, M. L., et al. (2014). CYP2J2-derived epoxyeicosatrienoic acids suppress endoplasmic reticulum stress in heart failure. *Molecular Pharmacology*, *85*(1), 105–115. http://dx.doi.org/10.1124/mol.113.087122.

Wassenaar, C. A., Ye, Y., Cai, Q., Aldrich, M. C., Knight, J., Spitz, M. R., et al. (2014). CYP2A6 reduced activity gene variants confer reduction in lung cancer risk in African

American smokers—Findings from two independent populations. *Carcinogenesis*. http://dx.doi.org/10.1093/carcin/bgu235.

Wei, X., Zhang, D., Dou, X., Niu, N., Huang, W., Bai, J., et al. (2014). Elevated 14,15-epoxyeicosatrienoic acid by increasing of cytochrome P450 2C8, 2C9 and 2J2 and decreasing of soluble epoxide hydrolase associated with aggressiveness of human breast cancer. *BMC Cancer*, *14*, 841. http://dx.doi.org/10.1186/1471-2407-14-841.

Wray, J. A., Sugden, M. C., Zeldin, D. C., Greenwood, G. K., Samsuddin, S., Miller-Degraff, L., et al. (2009). The epoxygenases CYP2J2 activates the nuclear receptor PPARalpha in vitro and in vivo. *PLoS One*. *4*(10). http://dx.doi.org/10.1371/journal.pone.0007421e7421.

Wu, S., Moomaw, C. R., Tomer, K. B., Falck, J. R., & Zeldin, D. C. (1996). Molecular cloning and expression of CYP2J2, a human cytochrome P450 arachidonic acid epoxygenase highly expressed in heart. *The Journal of Biological Chemistry*, *271*(7), 3460–3468.

Xiao, B., Li, X., Yan, J., Yu, X., Yang, G., Xiao, X., et al. (2010). Overexpression of cytochrome P450 epoxygenases prevents development of hypertension in spontaneously hypertensive rats by enhancing atrial natriuretic peptide. *The Journal of Pharmacology and Experimental Therapeutics*, *334*(3), 784–794. http://dx.doi.org/10.1124/jpet.110.167510.

Xu, M., Ju, W., Hao, H., Wang, G., & Li, P. (2013). Cytochrome P450 2J2: Distribution, function, regulation, genetic polymorphisms and clinical significance. *Drug Metabolism Reviews*, *45*(3), 311–352. http://dx.doi.org/10.3109/03602532.2013.806537.

Xu, D., Li, N., He, Y., Timofeyev, V., Lu, L., Tsai, H. J., et al. (2006). Prevention and reversal of cardiac hypertrophy by soluble epoxide hydrolase inhibitors. *Proceedings of the National Academy of Sciences of the United States of America*, *103*(49), 18733–18738. http://dx.doi.org/10.1073/pnas.0609158103.

Xu, X., Tu, L., Feng, W., Ma, B., Li, R., Zheng, C., et al. (2013). CYP2J3 gene delivery up-regulated adiponectin expression via reduced endoplasmic reticulum stress in adipocytes. *Endocrinology*, *154*(5), 1743–1753. http://dx.doi.org/10.1210/en.2012-2012.

Xu, X., Tu, L., Wang, L., Fang, X., & Wang, D. W. (2011). CYP2J3 gene delivery reduces insulin resistance via upregulation of eNOS in fructose-treated rats. *Cardiovascular Diabetology*, *10*, 114. http://dx.doi.org/10.1186/1475-2840-10-114.

Xu, X., Zhang, X. A., & Wang, D. W. (2011). The roles of CYP450 epoxygenases and metabolites, epoxyeicosatrienoic acids, in cardiovascular and malignant diseases. *Advanced Drug Delivery Reviews*, *63*(8), 597–609. http://dx.doi.org/10.1016/j.addr.2011.03.006.

Zandvliet, A. S., Huitema, A. D., Copalu, W., Yamada, Y., Tamura, T., Beijnen, J. H., et al. (2007). CYP2C9 and CYP2C19 polymorphic forms are related to increased indisulam exposure and higher risk of severe hematologic toxicity. *Clinical Cancer Research*, *13*(10), 2970–2976. http://dx.doi.org/10.1158/1078-0432.CCR-06-2978.

Zanger, U. M., & Schwab, M. (2013). Cytochrome P450 enzymes in drug metabolism: Regulation of gene expression, enzyme activities, and impact of genetic variation. *Pharmacology & Therapeutics*, *138*(1), 103–141. http://dx.doi.org/10.1016/j.pharmthera.2012.12.007.

Zeldin, D. C. (2001). Epoxygenase pathways of arachidonic acid metabolism. *The Journal of Biological Chemistry*, *276*(39), 36059–36062. http://dx.doi.org/10.1074/jbc.R100030200.

Zhang, L., Cui, Y., Geng, B., Zeng, X., & Tang, C. (2008). 11,12-Epoxyeicosatrienoic acid activates the L-arginine/nitric oxide pathway in human platelets. *Molecular and Cellular Biochemistry*, *308*(1–2), 51–56. http://dx.doi.org/10.1007/s11010-007-9611-6.

Zhao, G., Tu, L., Li, X., Yang, S., Chen, C., Xu, X., et al. (2012). Delivery of AAV2-CYP2J2 protects remnant kidney in the 5/6-nephrectomized rat via inhibition of apoptosis and fibrosis. *Human Gene Therapy*, *23*(7), 688–699. http://dx.doi.org/10.1089/hum.2011.135.

Zhao, G., Wang, J., Xu, X., Jing, Y., Tu, L., Li, X., et al. (2012). Epoxyeicosatrienoic acids protect rat hearts against tumor necrosis factor-alpha-induced injury. *The Journal of Lipid Research*, *53*(3), 456–466. http://dx.doi.org/10.1194/jlr.M017319.

Zheng, C., Wang, L., Li, R., Ma, B., Tu, L., Xu, X., et al. (2010). Gene delivery of cytochrome p450 epoxygenase ameliorates monocrotaline-induced pulmonary artery hypertension in rats. *American Journal of Respiratory Cell and Molecular Biology*, *43*(6), 740–749. http://dx.doi.org/10.1165/rcmb.2009-0161OC.

CHAPTER EIGHT

Cytochrome P450 ω-Hydroxylases in Inflammation and Cancer

Amanda L. Johnson*, Katheryne Z. Edson*,†, Rheem A. Totah*, Allan E. Rettie*,[1]

*Department of Medicinal Chemistry, School of Pharmacy, University of Washington, Seattle, Washington, USA

†Amgen Inc., Thousand Oaks, California, USA

[1]Corresponding author: e-mail address: rettie@uw.edu

Contents

Advances in Pharmacology, Volume 74
ISSN 1054-3589
http://dx.doi.org/10.1016/bs.apha.2015.05.002

Abstract

Cytochrome P450-dependent ω-hydroxylation is a prototypic metabolic reaction of CYP4 family members that is important for the elimination and bioactivation of not only therapeutic drugs, but also endogenous compounds, principally fatty acids. Eicosanoids, derived from arachidonic acid, are key substrates in the latter category. Human CYP4 enzymes, mainly CYP4A11, CYP4F2, and CYP4F3B, hydroxylate arachidonic acid at the omega position to form 20-HETE, which has important effects in tumor progression and angiogenesis and blood pressure regulation in the vasculature and kidney. CYP4F3A in myeloid tissue catalyzes the ω-hydroxylation of leukotriene B4 to 20-hydroxy leukotriene B4, an inactivation process that is critical for the regulation of the inflammatory response. Here, we review the enzymology, tissue distribution, and substrate selectivity of human CYP4 ω-hydroxylases and their roles as catalysts for the formation and termination of the biological effects of key eicosanoid metabolites in inflammation and cancer progression.

ABBREVIATIONS

17-ODYA 17-octadecynoic acid
Akt protein kinase B
AMP 5′ adenosine monophosphate
COX cyclooxygenase
CYP cytochrome P450
DDMS dibromo-dodecenyl-methylsulfimide
DHA docosahexaenoic acid
DHET dihydroxyeicosatrienoic acid
EET epoxyeicosatrienoic acid
EGF epidermal growth factor
EPA eicosapentaenoic acid
EPC endothelial progenitor cells
FGF fibroblast growth factor
HEET hydroxyepoxyeicosatrienoic acid
HET0016 *N*-hydroxy-*N*′-(4-*n*-butyl-2-methylphenyl)formamidine
HETE hydroxyeicosatetraenoic acid
HIF-1α hypoxia-inducible factor 1-alpha
HNF4α hepatocyte nuclear factor 4-alpha
HPETE hydroperoxyeicosatetraenoic acid
IL interleukin
LOX lipoxygenase
LPS lipopolysaccharide
LT leukotriene
LX lipoxin
MAPK mitogen-activated protein kinase
NF-κB nuclear factor kappa-B
NSCLC nonsmall cell lung cancer
PDGF platelet-derived growth factor

PG prostaglandin
PI3K phosphoinositide 3-kinase
PMNL polymorphonuclear leukocytes
PPAR peroxisome proliferator-activated receptor
PUFA polyunsaturated fatty acid
ROS reactive oxygen species
Rv resolvin
sEH soluble epoxide hydrolase
SREBP sterol regulatory element-binding protein
TNFα tissue necrosis factor alpha
VEGF vascular endothelial growth factor

1. INTRODUCTION

Omega (ω)-hydroxylation is an oxidation reaction catalyzed by cytochrome P450 (CYP) monooxygenases that transforms the terminal methyl group of a hydrophobic aliphatic chain into a more polar alcohol metabolite. Fatty acid ω-hydroxylation is the pivotal catalytic step that initiates formation of mono- and dicarboxylic acids that are then catabolized through the β-oxidation pathway. The biological ω-hydroxylation pathway was first described more than 80 years ago for medium-chain fatty acids that were metabolized to urinary dicarboxylic acids of the same chain-length (Verkade et al., 1933). These linked enzymatic processes may prevent toxic buildup of some fatty acids in the body (Hardwick, 2008).

In the early 1960s, ω-hydroxylation was localized to the microsomal fraction and shown to be dependent on NADPH and molecular oxygen (Preiss & Bloch, 1964; Wakabayashi & Shimazono, 1963). Confirmation that fatty acid ω-hydroxylation was catalyzed by CYP followed from the successful separation and reconstitution of the P450, reductase and lipid components of the enzyme system that supported lauric acid ω-hydroxylation (Lu & Coon, 1968).

Although ω-hydroxylation is relatively a minor route in the overall catabolism of fatty acids (Draye & Vamecq, 1989), the pathway is essential for both the anabolism and catabolism of critical lipid mediators such as 20-hydroxyeicosatetraenoic acid (20-HETE) and leukotriene B_4 (LTB_4), respectively (Fig. 1). P450-dependent metabolism of arachidonic acid and LTB_4 was established in the early 1980s (Bednar, Schwartzman, Ibraham, McGiff, & Mullane, 1984; Capdevila, Chacos, Werringloer, Prough, & Estabrook, 1981). Over the next 30 years, these ω-hydroxylation pathways

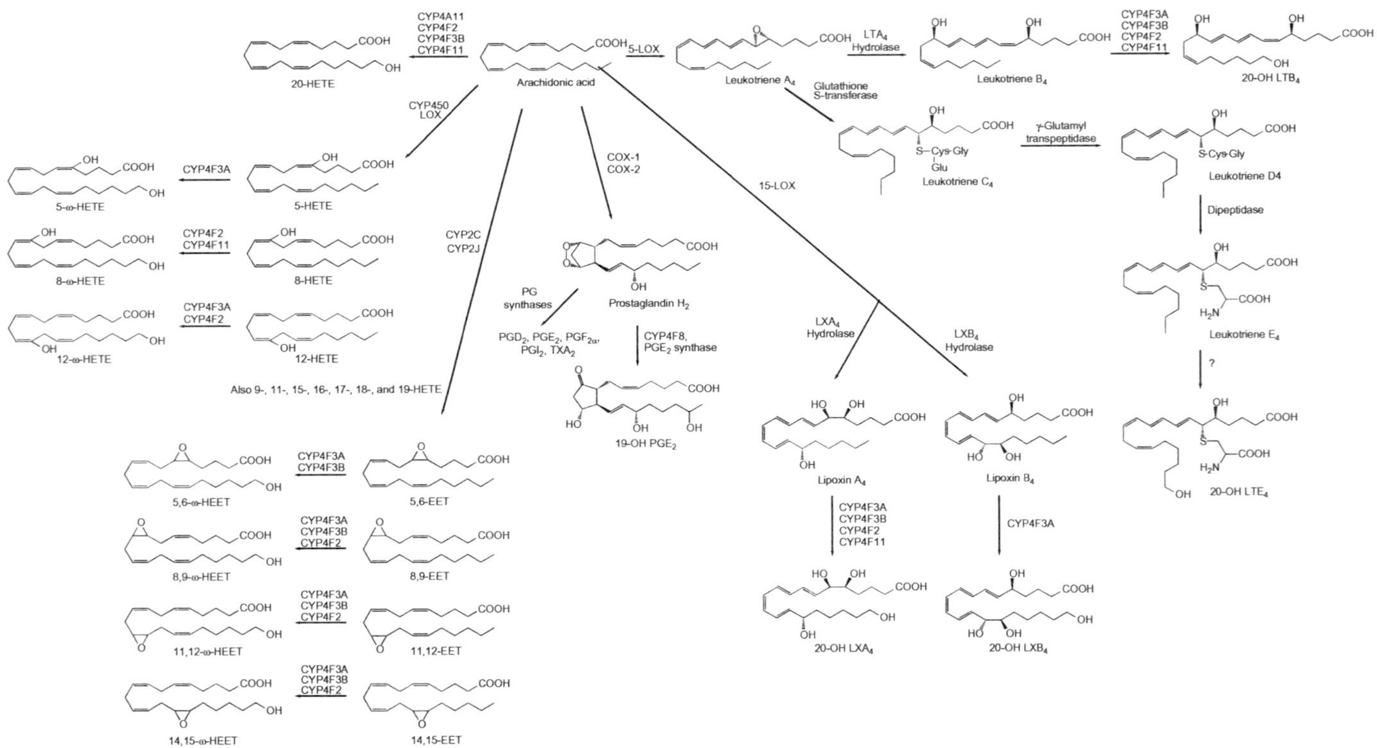

Figure 1 Eicosanoid pathways for bioactive lipid anabolism and catabolism.

have emerged as critical determinants of numerous disease processes, including inflammation and cancer progression, which are the focus of this review.

2. PHYSIOLOGICAL ROLES, MULTIPLICITY, TISSUE DISTRIBUTION, AND SUBSTRATE SPECIFICITIES OF THE CYP4 ω-HYDROXYLASES

The CYP4 enzyme family play a primary physiological role in the ω-hydroxylation of endogenous fatty acids, exemplified by the eicosanoids. Several CYP4A and CYP4F enzymes generate the active signaling compound, 20-HETE, by ω-hydroxylation of arachidonic acid (Kroetz & Xu, 2005; Lasker et al., 2000; Powell, Wolf, Jin, & Lasker, 1998). On the other hand, *inactivation* of LTB_4 via ω-hydroxylation is the defining function of neutrophil CYP4F3A (Kikuta et al., 1998). CYP4 enzymes also play important endogenous roles as ω-hydroxylases in vitamin E and vitamin K catabolism (Edson et al., 2013; McDonald, Rieder, Nakano, Hsia, & Rettie, 2009; Parker, Sontag, Swanson, & McCormick, 2004), and the metabolism of xenobiotics such as ebastine, terfenadine, pafuramidine, and fingolimod (Hashizume et al., 2002; Kovarik et al., 2009; Wang et al., 2006). The structural basis for selective ω-hydroxylation by CYP4 enzymes is thought to be a narrow channel near the heme (~4 Å for CYP4A1) that constrains access of only the terminal carbon to the active site iron-oxo species (He, Cryle, De Voss, & de Montellano, 2005; Lewis & Lake, 1999). A schematic of the proposed active site for CYP4A1 binding of lauric acid is shown in Fig. 2.

In humans, the cytochrome P450 4 (CYP4) family consists of 12 genes and 13 enzymes divided into 6 subfamilies: CYP4A, CYP4B, CYP4F, CYP4V, CYP4X, and CYP4Z (Table 1). At the gene level, *CYP4A, B, X,* and *Z* are clustered on chromosome 1, while the *CYP4F* and *CYP4V* genes reside on chromosome 19 and 4, respectively (Nelson et al., 2004). While the CYP4s represent one of the largest human P450 families, only a subset of these enzymes has significant ω-hydroxylase activity toward eicosanoids.

The most extensively studied human CYP4 fatty acid ω-hydroxylases are CYP4A11, CYP4F2, CYP4F3A, and CYP4F3B. CYP4A11 (ortholog of rat CYP4A1) is regulated by peroxisome proliferator-activated receptor-α (PPARα), expressed at high levels in liver and kidney, and catalyzes the metabolism of lauric and arachidonic acids with high regioselectivity at the ω-position (Johnson, Palmer, Griffin, & Hsu, 1996). CYP4F2 is also

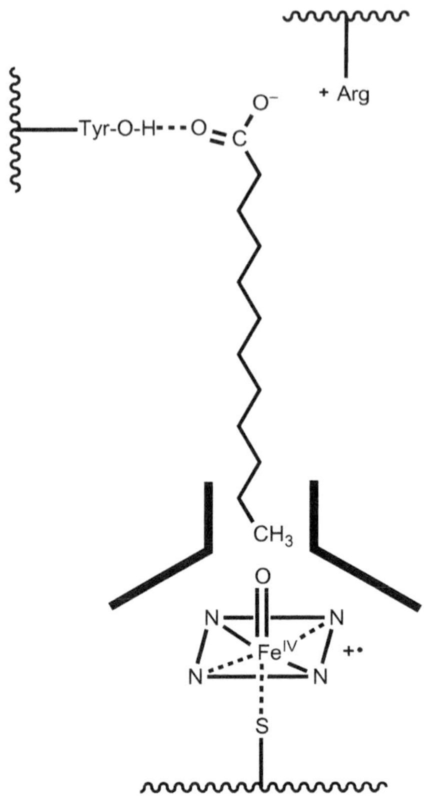

Figure 2 Proposed structural requirements for selective ω-hydroxylation of fatty acids by CYP4A1.

expressed at significant levels in liver and kidney, and although it is not regulated by PPARα, it is induced by some statins that signal through the sterol regulatory element-binding protein (SREBP) (Hsu, Savas, Griffin, & Johnson, 2007b).

CYP4F3 is an unusual human P450 gene in that it undergoes tissue-specific alternative splicing to CYP4F3A and CYP4F3B. The substrate specificities of CYP4F3A and CYP4F3B are determined by the incorporation of either exon 4 or exon 3, respectively, into the translated enzyme (Christmas, Ursino, Fox, & Soberman, 1999). CYP4F3A is expressed in neutrophils and has the highest affinity of all the CYP4F isoforms for LTB_4. In contrast, CYP4F3B is primarily expressed in human liver and kidney and has greater activity toward arachidonic acid and omega 3 polyunsaturated fatty acids (ω3 PUFAs) (Christmas et al., 2001; Fer et al., 2008). CYP4F3 enzymes are also

Table 1 Human CYP4 Enzymes: Tissue Distribution and Fatty Acid ω-Hydroxylase Activities

Enzyme	Tissue Distribution	Substrates
CYP4A11	Liver, kidney	AA, ω3 PUFA, MCFA, LCFA
CYP4A22	Liver, kidney	?
CYP4B1	Lung, bladder	Inactive in human, SCFA in other species
CYP4F2	Liver, kidney	LTB4, AA, MCFA, LCFA, VLCFA, LX, EET, HETE
CYP4F3A	Myeloid tissues	LTB4, ω3 PUFA, MCFA, LCFA, LX, EET, HETE
CYP4F3B	Liver, kidney	AA, ω3 PUFA, MCFA, LCFA, VLCFA, LTB4, EET
CYP4F8	Urogenital tissues	PGH1/2 (internal oxidations)
CYP4F11	Liver, kidney	MCFA, LCFA (3-hydroxy analogs), AA, LTB4, LX, HETE
CYP4F12	Liver, gut, kidney	AA, PGH1/2 (internal oxidations)
CYP4F22	Skin	Hepoxilins?, acylceramides?
CYP4V2	Retina	ω3 PUFA, AA, MCFA
CYP4X1	Brain, bronchial airways	No reported ω-hydroxylase activity
CYP4Z1	Mammary tissue	MCFA (internal oxidations), AA?

AA, arachidonic acid; LTB4, leukotriene B4; PGH1/2, prostaglandin H1/2; SCFA (C7–C9), short-chain fatty acid; MCFA (C10–C16), medium chain fatty acid; LCFA (C17–C21), long-chain fatty acid; VLCFA (C22–C26), very long-chain fatty acid; ω3 PUFA, omega 3 polyunsaturated fatty acid; LX, lipoxin; HETE, hydroxyeicosatrienoic acid; EET, epoxyeicosatrienoic acid.
Information taken from references throughout the text.

reported to be the major ω-hydroxylases of long-chain fatty acid epoxides (Le Quere, Plee-Gautier, Potin, Madec, & Salaun, 2004).

CYP4F8, CYP4F22, and CYP4V2 are all expressed primarily in extrahepatic tissues (Bylund, Hidestrand, Ingelman-Sundberg, & Oliw, 2000; Kelly, Nakano, Rohatgi, Yarov-Yarovoy, & Rettie, 2011), although only CYP4V2 exhibits clear fatty acid ω-hydroxylase activity (Nakano, Kelly, & Rettie, 2009; Nakano, Kelly, Wiek, Hanenberg, & Rettie, 2012). More information is available for CYP4F11 and CYP4F12, both of which are expressed primarily in drug clearance organs, including the liver. Whereas CYP4F12 can metabolize a diverse array of bulky xenobiotics, similar to

CYP3A4 (Eksterowicz, Rock, Rock, Wienkers, & Foti, 2014), CYP4F11 is associated with the ω-hydroxylation of endogenous substrates including longer chain fatty acids and vitamin K (Dhar, Sepkovic, Hirani, Magnusson, & Lasker, 2008; Edson et al., 2013).

CYP4X1 and CYP4Z1 are extrahepatic P450s, with sparse information about their biological function. Human CYP4X1 is highly expressed in the brain, skin, and airways, and although it is regulated by PPARα, the enzyme is not a recognized fatty acid ω-hydroxylase, generating only epoxide metabolites of arachidonic acid at low levels (Savas, Hsu, Griffin, Bell, & Johnson, 2005; Stark, Dostalek, & Guengerich, 2008). CYP4Z1 is expressed in mammary tissue and is inducible by glucocorticoids and progesterone (Rieger et al., 2004; Savas et al., 2005). The regioselectivity of CYP4Z1-dependent fatty acid metabolism is strikingly different from other ω-hydroxylases discussed so far in that the recombinant enzyme is a mid-chain hydroxylase of medium-chain fatty acids, adding a hydroxyl group to positions ω-2 to ω-5 (Zollner et al., 2009). Finally, CYP4A22 and CYP4B1 are expressed at very low levels and/or are inactive due to poor enzyme stability (Hsu, Savas, Griffin, & Johnson, 2007a; Wiek et al., 2015).

3. EICOSANOID PATHWAYS FOR BIOACTIVE LIPID ANABOLISM AND CATABOLISM

Eicosanoids are lipid mediators derived from arachidonic acid by the actions of cyclooxygenases (COXs), lipoxygenases (LOXs), and CYPs (Fig. 1) that together make up complex signaling networks of over 20 chemical messengers implicated in critical biological actions in practically every tissue, organ, and cell in the body (Funk, 2001).

3.1 COX Pathways

Prostaglandins (PGs) are autocrine and paracrine lipid mediators, synthesized from arachidonic acid by the dual-function cyclooxygenase enzymes, COX-1 and COX-2. COX-1 is constitutively expressed in most cell types, whereas COX-2 expression is generally induced in response to proinflammatory stimuli (Smith, DeWitt, & Garavito, 2000; Vane, Bakhle, & Botting, 1998), though both COX enzymes are involved in homeostatic- and inflammation-induced prostanoid synthesis (Seibert et al., 1997). COXs initially metabolize arachidonic acid, via their COX function, to unstable prostaglandin G_2 (PGG_2) that is then converted by their peroxidase function to prostaglandin H_2 (PGH_2). PGH_2 serves as a substrate for cell- and tissue-selective prostanoid synthases and isomerases that produce numerous bioactive prostanoids: prostaglandins D_2, E_2, $F_{2\alpha}$, I_2, and thromboxane A_2. Differential expression of these synthases determines the prostanoid profile of a given tissue (Naraba et al., 1998; Smith & Song, 2002). Prostanoids are widely distributed throughout the body and exert diverse physiological functions and responses. Their biosynthesis, mechanism of action, and role in health, inflammation and disease progression have been reviewed in detail elsewhere (Ricciotti & FitzGerald, 2011; Smyth, Grosser, Wang, Yu, & FitzGerald, 2009).

3.2 LOX Pathways

Of the various eicosanoids produced by the LOX pathway of arachidonic acid metabolism, the proinflammatory leukotrienes (LTs) are arguably the most significant in terms of their function in disease pathogenesis. Unlike PGs, which are synthesized by various cell types, LTs are predominantly formed in inflammatory cells such as polymorphonuclear leukocytes (PMNL), macrophages, and mast cells (Funk, 2001). 5-LOX catalyzes the epoxidation of arachidonic acid to leuktotriene A_4 (LTA_4), via the intermediacy of 5-hydroperoxyeicosatetraenoic acid (5-HPETE). LTA_4 then undergoes enzymatic hydrolysis to LTB_4 or conjugation with glutathione to form the cysteinyl-leukotrienes: LTC_4, LTD_4, and LTE_4 (Fig. 1). LTB_4 is a potent neutrophil chemotactic agent and a key mediator in inflammation, while the cysteinyl-LTs are involved in the contraction of smooth muscle to propagate the inflammatory response (Peters-Golden & Henderson, 2007; Samuelsson, Dahlen, Lindgren, Rouzer, & Serhan, 1987). Aside from LT formation, 5-, 12-, and 15-LOX can also convert arachidonic acid into the proinflammatory 5-, 12-, and 15-HETEs (Powell & Rokach, 2015). Additionally, 15-LOX synthesizes the lipoxins (LXs), arachidonic acid metabolites that act as important mediators in the resolution of inflammation (McMahon, Mitchell, Brady, & Godson, 2001; Serhan, Chiang, Dalli, & Levy, 2015). The biosynthesis, enzymology, and biological action of the LOX-derived arachidonic acid metabolites have been extensively reviewed elsewhere (Haeggstrom & Funk, 2011; Murphy & Gijon, 2007).

3.3 P450 Pathways

The CYP monooxygenases form two series of bioactive oxidized lipids from arachidonic acid (Capdevila & Falck, 2002; Konkel & Schunck, 2011); four regioisomeric epoxyeicosatrienoic acids (EETs), 14,15-, 11,12-, 8,9- and 5,6-EET; and eleven mono-HETEs, 5-, 8-, 9-, 11-, 12-, 15-, 16-, 17-, 18-, 19-, and 20-HETE (Fig. 1). Whereas human CYP2C and CYP2J enzymes are the principal catalysts of EETs formation, the internal HETEs are formed by a wide variety of P450s (Kaspera & Totah, 2009). CYP4 enzymes dominate 20-HETE formation and catabolism of LTB_4. The P450-derived eicosanoids exhibit a vast array of biological functions, although much of the research emphasis for the HETEs (and EETs) has been focused on the vascular (Fleming, 2008) and renal systems (Fan, Muroya, & Roman, 2015).

4. ω-HYDROXYLASES IN INFLAMMATION

Eicosanoid metabolites of arachidonic acid, including PGs, LTs, and 20-HETE, are key proinflammatory mediators of the inflammatory cascade (Khanapure, Garvey, Janero, & Letts, 2007). P450 ω-hydroxylases are linked to inflammation, largely through catabolism of the LTs and formation of the proinflammatory metabolite 20-HETE. It is well established that members of the CYP4F family function in the resolution phase of inflammation and in the ω-hydroxylation and deactivation of LTB_4 (Kalsotra & Strobel, 2006; Kikuta, Kusunose, & Kusunose, 2000). However, CYP4s may also play a role in the activation and amplification of inflammation through catabolism of other anti-inflammatory eicosanoids, such as the EETs and LXs (Kalsotra & Strobel, 2006; Kikuta, Kusunose, & Kusunose, 2002). In this section, we explore this dual role for P450 ω-hydroxylases in inflammation by reviewing eicosanoid metabolism catalyzed by the CYP4 family and examining what is known about the expression and regulation of *CYP4* genes during inflammation to better understand how these genes function in inflammatory diseases.

4.1 LT ω-Hydroxylases

LTB_4 and the cysteinyl-LTs, derived from arachidonic acid via the LOX pathway, are important proinflammatory mediators. LTs increase leukocyte tissue infiltration and amplify inflammation through production of cytokines and chemokines. LTB_4 is a potent chemotactic agent for neutrophils and other leukocytes and promotes expression of cell adhesion proteins. The cysteinyl-LTs mediate contraction of smooth muscle thereby increasing venopermeability to enhance recruitment of leukocytes. Numerous studies have demonstrated that elevated levels of LTB_4 and cysteinyl-LTs are linked to the pathogenesis of several inflammatory diseases, including asthma, Alzheimer's disease, rheumatoid arthritis, psoriasis, diabetes, cardiovascular disease, and cancer, among others (Peters-Golden & Henderson, 2007). Inactivation and elimination of LTs, especially LTB_4, is therefore expected to be a crucial part of the resolution of inflammation and return to normal homeostasis. While metabolism of LTB_4 is complex involving multiple enzymatic pathways, ω-hydroxylation is established as the primary route of inactivation. Conversion of LTB_4 to its 20-hydroxy metabolite (20-OH LTB_4) results in a dramatic loss of its chemotactic activity on leukocytes. 20-OH LTB_4 is rapidly oxidized to 20-COOH LTB_4 and targeted for β-oxidation or conjugation with a glucuronide (Kikuta et al., 1993).

ω-Hydroxylase function in inflammation can be traced back to the discovery of the CYP4F family in the 1980s through a series of studies aimed at identifying a deactivation pathway of LTB_4 in human PMNL. Experiments with $^{18}O_2$ revealed that the terminal hydroxyl group in a newly identified ω-hydroxylated metabolite of LTB_4 (20-OH LTB_4) was derived from molecular oxygen (Hansson, Lindgren, Dahlen, Hedqvist, & Samuelsson, 1981). Subsequently, it was shown that ω-hydroxylase activity in PMNL was confined to the microsomal fraction and could be inhibited by carbon monoxide (Shak & Goldstein, 1984). Inhibition of this activity by carbon monoxide and by antibodies against P450 reductase indicates catalysis by a P450 enzyme specific for LTB_4 ω-hydroxylation. This led to the cloning, expression, and identification of a novel LTB_4 ω-hydroxylase, CYP4F3A (Kikuta et al., 1993, 1998).

Human CYP4F2 and CYP4F3B, the major LTB_4 ω-hydroxylases expressed in human liver, play a central role in the inactivation and elimination of LTs from systemic circulation during the resolution of inflammation. The eicosanoid substrate selectivity and activity of the CYP4F subfamily as LTB_4 ω-hydroxylases varies considerably. Neutrophil-specific CYP4F3A has a much higher affinity for LTB_4 ($K_m = 0.64$ μM) than either CYP4F2 ($K_m = 47$ μM) or CYP4F3B ($K_m = 21$ μM) (Christmas et al., 2001; Kikuta et al., 1998). CYP4F11 also possesses LTB_4 ω-hydroxylase activity, though much less than that of the other CYP4Fs (Kalsotra, Turman, Kikuta, & Strobel, 2004). Early reports indicated that CYP4A11, purified from human liver microsomes, exhibited negligible ω-hydroxylase activity toward LTB_4 (Jin, Koop, Raucy, & Lasker, 1998), although CYP4A11 protein expression correlated with formation of 20-OH LTB_4. However, a more recent study demonstrated that CYP4A11 Supersomes™ generated 20-OH LTB_4 at a rate of 1.8 nmol/min/nmol P450 (Le Quere et al.,

2004), which is ~20-fold lower than catalysis by CYP4F3A (Kikuta et al., 1998). Other CYP4F family members also generate 17-, 18-, and 19-OH LTB_4 (Berry, Borgeat, Gosselin, Flamand, & Murphy, 2003). Further studies are needed to determine if these latter LT metabolites have any biological activity. It is still not known which P450s ω-hydroxylate and inactivate cysteinyl-LTs, although early work demonstrated ω-oxidation of LTE_4 in rat liver microsomes (Orning, 1987), and observed this metabolite in the urine of human subjects (Sala, Voelkel, Maclouf, & Murphy, 1990).

4.2 PG ω-Hydroxylases

PGs are proinflammatory metabolites of arachidonic acid, synthesized *de novo* at sites of injury by the COX pathway. PGs are largely responsible for the cardinal symptoms of inflammation: rubor (redness), calor (heat), tumor (swelling), and dolor (pain) (Ricciotti & FitzGerald, 2011). While PG metabolism is expected to be a key component of inflammation resolution, human P450 ω-hydroxylases do not appear to contribute significantly to this process. Several studies demonstrated that rabbit and rodent CYP4A isoforms can ω-hydroxylate PGs (Masters et al., 1993; Roman et al., 1993). However, human CYP4A11 does not exhibit PG ω-hydroxylase activity, although it does ω-hydroxylate PGH_2 analogs (Oliw, Stark, & Bylund, 2001). CYP4F8, largely expressed in prostate and seminal vesicles, is not a PG ω-hydroxylase but catalyzes the formation of 19-OH PGE_2 (Bylund et al., 2000).

4.3 Arachidonic Acid and ω-Hydroxylases

20-HETE plays a role in inflammation by stimulating the production of various proinflammatory mediators: PGE_2, cytokine tissue necrosis factor alpha (TNFα), and the chemokines IL-8, IL-12, IL-14 (Ishizuka et al., 2008). CYP4A11, CYP4F2, CYP4F3B, and CYP4F11 all catalyze formation of 20-HETE and are expressed in the liver and kidney, both major sites of 20-HETE synthesis (Christmas et al., 2001; Hsu et al., 2007a; Kalsotra et al., 2004). CYP4F2 is the major catalyst of 20-HETE formation in human liver and kidney microsomes (Lasker et al., 2000; Powell et al., 1998). Production of 20-HETE by the CYP4A and CYP4F enzymes therefore suggests a role for ω-hydroxylases in the activation of inflammation, which may exacerbate 20-HETE's role in tumor progression (see Section 5). In addition to CYP4Fs catalyzing the formation of 20-HETE, these enzymes also ω-hydroxylate and deactivate proinflammatory 5-, 8-, and 12-HETE

(Kalsotra & Strobel, 2006; Kikuta et al., 2002), again suggestive of dual roles for the P450 ω-hydroxylases in both the initiation and resolution phases of inflammation.

4.4 EET ω-Hydroxylases

The EETs mediate vasodilation of the vasculature and are endowed with anti-inflammatory properties, including prevention of cytokine-induced leukocyte adhesion to endothelial cells through inhibition of NF-κB, reducing platelet aggregation, and reducing smooth muscle cell migration (Fleming, 2008; Node et al., 1999). Catabolism of EETs proceeds primarily through soluble epoxide hydrolase (sEH) which catalyzes metabolism to the corresponding less active vicinal diol, dihydroxyeicosatrienoic acid (DHET) (Kaspera & Totah, 2009). The ratio of circulating 20-HETE:EETs + DHETs in plasma is usually used as an indicator of the relative rates of hydroxylation versus epoxidation of arachidonic acid, and therefore a gauge for inflammatory status (Theken et al., 2011).

EETs themselves can be ω-hydroxylated by CYP4 enzymes to their respective 20-hydroxyepoxyeicosatrienoic acids (HEETs) (Le Quere et al., 2004). Interestingly, human liver microsomes exhibit the highest ω-hydroxylase activity toward 11,12-EET, whereas 8,9-EET was the best substrate for recombinant CYP4F2, CYP4F3A, and CYP4F3B enzymes. It is not known whether other CYP4Fs or CYP4A11 can catalyze formation of ω-HEETs, although several rat CYP4A isoforms have demonstrated ω-hydroxylation activity toward the various EETs (Cowart et al., 2002), which is consistent with studies showing that ω-HEETs have high affinity toward PPARα and induce peroxisome proliferation in rodents (Gatica et al., 2007). It remains unclear whether ω-hydroxylation of EETs in humans is another deactivation process or a means of synthesizing biologically active compounds, as it is in rodents, as peroxisome proliferation is mediated through suppression of HNF4α and not by PPARα in humans (Hertz, Sheena, Kalderon, Berman, & Bar-Tana, 2001).

4.5 LX ω-Hydroxylases

LXs are anti-inflammatory metabolites of arachidonic acid, whose formation is catalyzed by 15-LOX, although they can also be synthesized from LTA_4 through transcellular metabolism (Gronert, Clish, Romano, & Serhan, 1999). LXs play an important role in the resolution phase of inflammation by decreasing leukocyte infiltration and cell adhesion, promoting macrophage clearance of apoptotic cells, and modulating the production of proinfammatory cytokines (Haeggstrom & Funk, 2011; McMahon et al., 2001). While the K_m values for LXs are several fold higher than that for LTB_4, CYP4F3A quite efficiently catalyzes ω-hydroxylation of both LXA_4 and LXB_4 but shows much higher affinity toward LXB_4 as a substrate (Kikuta et al., 1998). In contrast, CYP4F2 can ω-hydroxylate only LXA_4, with similar affinity and rate of formation as LTB_4 (Kikuta et al., 2000). The resolution of inflammation is an active process that involves the biosynthesis of proresolving mediators, in addition to catabolism of proinflammatory mediators. Failure of acute inflammation to resolve may lead to excessive tissue damage, chronic inflammation, and propagation of the autoimmune response. The ω-hydroxylation of the LXs by members of the CYP4F family suggests differential roles for these enzymes: promotion of inflammation resolution through deactivation of LTB_4 and prolongation of inflammation through LX catabolism.

4.6 ω3 PUFA ω-Hydroxylases

ω3 PUFAs, especially eicosapentaenoic acid (EPA) and docosahexaenoic acid (DHA), have long been touted as human dietary supplements for their beneficial impact in a wide range of diseases with an inflammatory etiology (Simopoulos, 2002). However, the molecular mechanisms by which they exert their anti-inflammatory and cardioprotective effects are poorly understood. EPA and DHA are precursors for resolvins (Rvs), protectins and maresins (Fig. 3), lipid mediators with potent anti-inflammatory and proresolving properties. Similar to the LXs, Rvs and protectins block leukocytic infiltration, modulate production of cytokines and other proinflammatory mediators, and promote macrophage uptake of apoptotic cells and clearance of cellular debris. Several recent reviews are available detailing the biosynthesis, regulation, and biological actions of this class of compounds (Serhan & Chiang, 2013; Serhan et al., 2015; Serhan & Petasis, 2011).

Direct effects of the CYP4 enzymes on protectins and Rvs appear to be minimal. There is an isolated report that human neutrophils can ω-hydroxylate RvE1 to 20-OHRvE1 (Arita et al., 2006), but the enzyme that catalyzes the reaction was not identified. It is also not clear if this is a true inactivation pathway, as 20-OHRvE1 retained some of the original RvE1 activity. However, P450 ω-hydroxylases may exert an indirect (proinflammatory) effect through reducing the biosynthesis of Rvs and

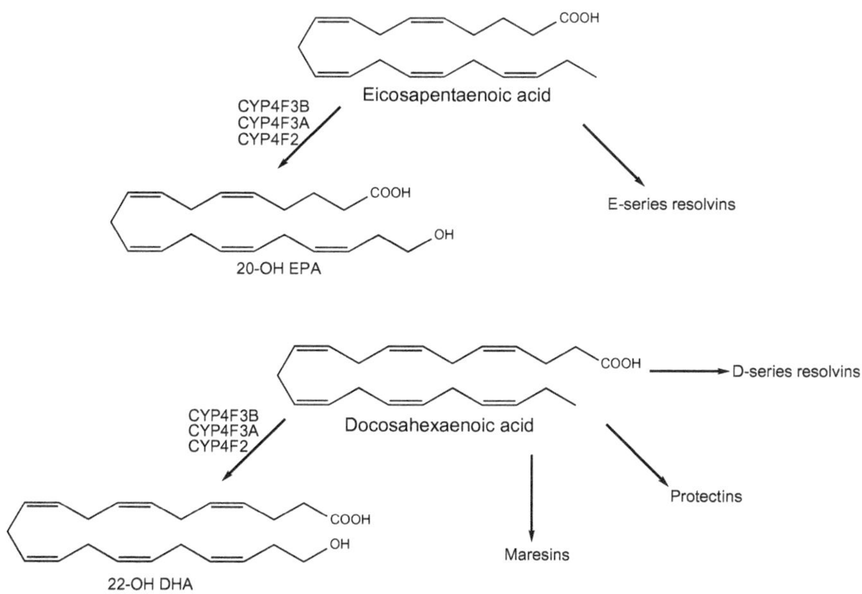

Figure 3 Metabolism of ω3 PUFAs by P450 ω-hydroxylases.

protectins through ω-hydroxylation of the parent compounds EPA and DHA (Fer et al., 2008). CYP4F3B can ω-hydroxylate EPA and DHA to their 20-hydroxy and 22-hydroxy metabolites, respectively (Harmon et al., 2006). Evidence was presented in the same study that 20-OH EPA and 22-OH DHA may function as lipid mediators rather than degradation products, much like 20-HETE, as these two metabolites were able to bind and activate PPARα in Cos-7 cells. However, the biological actions of these two ω-hydroxylated products are not well characterized.

Addition of EPA or DHA to incubations of arachidonic acid with each of the CYP4F enzymes resulted in decreased 20-HETE formation, suggesting that EPA and DHA compete with arachidonic acid as substrates for ω-oxidation (Fer et al., 2008; Harmon et al., 2006). In rats supplemented with dietary EPA/DHA, substantial replacement of arachidonic acid by EPA and DHA in membrane phospholipids of various organs was observed, providing *in vivo* evidence that EPA and DHA may compete with arachidonic acid for ω-hydroxylation by P450 (Arnold et al., 2010). In humans, supplementation with DHA/EPA leads to a change in the erythrocyte membrane fatty acid composition with DHA and EPA replacing arachidonic acid in phospholipids (Dawczynski et al., 2013). Conceivably, incorporation of EPA and DHA into membrane phospholipids may prevent conversion of arachidonic acid to the proinflammatory LTs and PGs by serving as alternative substrates for the COX and LOX enzymes to form docosanoids and related products (Fig. 3). Inhibition of arachidonic acid metabolism to proinflammatory mediators is potentially another mechanism by which ω3 PUFAs may exert their beneficial effects.

4.7 Gene Regulation of ω-Hydroxylases in Inflammation

CYP4F2, CYP4F3A, and CYP4F3B enzymes appear to be the primary P450 ω-hydroxylases involved in the inflammatory response, therefore understanding how these *CYP4F* genes are regulated during inflammation may provide insight into mechanisms involved in the activation and progression of inflammatory diseases. While regulation of the major drug-metabolizing P450s during inflammation and disease has been studied extensively, leading to reduced expression of the majority of P450s (Harvey & Morgan, 2014), little is known regarding the regulation of human *CYP4F* genes during inflammation. Studies in rat hepatocytes demonstrated induction of *CYP4F* genes after treatment with proinflammatory cytokines, TNFα, interleukin (IL)-1β, and IL-6, but downregulation of *CYP4Fs* by an anti-inflammatory cytokine IL-10 (Kalsotra et al., 2007). This is consistent with the view that cytokine-mediated suppression or induction of gene transcription is the primary mechanism for P450 regulation. However, identification of the specific nuclear receptors and transcription factors that mediate these effects has remained elusive, and analogous studies with human CYP4Fs are absent from the literature.

Multiple studies showed that retinoids, specifically 9-*cis*- and *all-trans*-retinoic acid (*at*RA), induce expression of CYP4F2 and 20-OH LTB_4 formation in HepG2 cells through the retinoid X receptor (RXR) (Zhang, Chen, & Hardwick, 2000; Zhang & Hardwick, 2000). Strobel and coworkers found that human keratinocytes can ω-hydroxylate LTB_4 and that this activity is increased following induction of CYP4F2, CYP4F3A, and CYP4F3B at both the transcript and protein levels by *at*RA activation of RXR, although the nuclear receptor partner of RXR remains unidentified (Kalsotra et al., 2008). A later study also confirmed that topical administration of retinoids to human skin samples enhanced CYP4F-mediated LTB_4 ω-hydroxylation, suggesting one way by which retinoid therapy might exert its anti-inflammatory effects in disorders such as psoriasis and atopic dermatitis (Du, Yin, Morrow, Strobel, & Keeney, 2009).

Induction of CYP4F2 has been found to proceed through other pathways in human hepatocytes and HepG2 cells, as well. *CYP4F2*-specific expression is induced by lovastatin through SREBP (Hsu et al., 2007b) and through an AMP-activated protein kinase pathway by genistein and resveratrol (Hsu, Savas, Lasker, & Johnson, 2011). Understanding how proinflammatory cytokines affect these pathways may help explain how CYP4F2, at least, is regulated during inflammation.

Much of the information regarding *CYP4F* gene regulation during inflammation has evolved from the use of mouse and animal models administered bacterial endotoxin lipopolysaccharide (LPS) or barium sulfate to induce an acute inflammatory response. However, the use of animal models in understanding human *CYP4F* gene expression during inflammation is limited by the fact that it is difficult to assign rodent orthologs of human CYP4F isoforms. Moreover, mechanisms of regulation by inflammatory mediators appear to be species specific. For instance, in a murine LPS-induced neuroinflammation model, pretreatment with the peroxisomal proliferator fenofibrate attenuated the inflammatory response and offered neuroprotection through induction of specific mouse *Cyp4fs*, an effect that was significantly reduced by the CYP4 inhibitor 17-ODYA (17-octadecynoic acid) (Sehgal et al., 2011). In contrast, *CYP4F2* expression in HepG2 cells was suppressed by clofibrate, ciprofibrate, and WY14,643 (Zhang et al., 2000). This complements *in vivo* evidence demonstrating that TNFα and IL-6 increase in response to fibrates in rodents (Rose et al., 1999; Rusyn, Tsukamoto, & Thurman, 1998), while levels of these proinflammatory cytokines decrease in human serum (Madej et al., 1998; Staels et al., 1998).

Additionally, CYP4F enzymes may be differentially regulated by the *type* of inflammatory insult, as evidenced by studies with murine Cyp4a enzymes. Murine hepatic *Cyp4a* mRNAs are typically downregulated in response to both LPS and *Citrobacter rodentium* infection models of inflammation. However, *Cyp4a10* and *Cyp4a14* knockout mice afforded protection and anti-inflammatory effects, such as reduced proinflammatory cytokine production, only in response to *C. rodentium* infection but not to LPS injection compared to wild-type mice (Nyagode, Williams, & Morgan, 2014). Similar gene-deletion studies of murine Cyp4f isoforms have not yet been performed. In a rodent model of septic shock, administration of a synthetic 20-HETE mimetic to rats following LPS injection, resulted in protection against inflammation, tachycardia, and hypotension, with attenuation of nitric oxide synthase activity in the heart and kidney being one of the most marked observations (Sari et al., 2014; Tunctan et al., 2013). During inflammation, there is often enhanced release of nitric oxide, which can inhibit 20-HETE formation by CYP ω-hydroxylases (Wang et al., 2003), leading to the resultant hypotension in septic shock. Thus, while CYP4F involvement was not directly investigated in the foregoing examples, study of the tissue-specific regulation of CYP4F enzymes in response to different inflammatory insults appears warranted.

Nonetheless, animal models have proven useful for establishing that individual *CYP4F* genes are differentially regulated in a tissue- and time-dependent manner. Multiple studies that examined the inflammatory response and *CYP4F* expression in skin, brain, and lung tissue showed an initial suppression of specific CYP4F isoforms for 12–24 h, then induction, followed by a gradual return to basal levels over a period ranging from hours to days to weeks (Birnie, Morrison, Camara, & Strauss, 2013; Du et al., 2009; Kalsotra et al., 2008; Stoilov et al., 2006; Wang et al., 2008). These changes in CYP4F enzyme expression coincided with an initial increase of LTB_4 in the tissues analyzed, consistent with activation and amplification of inflammatory response, followed by a decrease in LTB_4 as CYP4F expression increased in the resolution phase. In mice, *Cyp4f* regulation was found to be tissue- and isoform specific, in addition to being time dependent. P450 transcripts were analyzed in various mouse tissues for up to 48 h following LPS administration. While induction of *Cyp4f16* mRNA expression was observed at 24 h in mouse liver, kidney, and heart, *Cyp4f13* expression was induced in kidney and heart but suppressed in liver (Theken et al., 2011). Moreover, in mouse liver, the time course of *Cyp4f13* expression corresponded to 20-HETE levels, suggesting that *Cyp4f13* functions in the initiation phase of inflammation (20-HETE formation), while *Cyp4f16* is active in the resolution of inflammation (LTB_4 ω-hydroxylation). Down-regulation of *Cyp2c* and *Cyp2j* transcripts and suppression of EET and DHET formation during the initial 24 h following LPS treatment were observed as well, resulting in an increase in 20-HETE:EET + DHET compared to basal levels. These effects were attenuated at 48 h, indicating a time dependent, anti-inflammatory switch toward resolution of inflammation, suggestive of cross talk in the regulation pathways of epoxygenase and ω-hydroxylase enzymes throughout the inflammatory response. Collectively, these results indicate that CYP4Fs are temporally regulated in an isoform- and tissue-specific manner and have differential roles in the initiation and resolution phases of inflammation.

5. ω-HYDROXYLASES IN CANCER

In contrast to the inflammatory response where ω-hydroxylation of numerous eicosanoids may impinge on the processes of activation and resolution, consequences for eicosanoid ω-hydroxylation in cancer have largely been focused on 20-HETE. This arachidonic acid metabolite initially gained pharmacological interest for its potent vasoconstrictor activity in rat aorta (Schwartzman, Falck, Yadagiri, & Escalante, 1989) and inhibition of sodium transport in renal proximal tubules (Quigley, Baum, Reddy, Griener, & Falck, 2000). In recent years, new attention has been paid to the role of CYP4A/4F-derived arachidonic acid metabolites in cancer (Alexanian & Sorokin, 2013; Panigrahy, Kaipainen, Greene, & Huang, 2010). In this section, we examine the effects of cancer on P450 enyzme activity, CYP4 enzymes as potential cancer biomarkers and review the role for CYP4-mediated formation of 20-HETE in cancer progression.

5.1 P450 Expression and Activity in Cancer

P450-dependent drug elimination is typically reduced in cancer due to an increase in inflammatory cytokines that decrease transcription of the major drug-metabolizing enzyme, CYP3A4, in normal human liver (Harvey & Morgan, 2014; Rivory, Slaviero, & Clarke, 2002). In cancerous tissues, expression of most xenobiotic-metabolizing P450 enzymes, other than CYP1B1 and CYP2J2, is substantially reduced (or even absent) relative to neighboring noncancerous tissue (Forrester et al., 1990; Murray et al., 1997). In contrast, several CYP4 enzymes that are associated with the ω-hydroxylation of endogenous substrates are *increased* in cancer cells (Alexanian & Sorokin, 2013).

5.2 P450s as Biomarkers in Cancer

A relationship between P450 ω-hydroxylases and cancer can be traced back to the early 1980s when peroxisome proliferators were shown to cause liver cancer in rodents (Reddy, Azarnoff, & Hignite, 1980) with accompanying upregulation of other targets of PPARα, including microsomal CYP4A ω-hydroxylases and mitochondrial and peroxisomal enzymes that regulate fatty acid catabolism (Johnson et al., 1996). Pretreatment of rats with clofibrate facilitated the purification of the then novel "cytochrome P-452"

(Gibson, Orton, & Tamburini, 1982), now termed CYP4A1, which preferentially hydroxylates lauric and palmitic acids at their ω-terminus (Aoyama et al., 1990), very similar to the regioselectivity exhibited by human CYP4A11 (Hardwick, 2008). In subsequent years, much effort was devoted to investigating the relationship between target gene induction and genotoxicity, especially given that P450 metabolites may conceivably play a role in the pathophysiology of liver cancer. However, today, we recognize that while peroxisome proliferator-mediated gene induction and liver cancer in rodents is critically dependent on PPAR transactivation (Gonzalez & Shah, 2008), the concomitant upregulation of CYP4A enzymes is merely a biomarker of exposure to these ubiquitous environmental agents. A search in the early 1990s for new *CYP4A* genes that could be involved in carcinogenesis led to the discovery of the *CYP4F* subfamily. In rat liver tumors induced by a variety of chemical carcinogens, *CYP4A* gene expression and associated laurate ω-hydroxylase activities were decreased, but Northern blotting revealed a new gene (*CYP4F1*) that was expressed at high levels, especially in tumors induced with aflatoxin B1 (Chen & Hardwick, 1993).

In human thyroid, ovarian, breast, and colon cancer tissues, CYP4F2 and CYP4A11 expression is generally upregulated at the mRNA level, although interindividual differences are large (Alexanian, Miller, Roman, & Sorokin, 2012). Higher levels of CYP4F2 protein were confirmed in several tumor samples by Western blotting. Upregulation of mRNA for CYP4A11, CYP4F2, and CYP4F3 has also been reported in pancreatic ductal adenocarcinoma, with the suggestion that these ω-hydroxylases might be used as distinguishing markers in pancreatic pathology (Gandhi et al., 2013).

CYP4Z1 is selectively expressed in mammary tissue and also appears to be upregulated in breast cancer tissue (Rieger et al., 2004). Recently, CYP4Z1 was identified as a potential biomarker of poor prognosis in prostate cancer (Tradonsky et al., 2012). In an extensive series of studies by Murray and colleagues, immunohistochemical analysis of cancer tissue microarrays highlighted CYP4Z1 expression in both breast and ovarian cancer (Downie et al., 2005; Murray, Patimalla, Stewart, Miller, & Heys, 2010), but not colon cancer (Kumarakulasingham et al., 2005). CYP4V2 and CYP4X1 expression has also been correlated with breast tumor grade (Murray et al., 2010). These immunohistochemical studies employed highly selective P450 antibodies against more than 20 P450 enzymes, but unfortunately, the panel did not include antibodies for CYP4A and CYP4F enzymes.

5.3 ω-Hydroxylases and Cancer Progression

Local and distant metastases are the cause of 90% of human mortality to cancer (Weigelt, Peterse, & van 't Veer, 2005). A key event in tumor metastasis is neovascularization and associated angiogenesis (Folkman, 1971). Angiogenesis—the growth of new blood vessels from vascular endothelial cells—is critical to the tumor microenvironment that must provide nutrients and oxygen to support metastatic progression. Angiogenesis results from an imbalance between antiangiogenic and proangiogenic factors such as cytokines and proteases as well as a variety of growth factors, notably vascular endothelial growth factor (VEGF) (Nishida, Yano, Nishida, Kamura, & Kojiro, 2006). Tumor angiogenesis requires binding of VEGF to the VEGF-2 receptor, which activates a proliferative signaling cascade through the mitogen-activated protein kinase (MAPK) pathway (D'Angelo, Struman, Martial, & Weiner, 1995). Over the last decade, 20-HETE has become recognized as a critical modulator of cancer progression, operating together with VEGF (and probably other growth factors) to promote cellular proliferation, neovascularization and angiogenesis, cell migration, tumor growth, and metastasis. The evidence in support of these functions and the role of specific CYP4 ω-hydroxylases is discussed in the following sections.

5.4 Chemical Tools for Dissecting CYP4 Enzymology and Participation of 20-HETE in Cancer Progression

A cellular receptor for 20-HETE has not been identified, although a GPCR that binds 12(S)-HETE with high affinity has been described (Guo, Zhang, et al., 2011). Therefore, the current lexicon of 20-HETE modulators is restricted to chemical inhibitors of CYP ω-hydroxylases and chemical analogs of 20-HETE that act as agonists or antagonists (Fig. 4).

The availability of chemical inhibitors for the CYP4 ω-hydroxylases has been critical for elucidating the physiological roles of eicosanoids metabolized by these enzymes. The most widely used chemical inhibitors are the mechanism-based inhibitor, 17-ODYA and the tight-binding inhibitor, *N*-hydroxy-*N'*-(4-*n*-butyl-2-methylphenyl)formamidine (HET0016). 17-ODYA is one of a group of terminal acetylenes and halogenated olefins (e.g., DDMS) that destroy ω-hydroxylases upon metabolic turnover to reactive intermediates that bind to the enzyme (Kroetz & Xu, 2005). However, 17-ODYA is not specific for ω-hydroxylases and inhibits renal microsomal 20-HETE and EET formation to an equal extent (Zou et al., 1994).

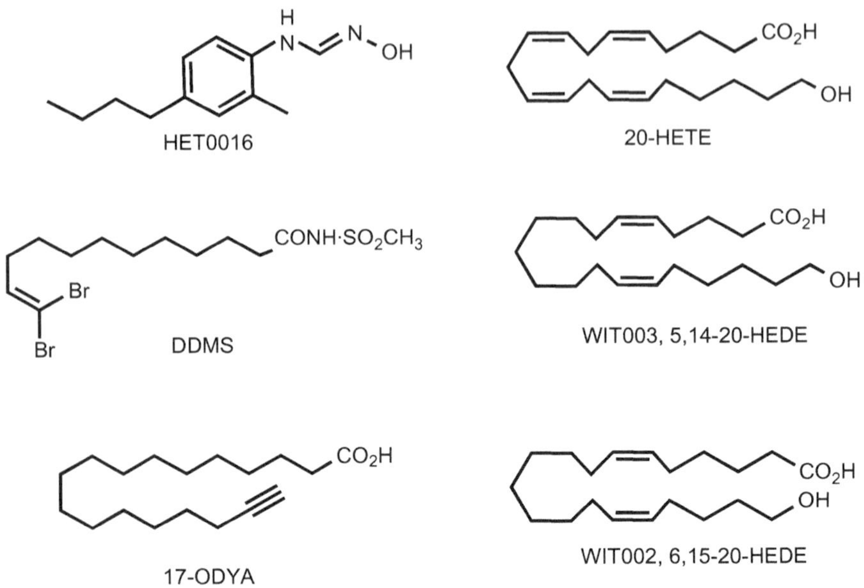

Figure 4 Structures of chemical tools used widely for dissecting ω-hydroxylase actions in inflammation and cancer.

HET0016 has emerged as both a potent *and* selective CYP4 chemical inhibitor that has proven exceptionally useful for studying 20-HETE pathways *in vitro* and *in vivo*. HET0016 was discovered in a screen to find a selective ω-hydroxylase inhibitor for the treatment of renal hypertension. In rat kidney microsomes, the IC_{50} for 20-HETE formation was 35 nM, while the IC_{50} for EET formation in this system was ~100 times higher (Miyata et al., 2001; Nakamura et al., 2003; Sato et al., 2001). Although the older literature tends to refer to HET0016 as a selective CYP4A inhibitor, it has the properties of a pan-CYP4 inhibitor, with IC_{50} values for CYP4A11, CYP4F2, CYP4F3B, CYP4V2, and CYP4B1 all in the 100 nM region or lower (Kehl et al., 2002; Nakano et al., 2009; Parkinson, Kelly, Bezabih, Whittington, & Rettie, 2012). High selectivity for human CYP4 ω-hydroxylases is clearly evident because IC_{50} values for the other major human liver P450s, CYP3A4, CYP2C9, and CYP2D6, are in the range 4–85 μM (Nakamura et al., 2003). Therefore, HET0016 is a more selective inhibitor of CYP4-dependent ω-hydroxylase activity than is 17-ODYA.

WIT002 and WIT003 are eicosanoid analogs that represent moderately potent (low μM) 20-HETE antagonist and agonists, respectively (Yu et al., 2004). WIT003 and HET0016 are often employed in study designs for evaluating 20-HETE signaling because of their complementary nature; i.e., demonstration of any WIT003-mediated reversal of HET0016's biological effects provides a powerful argument for the intermediacy of a CYP4 ω-hydroxylase in the pathway under study.

5.5 Cellular Proliferation and Tumor Growth

20-HETE has long been implicated in cellular proliferation (Roman, 2002), often invoking the participation of growth factors, such as VEGF, epidermal growth factor (EGF), fibroblast growth factor (FGF), and platelet-derived growth factor (PDGF). Lin et al. provided an early demonstration that 20-HETE, specifically, was a mitogen in rat proximal tubule cells (Lin, Rios, Falck, Belosludtsev, & Schwartzman, 1995). Notably, this effect was not observed with addition of other ω/ω-1 metabolites, such as 20-carboxy arachidonic acid, 19(S)- or 19(R)-HETE. In the same studies, EGF stimulated both cell growth and endogenous 20-HETE formation, while 17-ODYA attenuated the proliferative effects. In a complementary fashion, HET0016 inhibited the basal and EGF-/PDGF-stimulated growth of 9L rat gliosarcoma cells in culture, while WIT003 increased cellular proliferation (Guo et al., 2006), thereby providing strong evidence for a role in the CYP4:20-HETE pathway. Interestingly, 9L cells did not generate detectable quantities of 20-HETE when incubated with radiolabeled arachidonic acid. Similar effects were seen *in vitro* with U251 glioma cells (Guo, Roman, Falck, Edwards, & Scicli, 2005). These observations led the authors to question whether HET0016 might have antiproliferative effects unrelated to the compound's ability to inhibit 20-HETE formation—a situation that remains to be resolved.

Modulation of the CYP4:20-HETE pathway has very pronounced effects on tumor size in animal models of brain, kidney, and breast cancer. For example, following implantation into normal rat forebrain of U251 glioma cells that had been transfected with CYP4A1, a 10-fold increase in tumor volume was observed compared with the nontransfected cells (Guo et al., 2008). Similarly, brain implantation of 9L gliosarcoma cells caused rapid tumor growth that caused death in about 17 days. However, chronic treatment with HET0016 increased survival time by 5 days, apparently through a combination of reduced mitosis and increased apoptosis (Guo et al., 2006).

These results on tumor growth are impressive because brain cancers are difficult to treat, as are renal adenocarcinomas, which proliferate and metastasize extensively. Similar to its effects on glioma cells, HET0016 (and WIT002) reduced proliferation of renal adenocarcinoma cells (Alexanian et al., 2009). In the same studies, WIT002 also suppressed the growth of advanced renal carcinoma in a mouse model. Formation of low levels of 20-HETE by renal cancer cells could be demonstrated upon addition of arachidonic acid, and this effect was abolished by HET0016.

Nonsmall cell lung cancer (NSCLC) is another treatment-refractory malignancy. Injection of mice with an NSCLC-derived cell line (A549) transfected with CYP4A11 increased the tumor size and growth rate, both of which were reduced with HET0016 or WIT002 (Yu et al., 2011). Similar results were obtained in a separate lung metastasis model. Collectively, these studies provide strong support for the CYP4:20-HETE pathway as a drug target for combating tumor growth and metastatic potential across a variety of difficult to treat cancers.

5.6 Angiogenesis

The effect of the CYP4:20-HETE system on angiogenesis has also been investigated extensively in recent years. The majority of these efforts evaluated VEGF expression and/or cellular responses to the growth factor because VEGF receptors and their ligands play essential roles in the regulation of angiogenesis (Shibuya, 2013). These studies are of particular interest because they have helped illuminate mechanistic details of the pathway(s) involved for 20-HETE signaling in cancer progression.

Several groups, applying independent experimental approaches, established a role for 20-HETE in angiogenesis in the early 2000s. In one of the first of these studies, angiogenesis induced in skeletal muscle by chronic electrical stimulation was accompanied by a 2.5-fold increase in 20-HETE formation that could be completely blocked by HET0016 (Amaral, Maier, Schippers, Roman, & Greene, 2003). However, immunochemically detectable increases in VEGF in electrically stimulated muscles were *not* reduced by ω-hydroxylase inhibitors, although a neutralizing VEGF antibody did block electrically induced increases in 20-HETE. These observations led the authors to conclude that 20-HETE acted downstream of VEGF in this pathway for muscle angiogenesis induced by electrical stimulation.

The foregoing study did not directly link angiogenesis involving 20-HETE to CYP4 enzyme expression, but this was subsequently evaluated by Lanaido-Schwartzman and coworkers upon overexpression of rat CYP4A1 in microdissected renal arteries (Jiang et al., 2004). Transduced arteries produced an ~threefold higher level of 20-HETE in the culture medium and ~eightfold increase in endothelial cell sprouting. Both effects were blocked by HET0016 whose antiangiogenic activity could be reversed upon addition of a 20-HETE agonist. These *ex vivo* studies were complemented with an *in vivo* approach that used sustained release polymer pellets containing growth factors (VEGF, EGF, or FGF), plus or minus HET0016 or DDMS, implanted into rat eyes. Each rat served as its own control for evaluation of the effect of these modulators on corneal neovascularization. In all combinations, the P450 inhibitor completely abrogated the angiogenic response that had been stimulated by the growth factor. Similar results were obtained by these investigators with an experimental design that replaced the growth factors in the implanted pellets with U251 glioma spheroids. Although changes in 20-HETE levels were not confirmed in these experiments, the authors concluded that the similarity in effects caused *in vivo* by two P450 inhibitors with distinct inhibitory mechanisms (HET0016 and DDMS) argued for a role for the CYP4:20-HETE pathway in growth factor-induced changes in angiogenesis *in vivo*. More recently, CYP4Z1 overexpression in breast cancer cells has been linked to increased VEGF expression, angiogenesis, cell proliferation, and migration *in vitro* and increased tumor weight in xenograft models (Yu et al., 2012). These effects were accompanied by increased 20-HETE production and inhibition by HET0016. This is interesting because CYP4Z1 has yet to be directly shown to function as an ω-hydroxylase of arachidonic acid.

In recent years, endothelial progenitor cells (EPCs) have emerged as a model system to investigate the CYP4:20-HETE system in angiogenesis. Circulating EPCs contribute to postnatal vascularization, neovascularization, and tissue damage repair and so have attracted attention as a potential therapeutic target in vascular disease (Kirton & Xu, 2010). EPCs express relatively high levels of CYP4A11 and form 20-HETE and numerous other metabolites when incubated with arachidonic acid (Guo, Janic, et al., 2011). 20-HETE has been shown to regulate EPC angiogenesis both *in vitro* and *in vivo* (Chen et al., 2014; Guo, Janic, et al., 2011) and so this appears to be a highly useful cell system for dissecting CYP4 signaling pathways in cancer progression.

5.7 Signaling Mechanisms in CYP4:20-HETE-Induced Cancer Progression

Many of the studies discussed above highlight the interdependency between 20-HETE and VEGF in promoting critical features of cancer progression, including proliferation, angiogenesis, and metastases. Whereas Amaral et al. concluded that 20-HETE acts downstream of VEGF in skeletal muscle angiogenesis (Amaral et al., 2003), others have provided data that favor the reverse situation (Guo, Janic, et al., 2011).

The MAPK pathway has been invoked in phospholipase A2 activity that could release arachidonic acid, perhaps in concert with VEGF, and thereby amplify 20-HETE formation (Muthalif et al., 1998). Other critical elements of CYP4:20-HETE cancer signaling include activation of PI3K/Akt (Yu et al., 2011), as well as roles for HIF-1α and reactive oxygen species (ROS) (Guo et al., 2007). Very recently, Zeldin and coworkers generated a CYP4F2 transgenic mouse with selective endothelial cell expression (Cheng et al., 2014). Endothelial cells from transgenic animals exhibited twofold increases in levels of 20-HETE, increased growth and tube formation with upregulation of VEGF and the prooxidant enzyme NADPH oxidase, further underlining the importance of the MAPK pathway and ROS in cellular proliferation. Figure 5 attempts to capture essential elements of the complex signaling pathways of CYP4:20-HETE in cancer progression. More detailed mechanistic proposals can be found in two excellent recent reviews (Chen et al., 2014; Hoopes, Garcia, Edin, Schwartzman, & Zeldin, 2015)

6. CONCLUSION

Considerable evidence supports the view that P450-dependent ω-hydroxylases play pivotal roles in the progression of cancer and in the resolution of inflammation. However, the CYP4 enzymes that dominate these processes do not act in isolation. Similar to CYP4 enzymes, expression of CYP2J2 is upregulated in tumor cells of human origin compared to adjacent normal cells corresponding to an increase in EET formation (Wang & Dubois, 2010). Interestingly, during cancer progression, EETs generated by CYP2C and CYP2J appear to play rather similar roles to 20-HETE, in that they also stimulate tumor cell proliferation, inhibition of apoptosis, angiogenesis, and metastases in a variety of mouse tumors (Tacconelli & Patrignani, 2014). EETs also appear to participate in signaling pathways that mirror several of those invoked for 20-HETE's protumorigenic effects including VEGF, MAPK, and PI3K/Akt. Administration of synthetic EET analogs or sEH inhibitors recapitulated these effects, whereas EET antagonists or chemical inhibitors reduced the effects on tumor size and metastases (Jiang et al., 2007; Panigrahy et al., 2012). In contrast, during inflammation, EETs appear to *oppose* 20-HETE effects, suppressing

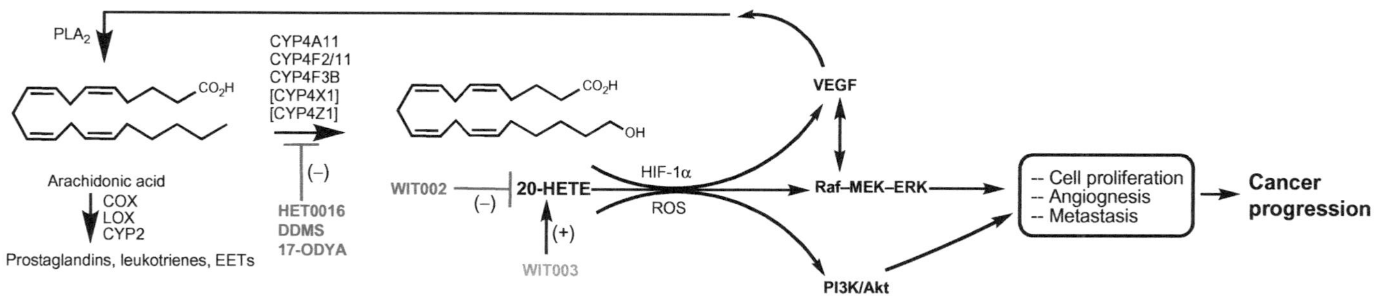

Figure 5 20-HETE signaling in cancer progression.

production of proinflammatory cytokines and chemokines and preventing leukocyte cell adhesion, thereby reducing release of LTB_4. There is also evidence that EET formation is decreased through downregulation of *CYP2C* and *CYP2J* genes during the initiation of acute inflammation, which creates favorable conditions for 20-HETE formation, and thus enhanced production of cytokines and LTB_4, resulting in amplification of inflammation (Theken et al., 2011). This suggests a functional balance between ω-hydroxylase and epoxygenase pathways throughout the different phases of the inflammatory response.

In general, it appears that investigators have not fully evaluated the ω-hydroxylase and epoxygenase pathways *together* in the same cell-based systems or animal models. This may reflect attempts to simplify the complex pharmacology or perhaps circumvent some of the analytical challenges in measuring 20-HETE and EETS in biological samples. Because EETs and 20-HETE are chemically unstable, subject to rapid oxidation, acid sensitive and present in low concentrations in blood and tissue samples, they present a serious analytical challenge. For EET quantification, determining whether downstream DHET analytes derive from the action of sEH or arise as artifacts of the extraction protocol can be quite difficult. Unfortunately, neither 20-HETE nor EETs are among the compounds included on traditional metabolomics or eicosanoid analytical platforms (Kofeler, Fauland, Rechberger, & Trotzmuller, 2012). In the cell, EETs and 20-HETE exist predominantly esterified at the *sn*-2 position in the phospholipid fraction of the membrane, with less than 5% present as the free acid (Hammond & O'Donnell, 2012). However, many studies only analyze the free EETs and 20-HETE by direct extraction into organic solvents, which will lead to variability in the level of lipids measured and make it difficult to observe changes in concentration following inhibition of CYP4 enzymes or treatment with sEH. There is also variability in analytical tools employed in EET and 20-HETE measurements. While LC–MS/MS is the instrumentation of choice for measuring EETs and 20-HETE owing to its sensitivity and rapid quantification of several lipids simultaneously, recent ELISA assays have emerged for 20-HETE and specific DHETs (Kroot et al., 2010). These varied methodologies render it even more difficult to interpret altered levels of lipid mediators in healthy versus cancerous tissue from different studies and determine how their concentrations change in cancer progression and metastases.

Another analytical challenge for helping understand the interplay between hydroxylation and epoxidation pathways is the quantitation of

CYP4 and CYP2 enzymes in the cell type or tissue under investigation. However, recent advances in mass spectrometry-based quantitation of P450 enzymes in whole and broken cell systems make this a realizable goal that does not rely on the availability of monospecific antibodies or the presumption that transcript profiling will always translate directly to protein levels (Aebersold, Burlingame, & Bradshaw, 2013). A wider array of P450 transgenic and knockout animal models can also be expected to provide valuable new tools for unveiling ω-hydroxylase biology *in vivo*. A few *Cyp4a* and *Cyp4f* knockout mice have been generated (Bardowell, Duan, Manor, Swanson, & Parker, 2012; Holla et al., 2001), but their utility for answering complex questions about disease in humans is limited by the expansion of these P450 subfamilies in mice relative to humans. In the absence of a clear identification of the mouse homolog for, say CYP4F2, either the entire *Cyp4* locus would need to be deleted or human CYP4F2 would need to be knocked in at high enough levels in a tissue-specific manner to facilitate data interpretation. As noted earlier, Zeldin and coworkers have generated a CYP4F2 transgenic mouse with selective endothelial cell expression (Cheng et al., 2014). Future development of animal models for other ω-hydroxylases will further expand the toolbox required for fully evaluating the biological roles of these enzymes in cancer and inflammation.

Finally, the development of drugs that selectively inhibit arachidonic acid pathways has been a seductive goal for many years, with clear therapeutic successes in aspirin, NSAIDs and zileuton, COX and LOX inhibitors, respectively, to draw on for inspiration. The CYP4F ω-hydroxylases also provide a potential drug target in the attenuation of the inflammatory response, as induction of these enzymes increases formation of 20-OH LTB_4. However, greater understanding of the differential tissue expression and regulation of the various CYP4Fs is needed before this becomes a truly viable option. Elucidating mechanisms by which EETs and ω3 PUFAs exert their anti-inflammatory benefits, and the extent to which ω-hydroxylases are involved, will also be an important focus in future development of novel anti-inflammatory therapies. sEH inhibitors have been developed and advanced into clinical trials with the intent of increasing *in vivo* EET levels for the potential treatment of hypertension, nociceptive pain, and diabetes (Kodani & Hammock, 2015). However, this approach has the potential downside that the undesirable protumorigenic features of the EETs might be unmasked in certain patients. Therefore, selective modulation of the 20-HETE pathway to combat cancer progression may be more tractable. Indeed, the tumor-shrinking properties of HET0016 are striking, although several studies have failed to link HET0016-mediated CYP4 inhibition to reductions in tumor levels of 20-HETE, raising the possibility that HET0016 might be acting through a different mechanism. Alternatively, extremely low 20-HETE levels, measuring only free 20-HETE or upregulated secondary metabolic pathways might explain these results. Continuing improvements in the sensitivity of mass spectrometry-based analysis of arachidonate metabolites (Tsikas & Zoerner, 2014) combined with a more thorough understanding of the metabolic fate of 20-HETE and EETs can be expected to further our understanding of the critical role these compounds play in the complex lipid signaling involved in inflammation and cancer.

CONFLICT OF INTEREST

The authors have no conflicts of interest.

ACKNOWLEDGMENTS

CYP4 research in the Rettie laboratory is currently supported by NIH Grant R01 GM109743. CYP2J2 and EET research in the Totah laboratory is supported by NIH Grant RHL096706.

REFERENCES

Aebersold, R., Burlingame, A. L., & Bradshaw, R. A. (2013). Western blots versus selected reaction monitoring assays: Time to turn the tables? *Molecular & Cellular Proteomics*, *12*, 2381–2382.

Alexanian, A., Miller, B., Roman, R. J., & Sorokin, A. (2012). 20-HETE-producing enzymes are up-regulated in human cancers. *Cancer Genomics Proteomics*, *9*, 163–169.

Alexanian, A., Rufanova, V. A., Miller, B., Flasch, A., Roman, R. J., & Sorokin, A. (2009). Down-regulation of 20-HETE synthesis and signaling inhibits renal adenocarcinoma cell proliferation and tumor growth. *Anticancer Research*, *29*, 3819–3824.

Alexanian, A., & Sorokin, A. (2013). Targeting 20-HETE producing enzymes in cancer—Rationale, pharmacology, and clinical potential. *Onco Targets and therapy*, *6*, 243–255.

Amaral, S. L., Maier, K. G., Schippers, D. N., Roman, R. J., & Greene, A. S. (2003). CYP4A metabolites of arachidonic acid and VEGF are mediators of skeletal muscle angiogenesis. *American Journal of Physiology. Heart and Circulatory Physiology*, *284*, H1528–H1535.

Aoyama, T., Hardwick, J. P., Imaoka, S., Funae, Y., Gelboin, H. V., & Gonzalez, F. J. (1990). Clofibrate-inducible rat hepatic P450s IVA1 and IVA3 catalyze the omega- and (omega-1)-hydroxylation of fatty acids and the omega-hydroxylation of prostaglandins E1 and F2 alpha. *Journal of Lipid Research*, *31*, 1477–1482.

Arita, M., Oh, S. F., Chonan, T., Hong, S., Elangovan, S., Sun, Y. P., et al. (2006). Metabolic inactivation of resolvin E1 and stabilization of its anti-inflammatory actions. *The Journal of Biological Chemistry*, *281*, 22847–22854.

Arnold, C., Markovic, M., Blossey, K., Wallukat, G., Fischer, R., Dechend, R., et al. (2010). Arachidonic acid-metabolizing cytochrome P450 enzymes are targets of {omega}-3 fatty acids. *The Journal of Biological Chemistry*, *285*, 32720–32733.

Bardowell, S. A., Duan, F., Manor, D., Swanson, J. E., & Parker, R. S. (2012). Disruption of mouse cytochrome p450 4f14 (Cyp4f14 gene) causes severe perturbations in vitamin E metabolism. *The Journal of Biological Chemistry*, *287*, 26077–26086.

Bednar, M. M., Schwartzman, M., Ibraham, N. G., McGiff, J. C., & Mullane, K. M. (1984). Conversion of arachidonic acid to two novel products by a cytochrome P450-dependent mixed-function oxidase in polymorphonuclear leukocytes. *Biochemical and Biophysical Research Communications*, *123*, 581–588.

Berry, K. A., Borgeat, P., Gosselin, J., Flamand, L., & Murphy, R. C. (2003). Urinary metabolites of leukotriene B4 in the human subject. *The Journal of Biological Chemistry*, *278*, 24449–24460.

Birnie, M., Morrison, R., Camara, R., & Strauss, K. I. (2013). Temporal changes of cytochrome P450 (Cyp) and eicosanoid-related gene expression in the rat brain after traumatic brain injury. *BMC Genomics*, *14*, 303.

Bylund, J., Hidestrand, M., Ingelman-Sundberg, M., & Oliw, E. H. (2000). Identification of CYP4F8 in human seminal vesicles as a prominent 19-hydroxylase of prostaglandin endoperoxides. *The Journal of Biological Chemistry*, *275*, 21844–21849.

Capdevila, J., Chacos, N., Werringloer, J., Prough, R. A., & Estabrook, R. W. (1981). Liver microsomal cytochrome P-450 and the oxidative metabolism of arachidonic acid. *Proceedings of the National Academy of Sciences of the United States of America*, *78*, 5362–5366.

Capdevila, J. H., & Falck, J. R. (2002). Biochemical and molecular properties of the cytochrome P450 arachidonic acid monooxygenases. *Prostaglandins & Other Lipid Mediators*, *68–69*, 325–344.

Chen, L., Ackerman, R., Saleh, M., Gotlinger, K. H., Kessler, M., Mendelowitz, L. G., et al. (2014). 20-HETE regulates the angiogenic functions of human endothelial progenitor cells and contributes to angiogenesis in vivo. *The Journal of Pharmacology and Experimental Therapeutics*, *348*, 442–451.

Chen, L., & Hardwick, J. P. (1993). Identification of a new P450 subfamily, CYP4F1, expressed in rat hepatic tumors. *Archives of Biochemistry and Biophysics*, *300*, 18–23.

Cheng, J., Edin, M. L., Hoopes, S. L., Li, H., Bradbury, J. A., Graves, J. P., et al. (2014). Vascular characterization of mice with endothelial expression of cytochrome P450 4F2. *The FASEB Journal*, *28*, 2915–2931.

Christmas, P., Jones, J. P., Patten, C. J., Rock, D. A., Zheng, Y., Cheng, S. M., et al. (2001). Alternative splicing determines the function of CYP4F3 by switching substrate specificity. *The Journal of Biological Chemistry*, *276*, 38166–38172.

Christmas, P., Ursino, S. R., Fox, J. W., & Soberman, R. J. (1999). Expression of the CYP4F3 gene. Tissue-specific splicing and alternative promoters generate high and low K(m) forms of leukotriene B(4) omega-hydroxylase. *The Journal of Biological Chemistry*, *274*, 21191–21199.

Cowart, L. A., Wei, S., Hsu, M. H., Johnson, E. F., Krishna, M. U., Falck, J. R., et al. (2002). The CYP4A isoforms hydroxylate epoxyeicosatrienoic acids to form high affinity peroxisome proliferator-activated receptor ligands. *The Journal of Biological Chemistry*, *277*, 35105–35112.

D'Angelo, G., Struman, I., Martial, J., & Weiner, R. I. (1995). Activation of mitogen-activated protein kinases by vascular endothelial growth factor and basic fibroblast growth factor in capillary endothelial cells is inhibited by the antiangiogenic factor 16-kDa N-terminal fragment of prolactin. *Proceedings of the National Academy of Sciences of the United States of America*, *92*, 6374–6378.

Dawczynski, C., Massey, K. A., Ness, C., Kiehntopf, M., Stepanow, S., Platzer, M., et al. (2013). Randomized placebo-controlled intervention with n-3 LC-PUFA-supplemented yoghurt: Effects on circulating eicosanoids and cardiovascular risk factors. *Clinical Nutrition*, *32*, 686–696.

Dhar, M., Sepkovic, D. W., Hirani, V., Magnusson, R. P., & Lasker, J. M. (2008). Omega oxidation of 3-hydroxy fatty acids by the human CYP4F gene subfamily enzyme CYP4F11. *Journal of Lipid Research*, *49*, 612–624.

Downie, D., McFadyen, M. C., Rooney, P. H., Cruickshank, M. E., Parkin, D. E., Miller, I. D., et al. (2005). Profiling cytochrome P450 expression in ovarian cancer: Identification of prognostic markers. *Clinical Cancer Research*, *11*, 7369–7375.

Draye, J. P., & Vamecq, J. (1989). The gluconeogenicity of fatty acids in mammals. *Trends in Biochemical Sciences*, *14*, 478–479.

Du, L., Yin, H., Morrow, J. D., Strobel, H. W., & Keeney, D. S. (2009). 20-Hydroxylation is the CYP-dependent and retinoid-inducible leukotriene B4 inactivation pathway in human and mouse skin cells. *Archives of Biochemistry and Biophysics*, *484*, 80–86.

Edson, K. Z., Prasad, B., Unadkat, J. D., Suhara, Y., Okano, T., Guengerich, F. P., et al. (2013). Cytochrome P450-dependent catabolism of vitamin K: Omega-hydroxylation catalyzed by human CYP4F2 and CYP4F11. *Biochemistry*, *52*, 8276–8285.

Eksterowicz, J., Rock, D. A., Rock, B. M., Wienkers, L. C., & Foti, R. S. (2014). Characterization of the active site properties of CYP4F12. *Drug Metabolism and Disposition*, *42*, 1698–1707.

Fan, F., Muroya, Y., & Roman, R. J. (2015). Cytochrome P450 eicosanoids in hypertension and renal disease. *Current Opinion in Nephrology and Hypertension*, *24*, 37–46.

Fer, M., Corcos, L., Dreano, Y., Plee-Gautier, E., Salaun, J. P., Berthou, F., et al. (2008). Cytochromes P450 from family 4 are the main omega hydroxylating enzymes in humans: CYP4F3B is the prominent player in PUFA metabolism. *Journal of Lipid Research*, *49*, 2379–2389.

Fleming, I. (2008). Vascular cytochrome p450 enzymes: Physiology and pathophysiology. *Trends in Cardiovascular Medicine*, *18*, 20–25.

Folkman, J. (1971). Tumor angiogenesis: Therapeutic implications. *The New England Journal of Medicine*, *285*, 1182–1186.

Forrester, L. M., Hayes, J. D., Millis, R., Barnes, D., Harris, A. L., Schlager, J. J., et al. (1990). Expression of glutathione S-transferases and cytochrome P450 in normal and tumor breast tissue. *Carcinogenesis*, *11*, 2163–2170.

Funk, C. D. (2001). Prostaglandins and leukotrienes: Advances in eicosanoid biology. *Science*, *294*, 1871–1875.

Gandhi, A. V., Saxena, S., Relles, D., Sarosiek, K., Kang, C. Y., Chipitsyna, G., et al. (2013). Differential expression of cytochrome P450 omega-hydroxylase isoforms and their association with clinicopathological features in pancreatic ductal adenocarcinoma. *Annals of Surgical Oncology*, *20*(Suppl. 3), S636–S643.

Gatica, A., Aguilera, M. C., Contador, D., Loyola, G., Pinto, C. O., Amigo, L., et al. (2007). P450 CYP2C epoxygenase and CYP4A omega-hydroxylase mediate ciprofibrate-induced PPARalpha-dependent peroxisomal proliferation. *Journal of Lipid Research*, *48*, 924–934.

Gibson, G. G., Orton, T. C., & Tamburini, P. P. (1982). Cytochrome P-450 induction by clofibrate. Purification and properties of a hepatic cytochrome P-450 relatively specific for the 12- and 11-hydroxylation of dodecanoic acid (lauric acid). *The Biochemical Journal*, *203*, 161–168.

Gonzalez, F. J., & Shah, Y. M. (2008). PPARalpha: Mechanism of species differences and hepatocarcinogenesis of peroxisome proliferators. *Toxicology*, *246*, 2–8.

Gronert, K., Clish, C. B., Romano, M., & Serhan, C. N. (1999). Transcellular regulation of eicosanoid biosynthesis. *Methods in Molecular Biology*, *120*, 119–144.

Guo, A. M., Arbab, A. S., Falck, J. R., Chen, P., Edwards, P. A., Roman, R. J., et al. (2007). Activation of vascular endothelial growth factor through reactive oxygen species mediates 20-hydroxyeicosatetraenoic acid-induced endothelial cell proliferation. *The Journal of Pharmacology and Experimental Therapeutics, 321*, 18–27.

Guo, A. M., Janic, B., Sheng, J., Falck, J. R., Roman, R. J., Edwards, P. A., et al. (2011). The cytochrome P450 4A/F-20-hydroxyeicosatetraenoic acid system: A regulator of endothelial precursor cells derived from human umbilical cord blood. *The Journal of Pharmacology and Experimental Therapeutics, 338*, 421–429.

Guo, M., Roman, R. J., Falck, J. R., Edwards, P. A., & Scicli, A. G. (2005). Human U251 glioma cell proliferation is suppressed by HET0016 [N-hydroxy-N'-(4-butyl-2-methylphenyl)formamidine], a selective inhibitor of CYP4A. *The Journal of Pharmacology and Experimental Therapeutics, 315*, 526–533.

Guo, M., Roman, R. J., Fenstermacher, J. D., Brown, S. L., Falck, J. R., Arbab, A. S., et al. (2006). 9L gliosarcoma cell proliferation and tumor growth in rats are suppressed by N-hydroxy-N'-(4-butyl-2-methylphenol) formamidine (HET0016), a selective inhibitor of CYP4A. *The Journal of Pharmacology and Experimental Therapeutics, 317*, 97–108.

Guo, A. M., Sheng, J., Scicli, G. M., Arbab, A. S., Lehman, N. L., Edwards, P. A., et al. (2008). Expression of CYP4A1 in U251 human glioma cell induces hyperproliferative phenotype in vitro and rapidly growing tumors in vivo. *The Journal of Pharmacology and Experimental Therapeutics, 327*, 10–19.

Guo, Y., Zhang, W., Giroux, C., Cai, Y., Ekambaram, P., Dilly, A. K., et al. (2011). Identification of the orphan G protein-coupled receptor GPR31 as a receptor for 12-(S)-hydroxyeicosatetraenoic acid. *The Journal of Biological Chemistry, 286*, 33832–33840.

Haeggstrom, J. Z., & Funk, C. D. (2011). Lipoxygenase and leukotriene pathways: Biochemistry, biology, and roles in disease. *Chemical Reviews, 111*, 5866–5898.

Hammond, V. J., & O'Donnell, V. B. (2012). Esterified eicosanoids: Generation, characterization and function. *Biochimica et Biophysica Acta, 1818*, 2403–2412.

Hansson, G., Lindgren, J. A., Dahlen, S. E., Hedqvist, P., & Samuelsson, B. (1981). Identification and biological activity of novel omega-oxidized metabolites of leukotriene B4 from human leukocytes. *FEBS Letters, 130*, 107–112.

Hardwick, J. P. (2008). Cytochrome P450 omega hydroxylase (CYP4) function in fatty acid metabolism and metabolic diseases. *Biochemical Pharmacology, 75*, 2263–2275.

Harmon, S. D., Fang, X., Kaduce, T. L., Hu, S., Raj Gopal, V., Falck, J. R., et al. (2006). Oxygenation of omega-3 fatty acids by human cytochrome P450 4F3B: Effect on 20-hydroxyeicosatetraenoic acid production. *Prostaglandins, Leukotrienes, and Essential Fatty Acids, 75*, 169–177.

Harvey, R. D., & Morgan, E. T. (2014). Cancer, inflammation, and therapy: Effects on cytochrome p450-mediated drug metabolism and implications for novel immunotherapeutic agents. *Clinical Pharmacology and Therapeutics, 96*, 449–457.

Hashizume, T., Imaoka, S., Mise, M., Terauchi, Y., Fujii, T., Miyazaki, H., et al. (2002). Involvement of CYP2J2 and CYP4F12 in the metabolism of ebastine in human intestinal microsomes. *The Journal of Pharmacology and Experimental Therapeutics, 300*, 298–304.

He, X., Cryle, M. J., De Voss, J. J., & de Montellano, P. R. (2005). Calibration of the channel that determines the omega-hydroxylation regiospecificity of cytochrome P4504A1: Catalytic oxidation of 12-HALODOdecanoic acids. *The Journal of Biological Chemistry, 280*, 22697–22705.

Hertz, R., Sheena, V., Kalderon, B., Berman, I., & Bar-Tana, J. (2001). Suppression of hepatocyte nuclear factor-4alpha by acyl-CoA thioesters of hypolipidemic peroxisome proliferators. *Biochemical Pharmacology, 61*, 1057–1062.

Holla, V. R., Adas, F., Imig, J. D., Zhao, X., Price, E., Jr., Olsen, N., et al. (2001). Alterations in the regulation of androgen-sensitive Cyp 4a monooxygenases cause hypertension. *Proceedings of the National Academy of Sciences of the United States of America, 98*, 5211–5216.

Hoopes, S. L., Garcia, V., Edin, M. L., Schwartzman, M. L., & Zeldin, D. C. (2015). Vascular actions of 20-HETE. *Prostaglandins & Other Lipid Mediators*. (Epub ahead of print).

Hsu, M. H., Savas, U., Griffin, K. J., & Johnson, E. F. (2007a). Human cytochrome p450 family 4 enzymes: Function, genetic variation and regulation. *Drug Metabolism Reviews*, *39*, 515–538.

Hsu, M. H., Savas, U., Griffin, K. J., & Johnson, E. F. (2007b). Regulation of human cytochrome P450 4F2 expression by sterol regulatory element-binding protein and lovastatin. *The Journal of Biological Chemistry*, *282*, 5225–5236.

Hsu, M. H., Savas, U., Lasker, J. M., & Johnson, E. F. (2011). Genistein, resveratrol, and 5-aminoimidazole-4-carboxamide-1-beta-D-ribofuranoside induce cytochrome P450 4F2 expression through an AMP-activated protein kinase-dependent pathway. *The Journal of Pharmacology and Experimental Therapeutics*, *337*, 125–136.

Ishizuka, T., Cheng, J., Singh, H., Vitto, M. D., Manthati, V. L., Falck, J. R., et al. (2008). 20-Hydroxyeicosatetraenoic acid stimulates nuclear factor-kappaB activation and the production of inflammatory cytokines in human endothelial cells. *The Journal of Pharmacology and Experimental Therapeutics*, *324*, 103–110.

Jiang, M., Mezentsev, A., Kemp, R., Byun, K., Falck, J. R., Miano, J. M., et al. (2004). Smooth muscle-specific expression of CYP4A1 induces endothelial sprouting in renal arterial microvessels. *Circulation Research*, *94*, 167–174.

Jiang, J. G., Ning, Y. G., Chen, C., Ma, D., Liu, Z. J., Yang, S., et al. (2007). Cytochrome p450 epoxygenase promotes human cancer metastasis. *Cancer Research*, *67*, 6665–6674.

Jin, R., Koop, D. R., Raucy, J. L., & Lasker, J. M. (1998). Role of human CYP4F2 in hepatic catabolism of the proinflammatory agent leukotriene B4. *Archives of Biochemistry and Biophysics*, *359*, 89–98.

Johnson, E. F., Palmer, C. N., Griffin, K. J., & Hsu, M. H. (1996). Role of the peroxisome proliferator-activated receptor in cytochrome P450 4A gene regulation. *The FASEB Journal*, *10*, 1241–1248.

Kalsotra, A., Anakk, S., Brommer, C. L., Kikuta, Y., Morgan, E. T., & Strobel, H. W. (2007). Catalytic characterization and cytokine mediated regulation of cytochrome P450 4Fs in rat hepatocytes. *Archives of Biochemistry and Biophysics*, *461*, 104–112.

Kalsotra, A., Du, L., Wang, Y., Ladd, P. A., Kikuta, Y., Duvic, M., et al. (2008). Inflammation resolved by retinoid X receptor-mediated inactivation of leukotriene signaling pathways. *The FASEB Journal*, *22*, 538–547.

Kalsotra, A., & Strobel, H. W. (2006). Cytochrome P450 4F subfamily: At the crossroads of eicosanoid and drug metabolism. *Pharmacology & Therapeutics*, *112*, 589–611.

Kalsotra, A., Turman, C. M., Kikuta, Y., & Strobel, H. W. (2004). Expression and characterization of human cytochrome P450 4F11: Putative role in the metabolism of therapeutic drugs and eicosanoids. *Toxicology and Applied Pharmacology*, *199*, 295–304.

Kaspera, R., & Totah, R. A. (2009). Epoxyeicosatrienoic acids: Formation, metabolism and potential role in tissue physiology and pathophysiology. *Expert Opinion on Drug Metabolism & Toxicology*, *5*, 757–771.

Kehl, F., Cambj-Sapunar, L., Maier, K. G., Miyata, N., Kametani, S., Okamoto, H., et al. (2002). 20-HETE contributes to the acute fall in cerebral blood flow after subarachnoid hemorrhage in the rat. *American Journal of Physiology Heart and Circulatory Physiology*, *282*, H1556–H1565.

Kelly, E. J., Nakano, M., Rohatgi, P., Yarov-Yarovoy, V., & Rettie, A. E. (2011). Finding homes for orphan cytochrome P450s: CYP4V2 and CYP4F22 in disease states. *Molecular Interventions*, *11*, 124–132.

Khanapure, S. P., Garvey, D. S., Janero, D. R., & Letts, L. G. (2007). Eicosanoids in inflammation: Biosynthesis, pharmacology, and therapeutic frontiers. *Current Topics in Medicinal Chemistry*, 7, 311–340.

Kikuta, Y., Kusunose, E., Endo, K., Yamamoto, S., Sogawa, K., Fujii-Kuriyama, Y., et al. (1993). A novel form of cytochrome P-450 family 4 in human polymorphonuclear leukocytes. cDNA cloning and expression of leukotriene B4 omega-hydroxylase. *The Journal of Biological Chemistry*, *268*, 9376–9380.

Kikuta, Y., Kusunose, E., & Kusunose, M. (2000). Characterization of human liver leukotriene B(4) omega-hydroxylase P450 (CYP4F2). *Journal of Biochemistry*, *127*, 1047–1052.

Kikuta, Y., Kusunose, E., & Kusunose, M. (2002). Prostaglandin and leukotriene omega-hydroxylases. *Prostaglandins & Other Lipid Mediators*, *68–69*, 345–362.

Kikuta, Y., Kusunose, E., Sumimoto, H., Mizukami, Y., Takeshige, K., Sakaki, T., et al. (1998). Purification and characterization of recombinant human neutrophil leukotriene B4 omega-hydroxylase (cytochrome P450 4F3). *Archives of Biochemistry and Biophysics*, *355*, 201–205.

Kirton, J. P., & Xu, Q. (2010). Endothelial precursors in vascular repair. *Microvascular Research*, *79*, 193–199.

Kodani, S. D., & Hammock, B. D. (2015). The 2014 bernard B. Brodie award lecture-epoxide hydrolases: Drug metabolism to therapeutics for chronic pain. *Drug Metabolism and Disposition*, *43*, 788–802.

Kofeler, H. C., Fauland, A., Rechberger, G. N., & Trotzmuller, M. (2012). Mass spectrometry based lipidomics: An overview of technological platforms. *Metabolites*, *2*, 19–38.

Konkel, A., & Schunck, W. H. (2011). Role of cytochrome P450 enzymes in the bioactivation of polyunsaturated fatty acids. *Biochimica et Biophysica Acta*, *1814*, 210–222.

Kovarik, J. M., Dole, K., Riviere, G. J., Pommier, F., Maton, S., Jin, Y., et al. (2009). Ketoconazole increases fingolimod blood levels in a drug interaction via CYP4F2 inhibition. *Journal of Clinical Pharmacology*, *49*, 212–218.

Kroetz, D. L., & Xu, F. (2005). Regulation and inhibition of arachidonic acid omega-hydroxylases and 20-HETE formation. *Annual Review of Pharmacology and Toxicology*, *45*, 413–438.

Kroot, J. J., Laarakkers, C. M., Geurts-Moespot, A. J., Grebenchtchikov, N., Pickkers, P., van Ede, A. E., et al. (2010). Immunochemical and mass-spectrometry-based serum hepcidin assays for iron metabolism disorders. *Clinical Chemistry*, *56*, 1570–1579.

Kumarakulasingham, M., Rooney, P. H., Dundas, S. R., Telfer, C., Melvin, W. T., Curran, S., et al. (2005). Cytochrome p450 profile of colorectal cancer: Identification of markers of prognosis. *Clinical Cancer Research*, *11*, 3758–3765.

Lasker, J. M., Chen, W. B., Wolf, I., Bloswick, B. P., Wilson, P. D., & Powell, P. K. (2000). Formation of 20-hydroxyeicosatetraenoic acid, a vasoactive and natriuretic eicosanoid, in human kidney. Role of Cyp4F2 and Cyp4A11. *The Journal of Biological Chemistry*, *275*, 4118–4126.

Le Quere, V., Plee-Gautier, E., Potin, P., Madec, S., & Salaun, J. P. (2004). Human CYP4F3s are the main catalysts in the oxidation of fatty acid epoxides. *Journal of Lipid Research*, *45*, 1446–1458.

Lewis, D. F., & Lake, B. G. (1999). Molecular modelling of CYP4A subfamily members based on sequence homology with CYP102. *Xenobiotica; The Fate of Foreign Compounds in Biological Systems*, *29*, 763–781.

Lin, F., Rios, A., Falck, J. R., Belosludtsev, Y., & Schwartzman, M. L. (1995). 20-Hydroxyeicosatetraenoic acid is formed in response to EGF and is a mitogen in rat proximal tubule. *The American Journal of Physiology*, *269*, F806–F816.

Lu, A. Y., & Coon, M. J. (1968). Role of hemoprotein P-450 in fatty acid omega-hydroxylation in a soluble enzyme system from liver microsomes. *The Journal of Biological Chemistry*, *243*, 1331–1332.

Madej, A., Okopien, B., Kowalski, J., Zielinski, M., Wysocki, J., Szygula, B., et al. (1998). Effects of fenofibrate on plasma cytokine concentrations in patients with atherosclerosis and hyperlipoproteinemia IIb. *International Journal of Clinical Pharmacology and Therapeutics, 36*, 345–349.

Masters, B. S., Clark, J. E., Roman, L. J., Nishimoto, M., McCabe, T. J., Ortiz de Montellano, P. R., et al. (1993). Functional aspects of eicosanoid hydroxylation by lung and kidney cytochromes P450. Expression of cDNAs in mammalian cells and E. coli. *Journal of Lipid Mediators, 6*, 353–360.

McDonald, M. G., Rieder, M. J., Nakano, M., Hsia, C. K., & Rettie, A. E. (2009). CYP4F2 is a vitamin K1 oxidase: An explanation for altered warfarin dose in carriers of the V433M variant. *Molecular Pharmacology, 75*, 1337–1346.

McMahon, B., Mitchell, S., Brady, H. R., & Godson, C. (2001). Lipoxins: Revelations on resolution. *Trends in Pharmacological Sciences, 22*, 391–395.

Miyata, N., Taniguchi, K., Seki, T., Ishimoto, T., Sato-Watanabe, M., Yasuda, Y., et al. (2001). HET0016, a potent and selective inhibitor of 20-HETE synthesizing enzyme. *British Journal of Pharmacology, 133*, 325–329.

Murphy, R. C., & Gijon, M. A. (2007). Biosynthesis and metabolism of leukotrienes. *The Biochemical Journal, 405*, 379–395.

Murray, G. I., Patimalla, S., Stewart, K. N., Miller, I. D., & Heys, S. D. (2010). Profiling the expression of cytochrome P450 in breast cancer. *Histopathology, 57*, 202–211.

Murray, G. I., Taylor, M. C., McFadyen, M. C., McKay, J. A., Greenlee, W. F., Burke, M. D., et al. (1997). Tumor-specific expression of cytochrome P450 CYP1B1. *Cancer Research, 57*, 3026–3031.

Muthalif, M. M., Benter, I. F., Karzoun, N., Fatima, S., Harper, J., Uddin, M. R., et al. (1998). 20-Hydroxyeicosatetraenoic acid mediates calcium/calmodulin-dependent protein kinase II-induced mitogen-activated protein kinase activation in vascular smooth muscle cells. *Proceedings of the National Academy of Sciences of the United States of America, 95*, 12701–12706.

Nakamura, T., Sato, M., Kakinuma, H., Miyata, N., Taniguchi, K., Bando, K., et al. (2003). Pyrazole and isoxazole derivatives as new, potent, and selective 20-hydroxy-5,8,11,14-eicosatetraenoic acid synthase inhibitors. *Journal of Medicinal Chemistry, 46*, 5416–5427.

Nakano, M., Kelly, E. J., & Rettie, A. E. (2009). Expression and characterization of CYP4V2 as a fatty acid omega-hydroxylase. *Drug Metabolism and Disposition, 37*, 2119–2122.

Nakano, M., Kelly, E. J., Wiek, C., Hanenberg, H., & Rettie, A. E. (2012). CYP4V2 in Bietti's crystalline dystrophy: Ocular localization, metabolism of omega-3-polyunsaturated fatty acids, and functional deficit of the p.H331P variant. *Molecular Pharmacology, 82*, 679–686.

Naraba, H., Murakami, M., Matsumoto, H., Shimbara, S., Ueno, A., Kudo, I., et al. (1998). Segregated coupling of phospholipases A2, cyclooxygenases, and terminal prostanoid synthases in different phases of prostanoid biosynthesis in rat peritoneal macrophages. *Journal of Immunology, 160*, 2974–2982.

Nelson, D. R., Zeldin, D. C., Hoffman, S. M., Maltais, L. J., Wain, H. M., & Nebert, D. W. (2004). Comparison of cytochrome P450 (CYP) genes from the mouse and human genomes, including nomenclature recommendations for genes, pseudogenes and alternative-splice variants. *Pharmacogenetics, 14*, 1–18.

Nishida, N., Yano, H., Nishida, T., Kamura, T., & Kojiro, M. (2006). Angiogenesis in cancer. *Vascular Health and Risk Management, 2*, 213–219.

Node, K., Huo, Y., Ruan, X., Yang, B., Spiecker, M., Ley, K., et al. (1999). Anti-inflammatory properties of cytochrome P450 epoxygenase-derived eicosanoids. *Science, 285*, 1276–1279.

Nyagode, B. A., Williams, I. R., & Morgan, E. T. (2014). Altered inflammatory responses to Citrobacter rodentium infection, but not bacterial lipopolysaccharide, in mice lacking the Cyp4a10 or Cyp4a14 genes. *Inflammation, 37*, 893–907.

Oliw, E. H., Stark, K., & Bylund, J. (2001). Oxidation of prostaglandin H(2) and prostaglandin H(2) analogues by human cytochromes P450: Analysis of omega-side chain hydroxy metabolites and four steroisomers of 5-hydroxyprostaglandin I(1) by mass spectrometry. *Biochemical Pharmacology, 62*, 407–415.

Orning, L. (1987). Omega-oxidation of cysteine-containing leukotrienes by rat-liver microsomes. Isolation and characterization of omega-hydroxy and omega-carboxy metabolites of leukotriene E4 and N-acetylleukotriene E4. *European Journal of Biochemistry, 170*, 77–85.

Panigrahy, D., Edin, M. L., Lee, C. R., Huang, S., Bielenberg, D. R., Butterfield, C. E., et al. (2012). Epoxyeicosanoids stimulate multiorgan metastasis and tumor dormancy escape in mice. *The Journal of Clinical Investigation, 122*, 178–191.

Panigrahy, D., Kaipainen, A., Greene, E. R., & Huang, S. (2010). Cytochrome P450-derived eicosanoids: The neglected pathway in cancer. *Cancer Metastasis Reviews, 29*, 723–735.

Parker, R. S., Sontag, T. J., Swanson, J. E., & McCormick, C. C. (2004). Discovery, characterization, and significance of the cytochrome P450 omega-hydroxylase pathway of vitamin E catabolism. *Annals of the New York Academy of Sciences, 1031*, 13–21.

Parkinson, O. T., Kelly, E. J., Bezabih, E., Whittington, D., & Rettie, A. E. (2012). Bioactivation of 4-Ipomeanol by a CYP4B enzyme in bovine lung and inhibition by HET0016. *Journal of Veterinary Pharmacology and Therapeutics, 35*, 402–405.

Peters-Golden, M., & Henderson, W. R., Jr. (2007). Leukotrienes. *The New England Journal of Medicine, 357*, 1841–1854.

Powell, W. S., & Rokach, J. (2015). Biosynthesis, biological effects, and receptors of hydroxyeicosatetraenoic acids (HETEs) and oxoeicosatetraenoic acids (oxo-ETEs) derived from arachidonic acid. *Biochimica et Biophysica Acta, 1851*, 340–355.

Powell, P. K., Wolf, I., Jin, R., & Lasker, J. M. (1998). Metabolism of arachidonic acid to 20-hydroxy-5,8,11, 14-eicosatetraenoic acid by P450 enzymes in human liver: Involvement of CYP4F2 and CYP4A11. *The Journal of Pharmacology and Experimental Therapeutics, 285*, 1327–1336.

Preiss, B., & Bloch, K. (1964). Omega-oxidation of long chain fatty acids in rat liver. *The Journal of Biological Chemistry, 239*, 85–88.

Quigley, R., Baum, M., Reddy, K. M., Griener, J. C., & Falck, J. R. (2000). Effects of 20-HETE and 19(S)-HETE on rabbit proximal straight tubule volume transport. *American Journal of Physiology. Renal Physiology, 278*, F949–F953.

Reddy, J. K., Azarnoff, D. L., & Hignite, C. E. (1980). Hypolipidaemic hepatic peroxisome proliferators form a novel class of chemical carcinogens. *Nature, 283*, 397–398.

Ricciotti, E., & FitzGerald, G. A. (2011). Prostaglandins and inflammation. *Arteriosclerosis, Thrombosis, and Vascular Biology, 31*, 986–1000.

Rieger, M. A., Ebner, R., Bell, D. R., Kiessling, A., Rohayem, J., Schmitz, M., et al. (2004). Identification of a novel mammary-restricted cytochrome P450, CYP4Z1, with over-expression in breast carcinoma. *Cancer Research, 64*, 2357–2364.

Rivory, L. P., Slaviero, K. A., & Clarke, S. J. (2002). Hepatic cytochrome P450 3A drug metabolism is reduced in cancer patients who have an acute-phase response. *British Journal of Cancer, 87*, 277–280.

Roman, R. J. (2002). P-450 metabolites of arachidonic acid in the control of cardiovascular function. *Physiological Reviews, 82*, 131–185.

Roman, L. J., Palmer, C. N., Clark, J. E., Muerhoff, A. S., Griffin, K. J., Johnson, E. F., et al. (1993). Expression of rabbit cytochromes P4504A which catalyze the omega-hydroxylation of arachidonic acid, fatty acids, and prostaglandins. *Archives of Biochemistry and Biophysics, 307*, 57–65.

Rose, M. L., Rivera, C. A., Bradford, B. U., Graves, L. M., Cattley, R. C., Schoonhoven, R., et al. (1999). Kupffer cell oxidant production is central to the mechanism of peroxisome proliferators. *Carcinogenesis, 20*, 27–33.

Rusyn, I., Tsukamoto, H., & Thurman, R. G. (1998). WY-14 643 rapidly activates nuclear factor kappaB in Kupffer cells before hepatocytes. *Carcinogenesis, 19*, 1217–1222.

Sala, A., Voelkel, N., Maclouf, J., & Murphy, R. C. (1990). Leukotriene E4 elimination and metabolism in normal human subjects. *The Journal of Biological Chemistry*, *265*, 21771–21778.

Samuelsson, B., Dahlen, S. E., Lindgren, J. A., Rouzer, C. A., & Serhan, C. N. (1987). Leukotrienes and lipoxins: Structures, biosynthesis, and biological effects. *Science*, *237*, 1171–1176.

Sari, A. N., Korkmaz, B., Serin, M. S., Kacan, M., Unsal, D., Buharalioglu, C. K., et al. (2014). Effects of 5,14-HEDGE, a 20-HETE mimetic, on lipopolysaccharide-induced changes in MyD88/TAK1/IKKbeta/IkappaB-alpha/NF-kappaB pathway and circulating miR-150, miR-223, and miR-297 levels in a rat model of septic shock. *Inflammation Research: Official Journal of the European Histamine Research Society ... [et al]*, *63*, 741–756.

Sato, M., Ishii, T., Kobayashi-Matsunaga, Y., Amada, H., Taniguchi, K., Miyata, N., et al. (2001). Discovery of a N'-hydroxyphenylformamidine derivative HET0016 as a potent and selective 20-HETE synthase inhibitor. *Bioorganic & Medicinal Chemistry Letters*, *11*, 2993–2995.

Savas, U., Hsu, M. H., Griffin, K. J., Bell, D. R., & Johnson, E. F. (2005). Conditional regulation of the human CYP4X1 and CYP4Z1 genes. *Archives of Biochemistry and Biophysics*, *436*, 377–385.

Schwartzman, M. L., Falck, J. R., Yadagiri, P., & Escalante, B. (1989). Metabolism of 20-hydroxyeicosatetraenoic acid by cyclooxygenase. Formation and identification of novel endothelium-dependent vasoconstrictor metabolites. *The Journal of Biological Chemistry*, *264*, 11658–11662.

Sehgal, N., Agarwal, V., Valli, R. K., Joshi, S. D., Antonovic, L., Strobel, H. W., et al. (2011). Cytochrome P4504f, a potential therapeutic target limiting neuroinflammation. *Biochemical Pharmacology*, *82*, 53–64.

Seibert, K., Zhang, Y., Leahy, K., Hauser, S., Masferrer, J., & Isakson, P. (1997). Distribution of COX-1 and COX-2 in normal and inflamed tissues. *Advances in Experimental Medicine and Biology*, *400A*, 167–170.

Serhan, C. N., & Chiang, N. (2013). Resolution phase lipid mediators of inflammation: Agonists of resolution. *Current Opinion in Pharmacology*, *13*, 632–640.

Serhan, C. N., Chiang, N., Dalli, J., & Levy, B. D. (2015). Lipid mediators in the resolution of inflammation. *Cold Spring Harbor Perspectives in Biology*, 7, a016311.

Serhan, C. N., & Petasis, N. A. (2011). Resolvins and protectins in inflammation resolution. *Chemical Reviews*, *111*, 5922–5943.

Shak, S., & Goldstein, I. M. (1984). Omega-oxidation is the major pathway for the catabolism of leukotriene B4 in human polymorphonuclear leukocytes. *The Journal of Biological Chemistry*, *259*, 10181–10187.

Shibuya, M. (2013). Vascular endothelial growth factor and its receptor system: Physiological functions in angiogenesis and pathological roles in various diseases. *Journal of Biochemistry*, *153*, 13–19.

Simopoulos, A. P. (2002). Omega-3 fatty acids in inflammation and autoimmune diseases. *Journal of the American College of Nutrition*, *21*, 495–505.

Smith, W. L., DeWitt, D. L., & Garavito, R. M. (2000). Cyclooxygenases: Structural, cellular, and molecular biology. *Annual Review of Biochemistry*, *69*, 145–182.

Smith, W. L., & Song, I. (2002). The enzymology of prostaglandin endoperoxide H synthases-1 and -2. *Prostaglandins & Other Lipid Mediators*, *68–69*, 115–128.

Smyth, E. M., Grosser, T., Wang, M., Yu, Y., & FitzGerald, G. A. (2009). Prostanoids in health and disease. *Journal of Lipid Research*, *50*(Suppl.), S423–S428.

Staels, B., Koenig, W., Habib, A., Merval, R., Lebret, M., Torra, I. P., et al. (1998). Activation of human aortic smooth-muscle cells is inhibited by PPARalpha but not by PPARgamma activators. *Nature*, *393*, 790–793.

Stark, K., Dostalek, M., & Guengerich, F. P. (2008). Expression and purification of orphan cytochrome P450 4X1 and oxidation of anandamide. *The FEBS Journal, 275*, 3706–3717.

Stoilov, I., Krueger, W., Mankowski, D., Guernsey, L., Kaur, A., Glynn, J., et al. (2006). The cytochromes P450 (CYP) response to allergic inflammation of the lung. *Archives of Biochemistry and Biophysics, 456*, 30–38.

Tacconelli, S., & Patrignani, P. (2014). Inside epoxyeicosatrienoic acids and cardiovascular disease. *Frontiers in Pharmacology, 5*, 239.

Theken, K. N., Deng, Y., Kannon, M. A., Miller, T. M., Poloyac, S. M., & Lee, C. R. (2011). Activation of the acute inflammatory response alters cytochrome P450 expression and eicosanoid metabolism. *Drug Metabolism and Disposition, 39*, 22–29.

Tradonsky, A., Rubin, T., Beck, R., Ring, B., Seitz, R., & Mair, S. (2012). A search for reliable molecular markers of prognosis in prostate cancer: A study of 240 cases. *American Journal of Clinical Pathology, 137*, 918–930.

Tsikas, D., & Zoerner, A. A. (2014). Analysis of eicosanoids by LC-MS/MS and GC-MS/MS: A historical retrospect and a discussion. *Journal of Chromatography B, Analytical Technologies in the Biomedical and Life Sciences, 964*, 79–88.

Tunctan, B., Korkmaz, B., Sari, A. N., Kacan, M., Unsal, D., Serin, M. S., et al. (2013). Contribution of iNOS/sGC/PKG pathway, COX-2, CYP4A1, and gp91(phox) to the protective effect of 5,14-HEDGE, a 20-HETE mimetic, against vasodilation, hypotension, tachycardia, and inflammation in a rat model of septic shock. *Nitric Oxide: Biology and Chemistry/Official Journal of the Nitric Oxide Society, 33*, 18–41.

Vane, J. R., Bakhle, Y. S., & Botting, R. M. (1998). Cyclooxygenases 1 and 2. *Annual Review of Pharmacology and Toxicology, 38*, 97–120.

Verkade, P. E., Elzas, M., van der Lee, J., de Wolff, H. H., Verkade-Sandbergen, A., & van der Sande, D. (1933). Untersuchungen über den Fettstoffwechsel I. *Hoppe-Seyler's Zeitschrift für physiologische Chemie, 215*, 225–257.

Wakabayashi, K., & Shimazono, N. (1963). Studies on omega-oxidation of fatty acids in vitro. I. Overall reaction and intermediate. *Biochimica et Biophysica Acta, 70*, 132–142.

Wang, D., & Dubois, R. N. (2010). Eicosanoids and cancer. *Nature Reviews. Cancer, 10*, 181–193.

Wang, M. Z., Saulter, J. Y., Usuki, E., Cheung, Y. L., Hall, M., Bridges, A. S., et al. (2006). CYP4F enzymes are the major enzymes in human liver microsomes that catalyze the O-demethylation of the antiparasitic prodrug DB289 [2,5-bis(4-amidinophenyl)furan-bis-O-methylamidoxime]. *Drug Metabolism and Disposition: The Biological Fate of Chemicals, 34*, 1985–1994.

Wang, M. H., Wang, J., Chang, H. H., Zand, B. A., Jiang, M., Nasjletti, A., et al. (2003). Regulation of renal CYP4A expression and 20-HETE synthesis by nitric oxide in pregnant rats. *American Journal of Physiology Renal Physiology, 285*, F295–F302.

Wang, Y., Zhao, J., Kalsotra, A., Turman, C. M., Grill, R. J., Dash, P. K., et al. (2008). CYP4Fs expression in rat brain correlates with changes in LTB4 levels after traumatic brain injury. *Journal of Neurotrauma, 25*, 1187–1194.

Weigelt, B., Peterse, J. L., & van 't Veer, L. J. (2005). Breast cancer metastasis: Markers and models. *Nature Reviews Cancer, 5*, 591–602.

Wiek, C., Schmidt, E. M., Roellecke, K., Freund, M., Nakano, M., Kelly, E. J., et al. (2015). Identification of amino acid determinants in CYP4B1 for optimal catalytic processing of 4-ipomeanol. *The Biochemical Journal, 465*, 103–114.

Yu, M., Cambj-Sapunar, L., Kehl, F., Maier, K. G., Takeuchi, K., Miyata, N., et al. (2004). Effects of a 20-HETE antagonist and agonists on cerebral vascular tone. *European Journal of Pharmacology, 486*, 297–306.

Yu, W., Chai, H., Li, Y., Zhao, H., Xie, X., Zheng, H., et al. (2012). Increased expression of CYP4Z1 promotes tumor angiogenesis and growth in human breast cancer. *Toxicology and Applied Pharmacology, 264*, 73–83.

Yu, W., Chen, L., Yang, Y. Q., Falck, J. R., Guo, A. M., Li, Y., et al. (2011). Cytochrome P450 omega-hydroxylase promotes angiogenesis and metastasis by upregulation of VEGF and MMP-9 in non-small cell lung cancer. *Cancer Chemotherapy and Pharmacology*, *68*, 619–629.

Zhang, X., Chen, L., & Hardwick, J. P. (2000). Promoter activity and regulation of the CYP4F2 leukotriene B(4) omega-hydroxylase gene by peroxisomal proliferators and retinoic acid in HepG2 cells. *Archives of Biochemistry and Biophysics*, *378*, 364–376.

Zhang, X., & Hardwick, J. P. (2000). Regulation of CYP4F2 leukotriene B4 omega-hydroxylase by retinoic acids in HepG2 cells. *Biochemical and Biophysical Research Communications*, *279*, 864–871.

Zollner, A., Dragan, C. A., Pistorius, D., Muller, R., Bode, H. B., Peters, F. T., et al. (2009). Human CYP4Z1 catalyzes the in-chain hydroxylation of lauric acid and myristic acid. *Biological Chemistry*, *390*, 313–317.

Zou, A. P., Ma, Y. H., Sui, Z. H., Ortiz de Montellano, P. R., Clark, J. E., Masters, B. S., et al. (1994). Effects of 17-octadecynoic acid, a suicide-substrate inhibitor of cytochrome P450 fatty acid omega-hydroxylase, on renal function in rats. *The Journal of Pharmacology and Experimental Therapeutics*, *268*, 474–481.

CHAPTER NINE

Bile Acid Metabolism and Signaling in Cholestasis, Inflammation, and Cancer

Tiangang Li[1], Udayan Apte
Department of Pharmacology, Toxicology and Therapeutics, The University of Kansas Medical Center, Kansas City, Kansas, USA
[1]Corresponding author: e-mail address: tli@kumc.edu

Contents

Abstract

Bile acids are synthesized from cholesterol in the liver. Some cytochrome P450 (CYP) enzymes play key roles in bile acid synthesis. Bile acids are physiological detergent molecules, so are highly cytotoxic. They undergo enterohepatic circulation and play important roles in generating bile flow and facilitating biliary secretion of endogenous metabolites and xenobiotics and intestinal absorption of dietary fats and lipid-soluble vitamins. Bile acid synthesis, transport, and pool size are therefore tightly regulated under physiological conditions. In cholestasis, impaired bile flow leads to accumulation of bile acids in the liver, causing hepatocyte and biliary injury and inflammation. Chronic cholestasis is associated with fibrosis, cirrhosis, and eventually liver failure. Chronic

Advances in Pharmacology, Volume 74
ISSN 1054-3589
http://dx.doi.org/10.1016/bs.apha.2015.04.003

cholestasis also increases the risk of developing hepatocellular or cholangiocellular carcinomas. Extensive research in the last two decades has shown that bile acids act as signaling molecules that regulate various cellular processes. The bile acid-activated nuclear receptors are ligand-activated transcriptional factors that play critical roles in the regulation of bile acid, drug, and xenobiotic metabolism. In cholestasis, these bile acid-activated receptors regulate a network of genes involved in bile acid synthesis, conjugation, transport, and metabolism to alleviate bile acid-induced inflammation and injury. Additionally, bile acids are known to regulate cell growth and proliferation, and altered bile acid levels in diseased conditions have been implicated in liver injury/regeneration and tumorigenesis. We will cover the mechanisms that regulate bile acid homeostasis and detoxification during cholestasis, and the roles of bile acids in the initiation and regulation of hepatic inflammation, regeneration, and carcinogenesis.

ABBREVIATIONS

ABC ATP-binding cassette
APAP acetaminophen
ASBT apical sodium-dependent bile salt transporter
BSEP bile salt export pump
CA cholic acid
CCK cholecystokinin
CDCA chenodeoxycholic acid
CYP27A1 sterol 27-hydroxylase
CYP7A1 cholesterol 7α-hydroxylase
CYP7B1 oxysterol 7α-hydroxylase
CYP8B1 sterol 12α-hydroxylase
DCA deoxycholic acid
FGF15 fibroblast growth factor 15
FGFR4 FGF receptor 4
FXR farnesoid X receptor
HCA hyocholic acid
HCC hepatocellular carcinoma
HNF4α hepatocyte nuclear factor 4α
I-BABP intestinal bile acid-binding protein
ICP intrahepatic cholestasis of pregnancy
LCA lithocholic acid
LPS lipopolysaccharide
LRH-1 liver receptor homolog-1
MCA muricholic acid
mEH microsomal epoxide hydrolase
MRP multidrug resistance-associated protein
NF-κB nuclear factor κB
norUDCA nor-ursodeoxycholic acid
NRs nuclear receptors

NTCP Na^+-dependent taurocholate transporter
OATP organic anion transporter
OCA obeticholic acid
OST organic solute transporter
PBC primary biliary cirrhosis
PFIC progressive familial intrahepatic cholestasis
PHX partial hepatectomy
PSC primary sclerosing cholangitis
PXR pregnane X receptor
SHP small heterodimer partner
UDCA ursodeoxycholic acid
VDR vitamin D receptor
VSMCs vascular smooth muscle cells

1. INTRODUCTION

Bile acids are physiological detergent molecules synthesized from cholesterol exclusively in the liver (Li & Chiang, 2014; Russell & Setchell, 1992). Under physiological conditions, most bile acids exist as glycine or taurine conjugates and are therefore referred to as "bile salts." Bile acid synthesis has several important functions in the liver. Conversion of cholesterol into bile acids in the liver accounts for a major fraction of daily cholesterol turnover in humans (Chiang, 2009). Biliary bile acid secretion generates bile flow and facilitates hepatobiliary secretion of various endogenous metabolites and xenobiotics (Trauner & Boyer, 2003). In the gallbladder, bile acids form mixed micelles with phospholipids and cholesterol to increase cholesterol solubility and decrease bile acid toxicity. Once released into the small intestine, bile acids facilitate the intestinal digestion and absorption of dietary cholesterol, fat, and other lipophilic nutrients. In addition, bile acids are signaling molecules that activate intracellular ligand-activated nuclear receptors (NRs) and cell surface G protein-coupled receptors to regulate various cellular processes from lipid and glucose metabolism, drug metabolism to immunity (Li & Chiang, 2014). Impaired bile flow, which can be caused by both genetic and environmental factors, can lead to cholestasis, in which accumulation of bile acids in the liver causes hepatic inflammation and injury (Zollner & Trauner, 2008). Patients with cholestasis over time develop fibrosis, cirrhosis, and eventually liver failure and increased risk of hepatocellular or cholangiocellular carcinomas. In cholestasis, a number of detoxification mechanisms, mainly mediated by bile acid-activated NRs, are activated to alleviate bile acid-induced inflammation and injury. In addition, bile acids are known to regulate cell growth and proliferation, and dysregulation of bile acid homeostasis and signaling can have an impact on liver regeneration and tumorigenesis (Fan, Wang, Xu, Yan, & Huang, 2015; Vacca et al., 2013). In this chapter, we will cover the roles of major cytochrome P450 (CYP) enzymes involved in hepatic bile acid synthesis, the mechanisms that regulate bile acid homeostasis, and the roles of bile acid metabolism and signaling in the regulation of hepatic inflammation, regeneration, and carcinogenesis.

2. BILE ACID SYNTHESIS AND TRANSPORT

2.1 Primary Bile Acid Synthesis

Primary bile acids are synthesized from cholesterol in the liver (Li & Chiang, 2014). Some primary bile acids are converted to secondary bile acids by bacterial enzymes in small and large intestine. In humans, the bile acid pool consists of the primary bile acids cholic acid (CA) and chenodeoxycholic acid (CDCA), and the secondary bile acid deoxycholic acid (DCA). In the liver, two pathways, namely the classic pathway and the alternative pathway, are responsible for the synthesis of primary bile acids (Fig. 1). The conversion of cholesterol to bile acids involves hydroxylation, saturation of the double bond at C5–C6, epimerization of the 3-hydroxyl group, and oxidative cleavage of a 3-carbon unit, and these reactions are catalyzed by a number of CYP enzymes localized in the endoplasmic reticulum, mitochondria, cytosol, and peroxisomes (Russell, 2003). All bile acids possess a 5β-hydrogen group and a *cis*-configuration along the plane of the fused A and B ring. Majority of the bile acids exist as glycine or taurine conjugates and are thus referred to as "sodium salts." Conjugated bile acids are negatively charged molecules with increased solubility under most physiological pH ranges. Bile acids usually have a few hydroxyl groups and the carboxyl group on one side of the carbon skeleton, which is the structural basis for bile acids to act as amphipathic detergent molecules in facilitating lipid absorption and cholesterol dissolution.

2.1.1 The Classic Pathway

The classic bile acid synthetic pathway is the major bile acid biosynthetic pathway in humans (Fig. 1). It accounts for about 90% of total bile acid

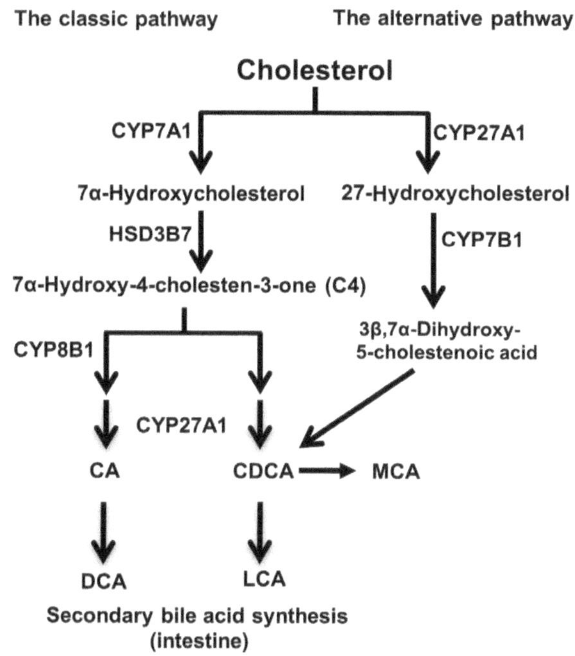

Figure 1 Bile acid synthetic pathways and bile acid structure. Cholesterol is the common precursor for bile acid synthesis via two major bile acid biosynthetic pathways. In the classic pathway, the rate-limiting enzyme cholesterol 7α-hydroxylase (CYP7A1) in the endoplasmic reticulum converts cholesterol into 7α-hydroxycholesterol. The 3β-hydroxysteroid dehydrogenase (3βHSD, HSD3B7) converts 7α-hydroxycholesterol to 7α-hydroxy-4 cholesten-3-one (C4). C4 can be converted to cholic acid (CA) which requires the sterol 12α-hydroxylase (CYP8B1). Without 12α-hydroxylation by CYP8B1, C4 is eventually converted to chenodeoxycholic acid (CDCA). In the classic pathway, the mitochondrial sterol 27-hydroxylase (CYP27A1) catalyzes the steroid side-chain oxidation in both CA and CDCA synthesis. In the alternative pathway, CYP27A1 catalyzes the first step to convert cholesterol to 27-hydroxycholesterol. Oxysterol 7α-hydroxylase (CYP7B1) catalyzes hydroxylation of 27-hydroxycholesterol to 3β,7α-dihydroxy-5-cholestenoic acid, which eventually is converted to CDCA. In the large intestine, bacterial 7α-dehydroxylase removes a hydroxyl group from C-7 and converts CA to deoxycholic acid (DCA) and CDCA to lithocholic acid (LCA). In mouse liver, most of CDCA is converted to α- and β-muricholic acid (MCA).

production in the liver. This pathway produces CA and CDCA in approximately equal amounts. The cholesterol 7α-hydroxylase (CYP7A1), a microsomal CYP enzyme, catalyzes the first and rate-limiting step to convert cholesterol to 7α-hydroxycholesterol (Chiang, 2009; Myant & Mitropoulos, 1977). The intermediate product 7α-hydroxy-4-cholestene-

3-one (C4) is the common precursor for CA and CDCA. Another microsomal CYP enzyme sterol 12α-hydroxylase (CYP8B1) catalyzes the addition of a hydroxyl group at the C-12 position on C4, and the resulting intermediate is eventually converted to CA. The C4 that escapes from this 12-hydroxylation reaction catalyzed by CYP8B1 is converted to CDCA. The synthesis of all bile acids involves the steroid side-chain cleavage, which is catalyzed by the mitochondrial sterol 27-hydroxylase (CYP27A1). The hepatic bile acid synthesis rate is mainly controlled via the transcriptional regulation of the CYP7A1 gene (Chiang, 2009). Hepatic CYP8B1 activity may regulate the CA:CDCA ratio in the bile acid pool. However, other mechanisms such as intestine bile acid transformation and transport may also have a significant impact on the bile acid pool composition. Studies have shown that the plasma level of C4, the common precursor for CA and CDCA, correlates well with hepatic bile acid synthesis and thus has been used as a surrogate serum marker for the rate of hepatic bile acid synthesis in humans (Axelson, Mork, & Sjovall, 1988).

2.1.2 The Alternative Pathway

This pathway is also called the acidic pathway due to the production of acidic intermediates (Fig. 1). In this pathway, the mitochondrial CYP27A1 catalyzes the first hydroxylation reaction to convert cholesterol to 27-hydroxycholesterol. The hydroxylation at the C-7 position is catalyzed by oxysterol 7α-hydroxylase (CYP7B1). In contrast to the classic pathway, the alternative pathway only produces CDCA, but not CA. In addition to cholesterol, oxysterol intermediates formed in the peripheral tissues can enter the liver and feed into this pathway for bile acid production. In humans, the alternative pathway is considered a minor bile acid synthetic pathway because it produces less than 10% of the total bile acids under normal physiological conditions. However, the alternative pathway seemed to be more active in rodents and can be responsible for the synthesis of about 50% of bile acids in rodents (Li & Chiang, 2014; Russell & Setchell, 1992).

2.2 Bile Acid Biotransformation in the Intestine

In the intestine, some bile acids undergo multistep biotransformation that is usually catalyzed by bacterial enzymes in the intestinal tract (Ridlon, Kang, & Hylemon, 2006). In the small and large intestine, a number of bacterial species have bile salt hydrolases. These are bile salt deconjugating enzymes that convert some taurine- and glycine-conjugated bile acids to free bile acids. Free bile acids can passively diffuse through the plasma membrane, but this process accounts for a very small fraction of bile acid reabsorption in the small and large intestine. Probably more importantly, bile acid deconjugation by bile salt hydrolases is a required step for subsequent conversion of primary bile acids into secondary bile acids in the intestine. For example, in the large intestine, bacterial 7α-dehydroxylase catalyzes C-7 dehydroxylation reaction on primary bile acids, which converts CA and CDCA to DCA and lithocholic acid (LCA), respectively (Bjorkhem, Einarsson, Melone, & Hylemon, 1989; Hylemon, Melone, Franklund, Lund, & Bjorkhem, 1991; Ridlon et al., 2006). These secondary bile acids are highly hydrophobic and toxic, and increased concentrations in the liver have been linked to inflammation, cholestasis, gallstone formation, and carcinogenesis (Low-Beer & Nutter, 1978; Marcus & Heaton, 1988; Ridlon & Hylemon, 2006). Both elevated levels of DCA and LCA have been implicated in the promotion of colon cancer (Hussaini, Pereira, Murphy, & Dowling, 1995). Elevated LCA in cholestasis contributes to liver injury and inflammation (Staudinger et al., 2001). Unconjugated DCA is reabsorbed in the large intestine via passive absorption and transported back to the liver, where it can be converted to conjugated form (Trauner & Boyer, 2003). In normal conditions, most LCA is sulfated in the intestine and excreted into feces. The trace amount of LCA that is transported to the liver can be efficiently sulfated and secreted into the circulation for renal excretion. In the intestine, CYP3A and CYP2 family enzymes can also metabolize LCA to more soluble and thus less toxic hyocholic acid (HCA) and ursodeoxycholic acid (UDCA) in humans (Fig. 2; Araya & Wikvall, 1999; Teixeira & Gil, 1991).

It should be noted that although rodents, especially mice and rats, have been widely used as a model for bile acid research, the bile acid pool composition and hydrophobicity in rats and mice are significantly different from that of humans. In mice and rats, the majority of CDCA is converted to more hydrophilic α-muricholic acid (α-MCA) and β-MCA in the liver, which does not occur in human livers. In humans, the highly hydrophobic bile acid pool consists of about 40% of CA, 40% of CDCA, and 20% DCA. Mouse bile acid pool consists of about equal amount of CA and MCAs with relatively low levels of DCA and CDCA, and thus is more hydrophilic. In addition, rats do not have gallbladders, and they secrete bile directly into the small intestine.

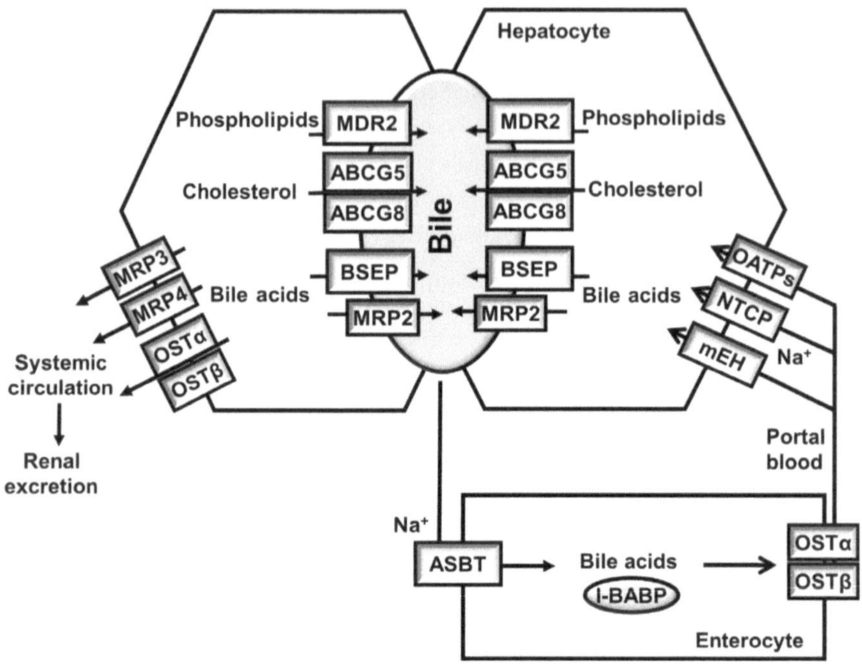

Figure 2 Bile acid transporters in the enterohepatic circulation. Bile acids, after synthesis, are secreted into the bile where bile acids, cholesterol, and phospholipids form micelles. Food intake stimulates the gallbladder to release bile acids into the small intestine. Conjugated bile acids are efficiently reabsorbed in the ileum by active transport systems in the ileum, and a small amount of unconjugated bile acids is reabsorbed by passive diffusion in the small and large intestine. Less than 5% of the bile acids are lost through fecal excretion, which is compensated by *de novo* synthesis in the liver. At the canalicular membrane of the hepatocytes, the bile salt export pump (BSEP) is the primary bile acid efflux transporter, while the multidrug resistance-associated protein-2 (MRP2) can also secrete organic substrates including bile acids, bilirubin, and glutathione. ABCG5 and ABCG8 form heterodimers and transport cholesterol into the bile, and multidrug resistance-2 (MDR2) is responsible for phospholipid secretion. At the basolateral membrane of the hepatocytes, the Na^+-dependent taurocholate transporter (NTCP) is mainly responsible for Na^+-dependent uptake of conjugated bile acids. The microsomal epoxide hydrolase (mEH) may also mediate Na^+-dependent uptake of conjugated bile acids. The organic anion transporters (OATPs) show substrate specificity for unconjugated bile acids. At the basolateral membrane of the hepatocytes, organic solute transporters OSTα and OSTβ heterodimers, MRP3, and MRP4 secrete bile acids into the systemic circulation. In cholestasis, this pathway is induced and leads to subsequent renal bile acid excretion. In the intestine, the apical sodium-dependent bile acid transport (ASBT) mediates bile acid uptake in the ileum. Intracellular bile acids are bound to the intestinal bile acid-binding protein (I-BABP) and are transported to the basolateral membrane where bile acids are secreted into the portal circulation by the OSTα and OSTβ heterodimer.

2.3 Genetic Defects of CYPs in Bile Acid Synthesis

Genetic mutations in *CYP7A1* gene are very rare and only one case has been reported so far. Individuals of homozygous CYP7A1 mutation showed decreased bile acid levels and developed hypercholesterolemia and premature gallstone disease, which is consistent with the key role of bile acid synthesis in overall cholesterol metabolism (Pullinger et al., 2002). Consistently, studies have linked *CYP7A1* polymorphisms to risk of gallstone, plasma LDL cholesterol levels, and cardiovascular risk (Hofman, Princen, Zwinderman, & Jukema, 2005; Jiang et al., 2004; Kajinami, Brousseau, Ordovas, & Schaefer, 2004). Mice lacking *cyp7a1* showed decreased postnatal survival rate and fat and vitamin malabsorption which could be partially corrected with vitamin and bile acid supplements (Ishibashi, Schwarz, Frykman, Herz, & Russell, 1996). Later studies also showed that *cyp7a1* knockout mice were hypercholesterolemia (Erickson et al., 2003). In contrast, *CYP7A1* transgenic mice had increased hepatic cholesterol catabolism and were resistant to diet-induced hypercholesterolemia and atherosclerosis (Miyake et al., 2001, 2002) and metabolic disorders (Li et al., 2011, 2010). *CYP7A1* transgenic mice had approximately two- to threefold increase in total bile acid pool size. Susceptibility to cholestatic liver injury has not been studied in these mice.

Humans with *CYP8B1* mutations have not been reported so far. *Cyp8b1* knockout mice lack CA production (Li-Hawkins et al., 2002). As mentioned early and also discussed in the later section, due to the efficient conversion of CDCA to MCAs, mice depend more on CA to inhibit *cyp7a1* and to facilitate intestine cholesterol absorption because MCAs neither inhibit *cyp7a1* nor facilitate cholesterol absorption. For these reasons, *cyp8b1* knockout mice had increased hepatic CYP7A1 levels and larger bile acid pool, and decreased intestinal fractional cholesterol absorption (Slatis et al., 2010). *Cyp8b1* knockout mice were also resistant to atherosclerosis development primarily due to decreased cholesterol absorption and increased hepatic CYP7A1 levels. In humans, CDCA is not converted to MCAs. Therefore, decreased CYP8B1 activity or defect in CA synthesis in humans may have different impact on hepatic bile acid synthesis and intestine cholesterol absorption compared to mice.

Because CYP27A1 mediates the sterol side-chain cleavage in both classic and alternative bile acid synthesis pathways, *CYP27A1* mutations in humans lead to decreased bile acid synthesis and increased levels of 7α-hydroxylated cholesterol metabolites, including C4 in the liver. Human mutations in

CYP27A1 cause a rare lipid storage disorder called cerebrotendinous xanthomatosis (CTX) (Leitersdorf et al., 1994). CTX is characterized by the accumulation of cholesterol and cholestanol, a derivative of C4, in xanthomas and brains, leading to xanthomatosis, premature atherosclerosis, and progressive neurological disorders. CTX patients are effectively treated with CDCA, which inhibits *CYP7A1* and thus decreases the production of 7α-hydroxylated cholesterol metabolites and therefore cholestanol. *CYP27A1* mutations in humans usually do not lead to hypercholesterolemia. *Cyp27a1* knockout mice had decreased bile acid synthesis and upregulation of hepatic *CYP7A1* gene, but were absent of CTX-related phenotypes (Rosen et al., 1998).

2.4 Bile Acid Transport

2.4.1 Enterohepatic Circulation of Bile Acids

Once synthesized, bile acids in the hepatocytes are efficiently secreted into the bile and stored in the gallbladder. Food intake stimulates the epithelial cells in the duodenum to secrete a peptide hormone called cholecystokinin (CCK), which stimulates gallbladder contraction to release bile acids into the intestinal tract (Otsuki, 2000). CCK also stimulates the pancreas to secrete digestive enzymes into the small intestine to facilitate digestion. Along the small intestine tract, concentrated bile acids form mixed micelles with dietary lipids, which are necessary for lipid digestion by pancreatic lipases. Bile acids are then reabsorbed in the ileum and transported back to the liver via portal blood for resecretion into the bile. The enterohepatic circulation of bile acids is a highly efficient process and only ~5% of the total bile acids are lost via excretion into the feces, which is replenished by the hepatic *de novo* synthesis of bile acid. The enterohepatic circulation of bile acids is illustrated in Fig. 2, and the bile acid transporters involved in this process are discussed below.

2.4.2 Hepatic Bile Acid Transport

Hepatocytes are polarized cells with basolateral (sinusoidal) and apical (canalicular) membrane domains. Bile acids in the portal circulation are taken up by hepatocytes across the basolateral membrane and then secreted into the bile canaliculi and subsequently drained into the bile. Hepatic bile acid uptake is highly efficient with about 90% first-pass extraction rate for conjugated bile acids. Conjugated bile acids cannot diffuse through the membrane and thus require active transport systems for cellular uptake (Meier, 1995). The Na^+-dependent taurocholate transporter (NTCP) is the major bile acid uptake transporter in the basolateral membrane of the hepatocytes (Fig. 2). Early studies conducted in cultured hepatocytes or perfused livers support the role of NTCP as the basolateral Na^+-dependent bile acid uptake transporter (Hagenbuch & Meier, 1994; Hagenbuch, Stieger, Foguet, Lubbert, & Meier, 1991; Meier & Stieger, 2002). Recently, a human patient with loss-of-function mutation of the *NTCP* gene was reported to show significantly elevated circulating bile acids without clinical symptoms of cholestasis, consistent with impaired basolateral bile acid uptake into the hepatocytes (Vaz et al., 2014). This is further supported by a recent study of *ntcp* knockout mice (Slijepcevic et al., 2015). The microsomal epoxide hydrolase (mEH), a bifunctional enzyme that is involved in xenobiotic metabolism, is another candidate basolateral Na^+-dependent bile acid transporter (Ananthanarayanan, von Dippe, & Levy, 1988; Fretland & Omiecinski, 2000; Von Dippe, Amoui, Alves, & Levy, 1993; von Dippe, Amoui, Stellwagen, & Levy, 1996). A human individual with significantly decreased mEH function due to a point mutation had hypercholanemia, a condition associated with highly elevated plasma bile acid levels in the absence of hepatocyte injury (Zhu et al., 2003). However, *mEH* knockout mice showed no apparent alteration in bile acid homeostasis (Miyata et al., 1999). Several organic anion transporter (OATP) isoforms, including human OATP1A2, OATP1B1, and OATP1B3, can mediate Na^+-independent bile acid uptake at the basolateral membrane of the hepatocytes. This pathway is mainly responsible for the uptake of unconjugated bile acids and accounts for about ~25% of bile acid uptake by the hepatocytes (Trauner & Boyer, 2003).

Bile acids, phospholipids, and cholesterol are three major organic solutes in the bile. They form mixed micelles to increase cholesterol solubility and decrease bile acid toxicity in the bile. Canalicular bile acid secretion into the bile against the concentration gradient is the rate-limiting step in bile formation (Boyer, 2013). The bile salt export pump (BSEP, ABCB11), an ATP-binding cassette (ABC) transporter also known as the sister of P-glycoprotein, is the major bile acid efflux transporter at the canalicular membrane (Childs, Yeh, Georges, & Ling, 1995). The multidrug resistance-associated protein-2 (MRP2, ABCC2) can also mediate the canalicular secretion of certain sulfated and conjugated bile acids into the bile. MRP2 also shows broad substrate specificity for bilirubin conjugates, glutathione, and drugs. Hepatobiliary free cholesterol secretion into the bile is a major route for cholesterol elimination from the body. The ABC transporters ABCG5 and ABCG8 form a heterodimer, which mediates cholesterol secretion into the bile (Berge et al., 2000). Phosphatidylcholine is the major phospholipid in the bile, and its secretion is mediated by the phospholipid flippase multidrug resistant 3 (MDR3, ABCB4) (Smit et al., 1993).

Bile acids in the hepatocytes can also be secreted across the basolateral membrane back into the systemic circulation for subsequent renal excretion. This process is mediated by several transporters including the organic solute transporters OSTα and OSTβ heterodimer, MRP3, and MRP4 (Ballatori et al., 2005; Kullak-Ublick, Stieger, Hagenbuch, & Meier, 2000; Trauner & Boyer, 2003). This basolateral bile acid efflux route in the hepatocytes is much less significant than biliary bile acid secretion under normal physiology. However, in cholestasis when biliary bile acid secretion is impaired, basolateral bile acid efflux is often induced as a protective mechanism (Ballatori et al., 2005; Boyer et al., 2006; Cui, Aleksunes, Tanaka, Goedken, & Klaassen, 2009). In the bile duct, some of the bile acids can be taken up by the cholangiocytes. Unconjugated bile acids can passively enter the cholangiocytes, and conjugated bile acids are taken up by the apical sodium-dependent bile salt transporter (ASBT, SLC10A1) (Xia, Francis, Glaser, Alpini, & LeSage, 2006). Bile acids in cholangiocytes are secreted into the peribiliary plexus via the OSTα/β heterodimer at the basolateral membrane. Cholangiocyte bile acid transport may play a role in bile formation. In addition, this process can also transport bile acids into the systemic circulation for renal elimination in cholestasis (Kullak-Ublick et al., 2000; Xia et al., 2006).

2.4.3 Intestinal Bile Acid Transport

Intestine bile acid reabsorption mainly occurs at the terminal ileum where bile acid transporters are highly expressed. This is a highly efficient process and only about ~5% of the bile acids are lost via fecal excretion. At the apical side, the ASBT mediates ileal bile acid uptake into the enterocytes (Shneider et al., 1995). In the enterocytes, bile acids are transported to the basolateral membrane by the intestinal bile acid-binding protein (I-BABP) (Gong, Everett, Schwartz, Norris, & Wilson, 1994). The OSTα and OSTβ form a functional heterodimer at the basolateral membrane and mediate bile acid efflux into the portal circulation (Ballatori et al., 2005; Rao et al., 2008). As most of the conjugated bile acids are efficiently reabsorbed in the small intestine via active transport systems, some unconjugated primary bile acids and secondary bile acids, mainly DCA and to a much less extent LCA, can also be reabsorbed in the colon via passive diffusion and returned to the liver via portal circulation.

3. REGULATION OF BILE ACID TOXICITY AND INFLAMMATION IN CHOLESTASIS

3.1 Cholestatic Liver Diseases

Cholestasis is resulted from disrupted bile flow from the liver to the intestine tract, leading to accumulation of toxic bile acids and other metabolites in the liver, decreased bile acids in the intestine, and increased bile acids in the systemic circulation. The accumulation of toxic bile acids in the hepatobiliary system damages bile duct epithelial cells and hepatocytes, causing liver injury and inflammation. Chronic cholestasis leads to fibrosis, cirrhosis, and eventually liver failure or hepatocellular or cholangiocellular carcinomas. Mechanical obstruction of extrahepatic bile ducts or major intrahepatic bile ducts, such as by tumors or gallstones, genetic mutations of bile acid transporter genes, and acquired dysregulation of bile transport system by drugs, pregnancy, and other pathophysiological conditions can cause either intra- or extrahepatic cholestasis (Srivastava, 2014; Zollner & Trauner, 2008). Congenital or acquired defects in canalicular membrane transporters are the major cause of intrahepatic cholestasis (Zollner & Trauner, 2008). Congenital cholestasis usually occurs very early in life. These patients show progressive liver damage, jaundice, pruritus, and slow growth due to malabsorption of fat-soluble vitamins. Progressive familial intrahepatic cholestasis (PFIC) and benign recurrent intrahepatic cholestasis are autosomal recessive diseases that are linked to genetic mutations in *ATP8B1* (Type 1, PFIC1), *BSEP* (Type 2, PFIC2), and *MDR3* (Type 3, PFIC3) (Srivastava, 2014). PFIC1, which is also known as Byler disease, is linked to mutations in the *ATP8B1* gene. This gene encodes a P-type ATPase that functions as an aminophospholipid flippase. Mutations in the *BSEP* gene are associated with PFIC subtype 2. Defective BSEP function causes hepatic bile acid accumulation and cholestasis. These patients showed markedly elevated plasma bile acid levels and extremely low biliary bile acid concentration (Strautnieks et al., 1997). In addition, *BSEP* polymorphisms have been linked to intrahepatic cholestasis of pregnancy (ICP) (Noe et al., 2005; Pauli-Magnus et al., 2004) and drug-induced liver injury (Lang et al., 2007). ICP is a reversible form of intrahepatic cholestasis associated with adverse pregnancy outcomes. PFIC3 patients, with defective phospholipid transporter MDR3, have high levels of γ-glutamyl transpeptidase activity, progressive cholestasis, bile duct damage, and may require liver transplant. As mentioned early, phospholipids in bile are required for mixed micelle formation. If not incorporated into mixed micelles, bile acids will damage the canalicular membrane and cholangiocytes, causing liver injury and cholestasis. Genetic polymorphisms and heterozygote mutations of the *PFIC1*, *PFIC2*, and *PFIC3* genes may increase susceptibility to acquired cholestasis in adults including ICP, drug-induced liver injury, primary biliary cirrhosis (PBC), and primary sclerosing cholangitis (PSC). MRP2 excretes conjugated bile acids, bilirubin, and other organic anions into the bile. Mutations in the *MRP2* gene have been linked to Dubin–Johnson syndrome, which is characterized by chronic hyperbilirubinemia (Keitel et al., 2003). These patients have elevated bile acids and cholestasis.

3.2 Bile Acid-Activated NR Regulation of Bile Acid Metabolism and Detoxification in Cholestasis

3.2.1 Nuclear Receptor

NRs are ligand-activated transcription factors. They play important roles in embryogenesis, development, and metabolism (Mangelsdorf & Evans, 1995; Mangelsdorf et al., 1995). A typical NR consists of an N-terminal DNA-binding domain (DBD) and a C-terminal ligand-binding domain (LBD) (Fig. 3). The DBD is a highly conserved region containing two Zinc finger motifs that mediate the NR binding to a consensus DNA sequence called hormone response element (HRE). Some NRs bind to the HRE as a homodimer, while some NRs form a heterodimer with another nuclear receptor retinoid X receptor to bind to DNA. A few NRs, such as the hepatocyte

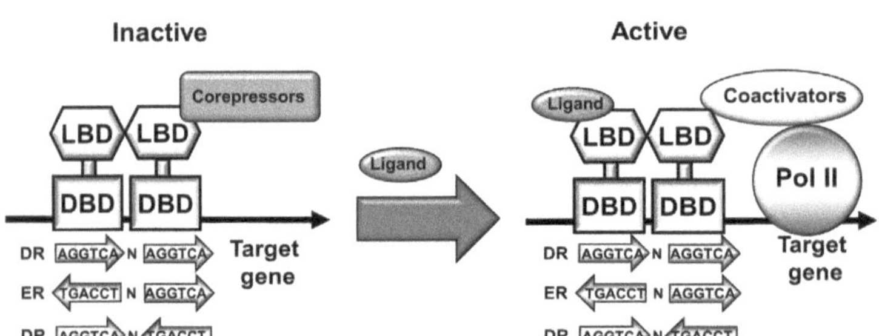

Figure 3 Nuclear receptors. The domain structure of a typical nuclear receptor contains a DNA-binding domain (DBD) and a ligand-binding domain (LBD). Nuclear receptors recognize consensus DNA sequence AGGTCA half-site arranged in direct repeat (DR), everted repeat (ER), and inverted repeat (IR). Ligand binding causes nuclear receptor LBD conformational change, which allows the nuclear receptor to recruit coactivators to replace corepressors. Coactivators facilitate chromatin remodeling and the assembly of general transcriptional machinery, leading to transactivation of the target gene.

nuclear factor 4α (HNF4α) and the liver receptor homolog-1 (LRH-1), bind DNA as monomers. In the LBD, a number of α-helices form a hydrophobic ligand-binding cavity where lipophilic small molecules bind as ligands. The LBD also contains motifs for dimerization and recruitment of coregulators (coactivators or corepressors). In general, ligand binding causes a conformational change in the LBD of a NR, which allows the NRs to recruit the coactivators and displace the corepressors (Fig. 3). This facilitates the assembly of general transcriptional machinery leading to transcriptional activation of the target genes. There are 48 NR genes in the human genome and 49 in the mouse genome (Mangelsdorf et al., 1995). Among these NRs, the farnesoid X receptor (FXR), pregnane X receptor (PXR), and vitamin D receptor (VDR) can be activated by bile acids (Chiang, 2003). Constitutive androstane receptor (CAR) is activated by drugs and xenobiotics, but not bile acids. However, CAR is activated in cholestasis likely by the accumulation of toxic metabolites. These NRs regulate a number of bile acid synthetic and metabolizing CYPs as well as bile acid conjugation enzymes and transporters. During cholestasis, activation of these NRs by bile acids and their metabolites plays a protective role against injury by decreasing hepatic bile acid synthesis and uptake, promoting hepatic bile acid efflux, and inducing phase-I bile acid metabolizing and phase-II bile acid conjugation enzymes. The bile acid-activated NRs and their target genes are summarized in Table 1. The role of major bile acid-activated receptors in the regulation of bile acid metabolism and detoxification are discussed below.

3.2.2 FXR Regulation of Bile Acid Feedback Inhibition of Bile Acid Synthesis

FXR is highly expressed in the hepatocytes and in the intestine, tissues that are exposed to high levels of bile acids. FXR can be activated by both free and conjugated bile acids (Makishima et al., 1999; Parks et al., 1999). The hydrophobic bile acid CDCA is the most efficacious ligand of FXR ($EC_{50} = \sim 10$ μmol/L), followed by LCA, DCA, and CA The hydrophilic bile acid MCAs and UDCA do not activate FXR. In fact, a recent study showed that elevated concentration of MCA in the bile acid pool may act as FXR antagonists to inhibit FXR activity (Sayin et al., 2013).

Hepatic bile acid synthesis is tightly regulated by bile acids through negative feedback mechanisms, which maintains a constant bile acid pool size in normal physiology. In cholestasis, accumulation of bile acids in the liver represses hepatic bile acid synthesis, which is deemed as a protective mechanism to alleviate bile acid damage to the liver. The activity of hepatic bile

Table 1 Bile Acid-Activated Nuclear Receptor Target Genes in Bile Acid Metabolism

Receptors	Target Genes	Regulation	Function and Pathways
FXR	CYP7A1	Down	Enzyme in bile acid synthesis
	CYP8B1	Down	Enzyme in bile acid synthesis
	CYP27A1	Down	Enzyme in bile acid synthesis
	SHP	Up	Negative regulator of bile acid synthesis
	FGF15/FGF19	Up	Negative regulator of bile acid synthesis
	BSEP	Up	Canalicular bile acid secretion
	OSTα/OSTβ	Up	Basolateral bile acid secretion
	NTCP	Down	Basolateral bile acid uptake
	UGT2B4/UGT2B7	Up	Bile acid conjugation
PXR	CYP3A	Up	Phase-I bile acid/drug metabolism
	CYP2B/CYP2C	Up	Phase-I bile acid/drug metabolism
	MRP1/MRP2/ MRP3	Up	Bile acid/drug transport
	UGT1A3	Up	Bile acid conjugation
	SULT2A1	Up	Bile acid conjugation
	OATP2	Up	Basolateral bile acid/drug uptake
	CYP7A1	Down	Enzyme in bile acid synthesis
CAR	CYP3A	Up	Phase-I bile acid/drug metabolism
	CYP2B, CYP2C	Up	Phase-I bile acid/drug metabolism
	UGT1A1	Up	Bile acid conjugation
	MRP2	Up	Bile acid/drug transport
	CYP7A1	Down	Enzyme in bile acid synthesis
VDR	CYP3A	Up	Phase-I bile acid/drug metabolism
	CYP2B/CYP2C	Up	Phase-I bile acid/drug metabolism
	SULT2A1	Up	Bile acid conjugation
	MRP3/MRP4	Up	Bile acid/drug transport
	ASBT	Up	Canalicular bile acid uptake
	CYP7A1	Down	Enzyme in bile acid synthesis

acid synthetic CYP enzymes, including CYP7A1, CYP8B1, and CYP27A1, is mainly controlled at the rate of gene transcription. Bile acid-activated FXR has been shown to cause transcriptional repression of *CYP7A1*, *CYP8B1*, and *CYP27A1*, and thus plays a central role in mediating the negative feedback regulation of bile acid synthesis (Li & Chiang, 2014). The mechanisms of FXR regulation of the *CYP7A1* gene have been extensively investigated in the past decades. FXR does not directly bind to the *CYP7A1* gene promoter. Instead, two indirect mechanisms mediate the FXR repression of *CYP7A1* gene (Fig. 4). First, activation of FXR induces the transcription of a NR small heterodimer partner (SHP). SHP is an atypical NR without a DBD and thus does not bind to DNA. Instead, it often acts as a transcriptional repressor and inhibits the trans-activating activity of a number of NRs and transcriptional factors, leading to the inhibition of their

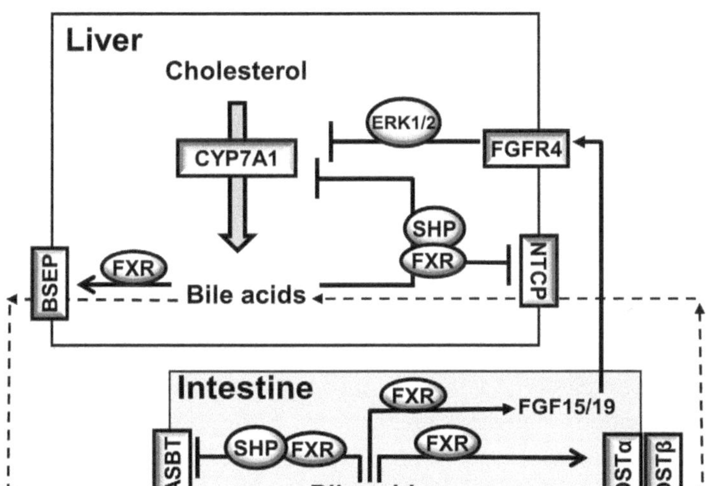

Figure 4 FXR regulation of bile acid feedback inhibition of bile acid synthesis and bile acid transport in the enterohepatic system. In hepatocytes, bile acid-activated FXR induces the repressor SHP, which interacts with and represses the trans-activating action of HNF4α and LRH-1, leading to CYP7A1 inhibition. Bile acid/FXR induces SHP to repress NTCP. FXR binds to *BSEP* gene promoter and induced BSEP and canalicular bile acid secretion. In the intestine, FXR activation inhibits ASBT and induces OSTα and OSTβ, and thus decreases bile acid absorption and promotes basolateral bile acid secretion. Bile acid-activated FXR induces FGF15 (FGF19 in humans). FGF15 binds and activates FGFR4 on the hepatocytes, leading to the inhibition of *CYP7A1* gene, a process that may involve ERK1/2 signaling. The downstream target of FGF15/19 has not been well characterized.

target genes. On the *CYP7A1* gene promoter, two NRs LRH-1 and HNF4α bind to their consensus DNA-binding sites and activate the basal *CYP7A1* gene transcription (Goodwin et al., 2000; Lu et al., 2000). Bile acid- and FXR-induced SHP interact with and inhibit LRH1 and HNF4α, leading to transcriptional repression of the *CYP7A1* gene. In normal physiology, hepatic bile acids are efficiently excreted by canalicular transporters into the bile to maintain a relatively low level of intrahepatic bile acids. Food intake subsequently triggers gallbladder bile acid release into the small intestine. Indeed, studies in mice revealed that intestine is a major bile acid storage site that can retain about 60–80% of the bile acid pool in the body (Li et al., 2012, 2010). Therefore, intestine bile acid sensing is very important in the bile acid pool maintenance. In 2005, Inagaki et al. reported that activation of FXR transcriptionally induces fibroblast growth factor 15 (FGF15) in mice. FGF15 then acts as an endocrine hormone, binds to the cell surface FGF receptor 4 (FGFR4) on the hepatocytes, and inhibits hepatic *CYP7A1* gene transcription (Inagaki et al., 2005). Studies have so far identified that β-Klotho, the SH2 domain containing protein tyrosine phosphatase (SHP-2), and FGF receptor substrate 2 are key components of the FGFR4 signaling complex at the plasma membrane, and that deletion of each of these signaling complex proteins led to increased hepatic bile acid synthesis and elevated bile acid pool size (Goetz et al., 2007; Ito et al., 2005; Li et al., 2014; Lin, Wang, Blackmore, & Desnoyers, 2007; Wang et al., 2014). Importantly, deletion of *shp*-2 resulted in a severe cholestasis phenotype in mice (Li et al., 2014). Human FGF19 shares ~51% amino acid sequence homology with mouse FGF15, and thus is the mouse FGF15 ortholog. Current evidence suggests that FGF19 also represses human *CYP7A1* in an FGFR4-dependent signaling mechanism (Song, Li, Owsley, Strom, & Chiang, 2009). However, different from FGF15 that is not expressed in mouse hepatocytes, FGF19 mRNA is detectable in human livers and primary human hepatocytes, and its expression can be induced by FXR (Song et al., 2009). One study showed that circulating FGF19 levels increased while CYP7A1 expression decreased in human patients with obstructive cholestasis, indicating that human hepatocytes produce FGF19 (Schaap, van der Gaag, Gouma, & Jansen, 2009). The *CYP8B1* and *CYP27A1* genes have been shown to be repressed by bile acid/FXR/SHP cascade, but not by the bile acid/FXR/FGF15 signaling axis (Chen & Chiang, 2003; Zhang & Chiang, 2001). Therefore, FXR activation represses key genes in the biosynthetic pathway to decrease bile acid output.

3.2.3 FXR Regulation of Bile Acid Transport

In addition to inhibiting hepatic bile acid synthesis, FXR activation by bile acids also inhibits hepatic bile acid uptake and promotes biliary bile acid secretion by regulating the expression of hepatic bile acid transporters, and thus prevents bile acid accumulation in the hepatocytes. FXR induces the apical bile acid efflux transporters BSEP (Ananthanarayanan, Balasubramanian, Makishima, Mangelsdorf, & Suchy, 2001), MRP2 (Kast et al., 2002), and the phospholipid transporter ABCB4 (Liu et al., 2003). At the basolateral membrane of the hepatocytes, FXR activation limits bile acid uptake by inhibiting the expression of bile acid uptake transporter NTCP via induction of SHP that represses the transactivation of NTCP by NRs such as retinoic acid receptor and HNF4α (Denson et al., 2001). In conditions associated with hepatic bile acid accumulation, there is usually a compensatory induction of transporters including OSTs and MRPs at the basolateral membrane of the hepatocytes to efflux bile acids into the systemic blood circulation for subsequent renal excretion (Fig. 2), resulting in elevated plasma bile acid concentration in cholestasis (Ballatori et al., 2005; Boyer et al., 2006; Cui et al., 2009). The *OSTα* and *OSTβ* genes are directly induced by bile acid-activated FXR, while MRP1, MRP3, and MRP4 are induced by bile acid-activated PXR during cholestasis, which will be discussed in the next section. In the intestine, FXR induces *OSTα* and *OSTβ* genes (Frankenberg et al., 2006) and represses *ASBT* gene (Chen et al., 2003; Neimark, Chen, Li, & Shneider, 2004). Therefore, similar to the role of FXR in the hepatocytes, FXR activation in the intestine also inhibits bile acid uptake and promotes basolateral bile acid secretion, which decreases cellular bile acid accumulation.

3.2.4 Bile Acid/Xenobiotic Receptors in Bile Acid Metabolism and Detoxification

Like drugs and xenobiotics, bile acid accumulation during cholestasis feedforward induces genes that are involved in phase-I and phase-II bile acid metabolism and detoxification. In the liver, CYP3A4 is the major enzyme that catalyzes the hydroxylation of both primary and secondary bile acids at various positions, converting bile acids into more hydrophilic and less toxic molecules (Araya & Wikvall, 1999; Bodin, Lindbom, & Diczfalusy, 2005; Chen, Zhao, & Chen, 2014). However, unlike bile acid transporters, genetic variation of *CYP3A4* has not been associated with the risk of cholestasis. Bile acids also undergo phase-II glucuronidation and sulfoconjugation reactions before biliary or renal excretion (Belanger, Pelletier, Labrie,

Barbier, & Chouinard, 2003; Kauffman, 2004). Significantly elevated levels of sulfonated and glucuronidated bile acids are often seen in patients with cholestasis (Takikawa, Otsuka, Beppu, Seyama, & Yamakawa, 1983). Sulfotransferase 2A isoforms (SULT2As) are major enzymes involved in the sulfoconjugation reactions of bile acids (Weinshilboum et al., 1997), which usually lead to the formation of 3-α-sulfated bile acids. As mentioned earlier, LCA in the liver can be efficiently sulfated and subsequently excreted in the kidney. The protective role of SULT enzymes against bile acid toxicity has been demonstrated in LCA-induced liver toxicity in mice (Kitada et al., 2003). Glucuronide conjugation of bile acids are catalyzed by UDP-glucuronosyltransferase (UGT) isoforms UGT2B4, UGT2B7, and UGT1A3 (Gall et al., 1999; Pillot et al., 1993; Trottier et al., 2006). These enzymes mediate the glucuronide conjugation of hydroxylated bile acids at different positions. For example, UGT2B4 is involved in the glucuronidation of 6α-hydroxylated bile acids, while UGT2B7 can mediate the glucuronidation of both 3α-hydroxylated bile acids and 6α-hydroxylated bile acids. Although some studies have shown that bile acid-activated FXR can regulate a few of the above-mentioned genes (Barbier et al., 2003; Lu et al., 2005), extensive studies suggest that the drug and bile acid/xenobiotic-activated NRs PXR, CAR, and VDR play a predominant role in inducing the phase-I and phase-II bile acid-metabolizing genes in cholestasis.

3.2.5 Pregnane X Receptor

PXR can be activated by a wide range of xenobiotics, endobiotics, and clinical drugs, and in turn induces a number of CYP3A and CYP2 family enzymes, conjugation enzymes, and transporters in drug metabolism (Kliewer, Goodwin, & Willson, 2002). In cholestasis, high levels of primary bile acids and secondary bile acids and some of the bile acid intermediate metabolites can activate PXR. Upon ligand activation, PXR usually induces gene transcription via binding to the xenobiotic response elements in its target gene promoter or enhancer regions. In phase-I bile acid metabolism, PXR induces *CYP3A4* and *CYP2B* genes (Staudinger et al., 2001). Hydroxylation of bile acids by CYP3A4 not only detoxifies bile acids but also increases their conjugation and subsequent excretion. Specifically, CYP3A4 can convert CDCA to HCA and 3α,7α-dihydroxy-3-oxo-5β-cholanoic acid, and CA to 3-dehydro-CA. CYP3A4 can convert the highly toxic and carcinogenic DCA into 3-dehydro-DCA and 1β,3α,12α-trihydroxy-5β-cholanoic acid, and LCA into 3-dehydro-LCA, hyodeoxycholic acid, and 1β-hydroxy-LCA. In phase-II and phase-III bile acid metabolism, PXR induces the bile acid conjugation enzymes, SULT2A1 and UGTs, the canalicular transporter MRP2 and the basolateral transporter OATP2 (Kliewer & Willson, 2002). These PXR target genes are induced in mice treated with LCA or subjected to bile duct ligation, while these adaptive responses were impaired in *pxr* knockout mice. PCN, a mouse PXR agonist, repressed hepatic CYP7A1 mRNA expression, enzyme activity, and biliary bile acid secretion in rodents (Mason & Boyd, 1978; Stahlberg, 1995; Staudinger et al., 2001), suggesting that PXR may regulate bile acid synthesis. Rifampicin-activated PXR was also shown to repress human *CYP7A1* gene transcription (Bhalla, Ozalp, Fang, Xiang, & Kemper, 2004; Li & Chiang, 2005). PXR does not directly bind to the *CYP7A1* gene promoter. Instead, it interacts with and represses the transactivation activity of HNF4α that binds the *CYP7A1* gene promoter. In addition, activation of PXR in the intestine was shown to induce FGF15 or FGF19 expression, and a PXR response element was identified in the promoter of the *FGF19* (Wang, Venkatesh, et al., 2011; Wistuba, Gnewuch, Liebisch, Schmitz, & Langmann, 2007). Mice lacking PXR were more susceptible to hepatotoxicity caused by LCA treatment or bile duct ligation (Staudinger et al., 2001; Stedman et al., 2005). Studies have also shown that pharmacological activation of PXR protected against bile acid-induced liver injury in experimental cholestasis models (Stedman et al., 2005). Rifampicin, the human PXR agonist, has been used to reduce pruritus associated with cholestasis in humans (Hofmann, 2002). The effectiveness of rifampicin in treating pruritus varied among individuals (Hofmann, 2002).

3.2.6 Constitutive Androstane Receptor

CAR is another key regulator of drug and bile acid-metabolizing genes in the liver (Stanley, Horsburgh, Ross, Scheer, & Wolf, 2006). CAR and PXR can bind to the same xenobiotic response elements in the target gene promoters and thus regulate an overlapping set of target genes including *CYP3A* and *CYP2Bs*. Phenobarbital and TCPOBOP are frequently used as CAR agonists in various experimental settings. Bile acids do not directly activate CAR, but CAR may be activated by toxic metabolites in cholestasis. CAR agonists have been shown to repress *CYP7A1* gene in hepatocytes (Miao, Fang, Bae, & Kemper, 2006). Studies have shown that activation of CAR was beneficial for protecting against bile acid toxicity during cholestasis in mice (Beilke et al., 2009; Guo et al., 2003; Saini et al., 2004; Stedman et al., 2005). The *car* knockout mice had higher degree of liver injury than wild-type mice upon LCA treatment or bile duct ligation (Stedman et al., 2005). CAR may play an important role in inducing sulfation of LCA because *car* transgenic mice had increased levels of sulfated LCA and were resistant to LCA toxicity (Saini et al., 2004). Treatment with CAR ligands phenobarbital or TCPOBOP in *fxr/pxr* double knockout mice protected these mice from CA feeding-induced bile acid toxicity, which was attributed to the induction of CAR target genes *Cpy2b*, *Cyp3a*, *Mrp2*, *Ugt1a1*, and *Gsta* (Guo et al., 2003).

3.2.7 Vitamin D_3 Receptor

VDR is highly expressed in the intestine, but is expressed at very low levels in human hepatocytes, and is not expressed in mouse liver. The role of VDR in the regulation of intestine bile acid metabolism is well documented (Makishima et al., 2002), but its role in hepatic bile acid metabolism is less clear. Among all bile acids, LCA and its metabolite 3 keto-LCA are the most potent activators of VDR. VDR acts as a bile acid sensor in the intestine to protect the gut from bile acid toxicity (Makishima et al., 2002; Nagpal, Na, & Rathnachalam, 2005). VDR can recognize the same xenobiotic response elements as PXR and CAR, and activation of VDR by 1α,25-dihydroxyvitamin D_3 induces CYP3A4, CYP2B, and CYP2C in drug and bile acid metabolism (Drocourt, Ourlin, Pascussi, Maurel, & Vilarem, 2002; Schmiedlin-Ren, Thummel, Fisher, Paine, & Watkins, 2001; Thummel et al., 2001). Activation of VDR also induced SULT2A1 and thus can stimulate bile acid sulfoconjugation (Chatterjee, Echchgadda, & Song, 2005). Furthermore, VDR induced two bile acid transporters MRP3 and ASBT in the intestine (Chen et al., 2006; McCarthy, Li, & Sinal, 2005). It has been suggested that high levels of LCA not only causes cholestasis but may also be involved in the promotion of colon cancer (Ajouz, Mukherji, & Shamseddine, 2014). When LCA levels increase in the gut, VDR may be activated as an adaptive response to convert LCA to less toxic intermediates for excretion (Makishima et al., 2002).

Unlike PXR and CAR, VDR seemed to be expressed at very low levels in primary human hepatocytes (Han, Li, Ellis, Strom, & Chiang, 2010). Treatment of 1α,25-dihydroxyvitamin D_3 induced CYP3A, CYP2B, and CYP2C in primary human hepatocytes (Drocourt et al., 2002). During cholestasis, LCA levels may increase significantly in the liver, which leads to VDR activation. A few studies have investigated the role of VDR in cholestasis in mice. These studies used 1α,25-dihydroxyvitamin D_3 as a VDR agonist but not bile acids which also activate PXR and FXR. One study showed that 1α,25-dihydroxyvitamin D_3 treatment did not affect hepatic or plasma bile acid levels in bile duct-ligated mice, suggesting a minimal role of VDR in modulating bile acid metabolism in cholestasis (Ogura et al., 2009). However, VDR activation by 1α,25-dihydroxyvitamin D_3 treatment in bile duct-ligated mice reduced proinflammatory cytokine expression, suggesting the anti-inflammatory properties of VDR may provide certain benefits during cholestasis (Nagpal et al., 2005). Another study showed that 1α,25-dihydroxyvitamin D_3 treatment increased renal MRP2, MRP3, and MRP4 mRNA expression and increased renal bile acid secretion (Nishida, Ozeki, & Makishima, 2009).

Some studies have shown that VDR agonists affected hepatic bile acid synthesis and bile acid pool size, but both the effects and the mechanisms are still somewhat controversial. Treating primary human hepatocytes with 1α,25-dihydroxyvitamin D_3 repressed CYP7A1 mRNA expression, suggesting VDR may inhibit bile acid synthesis in cholestasis (Han et al., 2010). A recent study showed that *vdr* knockout mice showed higher hepatic *cyp7a1* gene expression and a larger bile acid pool size, while 1α,25-dihydroxyvitamin D_3 treatment decreased hepatic *cyp7a1* gene expression in mice (Schmidt et al., 2010). VDR is not expressed in mouse hepatocytes. Instead, it was suggested that VDR may affect hepatic bile acid synthesis via regulation of intestine FGF15 transcription (Schmidt et al., 2010). In contrast, another study showed that injecting mice with 1α,25-dihydroxyvitamin D_3 increased hepatic CYP7A1 mRNA and lowered cholesterol levels (Chow et al., 2013). This was associated with decreased SHP expression in the liver upon VDR activation. These effects of VDR activation on bile acid synthesis are likely mediated by extrahepatic mechanisms.

3.3 Bile Acid Modulation of Hepatic Inflammation and Cholestatic Liver Injury

Both CDCA and UDCA have been used for effective gallstone dissolution in human patients for many years (Lioudaki, Ganotakis, & Mikhailidis, 2011). While CDCA can cause mild hepatotoxicity in some patients, UDCA is highly soluble and was found to be generally nontoxic to humans. UDCA (Ursodiol™) has also been approved by FDA for treating PBC, and has been shown to significantly improve liver tests and prolong the time needed for liver transplantation in these patients (Dyson et al., 2015). In contrast, UDCA is not effective in treating patient with PSC. Current evidence suggest that UDCA can provide multiple benefits including decreased hydrophobicity of the bile acid pool, increased hepatobiliary secretion, reduced inflammation, and cell death. Nor-ursodeoxycholic acid (norUDCA) is a side-chain-shortened C_{23} homolog of UDCA (Yeh et al., 1997; Yoon et al., 1986). It cannot be conjugated, and after administration it is secreted into the bile, reabsorbed by cholangiocytes, and returned to the liver. It has been shown that norUDCA increased bicarbonate in bile and thus hypercholeresis. It was shown that norUDCA improved sclerosing cholangitis in the *Mdr2−/−* model of cholangiopathy (Halilbasic et al., 2009).

As discussed earlier, activation of FXR provides multiple benefits in alleviating cholestatic liver injury. Based on these rationales, a potent FXR agonist obeticholic acid (OCA) has been tested for treating cholestasis in both experimental animal models and humans (Ali, Carey, & Lindor, 2015). OCA is a 6α-ethyl CDCA derivative that selectively activates FXR with a ~100-fold higher potency than CDCA (Pellicciari et al., 2004, 2002). In animal models of cholestasis, OCA effectively protected against cholestatic liver injury and inflammation (Fiorucci et al., 2005; Pellicciari et al., 2002). Recent clinical trials also showed that OCA significantly improved liver tests in patients with PBC (Hirschfield et al., 2015). In addition to decreasing bile acid synthesis, increasing bile flow, and promoting bile acid detoxification, FXR has recently been shown to directly modulate immune response in both hepatic and extrahepatic tissues. *Fxr* knockout mice showed increased liver inflammation, while FXR activation decreased lipopolysaccharide (LPS)-induced hepatic inflammation (Wang et al., 2008). Consistently, FXR activation protected against liver injury in *Mdr2* knockout mouse model of chronic cholangiopathy (Baghdasaryan et al., 2011). FXR has also been shown to play an anti-inflammatory role in extrahepatic tissues. For example, FXR modulates intestine immunity and FXR activation was shown to reduce inflammation in inflammatory bowel disease (Gadaleta et al., 2011; Vavassori, Mencarelli, Renga, Distrutti, & Fiorucci, 2009). FXR is expressed in vascular smooth muscle cells (VSMCs) and FXR agonists have been shown to inhibit inflammation in VSMC and slow the progression of atherosclerosis by decreasing inflammation of the vasculature (Bishop-Bailey, Walsh, & Warner, 2004; Hanniman, Lambert, McCarthy, & Sinal, 2005; Zhang, He, et al., 2008). The underlying molecular mechanism by which FXR modulates immune response is still not fully clear. FXR activation may antagonize nuclear factor κB (NF-κB) signaling to decrease proinflammatory cytokine production in the liver (Wang et al., 2008). Some studies have reported that FXR was expressed in macrophages and activation of FXR repressed LPS-induced proinflammatory cytokine expression, an effect that was abolished in *fxr−/−* macrophages (Mencarelli, Renga, Distrutti, & Fiorucci, 2009). In VSMC, FXR may induce SHP to inhibit the expression of cyclooxygenase 2 and inducible nitric oxide synthase, which are involved in vascular inflammation and VSMC migration. It is noted that beside cholestasis, FXR agonist OCA has also shown promise in treating nonalcoholic steatohepatitis (NASH) based on both animal studies and clinical trials (Ali et al., 2015; Neuschwander-Tetri et al., 2015). OCA improved lipid and glucose homeostasis, liver enzyme tests, and insulin sensitivity, which may be attributed to the role of FXR in the regulation of lipid and glucose homeostasis, inflammation, insulin sensitivity, and bile acid metabolism (Ali et al., 2015).

The G protein-coupled receptor TGR5 is a bile acid-activated membrane receptor (Kawamata et al., 2003; Maruyama et al., 2002). TGR5 activation stimulates adenylate cyclase, intracellular cAMP production, and PKA activation. Among all bile acids, LCA and 3-keto-LCA are the most potent TGR5 agonists with an EC_{50} of less than 1 μM. DCA, CDCA, and CA also activate TGR5 with an EC_{50} of ~1.0, 4.4, and 7.7 μM, respectively. Despite the liver being a major bile acid target organ, TGR5 is not expressed in hepatocytes. However, TGR5 is expressed in the liver sinusoidal endothelial cells (Keitel et al., 2007), gallbladder epithelial cells, and Kupffer cells (Keitel, Donner, Winandy, Kubitz, & Haussinger, 2008). TGR5 is highly expressed in the ileum and colon (Kawamata et al., 2003) and in nontraditional bile acid target organs including white and brown adipose, spleen, kidney, pancreas, lung, macrophages, and the central nervous system (Kawamata et al., 2003). Activation of TGR5 in adipose, muscle, and intestine has been shown to regulate lipid, glucose, and energy metabolism and thus improve metabolic homeostasis (Li & Chiang, 2014). TGR5 may be a potential therapeutic target for the treatment of diabetes and cardiovascular diseases. The metabolic regulation by TGR5 signaling will not be further discussed here.

How TGR5 regulates bile acid synthesis and metabolism under normal physiology is currently not very clear. However, it was reported that mice lacking TGR5 had reduced bile acid pool size (Maruyama et al., 2006), a more hydrophobic bile acid composition and showed more severe liver injury upon bile acid feeding or bile duct ligation (Pean et al., 2013). Studies have shown that pharmacological activation of TGR5 in macrophages may play an anti-inflammatory role in the immune system, which is supported by recent studies demonstrating a protective role of TGR5 activation in cholestasis and NASH (Kawamata et al., 2003; Keitel et al., 2008; McMahan et al., 2013; Pean et al., 2013). Activation of TGR5 reduced LPS-stimulated proinflammatory cytokine production (Keitel et al., 2008). *Tgr5* knockout mice challenged with LPS had higher plasma liver enzymes and elevated cytokine expression, while the selective TGR5 agonist 23(S)-mCDCA antagonized LPS-induced cytokine expression in mouse liver (Wang, Chen, Yu, Forman, & Huang, 2011). In the vasculature, TGR5 activation by 6-EMCA or INT-777 attenuated atherosclerosis in mice. Importantly, it was shown that INT-777 did not attenuate atherosclerosis in mice transplanted with bone marrow of *tgr5* knockout mice, proving the anti-inflammatory and antiatherogenic role of macrophage TGR5. In the intestine where TGR5 is highly expressed, a TGR5-selective agonist protected the integrity of intestinal barrier function, immune response, and proinflammatory cytokine production in experimental colitis models (Cipriani et al., 2011; Yoneno et al., 2013). Pruritus is commonly associated with cholestasis and treatment with bile acid derivatives. A recent study suggests that TGR5 mediates bile acid-induced itch and analgesia (Alemi et al., 2013). Bile acids activate TGR5 on sensory nerves and stimulate the release of neuropeptides in the spinal cord that transmits itch and analgesia.

4. ROLE OF BILE ACIDS IN LIVER INJURY, REGENERATION, AND CANCER

Patients with chronic and advanced-stage cholestasis including PBC and PSC may be at higher risk of developing hepatocellular carcinoma (HCC) and bile duct cancer (Eaton, Talwalkar, Lazaridis, Gores, & Lindor, 2013; Tomiyama et al., 2013). One of the unique characteristics of liver is its ability to regenerate after liver injury or surgical resection (Michalopoulos, 2013). Proper liver regeneration is an important determinant of final outcome after toxin or drug-induced liver injury (Mehendale, 2005). In cholestasis, both biliary epithelial cells and hepatocytes proliferate to compensate for liver cell death. On one hand, cholestasis is associated with reduced regenerating capability (Yokoyama, Nagino, & Nimura, 2007), and on the other hand, repeated injury and repair cycles during chronic cholestasis exacerbate liver fibrosis, inflammation, cirrhosis, and cancer. During cholestasis, bile acids not only act as toxins to cause chronic inflammation and cell death, bile acids also activate cellular mitogenic signaling pathways to regulate cell proliferation (Fan et al., 2015). Studies indicate that bile acids are tumor promoters and involved in the pathogenesis of HCC (Kim et al., 2007; Yang et al., 2007).

Liver can regenerate upon surgical resection. This is modeled using partial hepatectomy (PHX) procedures where about two-thirds of the liver was surgically removed in rodents to allow the study of the regeneration process. Studies in this model have obtained important information on the role and regulation of bile acid metabolism and signaling in the regulation of hepatocyte proliferation. In rats and mice, hepatic CYP7A1 enzyme activity and mRNA expression were significantly decreased immediately following PHX (Huang et al., 2006; Maeda et al., 2005; Nakano et al., 1995). This may reflect an increased bile acid influx and activation of the FXR signaling in the regenerating lobe. Overexpression of CYP7A1 in the liver impaired liver regeneration in mice, suggesting that repression of CYP7A1 and hepatic bile acid synthesis may be necessary for normal regeneration (Zhang et al., 2009). In addition, deletion of *fxr* gene in mice resulted in high mortality coupled with inhibited liver regeneration following PHX (Huang et al., 2006). Later studies also revealed a similar effect of FXR deletion in liver regeneration after CCl_4 and APAP-induced liver injury (Meng et al., 2010). Global deletion of FXR resulted in significantly induced CYP7A1, which caused enlarged bile acid pool and higher plasma bile acid levels in mice. Indeed, later studies showed that hepatocyte-specific *fxr* knockout mice displayed only a moderate delay in liver regeneration after PHX without any liver injury or necrosis in the regenerating lobes (Borude et al., 2012; Zhang et al., 2012). Hepatocyte-specific *fxr* knockout mice have intact intestine bile acids/FXR/FGF15 axis that limits the CYP7A1 expression and the expansion of bile acid pool. In addition, bile acid-induced FGF15 was recently shown to act as a secondary or auxiliary mitogen for hepatocytes and may enhance promitogenic effects of primary mitogens such as hepatocyte growth factors and epidermal growth factor (Limaye et al., 2008). Both *fgf15* knockout mice and *fgfr4* knockout mice had impaired liver regeneration and increased mortality after PHX (Chen et al., 2014; Uriarte et al., 2013). Current evidence suggests that intestine bile acid/FXR/FGF15 signaling axis is required to promote normal liver regeneration, while hepatic bile acid overload in cholestasis impairs normal liver regeneration.

The initial studies on the role of bile acids and liver cancers, specifically HCC, performed between 1970s and early 2000s showed that plasma bile acid concentrations are higher in HCC patients as compared to normal healthy subjects (Changbumrung, Tungtrongchitr, Migasena, & Chamroenngan, 1990; El-Mir, Badia, Luengo, Monte, & Marin, 2001; Hirayama & Irisa, 1976). These studies indicated that bile acid homeostasis was disturbed during HCC and cholangiocarcinoma development and could be potentially used as a biomarker. The first causative link between higher bile acids and HCC pathogenesis came in mid-2000s when two groups independently demonstrated that the *fxr* knockout mice develop spontaneous liver tumors (Kim et al., 2007; Yang et al., 2007). Further studies showed that genetic deletion of *shp* gene in mice also led to spontaneous HCC (Anakk et al., 2011; Katzenellenbogen et al., 2007; Zhang, Xu, et al., 2008). Deletion of *Mdr2* gene resulted in portal regurgitation of bile, extensive cholestasis resulting in a condition similar to PFIC. The *Mdr2−/−* mice developed biliary fibrosis and HCC (Katzenellenbogen et al., 2006). Similar condition was observed in humans where chronic cholestasis led to the development of HCC in many cases (Eaton et al., 2013; Tomiyama et al., 2013). In cholestasis, high levels of bile acids can lead to the generation of reactive oxygen species, disruption of cell membrane, impairment of mitochondrial function, and induction of DNA damage and mutation. Toxic bile acid-induced chronic inflammation and injury–repair response in the liver likely contribute to tumor promotion. In addition, bile acids, specifically the highly hydrophobic and highly toxic DCA, are known to activate cellular signaling pathways such as MAPK, STAT-3, and NF-κB to induce cytokines such as tumor necrosis factor α and interleukin 6. It is also known that bile acids can activate proinflammatory molecules such as Egr1 via an EGFR-dependent, FXR-independent mechanism (Allen, Jaeschke, & Copple, 2011). More recent studies revealed that conjugated bile acids can activate the sphingosine 1-phosphate receptor 2 that activates intracellular ERK1/2 and AKT signaling to promote the invasive growth of cholangiocarcinoma, which is commonly associated with chronic cholestasis (Liu et al., 2014; Studer et al., 2012). In summary, current studies suggest that dysregulation of hepatic bile acid synthetic and metabolizing enzymes and bile acid-activated receptors in cholestatic liver injury have an impact on hepatocyte proliferation, regeneration, and tumorigenesis.

5. CONCLUSION

Bile acids not only act as physiological detergent molecules that facilitate hepatic excretion of endogenous metabolites, xenobiotic and drugs, and intestine fat and nutrient absorption, but also act as signaling molecules that regulate various cellular processes involved in metabolism, immune response, and cell growth. Bile acids are highly toxic and accumulation of bile acids in cholestasis leads to tissue inflammation and injury, and increases the risk of liver cancer. CYP enzymes are involved in bile acid synthesis and detoxification. Bile acids, like drugs, undergo phase-II conjugation and phase-III excretion. Bile acid synthesis and transport are tightly regulated by bile acid and drug/xenobiotic sensing NRs to maintain bile acid homeostasis. These receptors regulate genes in phase-I, phase-II, and phase-III bile acid metabolism and play a critical role in the detoxification of bile acids. Knowledge on the mechanisms of the regulation of bile acid metabolism and signaling provides important molecular basis for the development of novel therapeutic approaches for the treatment of cholestasis and inflammation-related liver diseases.

CONFLICT OF INTEREST

The authors declare no conflict of interest.

ACKNOWLEDGMENTS

This work was supported by: The American Diabetes Association Junior Faculty Award (T.L.), NIH grant 1R01DK102487-01 (T.L.), the National Center for Research Resources (5P20RR021940-07), and the National Institute of General Medical Sciences (8 P20 GM103549-07) of the National Institutes of Health (T.L.). NIH grant 1R01DK098414 (U.A.). The American Association for the Study of Liver Diseases/ American Liver Foundation Liver Scholar Award (U.A.).

REFERENCES

Ajouz, H., Mukherji, D., & Shamseddine, A. (2014). Secondary bile acids: An under-recognized cause of colon cancer. *World Journal of Surgical Oncology*, *12*, 164.

Alemi, F., Kwon, E., Poole, D. P., Lieu, T., Lyo, V., Cattaruzza, F., et al. (2013). The TGR5 receptor mediates bile acid-induced itch and analgesia. *The Journal of Clinical Investigation*, *123*, 1513–1530.

Ali, A. H., Carey, E. J., & Lindor, K. D. (2015). Recent advances in the development of farnesoid X receptor agonists. *Annals of Translational Medicine*, *3*, 5.

Allen, K., Jaeschke, H., & Copple, B. L. (2011). Bile acids induce inflammatory genes in hepatocytes: A novel mechanism of inflammation during obstructive cholestasis. *The American Journal of Pathology*, *178*, 175–186.

Anakk, S., Watanabe, M., Ochsner, S. A., McKenna, N. J., Finegold, M. J., & Moore, D. D. (2011). Combined deletion of Fxr and Shp in mice induces Cyp17a1 and results in juvenile onset cholestasis. *The Journal of Clinical Investigation*, *121*, 86–95.

Ananthanarayanan, M., Balasubramanian, N., Makishima, M., Mangelsdorf, D. J., & Suchy, F. J. (2001). Human bile salt export pump promoter is transactivated by the farnesoid X receptor/bile acid receptor. *The Journal of Biological Chemistry*, *276*, 28857–28865.

Ananthanarayanan, M., von Dippe, P., & Levy, D. (1988). Identification of the hepatocyte Na+-dependent bile acid transport protein using monoclonal antibodies. *The Journal of Biological Chemistry*, *263*, 8338–8343.

Araya, Z., & Wikvall, K. (1999). 6alpha-hydroxylation of taurochenodeoxycholic acid and lithocholic acid by CYP3A4 in human liver microsomes. *Biochimica et Biophysica Acta*, *1438*, 47–54.

Axelson, M., Mork, B., & Sjovall, J. (1988). Occurrence of 3 beta-hydroxy-5-cholestenoic acid, 3 beta,7 alpha-dihydroxy-5-cholestenoic acid, and 7 alpha-hydroxy-3-oxo-4-cholestenoic acid as normal constituents in human blood. *The Journal of Lipid Research, 29*, 629–641.
Baghdasaryan, A., Claudel, T., Gumhold, J., Silbert, D., Adorini, L., Roda, A., et al. (2011). Dual farnesoid X receptor/TGR5 agonist INT-767 reduces liver injury in the Mdr2-/- (Abcb4-/-) mouse cholangiopathy model by promoting biliary HCO(-)(3) output. *Hepatology, 54*, 1303–1312.
Ballatori, N., Christian, W. V., Lee, J. Y., Dawson, P. A., Soroka, C. J., Boyer, J. L., et al. (2005). OSTalpha-OSTbeta: A major basolateral bile acid and steroid transporter in human intestinal, renal, and biliary epithelia. *Hepatology, 42*, 1270–1279.
Barbier, O., Torra, I. P., Sirvent, A., Claudel, T., Blanquart, C., Duran-Sandoval, D., et al. (2003). FXR induces the UGT2B4 enzyme in hepatocytes: A potential mechanism of negative feedback control of FXR activity. *Gastroenterology, 124*, 1926–1940.
Beilke, L. D., Aleksunes, L. M., Holland, R. D., Besselsen, D. G., Beger, R. D., Klaassen, C. D., et al. (2009). Constitutive androstane receptor-mediated changes in bile acid composition contributes to hepatoprotection from lithocholic acid-induced liver injury in mice. *Drug Metabolism and Disposition, 37*, 1035–1045.
Belanger, A., Pelletier, G., Labrie, F., Barbier, O., & Chouinard, S. (2003). Inactivation of androgens by UDP-glucuronosyltransferase enzymes in humans. *Trends in Endocrinology and Metabolism, 14*, 473–479.
Berge, K. E., Tian, H., Graf, G. A., Yu, L., Grishin, N. V., Schultz, J., et al. (2000). Accumulation of dietary cholesterol in sitosterolemia caused by mutations in adjacent ABC transporters. *Science, 290*, 1771–1775.
Bhalla, S., Ozalp, C., Fang, S., Xiang, L., & Kemper, J. K. (2004). Ligand-activated PXR interferes with HNF-4 signaling by targeting a common coactivator PGC-1alpha: Functional implications in hepatic cholesterol and glucose metabolism. *The Journal of Biological Chemistry, 279*, 45139–45147.
Bishop-Bailey, D., Walsh, D. T., & Warner, T. D. (2004). Expression and activation of the farnesoid X receptor in the vasculature. *Proceedings of the National Academy of Sciences of the United States of America, 101*, 3668–3673.
Bjorkhem, I., Einarsson, K., Melone, P., & Hylemon, P. (1989). Mechanism of intestinal formation of deoxycholic acid from cholic acid in humans: Evidence for a 3-oxo-delta 4-steroid intermediate. *The Journal of Lipid Research, 30*, 1033–1039.
Bodin, K., Lindbom, U., & Diczfalusy, U. (2005). Novel pathways of bile acid metabolism involving CYP3A4. *Biochimica et Biophysica Acta, 1687*, 84–93.
Borude, P., Edwards, G., Walesky, C., Li, F., Ma, X., Kong, B., et al. (2012). Hepatocyte-specific deletion of farnesoid X receptor delays but does not inhibit liver regeneration after partial hepatectomy in mice. *Hepatology, 56*, 2344–2352.
Boyer, J. L. (2013). Bile formation and secretion. *Comprehensive Physiology, 3*, 1035–1078.
Boyer, J. L., Trauner, M., Mennone, A., Soroka, C. J., Cai, S. Y., Moustafa, T., et al. (2006). Upregulation of a basolateral FXR-dependent bile acid efflux transporter OSTalpha-OSTbeta in cholestasis in humans and rodents. *American Journal of Physiology. Gastrointestinal and Liver Physiology, 290*, G1124–G1130.
Changbumrung, S., Tungtrongchitr, R., Migasena, P., & Chamroenngan, S. (1990). Serum unconjugated primary and secondary bile acids in patients with cholangiocarcinoma and hepatocellular carcinoma. *Journal of the Medical Association of Thailand = Chotmaihet thangphaet, 73*, 81–90.
Chatterjee, B., Echchgadda, I., & Song, C. S. (2005). Vitamin D receptor regulation of the steroid/bile acid sulfotransferase SULT2A1. *Methods in Enzymology, 400*, 165–191.
Chen, X., Chen, F., Liu, S., Glaeser, H., Dawson, P. A., Hofmann, A. F., et al. (2006). Transactivation of rat apical sodium-dependent bile acid transporter and increased bile

acid transport by 1alpha,25-dihydroxyvitamin D3 via the vitamin D receptor. *Molecular Pharmacology*, *69*, 1913–1923.
Chen, W., & Chiang, J. Y. (2003). Regulation of human sterol 27-hydroxylase gene (CYP27A1) by bile acids and hepatocyte nuclear factor 4alpha (HNF4alpha). *Gene*, *313*, 71–82.
Chen, F., Ma, L., Dawson, P. A., Sinal, C. J., Sehayek, E., Gonzalez, F. J., et al. (2003). Liver receptor homologue-1 mediates species- and cell line-specific bile acid-dependent negative feedback regulation of the apical sodium-dependent bile acid transporter. *The Journal of Biological Chemistry*, *278*, 19909–19916.
Chen, J., Zhao, K. N., & Chen, C. (2014). The role of CYP3A4 in the biotransformation of bile acids and therapeutic implication for cholestasis. *Annals of Translational Medicine*, *2*, 7.
Chiang, J. Y. (2003). Bile acid regulation of hepatic physiology: III. Bile acids and nuclear receptors. *American Journal of Physiology. Gastrointestinal and Liver Physiology*, *284*, G349–G356.
Chiang, J. Y. (2009). Bile acids: Regulation of synthesis. *The Journal of Lipid Research*, *50*, 1955–1966.
Childs, S., Yeh, R. L., Georges, E., & Ling, V. (1995). Identification of a sister gene to P-glycoprotein. *Cancer Research*, *55*, 2029–2034.
Chow, E. C., Magomedova, L., Quach, H. P., Patel, R. H., Durk, M. R., Fan, J., et al. (2013). Vitamin D receptor activation down-regulates small heterodimer partner and increases CYP7A1 to lower cholesterol. *Gastroenterology*, *146*, 1048–1059.
Cipriani, S., Mencarelli, A., Chini, M. G., Distrutti, E., Renga, B., Bifulco, G., et al. (2011). The bile acid receptor GPBAR-1 (TGR5) modulates integrity of intestinal barrier and immune response to experimental colitis. *PLoS One*, *6*, e25637.
Cui, Y. J., Aleksunes, L. M., Tanaka, Y., Goedken, M. J., & Klaassen, C. D. (2009). Compensatory induction of liver efflux transporters in response to ANIT-induced liver injury is impaired in FXR-null mice. *Toxicological Sciences: An Official Journal of the Society of Toxicology*, *110*, 47–60.
Denson, L. A., Sturm, E., Echevarria, W., Zimmerman, T. L., Makishima, M., Mangelsdorf, D. J., et al. (2001). The orphan nuclear receptor, shp, mediates bile acid-induced inhibition of the rat bile acid transporter, ntcp. *Gastroenterology*, *121*, 140–147.
Drocourt, L., Ourlin, J. C., Pascussi, J. M., Maurel, P., & Vilarem, M. J. (2002). Expression of CYP3A4, CYP2B6, and CYP2C9 is regulated by the vitamin D receptor pathway in primary human hepatocytes. *The Journal of Biological Chemistry*, *277*, 25125–25132.
Dyson, J. K., Hirschfield, G. M., Adams, D. H., Beuers, U., Mann, D. A., Lindor, K. D., et al. (2015). Novel therapeutic targets in primary biliary cirrhosis. *Nature Reviews. Gastroenterology & Hepatology*, *12*, 147–158.
Eaton, J. E., Talwalkar, J. A., Lazaridis, K. N., Gores, G. J., & Lindor, K. D. (2013). Pathogenesis of primary sclerosing cholangitis and advances in diagnosis and management. *Gastroenterology*, *145*, 521–536.
El-Mir, M. Y., Badia, M. D., Luengo, N., Monte, M. J., & Marin, J. J. (2001). Increased levels of typically fetal bile acid species in patients with hepatocellular carcinoma. *Clinical Science*, *100*, 499–508.
Erickson, S. K., Lear, S. R., Deane, S., Dubrac, S., Huling, S. L., Nguyen, L., et al. (2003). Hypercholesterolemia and changes in lipid and bile acid metabolism in male and female cyp7A1-deficient mice. *The Journal of Lipid Research*, *44*, 1001–1009.
Fan, M., Wang, X., Xu, G., Yan, Q., & Huang, W. (2015). Bile acid signaling and liver regeneration. *Biochimica et Biophysica Acta*, *1849*, 196–200.
Fiorucci, S., Clerici, C., Antonelli, E., Orlandi, S., Goodwin, B., Sadeghpour, B. M., et al. (2005). Protective effects of 6-ethyl chenodeoxycholic acid, a farnesoid X receptor ligand, in estrogen-induced cholestasis. *The Journal of Pharmacology and Experimental Therapeutics*, *313*, 604–612.

Frankenberg, T., Rao, A., Chen, F., Haywood, J., Shneider, B. L., & Dawson, P. A. (2006). Regulation of the mouse organic solute transporter alpha-beta, Ostalpha-Ostbeta, by bile acids. *American Journal of Physiology. Gastrointestinal and Liver Physiology*, *290*, G912–G922.
Fretland, A. J., & Omiecinski, C. J. (2000). Epoxide hydrolases: Biochemistry and molecular biology. *Chemico-Biological Interactions*, *129*, 41–59.
Gadaleta, R. M., van Erpecum, K. J., Oldenburg, B., Willemsen, E. C., Renooij, W., Murzilli, S., et al. (2011). Farnesoid X receptor activation inhibits inflammation and preserves the intestinal barrier in inflammatory bowel disease. *Gut*, *60*, 463–472.
Gall, W. E., Zawada, G., Mojarrabi, B., Tephly, T. R., Green, M. D., Coffman, B. L., et al. (1999). Differential glucuronidation of bile acids, androgens and estrogens by human UGT1A3 and 2B7. *The Journal of Steroid Biochemistry and Molecular Biology*, *70*, 101–108.
Goetz, R., Beenken, A., Ibrahimi, O. A., Kalinina, J., Olsen, S. K., Eliseenkova, A. V., et al. (2007). Molecular insights into the klotho-dependent, endocrine mode of action of fibroblast growth factor 19 subfamily members. *Molecular and Cellular Biology*, *27*, 3417–3428.
Gong, Y. Z., Everett, E. T., Schwartz, D. A., Norris, J. S., & Wilson, F. A. (1994). Molecular cloning, tissue distribution, and expression of a 14-kDa bile acid-binding protein from rat ileal cytosol. *Proceedings of the National Academy of Sciences of the United States of America*, *91*, 4741–4745.
Goodwin, B., Jones, S. A., Price, R. R., Watson, M. A., McKee, D. D., Moore, L. B., et al. (2000). A regulatory cascade of the nuclear receptors FXR, SHP-1, and LRH-1 represses bile acid biosynthesis. *Molecular Cell*, *6*, 517–526.
Guo, G. L., Lambert, G., Negishi, M., Ward, J. M., Brewer, H. B., Jr., Kliewer, S. A., et al. (2003). Complementary roles of farnesoid X receptor, pregnane X receptor, and constitutive androstane receptor in protection against bile acid toxicity. *The Journal of Biological Chemistry*, *278*, 45062–45071.
Hagenbuch, B., & Meier, P. J. (1994). Molecular cloning, chromosomal localization, and functional characterization of a human liver Na+/bile acid cotransporter. *The Journal of Clinical Investigation*, *93*, 1326–1331.
Hagenbuch, B., Stieger, B., Foguet, M., Lubbert, H., & Meier, P. J. (1991). Functional expression cloning and characterization of the hepatocyte Na+/bile acid cotransport system. *Proceedings of the National Academy of Sciences of the United States of America*, *88*, 10629–10633.
Halilbasic, E., Fiorotto, R., Fickert, P., Marschall, H. U., Moustafa, T., Spirli, C., et al. (2009). Side chain structure determines unique physiologic and therapeutic properties of norursodeoxycholic acid in Mdr2-/- mice. *Hepatology*, *49*, 1972–1981.
Han, S., Li, T., Ellis, E., Strom, S., & Chiang, J. Y. (2010). A novel bile acid-activated vitamin D receptor signaling in human hepatocytes. *Molecular Endocrinology*, *24*, 1151–1164.
Hanniman, E. A., Lambert, G., McCarthy, T. C., & Sinal, C. J. (2005). Loss of functional farnesoid X receptor increases atherosclerotic lesions in apolipoprotein E-deficient mice. *The Journal of Lipid Research*, *46*, 2595–2604.
Hirayama, C., & Irisa, T. (1976). Serum cholesterol and bile acid in primary hepatoma. *Clinica Chimica Acta; International Journal of Clinical Chemistry*, *71*, 21–25.
Hirschfield, G. M., Mason, A., Luketic, V., Lindor, K., Gordon, S. C., Mayo, M., et al. (2015). Efficacy of obeticholic acid in patients with primary biliary cirrhosis and inadequate response to ursodeoxycholic acid. *Gastroenterology*, *148*, 751–761, e758.
Hofman, M. K., Princen, H. M., Zwinderman, A. H., & Jukema, J. W. (2005). Genetic variation in the rate-limiting enzyme in cholesterol catabolism (cholesterol 7alpha-hydroxylase) influences the progression of atherosclerosis and risk of new clinical events. *Clinical Science*, *108*, 539–545.
Hofmann, A. F. (2002). Rifampicin and treatment of cholestatic pruritus. *Gut*, *51*, 756–757.

Huang, W., Ma, K., Zhang, J., Qatanani, M., Cuvillier, J., Liu, J., et al. (2006). Nuclear receptor-dependent bile acid signaling is required for normal liver regeneration. *Science*, *312*, 233–236.

Hussaini, S. H., Pereira, S. P., Murphy, G. M., & Dowling, R. H. (1995). Deoxycholic acid influences cholesterol solubilization and microcrystal nucleation time in gallbladder bile. *Hepatology*, *22*, 1735–1744.

Hylemon, P. B., Melone, P. D., Franklund, C. V., Lund, E., & Bjorkhem, I. (1991). Mechanism of intestinal 7 alpha-dehydroxylation of cholic acid: Evidence that allo-deoxycholic acid is an inducible side-product. *The Journal of Lipid Research*, *32*, 89–96.

Inagaki, T., Choi, M., Moschetta, A., Peng, L., Cummins, C. L., McDonald, J. G., et al. (2005). Fibroblast growth factor 15 functions as an enterohepatic signal to regulate bile acid homeostasis. *Cell Metabolism*, *2*, 217–225.

Ishibashi, S., Schwarz, M., Frykman, P. K., Herz, J., & Russell, D. W. (1996). Disruption of cholesterol 7alpha-hydroxylase gene in mice. I. Postnatal lethality reversed by bile acid and vitamin supplementation. *Journal of Biological Chemistry*, *271*, 18017–18023.

Ito, S., Fujimori, T., Furuya, A., Satoh, J., Nabeshima, Y., & Nabeshima, Y. (2005). Impaired negative feedback suppression of bile acid synthesis in mice lacking betaKlotho. *The Journal of Clinical Investigation*, *115*, 2202–2208.

Jiang, Z. Y., Han, T. Q., Suo, G. J., Feng, D. X., Chen, S., Cai, X. X., et al. (2004). Polymorphisms at cholesterol 7alpha-hydroxylase, apolipoproteins B and E and low density lipoprotein receptor genes in patients with gallbladder stone disease. *World Journal of Gastroenterology*, *10*, 1508–1512.

Kajinami, K., Brousseau, M. E., Ordovas, J. M., & Schaefer, E. J. (2004). Interactions between common genetic polymorphisms in ABCG5/G8 and CYP7A1 on LDL cholesterol-lowering response to atorvastatin. *Atherosclerosis*, *175*, 287–293.

Kast, H. R., Goodwin, B., Tarr, P. T., Jones, S. A., Anisfeld, A. M., Stoltz, C. M., et al. (2002). Regulation of multidrug resistance-associated protein 2 (ABCC2) by the nuclear receptors pregnane X receptor, farnesoid X-activated receptor, and constitutive androstane receptor. *The Journal of Biological Chemistry*, *277*, 2908–2915.

Katzenellenbogen, M., Mizrahi, L., Pappo, O., Klopstock, N., Olam, D., Jacob-Hirsch, J., et al. (2007). Molecular mechanisms of liver carcinogenesis in the mdr2-knockout mice. *Molecular Cancer Research: MCR*, *5*, 1159–1170.

Katzenellenbogen, M., Pappo, O., Barash, H., Klopstock, N., Mizrahi, L., Olam, D., et al. (2006). Multiple adaptive mechanisms to chronic liver disease revealed at early stages of liver carcinogenesis in the Mdr2-knockout mice. *Cancer Research*, *66*, 4001–4010.

Kauffman, F. C. (2004). Sulfonation in pharmacology and toxicology. *Drug Metabolism Reviews*, *36*, 823–843.

Kawamata, Y., Fujii, R., Hosoya, M., Harada, M., Yoshida, H., Miwa, M., et al. (2003). A G protein-coupled receptor responsive to bile acids. *The Journal of Biological Chemistry*, *278*, 9435–9440.

Keitel, V., Donner, M., Winandy, S., Kubitz, R., & Haussinger, D. (2008). Expression and function of the bile acid receptor TGR5 in Kupffer cells. *Biochemical and Biophysical Research Communications*, *372*, 78–84.

Keitel, V., Nies, A. T., Brom, M., Hummel-Eisenbeiss, J., Spring, H., & Keppler, D. (2003). A common Dubin-Johnson syndrome mutation impairs protein maturation and transport activity of MRP2 (ABCC2). *American Journal of Physiology. Gastrointestinal and Liver Physiology*, *284*, G165–G174.

Keitel, V., Reinehr, R., Gatsios, P., Rupprecht, C., Gorg, B., Selbach, O., et al. (2007). The G-protein coupled bile salt receptor TGR5 is expressed in liver sinusoidal endothelial cells. *Hepatology*, *45*, 695–704.

Kim, I., Morimura, K., Shah, Y., Yang, Q., Ward, J. M., & Gonzalez, F. J. (2007). Spontaneous hepatocarcinogenesis in farnesoid X receptor-null mice. *Carcinogenesis*, *28*, 940–946.

Kitada, H., Miyata, M., Nakamura, T., Tozawa, A., Honma, W., Shimada, M., et al. (2003). Protective role of hydroxysteroid sulfotransferase in lithocholic acid-induced liver toxicity. *The Journal of Biological Chemistry*, *278*, 17838–17844.

Kliewer, S. A., Goodwin, B., & Willson, T. M. (2002). The nuclear pregnane X receptor: A key regulator of xenobiotic metabolism. *Endocrine Reviews*, *23*, 687–702.

Kliewer, S. A., & Willson, T. M. (2002). Regulation of xenobiotic and bile acid metabolism by the nuclear pregnane X receptor. *The Journal of Lipid Research*, *43*, 359–364.

Kullak-Ublick, G. A., Stieger, B., Hagenbuch, B., & Meier, P. J. (2000). Hepatic transport of bile salts. *Seminars in Liver Disease*, *20*, 273–292.

Lang, C., Meier, Y., Stieger, B., Beuers, U., Lang, T., Kerb, R., et al. (2007). Mutations and polymorphisms in the bile salt export pump and the multidrug resistance protein 3 associated with drug-induced liver injury. *Pharmacogenetics and Genomics*, *17*, 47–60.

Leitersdorf, E., Safadi, R., Meiner, V., Reshef, A., Bjorkhem, I., Friedlander, Y., et al. (1994). Cerebrotendinous xanthomatosis in the Israeli Druze: Molecular genetics and phenotypic characteristics. *The American Journal of Human Genetics*, *55*, 907–915.

Li, T., & Chiang, J. Y. (2005). Mechanism of rifampicin and pregnane X receptor inhibition of human cholesterol 7 alpha-hydroxylase gene transcription. *American Journal of Physiology. Gastrointestinal and Liver Physiology*, *288*, G74–G84.

Li, T., & Chiang, J. Y. (2014). Bile acid signaling in metabolic disease and drug therapy. *Pharmacological Reviews*, *66*, 948–983.

Li, T., Francl, J. M., Boehme, S., Ochoa, A., Zhang, Y., Klaassen, C. D., et al. (2012). Glucose and insulin induction of bile acid synthesis: Mechanisms and implication in diabetes and obesity. *The Journal of Biological Chemistry*, *287*, 1861–1873.

Li, S., Hsu, D. D., Li, B., Luo, X., Alderson, N., Qiao, L., et al. (2014). Cytoplasmic tyrosine phosphatase Shp2 coordinates hepatic regulation of bile acid and FGF15/19 signaling to repress bile acid synthesis. *Cell Metabolism*, *20*, 320–332.

Li, T., Matozel, M., Boehme, S., Kong, B., Nilsson, L. M., Guo, G., et al. (2011). Overexpression of cholesterol 7alpha-hydroxylase promotes hepatic bile acid synthesis and secretion and maintains cholesterol homeostasis. *Hepatology*, *53*, 996–1006.

Li, T., Owsley, E., Matozel, M., Hsu, P., Novak, C. M., & Chiang, J. Y. (2010). Transgenic expression of cholesterol 7alpha-hydroxylase in the liver prevents high-fat diet-induced obesity and insulin resistance in mice. *Hepatology*, *52*, 678–690.

Li-Hawkins, J., Gafvels, M., Olin, M., Lund, E. G., Andersson, U., Schuster, G., et al. (2002). Cholic acid mediates negative feedback regulation of bile acid synthesis in mice. *The Journal of Clinical Investigation*, *110*, 1191–1200.

Limaye, P. B., Bowen, W. C., Orr, A. V., Luo, J., Tseng, G. C., & Michalopoulos, G. K. (2008). Mechanisms of hepatocyte growth factor-mediated and epidermal growth factor-mediated signaling in transdifferentiation of rat hepatocytes to biliary epithelium. *Hepatology*, *47*, 1702–1713.

Lin, B. C., Wang, M., Blackmore, C., & Desnoyers, L. R. (2007). Liver-specific activities of FGF19 require Klotho beta. *The Journal of Biological Chemistry*, *282*, 27277–27284.

Lioudaki, E., Ganotakis, E. S., & Mikhailidis, D. P. (2011). Lipid lowering drugs and gallstones: A therapeutic option? *Current Pharmaceutical Design*, *17*, 3622–3631.

Liu, Y., Binz, J., Numerick, M. J., Dennis, S., Luo, G., Desai, B., et al. (2003). Hepatoprotection by the farnesoid X receptor agonist GW4064 in rat models of intra- and extrahepatic cholestasis. *The Journal of Clinical Investigation*, *112*, 1678–1687.

Liu, R., Zhao, R., Zhou, X., Liang, X., Campbell, D. J., Zhang, X., et al. (2014). Conjugated bile acids promote cholangiocarcinoma cell invasive growth through activation of sphingosine 1-phosphate receptor 2. *Hepatology*, *60*, 908–918.

Low-Beer, T. S., & Nutter, S. (1978). Colonic bacterial activity, biliary cholesterol saturation, and pathogenesis of gallstones. *Lancet*, *2*, 1063–1065.

Lu, Y., Heydel, J. M., Li, X., Bratton, S., Lindblom, T., & Radominska-Pandya, A. (2005). Lithocholic acid decreases expression of UGT2B7 in Caco-2 cells: A potential role for a negative farnesoid X receptor response element. *Drug Metabolism and Disposition*, *33*, 937–946.

Lu, T. T., Makishima, M., Repa, J. J., Schoonjans, K., Kerr, T. A., Auwerx, J., et al. (2000). Molecular basis for feedback regulation of bile acid synthesis by nuclear receptors. *Molecular Cell*, *6*, 507–515.

Maeda, Y., Nagatomo, H., Kuroki, N., Nagatomo, J., Uchiyama, F., Shimozono, K., et al. (2005). Bile acid biosynthesis during liver regeneration: Enzyme activities of cholesterol 7alpha-hydroxylase and 3beta-hydroxy-delta5-C27-steroid dehydrogenase in rats. *Annals of Clinical and Laboratory Science*, *35*, 323–328.

Makishima, M., Lu, T. T., Xie, W., Whitfield, G. K., Domoto, H., Evans, R. M., et al. (2002). Vitamin D receptor as an intestinal bile acid sensor. *Science*, *296*, 1313–1316.

Makishima, M., Okamoto, A. Y., Repa, J. J., Tu, H., Learned, R. M., Luk, A., et al. (1999). Identification of a nuclear receptor for bile acids. *Science*, *284*, 1362–1365.

Mangelsdorf, D. J., & Evans, R. M. (1995). The RXR heterodimers and orphan receptors. *Cell*, *83*, 841–850.

Mangelsdorf, D. J., Thummel, C., Beato, M., Herrlich, P., Schutz, G., Umesono, K., et al. (1995). The nuclear receptor superfamily: The second decade. *Cell*, *83*, 835–839.

Marcus, S. N., & Heaton, K. W. (1988). Deoxycholic acid and the pathogenesis of gall stones. *Gut*, *29*, 522–533.

Maruyama, T., Miyamoto, Y., Nakamura, T., Tamai, Y., Okada, H., Sugiyama, E., et al. (2002). Identification of membrane-type receptor for bile acids (M-BAR). *Biochemical and Biophysical Research Communications*, *298*, 714–719.

Maruyama, T., Tanaka, K., Suzuki, J., Miyoshi, H., Harada, N., Nakamura, T., et al. (2006). Targeted disruption of G protein-coupled bile acid receptor 1 (Gpbar1/M-Bar) in mice. *The Journal of Endocrinology*, *191*, 197–205.

Mason, J. I., & Boyd, G. S. (1978). The suppressive effect of the catatoxic steroid, pregnenolone-16alpha-carbonitrile, on liver microsomal cholesterol-7alpha-hydroxlyase. *Steroids*, *31*, 849–854.

McCarthy, T. C., Li, X., & Sinal, C. J. (2005). Vitamin D receptor-dependent regulation of colon multidrug resistance-associated protein 3 gene expression by bile acids. *The Journal of Biological Chemistry*, *280*, 23232–23242.

McMahan, R. H., Wang, X. X., Cheng, L. L., Krisko, T., Smith, M., El Kasmi, K., et al. (2013). Bile acid receptor activation modulates hepatic monocyte activity and improves nonalcoholic fatty liver disease. *The Journal of Biological Chemistry*, *288*, 11761–11770.

Mehendale, H. M. (2005). Tissue repair: An important determinant of final outcome of toxicant-induced injury. *Toxicologic Pathology*, *33*, 41–51.

Meier, P. J. (1995). Molecular mechanisms of hepatic bile salt transport from sinusoidal blood into bile. *The American Journal of Physiology*, *269*, G801–G812.

Meier, P. J., & Stieger, B. (2002). Bile salt transporters. *Annual Review of Physiology*, *64*, 635–661.

Mencarelli, A., Renga, B., Distrutti, E., & Fiorucci, S. (2009). Antiatherosclerotic effect of farnesoid X receptor. *American Journal of Physiology. Heart and Circulatory Physiology*, *296*, H272–H281.

Meng, Z., Wang, Y., Wang, L., Jin, W., Liu, N., Pan, H., et al. (2010). FXR regulates liver repair after CCl4-induced toxic injury. *Molecular Endocrinology*, *24*, 886–897.

Miao, J., Fang, S., Bae, Y., & Kemper, J. K. (2006). Functional inhibitory cross-talk between constitutive androstane receptor and hepatic nuclear factor-4 in hepatic lipid/glucose metabolism is mediated by competition for binding to the DR1 motif and to the

common coactivators, GRIP-1 and PGC-1alpha. *The Journal of Biological Chemistry, 281*, 14537–14546.
Michalopoulos, G. K. (2013). Principles of liver regeneration and growth homeostasis. *Comprehensive Physiology, 3*, 485–513.
Miyake, J. H., Doung, X. D., Strauss, W., Moore, G. L., Castellani, L. W., Curtiss, L. K., et al. (2001). Increased production of apolipoprotein B-containing lipoproteins in the absence of hyperlipidemia in transgenic mice expressing cholesterol 7alpha-hydroxylase. *The Journal of Biological Chemistry, 276*, 23304–23311.
Miyake, J. H., Duong-Polk, X. T., Taylor, J. M., Du, E. Z., Castellani, L. W., Lusis, A. J., et al. (2002). Transgenic expression of cholesterol-7-alpha-hydroxylase prevents atherosclerosis in C57BL/6J mice. *Arteriosclerosis, Thrombosis, and Vascular Biology, 22*, 121–126.
Miyata, M., Kudo, G., Lee, Y. H., Yang, T. J., Gelboin, H. V., Fernandez-Salguero, P., et al. (1999). Targeted disruption of the microsomal epoxide hydrolase gene. Microsomal epoxide hydrolase is required for the carcinogenic activity of 7,12-dimethylbenz[a] anthracene. *The Journal of Biological Chemistry, 274*, 23963–23968.
Myant, N. B., & Mitropoulos, K. A. (1977). Cholesterol 7 alpha-hydroxylase. *The Journal of Lipid Research, 18*, 135–153.
Nagpal, S., Na, S., & Rathnachalam, R. (2005). Noncalcemic actions of vitamin D receptor ligands. *Endocrine Reviews, 26*, 662–687.
Nakano, K., Chijiiwa, K., Okamoto, S., Yamashita, H., Kuroki, S., & Tanaka, M. (1995). Hepatic cholesterol 7 alpha-hydroxylase activity and serum 7 alpha-hydroxy-cholesterol level during liver regeneration after partial hepatectomy in rats. *European Surgical Research Europaische Chirurgische Forschung Recherches Chirurgicales Europeennes, 27*, 389–395.
Neimark, E., Chen, F., Li, X., & Shneider, B. L. (2004). Bile acid-induced negative feedback regulation of the human ileal bile acid transporter. *Hepatology, 40*, 149–156.
Neuschwander-Tetri, B. A., Loomba, R., Sanyal, A. J., Lavine, J. E., Van Natta, M. L., Abdelmalek, M. F., et al. (2015). Farnesoid X nuclear receptor ligand obeticholic acid for non-cirrhotic, non-alcoholic steatohepatitis (FLINT): A multicentre, randomised, placebo-controlled trial. *Lancet, 385*, 956–965.
Nishida, S., Ozeki, J., & Makishima, M. (2009). Modulation of bile acid metabolism by 1alpha-hydroxyvitamin D3 administration in mice. *Drug Metabolism and Disposition, 37*, 2037–2044.
Noe, J., Kullak-Ublick, G. A., Jochum, W., Stieger, B., Kerb, R., Haberl, M., et al. (2005). Impaired expression and function of the bile salt export pump due to three novel ABCB11 mutations in intrahepatic cholestasis. *Journal of Hepatology, 43*, 536–543.
Ogura, M., Nishida, S., Ishizawa, M., Sakurai, K., Shimizu, M., Matsuo, S., et al. (2009). Vitamin D3 modulates the expression of bile acid regulatory genes and represses inflammation in bile duct-ligated mice. *The Journal of Pharmacology and Experimental Therapeutics, 328*, 564–570.
Otsuki, M. (2000). Pathophysiological role of cholecystokinin in humans. *Journal of Gastroenterology and Hepatology, 15*(Suppl.), D71–D83.
Parks, D. J., Blanchard, S. G., Bledsoe, R. K., Chandra, G., Consler, T. G., Kliewer, S. A., et al. (1999). Bile acids: Natural ligands for an orphan nuclear receptor. *Science, 284*, 1365–1368.
Pauli-Magnus, C., Lang, T., Meier, Y., Zodan-Marin, T., Jung, D., Breymann, C., et al. (2004). Sequence analysis of bile salt export pump (ABCB11) and multidrug resistance p-glycoprotein 3 (ABCB4, MDR3) in patients with intrahepatic cholestasis of pregnancy. *Pharmacogenetics, 14*, 91–102.
Pean, N., Doignon, I., Garcin, I., Besnard, A., Julien, B., Liu, B., et al. (2013). The receptor TGR5 protects the liver from bile acid overload during liver regeneration in mice. *Hepatology, 58*, 1451–1460.

Pellicciari, R., Costantino, G., Camaioni, E., Sadeghpour, B. M., Entrena, A., Willson, T. M., et al. (2004). Bile acid derivatives as ligands of the farnesoid X receptor. Synthesis, evaluation, and structure-activity relationship of a series of body and side chain modified analogues of chenodeoxycholic acid. *Journal of Medicinal Chemistry*, *47*, 4559–4569.

Pellicciari, R., Fiorucci, S., Camaioni, E., Clerici, C., Costantino, G., Maloney, P. R., et al. (2002). 6alpha-ethyl-chenodeoxycholic acid (6-ECDCA), a potent and selective FXR agonist endowed with anticholestatic activity. *Journal of Medicinal Chemistry*, *45*, 3569–3572.

Pillot, T., Ouzzine, M., Fournel-Gigleux, S., Lafaurie, C., Radominska, A., Burchell, B., et al. (1993). Glucuronidation of hyodeoxycholic acid in human liver. Evidence for a selective role of UDP-glucuronosyltransferase 2B4. *The Journal of Biological Chemistry*, *268*, 25636–25642.

Pullinger, C. R., Eng, C., Salen, G., Shefer, S., Batta, A. K., Erickson, S. K., et al. (2002). Human cholesterol 7alpha-hydroxylase (CYP7A1) deficiency has a hypercholesterolemic phenotype. *The Journal of Clinical Investigation*, *110*, 109–117.

Rao, A., Haywood, J., Craddock, A. L., Belinsky, M. G., Kruh, G. D., & Dawson, P. A. (2008). The organic solute transporter alpha-beta, Ostalpha-Ostbeta, is essential for intestinal bile acid transport and homeostasis. *Proceedings of the National Academy of Sciences of the United States of America*, *105*, 3891–3896.

Ridlon, J. M., & Hylemon, P. B. (2006). A potential role for resistant starch fermentation in modulating colonic bacterial metabolism and colon cancer risk. *Cancer Biology & Therapy*, *5*, 273–274.

Ridlon, J. M., Kang, D. J., & Hylemon, P. B. (2006). Bile salt biotransformations by human intestinal bacteria. *The Journal of Lipid Research*, *47*, 241–259.

Rosen, H., Reshef, A., Maeda, N., Lippoldt, A., Shpizen, S., Triger, L., et al. (1998). Markedly reduced bile acid synthesis but maintained levels of cholesterol and vitamin D metabolites in mice with disrupted sterol 27-hydroxylase gene. *The Journal of Biological Chemistry*, *273*, 14805–14812.

Russell, D. W. (2003). The enzymes, regulation, and genetics of bile acid synthesis. *Annual Review of Biochemistry*, *72*, 1370174.

Russell, D. W., & Setchell, K. D. (1992). Bile acid biosynthesis. *Biochemistry*, *31*, 4737–4749.

Saini, S. P., Sonoda, J., Xu, L., Toma, D., Uppal, H., Mu, Y., et al. (2004). A novel constitutive androstane receptor-mediated and CYP3A-independent pathway of bile acid detoxification. *Molecular Pharmacology*, *65*, 292–300.

Sayin, S. I., Wahlstrom, A., Felin, J., Jantti, S., Marschall, H. U., Bamberg, K., et al. (2013). Gut microbiota regulates bile acid metabolism by reducing the levels of tauro-beta-muricholic acid, a naturally occurring FXR antagonist. *Cell Metabolism*, *17*, 225–235.

Schaap, F. G., van der Gaag, N. A., Gouma, D. J., & Jansen, P. L. (2009). High expression of the bile salt-homeostatic hormone fibroblast growth factor 19 in the liver of patients with extrahepatic cholestasis. *Hepatology*, *49*, 1228–1235.

Schmidt, D. R., Holmstrom, S. R., Fon Tacer, K., Bookout, A. L., Kliewer, S. A., & Mangelsdorf, D. J. (2010). Regulation of bile acid synthesis by fat-soluble vitamins A and D. *The Journal of Biological Chemistry*, *285*, 14486–14494.

Schmiedlin-Ren, P., Thummel, K. E., Fisher, J. M., Paine, M. F., & Watkins, P. B. (2001). Induction of CYP3A4 by 1 alpha,25-dihydroxyvitamin D3 is human cell line-specific and is unlikely to involve pregnane X receptor. *Drug Metabolism and Disposition*, *29*, 1446–1453.

Shneider, B. L., Dawson, P. A., Christie, D. M., Hardikar, W., Wong, M. H., & Suchy, F. J. (1995). Cloning and molecular characterization of the ontogeny of a rat ileal sodium-dependent bile acid transporter. *The Journal of Clinical Investigation*, *95*, 745–754.

Slatis, K., Gafvels, M., Kannisto, K., Ovchinnikova, O., Paulsson-Berne, G., Parini, P., et al. (2010). Abolished synthesis of cholic acid reduces atherosclerotic development in apolipoprotein E knockout mice. *The Journal of Lipid Research*, *51*, 3289–3298.

Slijepcevic, D., Kaufman, C., Wichers, C. G., Gilglioni, E. H., Lempp, F. A., Duijst, S., et al. (2015). Impaired uptake of conjugated bile acids and Hepatitis B Virus preS1-binding in Na -taurocholate cotransporting polypeptide knockout mice. *Hepatology*, [Epub ahead of print].

Smit, J. J., Schinkel, A. H., Oude Elferink, R. P., Groen, A. K., Wagenaar, E., van Deemter, L., et al. (1993). Homozygous disruption of the murine mdr2 P-glycoprotein gene leads to a complete absence of phospholipid from bile and to liver disease. *Cell*, *75*, 451–462.

Song, K. H., Li, T., Owsley, E., Strom, S., & Chiang, J. Y. (2009). Bile acids activate fibroblast growth factor 19 signaling in human hepatocytes to inhibit cholesterol 7alpha-hydroxylase gene expression. *Hepatology*, *49*, 297–305.

Srivastava, A. (2014). Progressive familial intrahepatic cholestasis. *Journal of Clinical and Experimental Hepatology*, *4*, 25–36.

Stahlberg, D. (1995). Effects of pregnenolone-16 alpha-carbonitrile on the metabolism of cholesterol in rat liver microsomes. *Lipids*, *30*, 361–364.

Stanley, L. A., Horsburgh, B. C., Ross, J., Scheer, N., & Wolf, C. R. (2006). PXR and CAR: Nuclear receptors which play a pivotal role in drug disposition and chemical toxicity. *Drug Metabolism Reviews*, *38*, 515–597.

Staudinger, J. L., Goodwin, B., Jones, S. A., Hawkins-Brown, D., MacKenzie, K. I., LaTour, A., et al. (2001). The nuclear receptor PXR is a lithocholic acid sensor that protects against liver toxicity. *Proceedings of the National Academy of Sciences of the United States of America*, *98*, 3369–3374.

Stedman, C. A., Liddle, C., Coulter, S. A., Sonoda, J., Alvarez, J. G., Moore, D. D., et al. (2005). Nuclear receptors constitutive androstane receptor and pregnane X receptor ameliorate cholestatic liver injury. *Proceedings of the National Academy of Sciences of the United States of America*, *102*, 2063–2068.

Strautnieks, S. S., Kagalwalla, A. F., Tanner, M. S., Knisely, A. S., Bull, L., Freimer, N., et al. (1997). Identification of a locus for progressive familial intrahepatic cholestasis PFIC2 on chromosome 2q24. *The American Journal of Human Genetics*, *61*, 630–633.

Studer, E., Zhou, X., Zhao, R., Wang, Y., Takabe, K., Nagahashi, M., et al. (2012). Conjugated bile acids activate the sphingosine-1-phosphate receptor 2 in primary rodent hepatocytes. *Hepatology*, *55*, 267–276.

Takikawa, H., Otsuka, H., Beppu, T., Seyama, Y., & Yamakawa, T. (1983). Serum concentrations of bile acid glucuronides in hepatobiliary diseases. *Digestion*, *27*, 189–195.

Teixeira, J., & Gil, G. (1991). Cloning, expression, and regulation of lithocholic acid 6 beta-hydroxylase. *The Journal of Biological Chemistry*, *266*, 21030–21036.

Thummel, K. E., Brimer, C., Yasuda, K., Thottassery, J., Senn, T., Lin, Y., et al. (2001). Transcriptional control of intestinal cytochrome P-4503A by 1alpha,25-dihydroxy vitamin D3. *Molecular Pharmacology*, *60*, 1399–1406.

Tomiyama, Y., Takenaka, K., Kodama, T., Kawanaka, M., Sasaki, K., Nishina, S., et al. (2013). Risk factors for survival and the development of hepatocellular carcinoma in patients with primary biliary cirrhosis. *Internal Medicine*, *52*, 1553–1559.

Trauner, M., & Boyer, J. L. (2003). Bile salt transporters: Molecular characterization, function, and regulation. *Physiological Reviews*, *83*, 633–671.

Trottier, J., Verreault, M., Grepper, S., Monte, D., Belanger, J., Kaeding, J., et al. (2006). Human UDP-glucuronosyltransferase (UGT)1A3 enzyme conjugates chenodeoxycholic acid in the liver. *Hepatology*, *44*, 1158–1170.

Uriarte, I., Fernandez-Barrena, M. G., Monte, M. J., Latasa, M. U., Chang, H. C., Carotti, S., et al. (2013). Identification of fibroblast growth factor 15 as a novel mediator

of liver regeneration and its application in the prevention of post-resection liver failure in mice. *Gut*, *62*, 899–910.
Vacca, M., Degirolamo, C., Massafra, V., Polimeno, L., Mariani-Costantini, R., Palasciano, G., et al. (2013). Nuclear receptors in regenerating liver and hepatocellular carcinoma. *Molecular and Cellular Endocrinology*, *368*, 108–119.
Vavassori, P., Mencarelli, A., Renga, B., Distrutti, E., & Fiorucci, S. (2009). The bile acid receptor FXR is a modulator of intestinal innate immunity. *Journal of Immunology*, *183*, 6251–6261.
Vaz, F. M., Paulusma, C. C., Huidekoper, H., de Ru, M., Lim, C., Koster, J., et al. (2014). Sodium taurocholate cotransporting polypeptide (SLC10A1) deficiency: Conjugated hypercholanemia without a clear clinical phenotype. *Hepatology*, *61*, 260–267.
Von Dippe, P., Amoui, M., Alves, C., & Levy, D. (1993). Na(+)-dependent bile acid transport by hepatocytes is mediated by a protein similar to microsomal epoxide hydrolase. *The American Journal of Physiology*, *264*, G528–G534.
von Dippe, P., Amoui, M., Stellwagen, R. H., & Levy, D. (1996). The functional expression of sodium-dependent bile acid transport in Madin-Darby canine kidney cells transfected with the cDNA for microsomal epoxide hydrolase. *The Journal of Biological Chemistry*, *271*, 18176–18180.
Wang, Y. D., Chen, W. D., Wang, M., Yu, D., Forman, B. M., & Huang, W. (2008). Farnesoid X receptor antagonizes nuclear factor kappaB in hepatic inflammatory response. *Hepatology*, *48*, 1632–1643.
Wang, Y. D., Chen, W. D., Yu, D., Forman, B. M., & Huang, W. (2011). The G-protein-coupled bile acid receptor, Gpbar1 (TGR5), negatively regulates hepatic inflammatory response through antagonizing nuclear factor kappa light-chain enhancer of activated B cells (NF-kappaB) in mice. *Hepatology*, *54*, 1421–1432.
Wang, H., Venkatesh, M., Li, H., Goetz, R., Mukherjee, S., Biswas, A., et al. (2011). Pregnane X receptor activation induces FGF19-dependent tumor aggressiveness in humans and mice. *The Journal of Clinical Investigation*, *121*, 3220–3232.
Wang, C., Yang, C., Chang, J. Y., You, P., Li, Y., Jin, C., et al. (2014). Hepatocyte FRS2alpha is essential for the endocrine fibroblast growth factor to limit the amplitude of bile acid production induced by prandial activity. *Current Molecular Medicine*, *14*, 703–711.
Weinshilboum, R. M., Otterness, D. M., Aksoy, I. A., Wood, T. C., Her, C., & Raftogianis, R. B. (1997). Sulfation and sulfotransferases 1: Sulfotransferase molecular biology: cDNAs and genes. *The FASEB Journal*, *11*, 3–14.
Wistuba, W., Gnewuch, C., Liebisch, G., Schmitz, G., & Langmann, T. (2007). Lithocholic acid induction of the FGF19 promoter in intestinal cells is mediated by PXR. *World Journal of Gastroenterology*, *13*, 4230–4235.
Xia, X., Francis, H., Glaser, S., Alpini, G., & LeSage, G. (2006). Bile acid interactions with cholangiocytes. *World Journal of Gastroenterology*, *12*, 3553–3563.
Yang, F., Huang, X., Yi, T., Yen, Y., Moore, D. D., & Huang, W. (2007). Spontaneous development of liver tumors in the absence of the bile acid receptor farnesoid X receptor. *Cancer Research*, *67*, 863–867.
Yeh, H. Z., Schteingart, C. D., Hagey, L. R., Ton-Nu, H. T., Bolder, U., Gavrilkina, M. A., et al. (1997). Effect of side chain length on biotransformation, hepatic transport, and choleretic properties of chenodeoxycholyl homologues in the rodent: Studies with dinorchenodeoxycholic acid, norchenodeoxycholic acid, and chenodeoxycholic acid. *Hepatology*, *26*, 374–385.
Yokoyama, Y., Nagino, M., & Nimura, Y. (2007). Mechanism of impaired hepatic regeneration in cholestatic liver. *Journal of Hepato-Biliary-Pancreatic Surgery*, *14*, 159–166.
Yoneno, K., Hisamatsu, T., Shimamura, K., Kamada, N., Ichikawa, R., Kitazume, M. T., et al. (2013). TGR5 signalling inhibits the production of pro-inflammatory cytokines by

in vitro differentiated inflammatory and intestinal macrophages in Crohn's disease. *Immunology*, *139*, 19–29.

Yoon, Y. B., Hagey, L. R., Hofmann, A. F., Gurantz, D., Michelotti, E. L., & Steinbach, J. H. (1986). Effect of side-chain shortening on the physiologic properties of bile acids: Hepatic transport and effect on biliary secretion of 23-nor-ursodeoxycholate in rodents. *Gastroenterology*, *90*, 837–852.

Zhang, M., & Chiang, J. Y. (2001). Transcriptional regulation of the human sterol 12a-hydroxylase gene (CYP8B1): Roles of hepatocyte nuclear factor 4a (HNF4a) in mediating bile acid repression. *The Journal of Biological Chemistry*, *276*, 41690–41699.

Zhang, Q., He, F., Kuruba, R., Gao, X., Wilson, A., Li, J., et al. (2008). FXR-mediated regulation of angiotensin type 2 receptor expression in vascular smooth muscle cells. *Cardiovascular Research*, *77*, 560–569.

Zhang, L., Huang, X., Meng, Z., Dong, B., Shiah, S., Moore, D. D., et al. (2009). Significance and mechanism of CYP7a1 gene regulation during the acute phase of liver regeneration. *Molecular Endocrinology*, *23*, 137–145.

Zhang, L., Wang, Y. D., Chen, W. D., Wang, X., Lou, G., Liu, N., et al. (2012). Promotion of liver regeneration/repair by farnesoid X receptor in both liver and intestine. *Hepatology*, *56*, 2336–2343.

Zhang, Y., Xu, P., Park, K., Choi, Y., Moore, D. D., & Wang, L. (2008). Orphan receptor small heterodimer partner suppresses tumorigenesis by modulating cyclin D1 expression and cellular proliferation. *Hepatology*, *48*, 289–298.

Zhu, Q. S., Xing, W., Qian, B., von Dippe, P., Shneider, B. L., Fox, V. L., et al. (2003). Inhibition of human m-epoxide hydrolase gene expression in a case of hypercholanemia. *Biochimica et Biophysica Acta*, *1638*, 208–216.

Zollner, G., & Trauner, M. (2008). Mechanisms of cholestasis. *Clinics in Liver Disease*, *12*, 1–26, vii.

CHAPTER TEN

Translational Implications of the Alcohol-Metabolizing Enzymes, Including Cytochrome P450-2E1, in Alcoholic and Nonalcoholic Liver Disease

Byoung-Joon Song*[,1], Mohammed Akbar*, Inho Jo[†], James P. Hardwick[‡], Mohamed A. Abdelmegeed*

*Section of Molecular Pharmacology and Toxicology, Laboratory of Membrane Biochemistry and Biophysics, National Institute on Alcohol Abuse and Alcoholism, Bethesda, Maryland, USA

[†]Department of Molecular Medicine, Ewha Womans University School of Medicine, Seoul, South Korea

[‡]Biochemistry and Molecular Pathology in Department of Integrative Medical Sciences, Northeast Ohio Medical University, Rootstown, Ohio, USA

[1]Corresponding author: e-mail address: bj.song@nih.gov

Contents

Abstract

Fat accumulation (hepatic steatosis) in alcoholic and nonalcoholic fatty liver disease is a potentially pathologic condition which can progress to steatohepatitis (inflammation), fibrosis, cirrhosis, and carcinogenesis. Many clinically used drugs or some alternative medicine compounds are also known to cause drug-induced liver injury, which can

Advances in Pharmacology, Volume 74
ISSN 1054-3589
http://dx.doi.org/10.1016/bs.apha.2015.04.002

further lead to fulminant liver failure and acute deaths in extreme cases. During liver disease process, certain cytochromes P450 such as the ethanol-inducible cytochrome P450-2E1 (CYP2E1) and CYP4A isozymes can be induced and/or activated by alcohol and/or high-fat diets and pathophysiological conditions such as fasting, obesity, and diabetes. Activation of these P450 isozymes, involved in the metabolism of ethanol, fatty acids, and various drugs, can produce reactive oxygen/nitrogen species directly and/or indirectly, contributing to oxidative modifications of DNA/RNA, proteins and lipids. In addition, aldehyde dehydrogenases including the mitochondrial low K_m aldehyde dehydrogenase-2 (ALDH2), responsible for the metabolism of acetaldehyde and lipid aldehydes, can be inactivated by various hepatotoxic agents. These highly reactive acetaldehyde and lipid peroxides, accumulated due to ALDH2 suppression, can interact with cellular macromolecules DNA/RNA, lipids, and proteins, leading to suppression of their normal function, contributing to DNA mutations, endoplasmic reticulum stress, mitochondrial dysfunction, steatosis, and cell death. In this chapter, we specifically review the roles of the alcohol-metabolizing enzymes including the alcohol dehydrogenase, ALDH2, CYP2E1, and other enzymes in promoting liver disease. We also discuss translational research opportunities with natural and/or synthetic antioxidants, which can prevent or delay the onset of inflammation and liver disease.

ABBREVIATIONS

4-HNE 4-hydroxynonenal
ACR acrolein
ADME absorption, distribution, metabolism, and excretion
AFLD alcoholic fatty liver disease
AGE advanced glycation end product
ALD alcoholic liver disease
ALDH1 cytosolic aldehyde dehydrogenase
ALDH2 mitochondrial low K_m aldehyde dehydrogenase 2
AMPK AMP-activated protein kinase
APAP acetaminophen
BAC blood alcohol concentration
CMZ chlormethiazole
CNS central nerve system
complex I NADH-dependent ubiquinone oxidoreductase
complex III ubiquinone cytochrome bc1 oxidoreductase
complex IV cytochrome *c* oxidase
complex V ATP synthase
CYP2E1 ethanol-inducible cytochrome P450-2E1 isozyme
DAMP damage-associated molecular pattern
DILI drug-induced liver injury
ER endoplasmic reticulum
ERAD endoplasmic reticulum-associated degradation
ETC electron transport chain
EtG ethyl glucuronide
FAEE fatty acid ethyl ester
GSH glutathione

HETE 20-hydroxyeicosatetraenoic acid
HFCS high fructose corn syrup
HFD high-fat diet
HIF hypoxia-inducible factor
Hsp70 heat-shock protein 70
I/R ischemia–reperfusion
iNOS inducible nitric oxide synthase
IRS insulin receptor substrate protein
JNK c-Jun-N-terminal protein kinase
MAA malondialdehyde-acetaldehyde adducts
MAPK mitogen-activated protein kinase
MCD methionine and choline-deficient diet
MDA malondialdehyde
MDMA 3,4-methylenedioxymethamphetamine
MEOS microsomal ethanol-oxidizing enzyme system
Mito-CP mitochondria-targeted carboxy-proxyl
Mito-Q mitochondria-targeted ubiquinone
NAC *N*-acetylcysteine
NAFLD nonalcoholic fatty liver disease
NALD nonalcoholic liver disease
NASH nonalcoholic steatohepatitis
NF-κB nuclear factor-κB
NO nitric oxide
Nrf2 nuclear factor (erythroid-derived 2)-like 2
p38K p38 protein kinase
PAMP pathogen-associated molecular pattern
PGC-1α peroxisomal proliferator-activated receptor gamma coactivator-1α
PKC protein kinase C
PPARα peroxisome proliferator-activated receptor alpha
RNS reactive nitrogen species
ROS reactive oxygen species
SAMe *S*-adenosyl methionine
SOD superoxide dismutase
SREBP sterol-regulated element-binding protein
TCA tricarboxylic acid
UPR unfolded protein response
WT wild type
XDH xanthine dehydrogenase
XO xanthine oxidase

1. INTRODUCTION

Alcoholic and nonalcoholic fatty liver disease (AFLD and NAFLD, respectively) are major causes of morbidity and mortality in the world. Upon intake of alcohol or nonalcoholic substances such as high-fat diet (HFD) or soft drinks containing high fructose corn syrup (HFCS), triglycerides accumulate in the hepatocytes, leading to the development of fatty liver (hepatic steatosis), which is a reversible condition. Following continuous exposure or intake of these substances, hepatic steatosis can progress to inflammatory steatohepatitis, fibrosis, cirrhosis, and even hepatocellular carcinoma. In general, progression of benign fatty liver disease to more severe liver disease directly correlates with the amount, frequency of intake, and duration of stressors (e.g., alcohol or high fat) (Zakhari & Li, 2007). In addition, it is known that there are relatively large individual variations in the rate of ethanol elimination, possibly due to genetic and environmental factors (Li, Yin, Crabb, O'Connor, & Ramchandani, 2001). Furthermore, the liver disease progression can be exacerbated or facilitated especially in the presence of other comorbidity risk factors (Lieber, 2004a), such as hepatitis B or C virus (Mueller, Millonig, & Seitz, 2009; Otani et al., 2005; Rigamonti et al., 2003; Szabo, Saha, & Bukong, 2015; Szabo et al., 2010; Zakhari, 2013), HIV (Fan, Joshi, Koval, & Guidot, 2011; Persidsky et al., 2011), obesity (Cederbaum, 2012a; Hart, Morrison, Batty, Mitchell, & Davey, 2010; Loomba et al., 2013, 2010), diabetes (Hassan et al., 2002), smoking (Kuper et al., 2000; Purohit, Rapaka, Kwon, & Song, 2013; Salaspuro & Salaspuro, 2004), clinically used drugs (Boelsterli & Lee, 2014; McClain, Kromhout, Peterson, & Holtzman, 1980; Seeff, Cuccherini, Zimmerman, Adler, & Benjamin, 1986), or environmental contaminants such as benzene in gasoline (Kalf, Post, & Snyder, 1987). For instance, people who drink more than 60 g/day are more likely to develop fibrosis, cirrhosis, hepatocellular carcinoma, and ultimately liver failure (Lucey, Mathurin, & Morgan, 2009; Stickel & Seitz, 2010). In addition, simultaneous exposure to alcohol and one or two of these risk factors significantly increase the severity of liver disease with elevated morbidity and mortality (Neuman et al., 2014). In this review, we briefly describe the mechanisms of various types of liver disease caused by alcohol (ethanol), HFD, or other potentially hepatotoxic substances and the enzymes involved in the alcohol metabolism in promoting liver disease. Finally, we also describe translational research opportunities in preventing or treating various forms of liver disease.

2. PATHOLOGICAL MECHANISMS OF LIVER DISEASES

Excessive chronic intake of alcohol can promote alcoholic fatty liver, alcoholic steatohepatitis, fibrosis, cirrhosis, and hepatocarcinogenesis

(Gao & Bataller, 2011; Purohit, Gao, & Song, 2009; Szabo & Lippai, 2012). However, it is known that approximately 10–15% of steatotic individuals can develop more severe liver disease such as steatohepatitis and fibrosis/cirrhosis, indicating a requirement of secondary and/or tertiary hits (risk factors) for the progression (Day & James, 1998). In addition, 30–40% of all hepatic cancers are reported to be associated with chronic alcohol drinking (Morgan, Mandayam, & Jamal, 2004). Excessive amounts of alcohol intake, nonalcoholic substances such as high fat and HFCS-containing soft drinks, and potentially hepatotoxic agents including some dietary supplements can cause acute and chronic liver diseases, as reported (Jaeschke et al., 2002; McClain et al., 2004; McGill & Jaeschke, 2014; Navarro et al., 2014; Seth et al., 2013; Spruss et al., 2009; Stickel, Kessebohm, Weimann, & Seitz, 2011; Stickel & Shouval, 2015; Vos & Lavine, 2013). One of the common mechanisms by which hepatotoxic agents initiate pathophysiological conditions such as hypoxia-reoxygenation injury, obesity, and diabetes is by increasing oxidative/nitrosative/nitrative (i.e., nitroxidative) stress. It is well established that increased nitroxidative stress can be produced from impaired mitochondrial function (i.e., mitochondrial dysfunction), elevated levels of ethanol-inducible cytochrome P450-2E1 (CYP2E1) and other CYP isozymes, NADPH oxidases, inducible form of nitric oxide synthase (iNOS) and xanthine oxidase (XO) (Aubert, Begriche, Knockarert, Robin, & Fromenty, 2011; Lieber, 2004b; Spruss, Kanuri, Uebel, Bischoff, & Bergheim, 2011). On the other hand, the cellular levels of small-molecule antioxidants, such as glutathione (GSH), many vitamins including retinoic acid (vitamin A), thiamine (vitamin B1), ascorbic acid (vitamin C), and α-tocopherol (vitamin E), are known to be decreased following alcohol exposure (Cederbaum, 2012a; Leung & Nieto, 2013; Lieber, 1997; Nanji et al., 2003; Xiao et al., 2013) and/or under different pathophysiological conditions such as hypoxic liver injury and aging (Marí, Morales, Colell, García-Ruiz, & Fernández-Checa, 2009). Furthermore, recent data has shown that the activities of antioxidant enzymes such as glutathione peroxidase (Gpx), catalase, and superoxide dismutase (SOD) can be inhibited by alcohol and nonalcoholic substances (Abdelmegeed, Jang, Banerjee, Hardwick, & Song, 2013; Abdelmegeed, Moon, Chen, Gonzalez, & Song, 2010; Abdelmegeed & Song, 2014; Carmiel-Haggai, Cederbaum, & Nieto, 2005; Song et al., 2013). As a result of increased nitroxidative stress, many cellular macromolecules are covalently modified and these modifications including lipid peroxides can activate Kupffer cells, the liver resident immune cells, sinusoidal endothelial cells, and hepatic stellate cells to release inflammatory cytokines and profibrotic substances. Consequently, these changes elevate the ratio of pro-oxidants over antioxidant molecules, contributing to increased oxidative stress. In addition, iNOS, which produces nitric oxide (NO) at micromolar ranges (Dai et al., 2013), can be induced in AFLD and NAFLD, at least partly through the activation of a redox-sensitive transcription factor NF-κB, which also transcribes other downstream targets, i.e., TNFα, monocyte chemotactic protein-1 (MCP-1), and cyclooxygenase-2, involved in the initiation of inflammation (Nanji et al., 2003). The simultaneous production of reactive oxygen and nitrogen species (ROS/RNS) leads to production of more toxic peroxynitrite, which can nitrate Tyr residues as well as *S*-nitrosylate Cys residues of many target proteins (Abdelmegeed & Song, 2014; Song et al., 2013). Elevated nitroxidative stress can cause various types of post-translational modification (PTM) of some critical proteins in the endoplasmic reticulum (ER) and mitochondrial proteins accompanied by functional alterations and/or activity changes. These changes contribute to ER stress, resulting in unfolded protein responses (UPR) (Walter & Ron, 2011) and mitochondrial dysfunction with fat accumulation (Moon et al., 2006), respectively. In addition, increased nitroxidative stress can activate the cell-death-related protein kinases, such as c-Jun-N-terminal protein kinase (JNK) and p38 kinase (p38K), both of which can phosphorylate proapoptotic Bax and other substrate proteins to stimulate necroapoptotic cell death (Cederbaum, Lu, Wang, & Wu, 2015; Cnop, Foufelle, & Velloso, 2012; Gentile, Frye, & Pagliassotti, 2011; Hetz, Chevet, & Harding, 2013; Jaeschke, McGill, & Ramachandran, 2012; Kim, Ryu, & Song, 2006; Saberi et al., 2014; Seki, Brenner, & Karin, 2012; Song, Akbar, et al., 2014; Wang, 2014; Wu & Cederbaum, 2013).

Chronic alcohol intake is known to cause ER stress in the liver and many other tissues with elevation of three ER stress membrane proteins such as inositol-requiring enzyme-1 (IRE1α), transcription factor-6 (ATF6), and PKR-like eukaryotic initiation factor 2α kinase (PERK) (Chen, Budas, et al., 2008; Chen, Ma, et al., 2008; Ji, 2008; Kaplowitz & Ji, 2006; Malhi & Kaufman, 2014). HFD and pathological conditions, such as obesity and diabetes, are also known to be associated with elevated ER stress with accumulation of unfolded or misfolded proteins as well as aggregated proteins (Lee, Jeong, et al., 2014; Liu, Fan, Tang, & Ke, 2014; Lu et al., 2015; Ramirez et al., 2013). Under stressful conditions, ER-associated degradation (ERAD), a quality control system in ER, cannot remove all the unfolded proteins. Consequently, the cell undergoes an adaptive signaling process of UPR, which mainly promotes transcription of chaperone proteins with increased ERAD to efficiently remove damaged proteins. Petersen and other scientists reported that the main culprit for increased ER stress in alcohol-exposed rodents could be increased nitroxidative stress with decreased GSH levels and cystathionine β-synthase activity with elevated levels of homocysteine and protein glutathionylation (Barak, Beckenhauer, Kharbanda, & Tuma, 2001; Galligan, Smathers, Shearn, et al., 2012). Furthermore, the PERK, IRE1α, ATF6, and sterol-regulated element-binding protein (SREBP) pathways do not seem to play significant roles for the UPR pathways with ER stress and steatosis observed in alcohol-exposed mice. Consistent with the causal role of increased nitroxidative stress in ER stress (Galligan, Smathers, Shearn, et al., 2012), we had detected oxidatively modified ER proteins such as glucose-regulated proteins (e.g., Grp78, Grp75), heat-shock proteins (e.g., Hsp71, Hsp70, Hsp60), and protein disulfide isomerase (PDI) in alcohol-exposed hepatoma cells and rat livers (Kim et al., 2006; Moon et al., 2006; Suh et al., 2004). Similarly, oxidized chaperone proteins were also observed in 3,4-methylenedioxymethamphetamine (MDMA)-exposed rats (Moon, Upreti, et al., 2008) or mice following ischemia–reperfusion (I/R) injury (Moon, Hood, et al., 2008). Petersen and colleagues also reported that alcohol suppressed the activity of triacylglycerol hydrolase (carboxylesterase 3) via decreased glycosylation, contributing to fat (triglyceride) accumulation in alcohol-treated rodents (Galligan, Fritz, Tipney, et al., 2012). We believe that increased nitroxidative stress can promote various PTMs of ER-resident chaperone proteins, as recently reviewed (Song, Akbar, et al., 2014). Oxidative inactivation of these chaperone proteins is likely to result in the accumulation of unfolded/misfolded proteins of their client proteins, contributing to the UPR and ER stress, which stimulates fibrogenesis in hepatic stellate cells (Hernández-Gea et al., 2013) or cell death (Sano & Reed, 2013).

In addition to elevated ER stress, chronic and binge alcohol intake can cause mitochondrial dysfunction, leading to decreased energy supply, partly through direct suppression of the mitochondrial complexes in the electron transport chain (ETC) with decreased mitochondrial membrane potential (Bailey & Cunningham, 1999; Feldstein & Bailey, 2011; Hoek, Cahill, & Pastorino, 2002). In addition, alcohol-mediated redox change with decreased NAD^+ levels can interfere with the activities of various

NAD^+-dependent dehydrogenases in the mitochondria, leading to mitochondrial dysfunction (Cederbaum, 2012b; Lieber, 1997). In fact, mitochondrial dysfunction can be observed in experimental rodents exposed to alcohol (Moon et al., 2006), MDMA (Moon, Upreti, et al., 2008), and acetaminophen (APAP) (Abdelmegeed, Jang, et al., 2013; Abdelmegeed et al., 2010). Impaired mitochondrial function were also reported in pathological conditions such as obesity, diabetes (Dey & Swaminathan, 2010; Paradies, Paradies, Ruggiero, & Petrosillo, 2014), and I/R injury (Moon, Hood, et al., 2008) as well as in patients who consumed alcohol (Addolorato et al., 1998; Fromenty et al., 1995; Witschi, Mossi, Meyer, Junker, & Lauterburg, 1994) and NAFLD/NASH patients (Caldwell et al., 1999; Kojima et al., 2007; Sanyal et al., 2001). Increased nitroxidative stress can also serve as a major cause for the mitochondrial dysfunction in these cases. We and other laboratories consistently showed that many mitochondrial proteins were modified by different types of PTMs (e.g., oxidation, nitration, phosphorylation, acetylation, 4-HNE adduct formation, and others), as summarized (Fritz & Petersen, 2013; Song, Akbar, et al., 2014; Song et al., 2013). Oxidative inactivation of the modified mitochondrial proteins is likely to cause mitochondrial dysfunction, which can stimulate ER stress (Kozlov et al., 2009).

Along with stimulation of ER stress and mitochondrial dysfunction through PTMs, increased nitroxidative stress can directly and indirectly activate the apoptosis signaling pathways, contributing to cell death. For instance, binge alcohol can activate the cell-death-associated mitogen-activated protein kinases (MAPKs), including JNK and p38K (Brenner, Galluzzi, Kepp, & Kroemer, 2013; Lee & Shukla, 2005). In addition, increased nitroxidative stress in HFD and diabetic conditions or APAP can activate or alter JNK and other protein kinases such as PKCα, which stimulates JNK in a feed-forward manner (Saberi et al., 2014), contributing to insulin resistance and acute cell death. In fact, *JNK*-null mice or *iNOS*-null mice were resistant to alcohol- or HFD-induced insulin resistance and apoptosis. Specific deletion or suppression of JNK1 or iNOS, with either using a specific siRNA or an inhibitor, or in knockout mouse strain, can also lead to resistance to HFD-induced insulin resistance and hepatocytes death or drug-induced liver injury (DILI) (Charbonneau & Marette, 2010; Fujimoto et al., 2005; He et al., 2015; Kamanaka et al., 2003; Saberi et al., 2014; Seki et al., 2012; Singh et al., 2009; Song, Fu, Xia, Su, & Song, 2014; Tipoe et al., 2006; Zhang et al., 2011). The opposite case with overexpression of these pro-oxidant proteins would increase the tissue sensitivity toward apoptotic death. In many cases, elevated CYP2E1, NADPH oxidase, and iNOS play critical roles in causing increased nitroxidative stress, ER stress, mitochondrial dysfunction, and cell death (Fig. 1), as described later in this review.

Acetate, the final product of the oxidative alcohol metabolism, can be further converted to acetyl-CoA by acetyl-CoA synthase, before it is oxidized to carbon dioxide (CO_2) in the tricarboxylic acid (TCA) cycle. In addition, it can be transported to other extra-hepatic tissues such as heart,

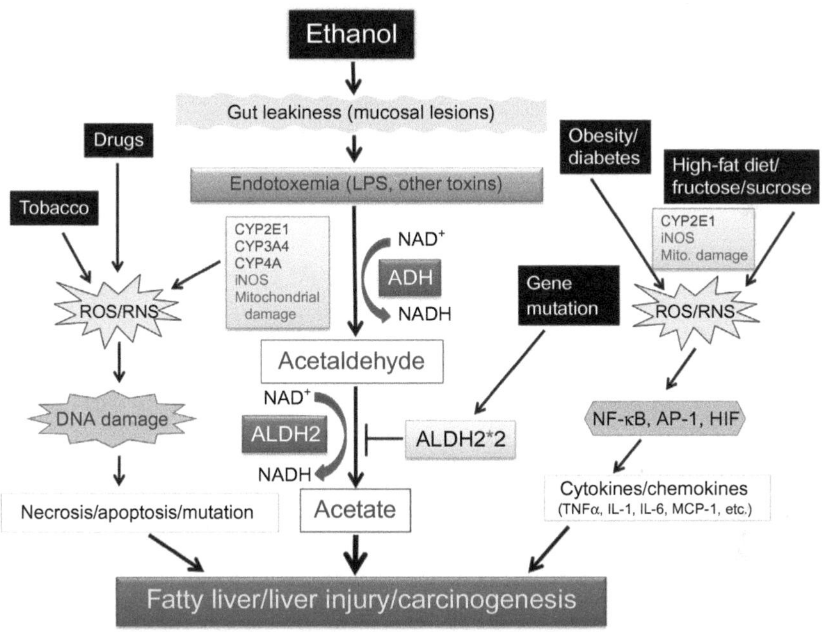

Figure 1 Schematic diagram for the pathological mechanisms of AFLD, NAFLD, DILI, and carcinogenesis. Known risk factors for various liver diseases are listed. Alcohol, HFD, fructose/sucrose, drugs, tobacco smoking, obesity, diabetes, and genetic polymorphisms with or without gut leakiness can increase cellular nitroxidative stress. Oxidative metabolism of alcohol and acetaldehyde by ADH and ALDH2, respectively, also causes a redox change with a decreased NAD^+/NADH ratio. These changes in the redox state with elevated nitroxidative stress can promote DNA mutations, lipid peroxidation, and protein modifications, leading to ER stress, mitochondrial dysfunction, and necroapoptosis of hepatocytes. Activation of a few redox-sensitive transcription factors such as NF-κB and HIF can increase the levels of proinflammatory cytokines and chemokines, which can further activate liver Kupffer and stellate cells, aggravating the liver disease conditions.

skeletal muscle, and brain to exert its effects, as reported (Pawlosky et al., 2010; Zakhari, 2013). In the peripheral tissues, alcohol or acetate can increase the portal blood flow into the liver, possibly through stimulating intestinal blood flow (Israel, Orrego, & Carmichael, 1994). The elevated intestinal and portal blood flow can be mediated by adenosine, since treatment with 8-phenyltheophylline, an adenosine receptor blocker, significantly blocked the alcohol- or acetate-mediated changes. In the heart, which depends on fatty acids as the major energy source, acetate can be converted to acetyl-CoA, which can be used as an alternative energy source (Kodde, van der Stok, Smolenski, & de Jong, 2007), especially after long-term alcohol ingestion (Lukivskaya & Buko, 1993) or under metabolically stressful conditions such as lipid-depleted cancer cells with low oxygen supply (Schug et al., 2015). However, acetate can suppress the brain function by decreasing glucose metabolism in the brains of rats (Pawlosky et al., 2010) and humans (Volkow et al., 2006), while acute alcohol intake increases acetate uptake in human brains (Volkow et al., 2013).

3. ENZYMES INVOLVED IN THE ALCOHOL METABOLISM

Alcohol is a water-soluble substance and thus can be easily distributed to virtually every organ in the body. Small amounts of alcohol intake can stimulate the central nerve system (CNS) with mood-enhancing euphoria (pleasant feeling), psychological relaxation, and outgoing behaviors. In fact, alcohol at low and moderate doses can exert many beneficial health effects such as increasing appetite with cardiovascular protection and neuroprotection (Collins et al., 2009; Gunzerath, Faden, Zakhari, & Warren, 2004; Katsiki, Tziomalos, & Mikhailidis, 2014). These beneficial effects of alcohol seem to be mediated by activation of protein kinase C epsilon (Chen, Gray, & Mochly-Rosen, 1999) and its translocation to mitochondria to activate ALDH2 through phosphorylation (Churchill, Disatnik, & Mochly-Rosen, 2009).

However, alcohol is an addictive substance. Habitual alcohol drinking can actually lead to physical and psychological dependency with alcoholism and alcohol abuse. Alcohol addiction alone, with or without other abused substances or other risk factors, can cause significant sociomedical problems to the alcoholic individuals, families, and societies. Consumption of excessive amounts of alcohol within a short period of time (i.e., binge alcohol) can suppress the CNS, resulting in impaired motor control and improper judgment. The decreased CNS function can lead to many unwarranted sociopathic behaviors such as motor vehicle-related accidents and alcohol-related violent criminal behaviors including physical and verbal attack, rape, homicide, and arson (Beck & Heinz, 2013). Volkow et al. (2008) reported that even moderate doses of alcohol can significantly disrupt the functional organization, which is accompanied with motor, cognition, behavior, and mood changes in humans. Long-term excessive amounts of alcohol intake can cause various medical problems with tissue damage to many organs including liver fibrosis/cirrhosis and carcinogenesis with alterations of endocrine and immune functions (Badger et al., 2003; Cederbaum, Lu, & Wu, 2009; Hoek et al., 2004; Lieber, 1997). The majority of these medical consequences are associated with the metabolism of alcohol, as briefly described below. Similar adverse health problems can be also observed with excessive intake of western HFD with high salt or soft drinks containing HFCS. In fact, these metabolic syndromes in the liver, heart, and other peripheral tissues can be frequently reported in experimental models and obese and diabetic people (Adkins et al., 2013; Marseglia et al., 2014; Mells et al., 2015).

Small amounts of alcohol can be cleared from the body through breath (lung), urine (kidney), and sweat (skin). However, more than 90% of alcohol consumed is oxidatively metabolized in the liver and stomach by cytosolic alcohol dehydrogenase (ADH; E.C. 1.1.1.1) to acetaldehyde, which is further metabolized to acetate by mitochondrial aldehyde dehydrogenase (ALDH2; E.C. 1.2.1.3) with a very low K_m toward acetaldehyde (Lieber, 1997, 2005; Zakhari & Li, 2007). After long-term alcohol exposure, large amounts of alcohol can also be metabolized through the microsomal ethanol-oxidizing system (MEOS) consisting of cytochrome P450 isozymes such as CYP2E1, CYP1A2, and CYP3A (Cederbaum, 2012b; Lieber, 2005). Alcohol can also be nonoxidatively metabolized via conjugation with fatty acids and fatty acyl-CoAs to produce fatty acid ethyl esters (FAEEs), which can be used as a marker for alcohol drinking (Deng & Deitrich, 2007; Laposata & Lange, 1986; Soderberg, Salem, Best, Cluette-Brown, & Laposata, 2003).

3.1 Role and Regulation of ADH Isozymes in Liver Disease

Ingested alcohol can be quickly transported from stomach to duodenum to be absorbed into blood stream, before being circulated to the liver for hepatic metabolism. Absorption, distribution, metabolism, and excretion (ADME) of alcohol can be influenced by genetic polymorphisms and ethnic backgrounds. In addition, the ADME of alcohol can be affected by gender, age, nutritional status, compositions of different diets (e.g., contents and composition of foods and fatty acids), biological clocks (e.g., day vs. night), and drugs or smoking, resulting in three- to fourfold differences among different individuals (Cederbaum, 2012b). Unlike many drugs and xenobiotic substances, which are usually excreted via the first-order kinetics in a concentration-dependent manner, alcohol is cleared from the body at a constant rate via pseudolinear near-zero-order kinetics, suggesting its removal in a concentration-independent manner (Cederbaum, 2012b; Lee, Liao, et al., 2013). There are many ADH isozymes expressed in a tissue-specific manner. The expression and activity of each ADH isozyme are also dependent on genetic polymorphisms (Agarwal, 2001; Chen, Peng, Wang, Tsao, & Yin, 2009). However, so far, the genetic polymorphisms of ADH isozymes do not appear to be related to alcoholic liver disease (ALD) or alcohol drinking pattern (Cederbaum, 2012b). In addition, each class of ADH isozyme exhibits different catalytic activities in the metabolism of alcohol (K_m for ADH is about 0.8–1 mM) or other substrates, including fatty alcohols (Agarwal, 2001; Cederbaum, 2012b; Lieber, 2005). For instance, orally consumed alcohol can be metabolized in the stomach by the classes I, III, and IV ADHs, although overall ADH activity in the stomach is smaller than that in the liver, partly due to the levels of their expression and catalytic efficiency. The ADH isozymes (a dimer with a 40 kDa monomer) are also responsible for the metabolism of a wide variety of substrates such as ethanol (including alcohol endogenously produced by gut bacteria), retinol, aliphatic alcohols, hydroxysteroids, and lipid aldehydes. The oxidized adenine dinucleotide (NAD^+) is needed as a cofactor for the ADH-mediated metabolism of these substrates. The oxidative metabolism of different alcohols by the ADH enzymes results in the production of their corresponding aldehydes and a reducing equivalent NADH, which can interfere with the activities of many NAD^+-dependent dehydrogenases in the liver (Ceni, Mello, & Galli, 2014). Because of the usage of NAD^+ preferentially for alcohol metabolism, other NAD^+-requiring activities are not properly executed (Lieber, 1997, 2005). For instance, some mitochondrial NAD^+-dependent dehydrogenases, involved in the fatty acid β-oxidation pathway (e.g., acyl-CoA dehydrogenase) or the TCA cycle (e.g., pyruvate dehydrogenase), can be inhibited. Decreased NAD^+/NADH can also suppress the activities of cytosolic NAD^+-dependent dehydrogenases, including glyceraldehyde 3-phosphate dehydrogenase (GAPDH) involved in the glycolysis for the breakdown of glucose. In fact, during oxidative stress, the active site cysteine of GAPDH can be oxidized to sulfenic acid, leading to decreased ATP levels (Cremers & Jakob, 2013). Oxidative inactivation of GAPDH can divert glucose from the glycolysis to the pentose monophosphate shunt, which generates NADPH, needed for the thioredoxin and glutaredoxin systems as well as glutathione reductase. NADPH can also serve as a cofactor for fatty acid synthesis. The alcohol-mediated redox change can also interfere with the gluconeogenesis process. Collectively, the suppression of these dehydrogenases is likely to decrease fat and carbohydrate metabolism with fat accumulation and decreased energy supply, respectively, as observed in AFLD and NAFLD.

The liver is the major organ for the ethanol metabolism. In fact, more than 90% of ingested alcohol is metabolized by the oxidative and nonoxidative pathways in the liver (Agarwal, 2001; Lieber, 2005). During ethanol metabolism mainly by ADH isozymes, reactive acetaldehyde is produced with a redox change (i.e., a decreased ratio of NAD^+/NADH). It is known that multiple ADH isozymes are expressed in the liver and other tissues. These ADH isozymes have distinct affinities (different K_m values for ethanol) and catalytic efficiencies (i.e., K_{cat}/K_m values) in the ethanol metabolism (Cederbaum, 2012b; Zakhari & Li, 2007). However, class I ADH is thought to be the major enzyme for the hepatic metabolism of ethanol. The ADH reaction is reversible, depending on the amounts of products—acetaldehyde and NADH. However, NADH-reoxidation step seems to be the rate-limiting process in alcohol metabolism (Israel, Khanna, & Lin, 1970). Decreased NAD^+/NADH is known to cause fat accumulation through increased NADPH level by mitochondrial transhydrogenase-mediated conversion of NADH to NADPH and thus promotes *de novo* fat synthesis in the cytoplasm and fat transport from adipose tissues. The activities of ADH isozymes are known to fluctuate, depending on the status of nutrition, growth hormone, epinephrine, and estrogens. However, the direct effects of alcohol or high fat on ADH activities have not been clearly elucidated, although a few reports suggested slight but significant suppression of ADH in alcohol-exposed models and individuals (Lieber, 2005; Salaspuro, Shaw, Jayatilleke, Ross, & Lieber, 1981; Ugarte, Pino, & Insunza, 1967). It is unclear whether ADH isozymes can be oxidatively modified under elevated nitroxidative stress in alcohol-exposed models and/or pathophysiological conditions since various ADH isozymes contain a highly conserved zinc-binding site with one histidine and three or four cysteine residues (Auld & Bergman, 2008). This area needs further studies.

The redox change with decreased NAD^+/NADH following alcohol metabolism can also negatively interfere with the activities of

NAD^+-dependent deacetylases such as cytosolic sirtuins 1 and 2 as well as mitochondrial sirtuin 3 and sirtuin 5, involved in cellular aging, lipid metabolism, and antioxidant defense. In fact, chronic alcohol consumption is known to suppress the activities and/or levels of sirtuins 1 and 3 (Lieber, Leo, Wang, & Decarli, 2008a, 2008b; You, Liang, Ajmo, & Ness, 2008) in a CYP2E1-independent manner (Picklo, 2008). Alcohol-mediated suppression of cytosolic and mitochondrial sirtuin proteins likely leads to elevated levels of many acetylated proteins in the liver and other tissues. Some of the hyper-acetylated proteins are SREBP and peroxisome proliferator-activated receptor gamma coactivator-1α (PGC-1α), involved in the fat synthesis and metabolism, respectively. Since acetylation of histone and other nuclear proteins is associated with epigenetic expression, alcohol-related decreased NAD^+/NADH ratio becomes important in liver disease, as reviewed (Shukla & Lim, 2013; Shukla et al., 2008). In addition, it is known that sirtuin 1 can be phosphorylated at Ser-46 by JNK, leading to ubiquitin-dependent degradation and fat accumulation in obese mice (Gao et al., 2011). The activities of mitochondrial sirtuin 3 and other isoforms were also suppressed by HFD (Valdecantos et al., 2012), thus increasing the number of acetylated mitochondrial proteins involved in the fatty acid metabolism, as similar to those hyper-acetylated proteins observed in *sirtuin 3*-null mice (Hirschey et al., 2011). In contrast, sirtuins can be activated by small-molecule polyphenols including resveratrol (Li, Wong, et al., 2014; Li, Zhao, et al., 2014; Yang & Lim, 2014) and green tea extracts (Wang, Moustaid-Moussa, et al., 2014), thereby decreasing the number of acetylated proteins, including PGC-1α. These changes can contribute to increased fat oxidation with improved insulin sensitivity and better outcome of the metabolic syndrome.

3.2 Role and Regulation of ALDH2 in Liver Disease

The ADH-mediated ethanol metabolism produces acetaldehyde, which is further metabolized to acetate by mitochondrial ALDH2 (homotetrameric enzyme with ~54 kDa monomer) and other ALDH isozymes. The blood alcohol concentration (BAC) (10–100 mM range in some alcoholics) is usually greater than those of acetaldehyde (10–100 μM range), although some exceptions exist. These results suggest that the catalytic activities of ALDH2 and isoforms, if any involved, are greater than those of ADH isozymes (Cederbaum, 2012b). However, markedly elevated amounts of acetaldehyde can be accumulated after inhibition of ALDH2 with its chemical inhibitors such as disulfiram (Antabuse®) and cyanamide. Alternatively, acetaldehyde can be elevated after suppression of mitochondrial ALDH2 through various forms of PTM following exposure to many hepatotoxic agents such as alcohol (Moon et al., 2006; Venkatraman, Landar, Davis, Ulasova, et al., 2004), APAP, high fat, anticancer drugs, toxic substances, and other pathological conditions, as reviewed (Song, Akbar, et al., 2014; Song et al., 2013, 2011). In fact, accumulated NADH and acetaldehyde caused by using either disulfiram (Antabuse®) or cyanamide caused fat accumulation in alcohol-exposed hepatoma cells H4EIIE cells or rodents (Kato, Kawase, Alderman, Inatomi, & Lieber, 1990; You, Fischer, Deeg, & Crabb, 2002; You, Matsumoto, Pacold, Cho, & Crabb, 2004). Mitochondrial ALDH2 is the major enzyme responsible for acetaldehyde metabolism in humans, although cytosolic ALDH1 may also be involved in rodents, due to its relatively low K_m value (11–18 μM for acetaldehyde) in comparison to that of human counterpart (>180 μM) (Klyosov, Rashkovetsky, Tahir, & Keung, 1996). In addition, ALDH2 is known to metabolize many other aldehydes including lipid peroxides which are produced from the degradation of numerous fatty acids. Inactivation of ALDH2 and its isozymes is likely to cause accumulation of toxic carbonyl compounds such as acetaldehyde, 4-hydroxynonenal (4-HNE), malondialdehyde (MDA), and acrolein (ACR), all of which can promote protein adduct formation and necroapoptotic cell death while activating the immune cells and hepatic stellate cells, further contributing to the development of fibrosis in the liver and other tissues (Fritz & Petersen, 2013; Lieber, 1997; Mottaran et al., 2002; Sutti, Rigamonti, Vidali, & Albano, 2014). One report suggested that acetaldehyde can be responsible for the onset and maintenance of fibrogenesis (Mello, Ceni, Surrenti, & Galli, 2008).

Under increased nitroxidative stress as observed in alcohol- or high-fat-exposed rodents and humans, the ALDH2 protein can undergo various forms of PTMs with suppressed activity, although its phosphorylation by PKCε or phosphatidylinositol-3-kinase can increase ALDH2 activity, as recently reviewed (Song et al., 2011). These PTMs include oxidation, disulfide formation, S-nitrosylation, nitration, phosphorylation, acetylation, carbonylation, protein adduct formation, and many others (Galligan, Smathers, Fritz, et al., 2012; Song, Akbar, et al., 2014). For instance, mitochondrial ALDH2 could be inactivated through various PTMs in experimental models exposed to alcohol (Doorn, Hurley, & Petersen, 2006; Moon et al., 2006; Venkatraman, Landar, Davis, Ulasova, et al., 2004) and other potentially toxic substances (Abdelmegeed, Jang, et al., 2013; Banfi et al., 1994;

Landin, Cohen, & Khairallah, 1996; Lee, Liao, et al., 2013; Mali et al., 2014; Méndez et al., 2014; Mitchell & Petersen, 1988; Moon, Kim, & Song, 2005; Moon, Lee, & Song, 2010; Moon, Upreti, et al., 2008). Decreased activity could result from decreased ALDH2 protein levels, as reported in alcohol-exposed rats (Venkatraman, Landar, Davis, Chamlee, et al., 2004) and high-fat-exposed mice (Eccleston et al., 2011). Suppressed ALDH2 activity can also be observed, as demonstrated in a few pathological conditions such as partially hepatotectomized rodents (Watanabe et al., 1985), I/R injury (Moon, Hood, et al., 2008), and cancer tissues (Kim et al., 2002; Oshita et al., 2010; Park, Cho, Kim, & Paik, 2002). Following ALDH2 suppression, the serum and hepatic levels of lipid peroxides such as acetaldehyde, 4-HNE, MDA, and malondialdehyde-acetaldehyde adducts (MAA) were markedly elevated, as shown in alcohol-exposed monkeys (Pawlosky, Flynn, & Salem, 1997). These results are consistent with elevated levels of acetaldehyde in alcohol-exposed *Aldh2*-null mice (Isse, Matsuno, Oyama, Kitagawa, & Kawamoto, 2005; Kwon et al., 2014), UChA rats containing *Aldh2* mutant genes (Quintanilla, Israel, Sapag, & Tampier, 2006; Quintanilla, Tampier, Sapag, & Israel, 2005), and rodents treated with disulfiram or cyanamide (Kato et al., 1990). Furthermore, elevated levels of reactive lipid peroxides were also observed in HFD-exposed animals (Abdelmegeed et al., 2012, 2011). These reactive carbonyl compounds can then interact with cellular macromolecules such as DNA and proteins, leading to their modifications with altered functions and apoptosis of the target cells. If the mutated DNA is not properly handled or removed by the repair enzymes or by autophagy, persistently elevated modified DNA can lead to DNA mutation and ultimately cancer, as observed in experimental models and humans exposed to alcohol and/or other hepatotoxic substances (Brooks, Enoch, Goldman, Li, & Yokoyama, 2009). Furthermore, mitochondrial DNA is known to be more sensitive to oxidative damage than the nuclear DNA possibly due to the absence or low levels of histone and DNA repair enzymes in the mitochondria, leading to deletion of mitochondrial DNA, as demonstrated with alcoholic individuals (Fromenty et al., 1995).

Because of the importance in intermediary metabolism and cellular defense (Alary, Gueraud, & Cravedi, 2003; Hartley, Ruth, & Petersen, 1995; Jin & Penning, 2007), tissue-specific expression, substrate specificity, biochemical characteristics, and functional role of each ALDH isoform including ADLH2 have been extensively studied (Marchitti, Brocker, Stagos, & Vasiliou, 2008). Large amounts of ALDH2 and other ALDH isoforms are generally expressed in the liver. A single nucleotide mutation (G to A nucleotide substitution) in human *ALDH2* gene can result in conversion of Glu487 to Lys487 with a dominant inactivation of ALDH2 activity through a markedly decreased NAD^+-binding affinity (Farres et al., 1994; Sheikh, Ni, Hurley, & Weiner, 1997; Yoshida, Huang, & Ikawa, 1984). The frequency of individuals with *ALDH2*2* mutant allele is abundant (30–50%) in East Asians. Individuals with either heterozygous or homozygous *ALDH2*2* mutant allele showed a markedly reduced ALDH2 activity with "flushing responses" such as red face, sweating, and uncomfortable feeling with difficulty in breathing and aberrant heart rates possibly due to elevated acetaldehyde upon alcohol drinking (Peng, Chen, Wang, Lai, & Yin, 2014). Therefore, the *ALDH2*2* variant is considered as a protective allele against alcoholism and serious tissue injury (Bosron, Ehrig, & Li, 1993; Day et al., 1991; Li, 2000). In fact, not a single individual with *ALDH2*2/2*2* homozygous alleles was identified after genetic screening of more than 1300 Japanese alcoholic individuals (Higuchi et al., 1994), further supporting the protective allele of *ALDH2*2* against alcoholism. However, if these individuals, with the heterozygous or homozygous *ALDH2*2* mutant allele, continue drinking alcohol despite uncomfortable feeling, they become more susceptible to alcohol- and acetaldehyde-related tissue injury and carcinogenesis especially in the oral-esophageal-gastrointestinal track (Brooks et al., 2009; Seitz & Cho, 2009; Setshedi, Wands, & de la Monte, 2010). Furthermore, many genetic studies reported that people with a dominant negative *ALDH2*2* mutant allele and a decreased ALDH2 activity are likely to have greater risks for cancer development in different tissues, myocardial infarct, ALD, and other pathological states, as reported (Jo et al., 2007; Muto et al., 2000; Takagi et al., 2002; Yokoyama et al., 1998, 2001). All these pathological conditions seem to be mediated by the suppressed ALDH2 and other ALDH isozymes with elevated DNA mutations (Minko et al., 2009) and protein modifications (Fritz & Petersen, 2013) by highly reactive carbonyl compounds such as acetaldehyde, 4-HNE, MDA, and ACR produced after alcohol intake and/or exposure to toxic substances including APAP and HFD. Similar pathological roles of ALDH2 were demonstrated with *Aldh2*-null mice with increased levels of acetaldehyde- or *N*-ethylidene-dG DNA adducts (Isse et al., 2005; Ogawa et al., 2007; Yu et al., 2010, 2012; Yukawa et al., 2014) and severe inflammatory liver disease (Kwon et al., 2014). Conversely, overexpression of *ALDH2* gene (Doser et al., 2009; Sun et al., 2014; Zhang & Ren, 2011; Zhang et al., 2014) or by using chemical ALDH2 activators such as alda-1 and alda-44 can protect cellular and tissue damage caused by alcohol, aging, and I/R (Chen, Budas, et al., 2008; Churchill et al., 2009; Perez-Miller et al., 2010).

The tissue and serum levels of acetaldehyde can be markedly elevated when ALDH2 is suppressed by a dominant negative mutation in the *ALDH2* gene (e.g., *ALDH2*2*) (Isse et al., 2005) or following exposure to the inhibitors of ALDH2 disulfiram and cyanamide (Kato et al., 1990) or other toxic agents such as APAP and CCL4 (Song et al., 2011). Accumulated acetaldehyde, which inhibits mitochondrial GSH transport (Lluis, Colell, García-Ruiz, Kaplowitz, & Fernández-Checa, 2003), can stimulate fat biosynthesis by directly activating the matured form of SREBP-1 (You et al., 2002). The latter results were demonstrated in alcohol-exposed hepatoma cells in the absence or presence of ADH inhibitor 4-methylpyrazole or ALDH2 inhibitor cyanamide. Similar results of SREBP activation with transcriptional activation of downstream lipogenic enzymes and hepatic triglyceride accumulation were also observed in C57BL mice fed with the low-fat containing alcohol liquid diet for 4 weeks (You et al., 2002). In addition, acetaldehyde can inhibit the phosphorylated (active) form of AMP-activated protein kinase (AMPK), a master regulatory protein in controlling metabolic syndrome, which negatively affects key proteins such as SREBP (You et al., 2004), acyl-CoA carboxylase, and malonyl-CoA decarboxylase (Purohit et al., 2009). AMPK-mediated phosphorylation inactivates acyl-CoA carboxylase but activates malonyl-CoA decarboxylase, contributing to decreased synthesis and increased degradation, respectively, of malonyl-CoA, which is a critical precursor for fat accumulation.

Acetaldehyde can also suppress the function of the peroxisome proliferator-activated receptor-α (PPARα), which is a key transcription factor in fat transport and oxidation as well as inflammatory function (Moraes, Piqueras, & Bishop-Bailey, 2006). Thus, suppression of PPARα or deletion of its gene, as seen in *Ppara*-null mice, can lead to fat accumulation and liver disease upon alcohol exposure (Nakajima et al., 2004) and HFD (Abdelmegeed et al., 2011). In contrast, activation of PPARα can decrease fat accumulation and improved liver functions, partly due to enhanced fat degradation and anti-inflammatory activity, as observed with many PPARα agonists such as clofibrate, adiponection (Xu et al., 2003; Yamauchi et al., 2002, 2001), and IL-6 (Hong et al., 2004). Experimental data showed that alcohol exposure significantly blunted transcriptional activation of PPARα of a reporter construct through impaired DNA-binding ability in hepatoma cells (Galli, Pinaire, Fischer, Dorris, & Crabb, 2001) and/or degradation of its binding partner RXRα (Fischer, You, Matsumoto, & Crabb, 2003). Consistently, alcohol administered with intragastric infusion downregulated PPARα with fat accumulation (Nanji, Dannenberg, Jokelainen, & Bass, 2004). In addition, our recent results showed that binge alcohol and APAP could decrease the level of PPARα, possibly through its nitration followed by proteasomal degradation (Yun et al., 2014). In this case, decreased levels of PPARα may lead to disrupted cell protection with a reduced supply of alternative energy ketone bodies, thereby contributing to acute death of hepatocytes in rodents and people. Acetaldehyde has been also shown to be conjugated with numerous proteins including tubulin (Israel, 1997; Jennett, Sorrell, Saffari-Fard, Ockner, & Tuma, 1989; Niemelä et al., 1998, 1994; Sutti et al., 2014) to activate immune reactions. Critical roles of ALDH2 and CYP2E1 in producing MAA or advanced glycation end product (AGE) adducts and other protein adducts have been suggested (Anderson et al., 2014; Duryee et al., 2005; Jeong et al., 2000; Kwon et al., 2014; Swaminathan, Clemens, & Dey, 2013; Swaminathan, Kumar, Clemens, & Dey, 2013; Thiele et al., 2001). In addition, acetaldehyde can interact with DNA to produce DNA adducts (Brooks et al., 2009; Seitz & Stickel, 2010; Yu et al., 2010, 2012). The acetaldehyde DNA adducts can promote cancer in many tissues including liver, mouth, esophagus, and gastrointestinal tract. Furthermore, acetaldehyde was demonstrated to be involved in promoting gut leakiness (Basuroy, Sheth, Mansbach, & Rao, 2005; Dunagan, Chaudhry, Samak, & Rao, 2012; Elamin, Masclee, Dekker, & Jonkers, 2013; Elamin, Masclee, Troost, Dekker, & Jonkers, 2014). In this system, treatment with an ADH inhibitor 4-methylpyrazole prevented gut leakiness, whereas treatment with cyanamide, an ALDH2 inhibitor, increased the rate of leakiness (permeability) in cultured Caco-2 cells and animal models. Last, some reports suggested that acetaldehyde produced and accumulated in the brain is involved in alcohol-seeking behavior, likely contributing to alcohol addiction (Deitrich, 2011; Karahanian et al., 2011, 2015; Rodd-Henricks et al., 2002).

3.3 Role and Regulation of CYP2E1 in Liver Disease

In addition to ADH, a small but significant amount of alcohol is metabolized by another enzyme system so-called the MEOS, where CYP2E1 is a major component (Cederbaum, 2012b; Lieber, 1997, 2005). The CYP2E1-related activity in the MEOS is similar to the combined total activities of CYP1A2 and CYP3A. Because CYP2E1 has approximately 10 times greater K_m value

(~10 mM for ethanol) compared to that of ADH (0.8–1 mM), it may play a minor role (less than 10%) in ethanol metabolism under physiological conditions. However, after binge and chronic alcohol drinking with higher BAC up to ~100 mM, as observed in some alcoholics (Lindblad & Olsson, 1976), induced CYP2E1 becomes important in ethanol metabolism. Unlike other P450 enzymes, CYP2E1, a loosely bound enzyme to the ER membrane, exhibits NADPH-oxidase activity, thus producing ROS during its catalytic cycle (Ekström & Ingelman-Sundberg, 1989; Terelius & Ingelman-Sundberg, 1988). The ROS include superoxide anion, hydroxyethyl radical, and hydrogen peroxide, depending on the local environment and pre-existing conditions. CYP2E1, present in both ER and mitochondria (Bansal et al., 2013, 2010; Knockaert, Fromenty, & Robin, 2011; Robin et al., 2002), is induced and activated by acute or chronic exposure to alcohol and other small molecules such as acetone and HFD or diabetes by different regulatory mechanisms (Koop, 1992; Roberts, Shoaf, Jeong, & Song, 1994; Roberts, Song, Soh, Park, & Shoaf, 1995; Song, Gelboin, Park, Yang, & Gonzalez, 1986; Song, Veech, Park, Gelboin, & Gonzalez, 1989; Song et al., 1987; Yun, Casazza, Sohn, Veech, & Song, 1992). Mitochondrial CYP2E1 can cause ethanol-induced oxidative stress and mitochondrial toxicity, leading to cell damage (Bansal et al., 2013, 2012; Robin et al., 2005). Moreover, its level and activity are elevated in experimental models of obese and hyperglycemic diabetic rodents and in humans (Dey & Kumar, 2011; Song, Veech, & Saenger, 1990; Song et al., 1987; Surapaneni, Priya, & Mallika, 2014; Weltman, Farrell, Hall, Ingelman-Sundberg, & Liddle, 1998; Weltman, Farrell, & Liddle, 1996; Yun et al., 1992). Because of different mechanisms of CYP2E1 induction (e.g., protein stabilization by ethanol or acetone and mRNA increase by HFD or diabetes), the overall levels of CYP2E1 can be increased in an additive or synergistic manner (Caro & Cederbaum, 2004; Song, Koop, Ingelman-Sundberg, Nanji, & Cederbaum, 1996). For instance, alcohol exposure in diabetic or obese rodents and people would markedly elevate CYP2E1 activity, thus producing greater amounts of ROS and increased oxidative stress (Cederbaum, 2012a). Another example of additive effect is interaction between alcohol drinking and other risk factors such as high-fat-induced obesity, smoking, infection with hepatitis virus or HIV, and certain drugs including APAP, halothane, and isoniazid, promoting acute hepatotoxicity or liver failure (Boelsterli & Lee, 2014; Jaeschke et al., 2002; Lu, Ward, & Cederbaum, 2013; McClain et al., 1980; Pessayre et al., 2012; Seeff et al., 1986; Yuan et al., 2014). The effect of each risk factor would add up toward greater levels of nitroxidative stress, contributing to increased sensitivity to cell death and inflammatory tissue injury (Fig. 2).

Enhanced alcohol oxidation by the MEOS including CYP2E1, CYP1A2, and CYP3A may be associated with metabolic tolerance for handling large amounts of alcohol due to no apparent change in the ADH activity (Videla & Israel, 1970). However, these P450 enzymes in the MEOS may stimulate alcohol-related toxicities with increased chances of tissue injury through alcohol and drug interactions, cytokine signaling, antigen presentation, and autophagy regulation (Osna & Donohue, 2013). These P450 isozymes, expressed in the liver and extra-hepatic tissues, are responsible for the metabolism of many potentially toxic substances including many FDA-approved drugs described below. Thus, their induction and catalytic activity are likely to cause greater production of ROS and

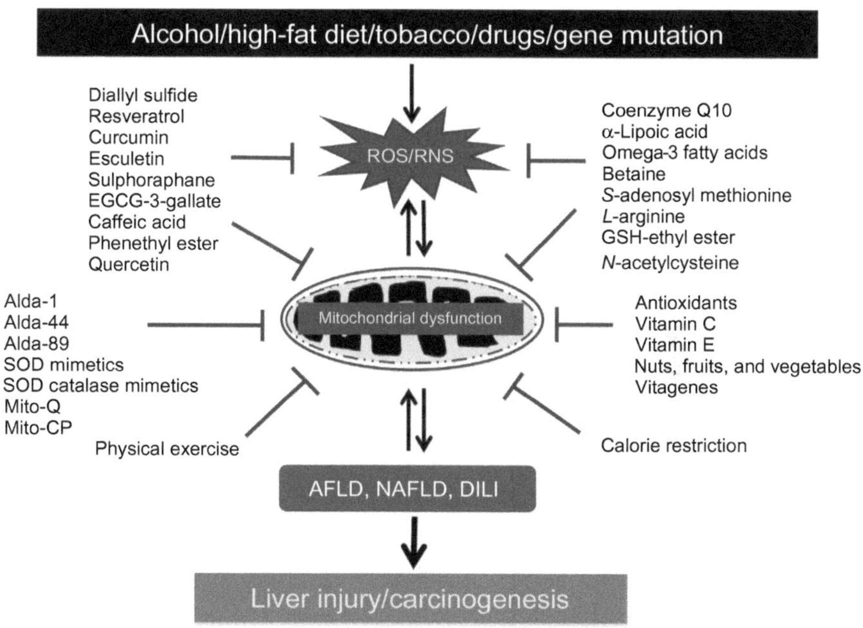

Figure 2 Potential prevention and protection against various liver diseases by natural or synthetic antioxidants with physical and behavioral modifications. Many small molecules including inhibitors of CYP2E1, activators of ALDH2, natural or synthetic antioxidants, and physical or behavioral modifications including exercise and calorie restrictions can be used to prevent or reduce the nitroxidative stress, mitochondrial dysfunction, and liver diseases. Unidirectional and bidirectional arrows indicate exclusive and mutual influences, respectively.

reactive metabolites of these substrates, which can interact with cellular macromolecules to cause cell/tissue injury and carcinogenesis in the liver and extra-hepatic tissues (Cederbaum, 2012a; Cederbaum et al., 2009; Lavandera, Ruspini, Batlle, & Buzaleh, 2015; Leung & Nieto, 2013; Pessayre et al., 2012; Zhang, Gao, et al., 2013). Increased ROS via CYP2E1-mediated alcohol metabolism can activate a redox-sensitive transcription factor nuclear factor-κB (NF-κB), which stimulates iNOS expression for production of large amounts of RNS (Dai et al., 2013; Wu, Xu, & Cederbaum, 2009). Concurrent presence of ROS/RNS can produce peroxynitrite, which can inhibit many mitochondrial complexes I, II, III, IV, and V, causing more ROS leaked out the ETC and suppression of ATP synthesis, contributing to lipid peroxidation, various PTMs of cellular proteins with ER stress or mitochondrial dysfunction, and necroapoptosis of hepatocytes. In addition, due to markedly elevated ROS/RNS following alcohol intake or chronic HFD, some of the cellular signaling pathways are affected. For instance, cell-death-related JNK and p38K are activated via phosphorylation through activation of the upstream kinases and suppression of phospho-protein phosphatases, as described (Cederbaum, Yang, Wang, & Wu, 2012; Heneberg & Dráber, 2005; Son, Kim, Chung, & Pae, 2013; Song, Akbar, et al., 2014). Combination of activated JNK/P38K and suppressed Akt or extracellular signal-regulated protein kinase (ERK) may contribute to hepatocyte cell death (Bae, Pie, & Song, 2001; Bae & Song, 2003; Gao et al., 2014; Soh et al., 2000). Based on the activation of JNK and p38K in HFD-exposed rodents (Czaja, 2010; Yang et al., 2014), similar patterns of hepatic injury may develop in NAFLD and diabetes.

In addition to alcohol metabolism, CYP2E1 is known to metabolize many small-molecule substrates, which serve as the inducers of CYP2E1. The exogenous compounds are APAP, halothane, isofluorane, isoniazid, solvents (e.g., carbon tetrachloride, chloroform, dichloromethane, benzene), various fatty acids, dimethylnitrosamine, diethylnitrosamine, bromodichloromethane, Vitamin A derivatives (retinol and retinoic acid), and others (Guengerich, Kim, & Iwasaki, 1991; Koop, 1992). Endogenous substrates can be acetaldehyde, acetone, ketone bodies, 4-HNE, and others including ethanol and acetaldehyde produced by gut bacteria (Casazza, Felver, & Veech, 1984; Cederbaum, 2012b; Koop, 1992; Kurkivuori et al., 2007). Metabolism of these substrates by CYP2E1 and relevant toxicities seem proportionally correlated with the induced levels of CYP2E1, despite the presence of a few exceptions such as APAP and CCL4-exposed models through proteolytic degradation of CYP2E1, as demonstrated

(Sohn, Yun, Park, Veech, & Song, 1991). For instance, clinically relevant doses of APAP or carbon tetrachloride (CCL4) can cause acute DILI via alcohol and drug interactions, especially in alcohol-exposed individuals or rodents with increased CYP2E1. The APAP- or CCL4-mediated hepatic (or kidney) injury is initiated through their metabolism by CYP2E1, since pretreatment with CYP2E1 inhibitors or *Cyp2e1*-null mice were fully protected from these types of DILI (Lee, Buters, Pineau, Fernandez-Salguero, & Gonzalez, 1996; Wong, Chan, & Lee, 1998). Based on these results, we can expect that similar cases of DILI could be observed in obese or diabetic individuals, who likely have elevated levels of CYP2E1. The metabolism of retinoic acid by CYP2E1 and competition by another substrate ethanol may also be related to increased carcinogenesis in alcohol-exposed rodents and alcoholic individuals (Leo & Lieber, 1999; Liu, Russell, Seitz, & Wang, 2001; Seitz & Wang, 2013). By using specific knockout mice, Rusyn and colleagues clearly showed that CYP2E1, but not NADPH oxidase, is important in promoting alcohol-mediated DNA damage (Bradford et al., 2005). The important role of CYP2E1 in producing carcinogenic etheno-DNA lesions was consistently reported in the experimental model and alcoholic individuals (Wang et al., 2009). In addition, abnormal retinoic acid metabolism is considered important in fetal alcohol syndrome or effects due to its importance in early development and differentiation (Feltes, de Faria Poloni, Nunes, & Bonatto, 2014; Kane, Folias, Wang, & Napoli, 2010; Keyte & Hutson, 2012).

Many reports suggest that CYP2E1-related metabolisms of alcohol and other substrates are directly and indirectly related to various PTMs of cellular proteins and DNA. These PTMs include acetaldehyde-protein adduct formation (Jeong et al., 2000; Niemelä et al., 1994), hydroxyethyl radical protein adducts (Albano et al., 1996; Clot et al., 1996; Moncada, Torres, Varghese, Albano, & Israel, 1994), oxidation (Suh et al., 2004), nitration (Abdelmegeed, Jang, et al., 2013; Abdelmegeed et al., 2010), proteasomal degradation (Bardag-Gorce, Li, French, & French, 2005), gamma-ketoaldehyde-protein adducts (Roychowdhury et al., 2009), AGE-adducts in high glucose exposed VL-17A cells (Swaminathan, Kumar, et al., 2013), MAA adduct in VL-17A cells (Swaminathan, Clemens, et al., 2013), etheno-DNA adduct formation (Linhart, Bartsch, & Seitz, 2014), and formation of CYP2E1 auto-antibodies detected in experimental models and alcoholic subjects (French et al., 1993; Sutti et al., 2014). In contrast, CYP2E1 was shown to suppress diethyl-1, 4-dihydro-2, 4, 6-trimethyl-3, 5-pyridinedicarboxylate (DDC)-induced Mallory Body formation in the liver (Bardag-Gorce, Wilson, et al., 2005). These modifications seem to depend on the amount and time of exposure to alcohol or other agents, because suppressed CYP2E1 activities by its specific inhibitors decreased the levels of these modifications. *Cyp2e1*-null mice were also resistant to these modifications compared to the wild-type (WT) counterparts, when they were exposed to the same agents including alcohol or other toxic substances. In addition, overexpression of CYP2E1 in E47-HepG2 hepatoma cells (Cederbaum, 2014) or transgenic mice (Morgan, French, & Morgan, 2002) or *Cyp2e1*-knockin mice (Cederbaum, 2012a; Wu & Cederbaum, 2013) revealed that these mice are more susceptible to oxidative stress, mitochondrial dysfunction, and liver injury by alcohol than the corresponding WT mice. These results clearly support the role of CYP2E1 in these PTMs, resulting in adverse responses and tissue injury in AFLD (Lakshman et al., 2013; Song, Akbar, et al., 2014). It is likely that similar types of PTMs and unhealthy outcomes can be observed in NAFLD, as reported with CYP2E1 transgenic mice (Kathirvel, Chen, Morgan, French, & Morgan, 2010; Kathirvel, Morgan, French, & Morgan, 2009).

Chronic or binge alcohol exposure can induce CYP2E1, which is predominantly expressed in the pericentral regions, where oxygen levels are lower compared to those of periportal regions based on the oxygen gradient in the liver. Liver injury, including ALD, nonalcoholic liver disease (NALD), or DILI, is often associated with suppressed vascular endothelial cell function, which is accompanied with restricted blood and oxygen supply (Doggett & Breslin, 2014; Ito, Abril, Bethea, & McCuskey, 2004; McCuskey et al., 2004; Tarnawski, Ahluwalia, & Jones, 2012). CYP2E1 also needs molecular oxygen for its catalytic activity, further lowering oxygen concentration in the pericentral regions. Combinations of these results lead to alcohol-induced hypoxia in animal models and human alcoholics (Arteel, Iimuro, Yin, Raleigh, & Thurman, 1997; Arteel, Raleigh, Bradford, & Thurman, 1996; Israel et al., 1975; Ji, Lemasters, Christenson, & Thurman, 1982; Wang, Wu, Yang, Gan, & Cederbaum, 2013; Yun et al., 2014). Cellular hypoxia can activate a transcription factor hypoxia-induced factor (HIF), which regulates the transcription of many downstream targets including iNOS (Nath & Szabo, 2012). As mentioned above, induction of iNOS with elevated levels of RNS and CYP2E1-mediated ROS can stimulate nitration of various cellular proteins including mitochondrial complexes I, III, and V, leading to greater ROS production and energy depletion, ultimately leading to p53-Bax-mediated cell death, as recently reported (Yun et al., 2014). Similar patterns of hypoxia-related liver injury has been reported with high-fat alone or with tobacco smoking-exposed mice (Eccleston et al., 2011; Mantena et al., 2009), and alcohol and tobacco smoking (Bailey et al., 2009). Furthermore, reoxygenation following hypoxic ischemia may also contribute to mitochondrial dysfunction and liver injury, as described (Moon, Hood, et al., 2008). Hypoxia-related liver injury due to CYP2E1-dependent metabolism of its potentially toxic substrates such as APAP, halothane, and CCL4 can be also observed in acute DILI, as reported (James, Donahower, Burke, McCullough, & Hinson, 2006; Noll & De Groot, 1984; Sparkenbaugh et al., 2011). However, Lieber (2005) suggested that APAP-induced liver injury could be worse during alcohol withdrawal, due to less competition by alcohol for the CYP2E1-dependent metabolism of APAP. It is also expected that hepatotoxicity by APAP can be enhanced or accelerated in obese or diabetic conditions with elevated levels of CYP2E1 and/or CYP3A (Michaut, Moreau, Robin, & Fromenty, 2014), as similar to those observed in alcoholic individuals (McClain et al., 1980; Seeff et al., 1986).

Autophagy is a cellular protection mechanism by which damaged cells are removed to produce an alternative energy especially during starvation. Removal of damaged mitochondria and lipids are called mitophagy and lipophagy, respectively. Recent studies indicated that ALD can be produced by inhibition of autophagy and mitophagy, as demonstrated in HepG2 hepatoma cells and WT mice in a CYP2E1-dependent manner (Wu, Wang, Zhou, Yang, & Cederbaum, 2012; Yang, Wu, Wang, & Cederbaum, 2012). CYP2E1-mediated elevated ROS/RNS likely negatively affect the components of mitophagy and autophagy with accumulation of damaged mitochondria and hepatocytes, finally contributing to ER stress and liver injury (Czaja, 2011; Ding, Manley, & Ni, 2011). The critical importance of mitophagy, mitochondrial fission/fusion and autophagy are also suggested in promoting NAFLD (Amir & Czaja, 2011; Brenner et al., 2013). It is also possible that some damaged hepatocytes in both ALD, NALD, and DILI, mediated at least partially by CYP2E1-related oxidative stress, may serve as cellular sources of damage-associated molecular pattern (DAMP) molecule or pathogen-associated molecular pattern (PAMP) molecule to activate immune responses for inflammatory liver disease (Szabo, 2015). For instance, high-mobility group box-1 (HMGB-1), DNA, and microRNAs, as DAMPs secreted from the nuclei of damaged or stressed cells (Beyer et al., 2012; Brenner et al., 2013; Eguchi, Wree, & Feldstein, 2014; Kubes & Mehal, 2012), may be involved in pathogenesis of ALD in a mouse model. Using the recombinant HMGB-1, a neutralizing antibody, or a specific siRNA to HMGB-1, Shah and colleagues showed that HMGB-1, released from hepatocytes, can contribute to alcohol-induced hepatotoxicity by recruiting hepatic stellate and endothelial cells to the site of parenchymal cell injury (Seo et al., 2013). Elevated levels of HMGB-1 were also observed in the liver biopsy samples from human alcoholic individuals compared to normal subjects (Ge et al., 2014). Elevated release of HMGB-1 and damage signaling were also observed in HFD-exposed rats (Zhang, Wang, et al., 2013), APAP-induced DILI (Cai et al., 2014), and I/R injury (Kamo et al., 2013), although a few exceptions exist. For instance, one report indicated that hepatocyte-specific deletion of HMGB-1 worsens I/R liver injury (Huang et al., 2014), suggesting that HMGB-1 may play a dual role in regulating the cellular immune function, depending on the cellular context.

Simple steatosis in AFLD and NAFLD can progress to more severe inflammatory hepatitis (steatohepatitis) and fibrosis/cirrhosis in the presence of a second hit such as increased nitroxidative stress or proinflammatory cytokines including TNFα. Alcohol-mediated increased bacteria leaked from the intestine can serve as a second hit, since gut leakage produces large amounts of endotoxin lipopolysaccharide (LPS), which can stimulate the production of proinflammatory cytokines and nitroxidative stress (Yoo, Abdelmegeed, & Song, 2013). In fact, many laboratories consistently reported that excessive amounts or binge alcohol can damage intestinal membrane and thus stimulate gut leakage in animal models (Keshavarzian, Jacyno, Urban, Winship, & Fields, 1996) and Caco-2 intestinal cells (Forsyth et al., 2013). Alcohol-induced gut leakage in experimental models (Abdelmegeed, Banerjee, et al., 2013; Forsyth et al., 2013) was also observed in healthy people with just one episode of acute binge drinking (Bala, Marcos, Gattu, Catalano, & Szabo, 2014) and in alcoholic individuals who suffer from cirrhosis (Bode, Kugler, & Bode, 1987; Pijls, Jonkers, Elamin, Masclee, & Koek, 2013). In animal models, exposure to 30% fructose in drinking water or HFD can increase gut leakage with elevated levels of endotoxin in the blood (Bergheim et al., 2008; Spruss, Kanuri, Stahl, Bischoff, & Bergheim, 2012; Spruss et al., 2009). Increased gut leakiness with significantly higher levels of plasma endotoxin and hepatic toll-like receptor 4 were also observed in patients with NAFLD compared to those in controls (Thuy et al., 2008). The mechanisms of gut leakiness by alcoholic and nonalcoholic substances have been studied. Acetaldehyde, NO produced by iNOS, and JNK-mediated phosphorylation of tight junction proteins seem to be important in gut leakiness and inflammatory ALD and

NALD (Basuroy et al., 2005; Dunagan et al., 2012; Keshavarzian et al., 1996; Lambert et al., 2003, 2004; Samak, Suzuki, Bhargava, & Rao, 2010; Spruss et al., 2011; Tang et al., 2009). Our recent results showed that binge alcohol increased the levels of intestinal CYP2E1 and iNOS, which produce ROS and RNS, respectively, contributing to increased nitroxidative stress with nitration of intestinal proteins. These results suggest that nitration of tight junction proteins seems critically important in gut leakiness since treatment with a CYP2E1 inhibitor chlormethiazole (CMZ) or an antioxidant *N*-acetylcysteine (NAC) significantly decreased the levels of CYP2E1 and iNOS with possibly decreased nitration of intestinal proteins. Furthermore, CMZ or NAC treatment ameliorated binge alcohol-mediated endotoxemia and liver inflammation. *Cyp2e1*-null mice were also resistant to binge alcohol-induced gut leakiness. These results further support the important role of CYP2E1 in binge alcohol-induced gut leakiness and inflammatory liver disease (Abdelmegeed, Banerjee, et al., 2013), as recently reviewed (Forsyth, Voigt, & Keshavarzian, 2014).

DILI caused by clinically used drugs, industrial solvents, and environmental agents represent another significant medical problem with acute fulminant liver failure and death (Boelsterli & Lee, 2014; Jaeschke et al., 2012; Xie et al., 2014; Yuan & Kaplowitz, 2013). APAP, isoniazid, halothane, isoflurane, troglitazone, CCL4, benzene, and bromodichloromethane can cause DILI. Many of these DILI-causing agents are substrates of CYP2E1. Thus, CYP2E1-mediated metabolism of these hepatotoxic compounds produces reactive intermediates, ROS, and lipid peroxidation with hypoxia in the pericentral regions. Depending on the concentrations of these agents, their reactive metabolites and lipid peroxide aldehydes (e.g., 4-HNE, MDA, and ACR) can bind to cellular proteins, producing various protein adducts, contributing to their inactivation. For instance, mitochondrial proteins such as ALDH2, enoyl-CoA hydratase, electron transfer flavoprotein-α, cytochrome *c* oxidase (complex IV), and sirtuin 3 are known to bind 4-HNE or reactive metabolites of APAP to produce protein adducts (Andringa, Udoh, Landar, & Bailey, 2014; Chen, Robinson, Schenker, Frosto, & Henderson, 1999; Fritz et al., 2011; Landin et al., 1996). Consequently, the activities of these proteins become suppressed, leading to mitochondrial dysfunction. In addition, the ROS produced from the CYP2E1-mediated metabolism of these substrates can activate the JNK/p38K-related cell-death signaling pathway, as demonstrated with ethanol, APAP, troglitazone, and CCL4 (Bae & Song, 2003; Bae et al., 2001; Kim, Ryu, & Song, 2006; Lee & Shukla, 2005; Nishitani & Matsumoto,

2006; Saberi et al., 2014; Schattenberg & Czaja, 2014; Soh et al., 2000; Wu & Cederbaum, 2013; Yang et al., 2012). *Cyp2e1*-null mice were more resistant while pretreatment with a specific siRNA to JNK1 prevented the hepatocytes from acute DILI caused by APAP. In contrast, it is expected that CYP2E1 transgenic mice or knockin mice or people with elevated levels of CYP2E1 are more likely susceptible to DILI, as observed with increased hepatotoxicity of APAP or halothane in alcoholic individuals. In these cases, HMGB-1 along with secreted DNA and microRNA, as DAMP and PAMP, may play important roles in promoting DILI, as discussed with APAP (Kubes & Mehal, 2012; Maher, 2009; Martin-Murphy, Holt, & Ju, 2010).

Many laboratories reported that the level and activity of CYP2E1 are increased by HFD, obesity, diabetes, and hyperglycemic conditions in cultured hepatoma cells, rodents, and people (Caro & Cederbaum, 2004; Chalasani et al., 2003; Lieber, 2004b; Purohit et al., 2009; Song et al., 1996; Weltman et al., 1998; Yun et al., 1992). However, so far, three genome-wide association studies indicated that genetic polymorphisms in *CYP2E1* gene do not appear to be related to the increased susceptibility to NAFLD (Daly, 2013). In addition, a few exceptions exist especially in leptin-deficient *ob/ob* mice or *fa/fa* Zucker rats with a defective leptin receptor (Carmiel-Haggai et al., 2005; Enriquez, Leclercq, Farrell, & Robertson, 1999; Leclercq, Field, Enriquez, Farrell, & Robertson, 2000). CYP2E1-mediated metabolism of fatty acids and ketones produces ROS and hypoxia in the pericentral regions in the liver. These changes indirectly produce RNS, possibly though induction of iNOS from the activation of transcription factors NF-κB and HIF. Concurrent presence of ROS/RNS can activate JNK/p38K, probably leading to hepatocyte apoptosis and insulin resistance through phosphorylation of the insulin receptor and its substrate proteins 1 (IRS1) and 2 (IRS2). In addition, IRS1 and IRS2 can undergo nitration of tyrosine (Tyr) residues, which can interfere with the regular Tyr phosphorylation, contributing to insulin resistance frequently observed in diet-induced obesity or diabetes models and hyperglycemic conditions. In fact, treatment with an inhibitor or a specific siRNA to CYP2E1, JNK1, or iNOS can prevent insulin resistance (Abdelmegeed et al., 2012; Schattenberg & Czaja, 2014). In addition, *Cyp2e1*-null mice or *iNOS*-null mice were resistant to high-fat- or fructose-mediated insulin resistance and nonalcoholic steatohepatitis (NASH) development (Abdelmegeed et al., 2012; Spruss et al., 2011; Zong, Armoni, Harel, Karnieli, & Pessin, 2012), although the detailed molecular mechanisms of insulin resistance by nitration or phosphorylation of IRS1 or IRS2 were not studied in these reports. Furthermore, CYP2E1 transgenic mice become more susceptible to insulin resistance and NASH following high-fat exposure (Kathirvel et al., 2009). In addition, chronic alcohol administration is known to cause insulin resistance and fat accumulation in mice (Carr, Dhir, Yin, Agarwal, & Ahima, 2013) while insulin secretion from pancreatic beta-cells can be suppressed by ethanol (Nguyen, Lee, & Nyomba, 2012), contributing to fatty liver disease. Collectively, many of these changes with adverse health outcomes, as briefly described here, can be prevented by CYP2E1 inhibition, suggesting that CYP2E1 can be an important therapeutic target.

3.4 Role of Nonoxidative Alcohol Metabolism in Liver Disease

A minor portion of ingested alcohol can be metabolized in a nonoxidative metabolic pathway such as conjugation with fatty acyl-CoA, producing FAEEs. Production of FAEEs, including ethyl myristate, ethyl palmitate, ethyl oleate, and ethyl stearate, can be catalyzed by an enzyme FAEE synthase (homodimer with a 26 kDa monomer) expressed in many tissues (Laposata, 1998; Laposata & Lange, 1986; Mogelson & Lange, 1984). Earlier reports showed that a FAEE synthase in myocardium could be a cholesterol esterase (sterol-ester acylhydrolase) (Lange, 1982) or glutathione *S*-transferase (Bora, Spilburg, & Lange, 1989). Despite being minor in the liver, it can become an important metabolic pathway with functional significance in some tissues such as pancreas, heart, and intestines (Elamin et al., 2013; Lange & Sobel, 1983; Werner et al., 1997). For instance, different laboratories reported toxic effects of FAEEs where they can cause mitochondrial dysfunction in heart through direct mitochondrial binding of FAEEs. Bound FAEEs can be accumulated and may serve as potential uncouplers of the mitochondrial oxidative phosphorylation, as demonstrated with ethyl oleate (Lange, 1982; Laposata, 1998; Lange & Sobel, 1983). Alternatively, FAEEs can stimulate apoptosis in pancreatic acinar cells with elevated calcium-related toxicity with decreased ATP synthesis (Criddle et al., 2006, 2004; Kaphalia et al., 2010). On the other hand, recent results demonstrated that exogenous ethyl esters of n-3 fatty acids including eicosatetrapentaenoic acid and docosahexaenoic acid show protective effects against obesity-related metabolic syndrome (Depner, Philbrick, & Jump, 2013; Mori et al., 1999; Pérez-Echarri et al., 2008, 2009; Pérez-Matute, Pérez-Echarri, Martínez, Marti, & Moreno-Aliaga, 2007; Spencer et al., 2013). Therefore, it appears that the levels of endogenously produced FAEEs are elevated, when the oxidative alcohol metabolism is suppressed or blocked. For instance, under normal physiological conditions, FAEE levels were less than 0.001 μmol/L in the heart. However, their levels can be markedly elevated up to 115 μmol/L in alcoholic individuals, as reviewed (Kodde et al., 2007). Since the half-lives of FAEEs are much longer than that of ethanol, their detection in the blood can be used as a useful marker for alcohol drinking. This detection of FAEEs can be practically important even when alcohol in blood no longer exists.

Alcohol can also be conjugated with glucuronic acid to produce ethyl glucuronide (EtG) catalyzed by UDP-glucuronosyltransferases (Schwab & Skopp, 2014). EtG is water soluble and thus easily gets excreted. EtG is a nonvolatile stable compound and can be detected in various body fluids and tissues. However, EtG is not detected in control, nonalcoholic individuals, and teetotalers, suggesting its specificity for alcohol intake. Therefore, detection of EtG in easily obtainable specimens such as hair and body fluids can become important in forensic legal medicine for determining alcohol drinking even in the absence of detectable levels of ethanol (Cappelle et al., 2015). Many methods have been developed for detecting EtG in hair from people. However, extra caution should be taken in using EtG as a marker for alcohol intake since the levels of EtG can be negatively affected by the presence of flavonoids or presence of ethanol-based hair conditioning products (Cappelle et al., 2015; Schwab & Skopp, 2014).

3.5 Role and Regulation of NADPH Oxidase in Liver Disease

Alcohol intake or exposure is known to suppress overall immune function in general, leading to increased infections by pathogenic bacteria and viruses (Jerrells, Peritt, Marietta, & Eckardt, 1989; Marietta et al., 1988). However, alcohol- or fructose-associated gut leakage and endotoxin can stimulate hepatic Kupffer cells with infiltration of neutrophils and monocytes into the liver. Activation of Kupffer cells and infiltrated immune cells can increase nitroxidative stress, HIF, and NF-κB, which produces cytokines, chemokines, and iNOS (Wang et al., 2013; Yun et al., 2014), where CYP2E1 is directly involved or at least plays a permissive role in activating immune cells (Barnes, Roychowdhury, & Nagy, 2014; Cao, Mak, & Lieber, 2005; Lieber, 1997). In addition, activated immune cells can stimulate membrane association of the subunits of NADPH oxidase, producing ROS (Kim, Nagy, & Park, 2014). In fact, Thurman and colleagues showed that NADPH oxidase is critically important in early ALD, since *NADPH-oxidase* knockout mice were resistant to early ALD (Kono et al., 1999). In this model by intragastric infusion of alcohol, CYP2E1 does not seem to be important, although many other laboratories reported the important role of CYP2E1 in AFLD, as described above. None-the-less, activation of NADPH oxidase is usually accompanied with increased levels of TNFα and other proinflammatory cytokines, contributing to inflammatory liver disease. The NADPH-oxidase activity in immune cells also plays an important role in diet-induced obesity, NAFLD, and NASH. For instance, activation of NADPH oxidase seems important in diet-induced NASH and fibrosis in HFD-exposed obese *fa/fa* Zucker rats compared to the lean *fa/?* rats (Carmiel-Haggai et al., 2005). In this model, the amount and activity levels of CYP2E1 and XO in the obese *fa/fa* rats with defective leptin receptor were significantly lower than those of the corresponding lean *fa/?* rats and unchanged even after HFD exposure. These results may indicate that CYP2E1 is not as important as NADPH oxidase. However, the expression of CYP2E1 may have a permissive role for priming the macrophages for their sensitization with elevated NADPH oxidase, hydrogen peroxide, NF-κB, and TNFα levels following exposure to LPS (Cao et al., 2005). Suppression of CYP2E1 with an inhibitor diallyl sulfide or NADPH oxidase by its inhibitor diphenyleneiodonium equally decreased the levels of hydrogen peroxide, suggesting both enzymes are involved in producing oxidative stress in this model. Furthermore, HFD-induced obesity and NALD are associated with increased activities of NADPH oxidase (Chatterjee et al., 2013; Chung, Park, Manautou, Koo, & Bruno, 2012; Gao et al., 2010; Sarna, Wu, Wang, Hwang, & Siow, 2012). All these reports suggest an important role of NADPH oxidase in ALD and NALD. In contrast, NADPH oxidase does not play an important role in promoting oxidative stress in a mouse model of NASH exposed to a methionine and choline-deficient diet (MCD) (dela Peña, Leclercq, Williams, & Farrell, 2007). Therefore, it would be prudent that both CYP2E1 and NADPH oxidase are likely important for providing oxidative stress for AFLD and NAFLD in a complementary manner because of a few exceptions for each protein, depending on the experimental model systems.

3.6 Role and Regulation of Xanthine Oxidase in Liver Disease

The oxidative metabolism of alcohol and acetaldehyde by ADH and ALDH2, respectively, requires NAD^+ as a cofactor, producing the reduced NADH. The altered redox state especially with a decreased NAD^+/NADH ratio can suppress the activity of a NAD^+-dependent xanthine dehydrogenase (XDH, D form), thereby converting it to oxygen-requiring XO (O form). Alcohol-mediated increased oxidative stress can also switch XDH to XO either by irreversible proteolysis or reversible oxidation of sulfhydryl groups. In fact, XDH became XO in a time-dependent manner after alcohol intake (Abbondanza, Battelli, Soffritti, & Cessi, 1989; Battelli, Abbondanza, & Stirpe, 1992; Sultatos, 1988). Subsequently, purine metabolism by XO may produce superoxide anion during its catalytic cycle. XO was also shown to mobilize iron from ferritin, thereby promoting conversion of superoxide to more toxic hydroxyl radical, which can cause oxidative organ damage during ischemia and inflammation (Biemond, Swaak, Beindorff, & Koster, 1986). Consequently, the levels of purine metabolites such as hypoxanthine, xanthine, and uric acid were significantly elevated in the liver and serum. These metabolic changes could be one mechanism responsible for the increased gout incidences observed in alcoholic individuals (Ka et al., 2006; Lieber, 2005; Yamamoto, Moriwaki, & Takahashi, 2005; Zakhari & Li, 2007). In addition, allopurinol, an inhibitor of XO, was shown to be effective in treating ALD with markedly decreased lipid peroxidation (Kato et al., 1990). The role of XO in oxidative stress and lipid peroxidation in extra-hepatic tissues such as heart, testes, and cerebellum has been described (Nordmann, Ribière, & Rouach, 1990). Administration of allopurinol significantly blunted alcohol-mediated abnormalities in these tissues, suggesting the role of XO in causing oxidative stress. Hypoxia is also known to conversion of NAD^+-dependent XDH to XO (Younes & Strubelt, 1987). Since chronic and acute alcohol exposure can produce local hypoxia in the pericentral regions of the liver (Israel et al., 1975), increased conversion of XO, by hypoxia due to the induced CYP2E1, may also contribute to hepatotoxicity in the pericentral regions. Administration of allopurinol (100 mg/kg) significantly blocked the combined toxic effects of ethanol and hypoxia. For instance, the inhibitory effects of ethanol on glycolysis and purine metabolism were prevented by allopurinol (Younes & Strubelt, 1987). Hyperuricemia produced by activated XO is positively correlated with individuals with NAFLD (Xu et al., 2015). Inhibition of XO by allopurinol significantly decreased the levels of steatosis in HFD-fed mice. Further, suppression of *XO* gene expression or its inactivation significantly decreased the uric acid production and fat accumulation in hepatoma HepG2 cells. These results suggest an important role of XO in HFD-induced NAFLD and hyperuricemia. Many other reports also showed the potential roles of XO and uric acid in causing toxicities in extra-hepatic tissues such as heart (Jia et al., 2015). Thus, XO, along with CYP2E1, could be a target for translational opportunities.

3.7 Role and Regulation of CYP2A5, CYP3A, and CYP4 Isozymes in Liver Disease

CYP4 enzyme family members have multiple functions in human biochemistry and physiology through not only the metabolism of potent signaling eicosanoids, but also their functional role in peroxisome-mediated fatty acids oxidation, vitamin, and steroid metabolism. These CYP4A enzymes also play pathophysiological roles in liver disease, hypertension, shock and sepsis, ischemic stroke adrenoleukodystrophy, Refsum disease, Bietti's crystalline dystrophy, and hyperkeratotic skin disease (See chapter "Cytochrome P450 ω-hydroxylases in inflammation and cancer" by Johnson et al.). For instance, CYP4A isozyme can be important in producing ROS and NASH, as shown in mice exposed to MCD, which was shown to elevate CYP2E1 mRNA and activity along with NASH-like inflammation (Chalasani et al., 2003; Weltman et al., 1996). In this case, CYP2E1 may be important in promoting NASH-related inflammatory changes in the pericentral regions. However, MCD-exposed *Cyp2e1*-null mice still developed NASH with lipid peroxidation, despite the absence of CYP2E1 (Leclercq et al., 2000; Robertson, Leclercq, Farrell, & Robertson, 2001). In this model, CYP4A becomes a major player in producing ROS and NADPH-dependent lipid peroxidation, which can cause liver injury. In fact, treatment with a specific antibody to CYP4A in *Cyp2e1*-null mice prevented the ROS production and lipid peroxidation, although treatment with the same CYP4A antibody did not prevent lipid peroxidation in wild-type mice. Similar to CYP2E1, CYP4A can metabolize various long-chain fatty acids at ω and ω-1 positions to produce shorter chain fatty acids (Hardwick, 2008). Through uncoupling during its catalytic cycle, CYP4A-mediated metabolism can produce ROS (Hardwick et al., 2013). In addition, it can produce dicarboxylic acids of long-chain fatty acids, which can inhibit the mitochondrial ETC, increased oxidative stress, and toxicity (Hardwick, 2008; Hardwick et al., 2013). One recent study showed that CYP4A, elevated in *db/db* mice, seems to play a major role in promoting high-fat-induced insulin resistance, ER stress, and apoptosis since inhibition of CYP4A with a specific inhibitor (HET0016) or intravenous injection of a small hairpin RNA specific to *CYP4A* mRNA efficiently blocked insulin resistance, ER stress, and apoptosis in diabetic *db/db* mice (Park et al., 2014).

Compared to CYP2E1, there are few studies on the role of human CYP4 family members in either NAFLD or AFLD. In humans with NAFLD, a fourfold increase in CYP4A11, which metabolizes arachidonic acid to 20-hydroxyeicosatetraenoic acid (HETE) (Nakamura et al., 2008), was observed with a slight increase in CYP2E1 during steatosis and a decrease of CYP2E1 in patients with NASH (Fisher et al., 2009). Because vitamin E improves liver histology in patients with NAFLD and that CYP4F2 is the major enzyme metabolizing vitamin E, participants in PIVENS and TONIC clinical trials were genotyped for CYP4F2 variants (V433M and W12G) (Athinarayanan et al., 2014). The results showed a significant decrease in plasma α-tocopherol in patients with CYP4F2 V433M genotype, but CYP4F2 polymorphisms likely play a minor or moderate role in the overall pharmacokinetics of vitamin E used as a therapeutic agent. A recent publication indicated that 20-HETE impairs endothelial insulin signaling by inducing the phosphorylation of IRS-1 (Li, Wong, et al., 2014; Li, Zhao, et al., 2014) with activation of SREBP-1α that induces the expression of mouse hepatic *CYP4A* genes, possibly leading to increased production of 20-HETE (Horton et al., 2003). These data indicate an important role of CYP4A/CYP4F produced 20-HETE in the regulation of insulin signaling in mice. However, the interplay of CYP4A11 and CYP4F2 P450s in the regulation of the fasting and feeding response in the progression of NAFLD needs further studies to identify the precise role of 20-HETE in insulin resistance and activation of AMPK by cellular stress.

Several reports indicated the differential expression of cytochrome P450 omega-hydroxylase isoforms in the clinic-pathological features of liver cirrhosis and cancer. The human CYP4F2 metabolizes the potent chemotactic eicosanoid leukotriene B4 to 20-hydroxy-leukotriene B4, which has less potent capabilities in recruiting immune cells. The induction of mouse CYP4A during hepatic steatosis along with fatty acid-induced uncoupling of the catalytic cycle can produce ROS. Increased ROS production and decreased levels of 20-hydroxy-leukotriene B4 due to suppressed CYP4F may be an important mechanism for providing the third hit, which promotes the progression of steatosis to steatohepatitis and eventually liver fibrosis, cirrhosis, and hepatocarcinogenesis. Decreased activity of CYP4F2 in the metabolism of arachidonic acid to 20-HETE due to Val433Met (1297C/T) substitution was strongly associated with rapid hepatic cirrhosis development ($OR = 6.0$, $CI = 0.28$, $p = 0.222$) (Vavilin et al., 2013). In contrast, the potent vasoconstrictive 20-HETE, which has strong mitogenic and angiogenic properties, is increased in tumors of liver, brain, kidney, and ovary with increased expression of *CYP4A/4F* genes compared to those in normal tissues (Alexanian, Miller, Roman, & Sorokin, 2012). Similarly, increased expression of CYP4A11, CYP4F2, and CYP4F3 isoforms were significantly expressed in pancreatic ductal adenocarcinoma (Gandhi et al., 2013), suggesting that 20-HETE, which increases expression of HIF and its downstream target vascular endothelial growth factor (VEGF), promotes blood vessel sprouting and metastasis by activation of metalloproteinases (MMPs) (Yu et al., 2011). Thus, selective inhibitors of 20-HETE synthesis by CYP4 omega hydroxylase have been demonstrated to reduce proliferation, angiogenesis, and invasion in lung, renal, and brain cancers (Edson & Rettie, 2013). Consistently, other reports indicated the utility of selective inhibitors of 20-HETE formation as potential therapeutic agents to inhibit tumor progression. In fact, the administration of HET0016 inhibited both 9L gliosarcoma and U251 glioma cell proliferation and tumor growth in a dose-dependent manner (Guo, Roman, Falck, Edwards, & Scicli, 2005), leading to increased mean survival time of the animals (Guo et al., 2006). Although these results and other reports suggest a promising role of 20-HETE antagonist as a therapeutic agent in the treatment of cancer, the development of isoform-selective antagonist may show increased efficacy without adverse drug reactions that may be present. Many of the presently used antagonists inhibit CYP4-mediated formation of 20-HETE in human microsomes with an IC50 value of less than 100 nM (Sato et al., 2001) although various CYP4A/4F isoforms can be differentially inhibited by broad-spectrum pan-CYP4 inhibitors (Miyata et al., 2001; Nakano, Kelly, & Rettie, 2009). These results suggest that careful cautions should be considered when using these pan-CYP4A inhibitors to define the role of 20-HETE CYP4A isoforms in the pathophysiological progression of disease. Thus, future efforts need to focus on the development of selective inducers and inhibitors of specific CYP4 subfamily members, and identification of major CYP4 isoforms in these widely diverse diseases.

Alcohol intake or nonalcoholic molecules can increase the levels of CYP3A and CYP2B, although the degree of their elevation is lower than that of CYP2E1 (Johansson et al., 1988; Niemelä et al., 2000). Since CYP3A is responsible for the metabolism of many drugs, it is likely that metabolic activation of some drugs by CYP3A may be directly related to drug disposition (Yin, Tomlinson, & Chow, 2010) or drug-induced cytotoxicity (Hosomi, Fukami, Iwamura, Nakajima, & Yokoi, 2011; Hosomi et al., 2010), especially after alcohol intake, as reported (Wolf et al., 2007). Examples of fat accumulation and DILI include APAP, isoniazid, valproate, tamoxifene, troglitazone, tacrin, rifampicin, and many others. The reactive metabolites of these drugs may be responsible for stimulating DILI (Jaeschke et al., 2012; Pessayre et al., 2012; Stachlewitz et al., 1997; Yuan & Kaplowitz, 2013). Alternatively, metabolism of these substrates may increase oxidative stress, which can activate the cell-death-associated JNK and/or p38K, leading to mitochondria-dependent apoptosis, as demonstrated with APAP (Bae et al., 2001) and troglitazone (Bae & Song, 2003). Furthermore, the levels of acetaldehyde and lipid peroxidation-protein adducts seem to correlate with the induced levels of CYP3A and CYP2E1 in alcohol or high-fat exposed rats, suggesting an important role of CYP3A in protein adducts formation (Niemelä et al., 1998).

Additive or synergistic interactions between alcohol and smoking can lead to increased hepatotoxicity and carcinogenesis in experimental animal models and human cases (Kuper et al., 2000; Purohit et al., 2013; Seitz & Cho, 2009). Chronic alcohol intake is known to increase the levels of CYP2A5, which can metabolize nicotine, a major ingredient of tobacco (Lu, Zhuge, Wu, & Cederbaum, 2011; Niemelä et al., 2000). In mice, alcohol feeding induces CYP2A5 in a CYP2E1-dependent manner (Lu et al., 2011), possibly through the CYP2E1-ROS-Nrf2 axis (Lu, Zhang, & Cederbaum, 2012). Elevated levels of hepatic CYP2A6 (the human ortholog of the mouse CYP2A5) were also observed in some patients with ALD or cirrhosis than the control, despite the small sample size (Lu et al., 2011). In the mouse model, ethanol-mediated CYP2A5 induction was dependent on the presence of CYP2E1, while ethanol induction of CYP2E1 was not CYP2A5 dependent. Ethanol-mediated CYP2A5 induction was not observed in *Cyp2e1*-null mice despite ethanol feeding. However, CYP2A5 induction was markedly elevated in the *Cyp2e1* knockin mice after treatment with ethanol but not with the dextrose-control. Furthermore, CYP2E1-dependent ROS was needed for CYP2A5 induction through activation of the NRF2 (Lu et al., 2012). Since CYP2A5 can also metabolize many cancer causing agents, such as aflatoxin B1 and nitrosamines, the induction of CYP2A5 in rodents by alcohol (Lu et al., 2011) or HFD (Choi et al., 2013) and CYP2A6 in alcoholic individuals is likely to contribute to increased oxidative stress and hepatic injury in ALD and NALD.

4. TRANSLATIONAL RESEARCH OPPORTUNITIES

As described above, increased activity of CYP2E1 with decreased ALDH2 activity can increase nitroxidative stress and hepatocyte apoptosis, as observed in AFLD, NAFLD, or acute DILI. Increased nitroxidative stress can promote multiple PTMs, as recently described (Song, Akbar, et al., 2014). The PTMs include the hydroxyethyl-adducts which were observed in alcohol-exposed rodents (Albano et al., 1996) and people (Clot et al., 1996). In addition, all these PTMs of ER and mitochondrial proteins are likely to contribute to increased ER stress and mitochondrial dysfunction, leading to accumulation of misfolded proteins with energy depletion, fat accumulation, altered metabolism, inflammation, and necrotic/apoptotic tissue damage. Because of recent developments in understanding the molecular mechanisms of ER stress and mitochondrial dysfunction in ALD, NALD, and DILI, it would be ideal to demonstrate or evaluate the efficacy of beneficial agents by studying the levels of oxidized, nitrated, phosphorylated proteins as novel approaches, as demonstrated (Song, Moon, Olsson, & Salem, 2008).

Due to the critical role of elevated nitroxidative stress in acute and chronic liver diseases, protective effects of many antioxidants from natural and synthetic origins have been evaluated in *in vitro* and *in vivo* models. These antioxidant agents include natural antioxidants [e.g., vitamin C, E, coenzyme Q10, alpha-lipoic acid, fish oil containing n-3 fatty acids, betaine, and *S*-adenosyl-methionine (SAMe)], L-arginine, small-molecule metabolites (e.g., GSH-ethyl ester and NAC), and plants polyphenols (silimarin in milk thistle, curcumin, esculetin, sulforaphane, resveratrol, quercetin, epigallocatechin-3-gallate, caffeic acid phenethyl ester, and many others) (Andringa et al., 2010; Bailey et al., 2006; Cao et al., 2013; Cederbaum, 2010; Choi et al., 2013; Chung et al., 2012; Esfandiari et al., 2007; Ji & Kaplowitz, 2003; Kharbanda et al., 2012; Kim, Nagy, et al., 2014; Kim, Quon, & Kim, 2014; Lee, Mcgregor, et al., 2013; Lee, Yun, Seo, Kim, & Lee, 2014; Lieber, 2002; Marcolin et al., 2012; Nanji et al., 2003; Powell et al., 2010; Rodrigues et al., 2013; Scorletti et al., 2014; Shin et al., 2014; Song, et al., 2008; Surapaneni et al., 2014). These antioxidants, contained in many fruits and vegetables, show beneficial effects on AFLD, NAFLD, and DILI, although some of these antioxidants also exhibit significant protection in other tissues such as heart, muscle, and brain (Pallauf, Giller, Huebbe, & Rimbach, 2013; Rodriguez et al., 2015). Some of these agents include inhibitors of CYP2E1 such as diallyl sulfide in garlic, phenethyl isothiocyanate (Stice et al., 2015; Yoshigae, Sridar, Kent, & Hollenberg, 2013) in crucible vegetables, or dioscin in edible plants (Xu et al., 2014), although these inhibitors seem less potent than the synthetic CYP2E1 inhibitor CMZ (Hu et al., 1994) or YH439 (Jeong et al.,

1996). In addition, general antioxidants such as SAMe and betaine were shown to preserve mitochondrial function and proteome in the animal models of AFLD (Bailey et al., 2006; Purohit et al., 2007) and NAFLD (Santamaria et al., 2003), partly through blocking the ROS/RNS production and restoring the physiological levels of GSH, SAMe, and *S*-adenosyl-homocysteine. Numerous other herbs and supplements have shown a bit of promise for protecting the liver, although some of them may cause harmful effects, as reported (Stickel & Shouval, 2015; Stickel et al., 2011). These protective agents include *Andrographis*, artichoke leaf, beet leaf, choline, dandelion, inositol, lecithin, licorice, lipoic acid, *Picrorhiza kurroa*, schisandra, taurine, and turmeric. In addition, probiotics, belonging to *Lactobacillus bifidus*, *L. acidophilus*, *L. bulgaricus*, and *S. thermophilus*, can play an important role for the prevention and/or treatment of liver disease possibly improving the gut microbiome. Based on these reports, it is expected that benefits of many other antioxidants will be described in the future.

Glucocorticosteroids represent the most widely accepted therapy in patients with severe ASH. In clinics, patients with severe ASH are treated with a short course of glucocorticoid therapy, as a first line of treatment agent, despite their side-effects. However, glucocorticoids cannot be used in patients with gastrointestinal bleeding, chronic hepatitis B virus infection, evidence of active infection, and probably in hepatorenal syndrome (Depew, Boyer, Omata, Redeker, & Reynolds, 1980). In these cases, the use of Pentoxifylline (PTX) is highly recommended (Frazier, 2011). PTX is a nonselective phosphodiesterase inhibitor that increases intracellular concentrations of adenosine 3′, 5′-cyclic monophosphate (cAMP) and guanosine 3′, 5′-cyclic monophosphate (cGMP), improves the alcoholic hepatitis via downregulation of proinflammatory cytokines such as TNFα. It has also been shown to have antifibrotic effects through the attenuation of both profibrogenic cytokines and procollagen I expression (Raetsch et al., 2002). In addition, some PPARα agonists, including Wy-14643, or PPARγ agonists, such as pioglitazone and rosiglitazone, have been effective against ALD and NAFLD in experimental models (Del Ben et al., 2014; Ip, Farrell, Hall, Robertson, & Leclercq, 2004; Lomonaco, Sunny, Bril, & Cusi, 2013). Therefore, drug repurposing of the already Food and Drug Administration-approved drugs should be carefully evaluated for the treatment and management of various liver diseases.

Mitochondrial ALDH2 is an important defensive enzyme against acetaldehyde and lipid peroxides such as 4-HNE and MDA. Many reports showed that ALDH2 and its isozymes are inactivated in alcohol-exposed animals and human alcoholics, as recently reviewed (Song et al., 2011). In fact, ALDH2 activity was shown to be suppressed by many hepatotoxic substances such as APAP, CCL4, cyanamide, disulfiram, MDMA, HFD, smoking, daunomycin, and others. The ALDH2 activity is also depressed in many pathological conditions such as I/R injury, obesity/diabetes, and cancer, possibly via oxidative modifications (e.g., oxidation, nitration, phosphorylation, acetylation, adduct formation, etc.) of the critical amino acids including the active site Cys residue, which is highly conserved among many ALDH isozymes (Moon et al., 2010). Decreased ALDH2 due to genetic mutation or oxidative inactivation seems to be a major risk factor for various disease states including alcoholic organ damage, cancer, and cardiovascular diseases (Song et al., 2011). Due to inactivated ALDH2 and other ALDH isozymes, the serum and tissue levels of toxic lipid peroxidation products such as acetaldehyde, 4-HNE, and MDA were consequently elevated, as reported earlier (Isse et al., 2005; Pawlosky et al., 1997). Elevation of these reactive compounds would cause DNA damage and promote acute cell death with the activation of immune cells. Therefore, restoration of the suppressed ALDH2 by antioxidants, as demonstrated with antioxidants dithiothreitol (DTT) (Moon et al., 2006) and lipoic acid (Wenzel et al., 2007) or small-molecule synthetic activators would be an ideal approach to protect various cells or organs from oxidative damage in ALD, NALD, and DILI. Furthermore, ALDH2 deletion was associated with increased cardiovascular problems (Wenzel et al., 2008). In fact, Mochly-Rosen and her colleagues identified ALDH2 activators through screening chemical libraries and demonstrated the protective effects of synthetic ALDH2 activators such as alda-1 [*N*-(1,3-benzodioxol-5-ylmethyl)-2,6-dichlorobenzamide], alda-44, and alda-89, which not only restored the suppressed ALDH2 activity but also significantly protected the heart under I/R condition (Budas, Disatnik, Chen, & Mochly-Rosen, 2010; Chen, Budas, et al., 2008). More recent reports showed that alda-1 is cardioprotective in post-myocardial infarction (Gomes et al., 2015) or in aged mice (Zhang et al., 2014) as well as hepatoprotective against alcohol-induced steatosis and cell death (Zhong et al., 2014). These results clearly demonstrate that ALDH2 could become a new, emerging target for developing medicines not only for treating cardiovascular diseases but also for other tissues including the liver and brain (Luo, Liu, Ma, & Peng, 2014). Due to the critical role of ALDH2 in serious liver disease (Day et al., 1991; Li, 2000; Li et al., 2001), we expect that small-molecule ALDH2 activators can protect the liver disease caused by alcohol, as recently reported (Zhong et al., 2014) and nonalcoholic substances and in DILI. For instance, alda-1 and its structural analogs could be used to support many East Asian people with the dominant negative mutant (i.e., *ALDH2*1/2* or *ALDH2*2/2)* gene.

Furthermore, many naturally occurring antioxidants from fruits and vegetables may have a weakness of poor quality control. For instance, the amounts of an active ingredient can be variable depending on the location of cultivation, seasonal weather, extraction process, being a minor component, and unacceptable levels of residual herbicides/pesticides. Other problems can be poor bioavailability with low solubility, stability, and little mitochondrial transport, as discussed (Song, Akbar, et al., 2014). To overcome these weaknesses, many antioxidants with improved mitochondrial targeting properties have been synthesized and evaluated to block increased nitroxidative stress and mitochondrial dysfunction in many pathological states including the experimental models of liver disease, as discussed above. For instance, various antioxidants including SOD-mimetics were developed to scavenge or remove peroxynitrite and thus prevent tissue injury. In fact, the analogs of SOD-catalase mimetics also showed beneficial effects on various disease states in many tissues (Melov et al., 2001). In addition, a peroxynitrite scavenger/SOD mimetic MnTMPyP were effective in preventing oxidative stress, mitochondrial dysfunction, and liver injury following I/R in mice (Moon, Hood, et al., 2008).

To improve the intracellular delivery of target molecules, triphenyl phosphonium (TPP^+, a cell-permeable lipophilic cation) has been developed for conjugation with various drugs and antioxidants, as reviewed (Murphy, 2014; Reily et al., 2013). The results with mitochondria-targeted ubiquinone (Mito-Q) or mitochondria-targeted carboxy-proxyl (Mito-CP) so far have shown promising results in preventing mitochondrial abnormalities and nitroxidative liver damage in mice following hepatic I/R procedure (Mukhopadhyay et al., 2012). In a dose-dependent manner, Mito-Q and Mito-CP significantly attenuated the nitroxidative stress markers (e.g., HNE/carbonyl adducts, MDA, 8-OH-dG, and 3-nitrotyrosine levels), mitochondrial dysfunction, and histopathological signs of liver injury as well as delayed inflammatory cell infiltration and cell death. Similarly, Mito-Q was also effective in preventing micro- and macro-vesicular steatosis in AFLD (Chacko et al., 2011). However, in this model, Mito-Q did neither change the levels of CYP2E1 and ALDH2 nor the mitochondrial respiratory abnormality caused by ethanol exposure. Therefore, the results strongly suggested that Mito-Q decreased the levels of protein nitration and

HIF-1α stabilization, possibly through suppressing the production of ROS/RNS. These beneficial results suggest that some synthetic antioxidants are far more effective than the untargeted natural antioxidants in neutralizing elevated nitroxidative stress and metabolic syndrome features, as reported (Feillet-Coudray et al., 2014). Because of the recent clinical testing, we expect to see approval of some of these antioxidants in treating oxidative stress-mediated various forms of liver disease.

5. CONCLUSION

We have briefly described the properties of the major alcohol-metabolizing enzymes, namely, ADH, ALDH2, and CYP2E1 in the liver. We also described their functional roles in promoting ER stress, mitochondrial dysfunction, apoptosis, fat accumulation, inflammation, DILI, fibrosis/cirrhosis, and hepatocarcinogenesis by alcoholic and nonalcoholic substances. These pathological conditions are likely mediated through a variety of PTMs of many ER and mitochondrial proteins under increased nitroxidative stress. In this review, we have described the opposite roles of ALDH2 and CYP2E1 in producing ROS/RNS, hypoxia, autophagy, protein modifications, JNK/p38K-related cell-death signaling to promote ER stress, mitochondrial dysfunction, and necroapoptosis in AFLD, NAFLD, and DILI. We also discussed the potential protections by using natural and synthetic antioxidants in preventing various liver diseases. Finally, further development of highly specific inhibitors of CYP2E1 and activators of ALDH2 without toxicities would provide opportunities and challenges in future translational research.

CONFLICT OF INTEREST

All authors declared no conflict of interest.

ACKNOWLEDGMENTS

This research was supported by the Intramural Program Fund at the National Institute on Alcohol Abuse and Alcoholism. The authors thank Dr. Klaus Gawrisch for his support.

REFERENCES

Abbondanza, A., Battelli, M. G., Soffritti, M., & Cessi, C. (1989). Xanthine oxidase status in ethanol-intoxicated rat liver. *Alcoholism: Clinical and Experimental Research*, *13*(6), 841–844.

Abdelmegeed, M. A., Banerjee, A., Jang, S., Yoo, S. H., Yun, J. W., Gonzalez, F. J., et al. (2013). CYP2E1 potentiates binge alcohol-induced gut leakiness, steatohepatitis, and apoptosis. *Free Radical Biology and Medicine*, *65*, 1238–1245.

Abdelmegeed, M. A., Banerjee, A., Yoo, S. H., Jang, S., Gonzalez, F. J., & Song, B. J. (2012). Critical role of cytochrome P450 2E1 (CYP2E1) in the development of high fat-induced non-alcoholic steatohepatitis. *Journal of Hepatology*, *57*(4), 860–866.

Abdelmegeed, M. A., Jang, S., Banerjee, A., Hardwick, J. P., & Song, B. J. (2013). Robust protein nitration contributes to acetaminophen-induced mitochondrial dysfunction and acute liver injury. *Free Radical Biology and Medicine*, *60*, 211–222.

Abdelmegeed, M. A., Moon, K. H., Chen, C., Gonzalez, F. J., & Song, B. J. (2010). Role of cytochrome P450 2E1 in protein nitration and ubiquitin-mediated degradation during acetaminophen toxicity. *Biochemical Pharmacology*, *79*(1), 57–66.

Abdelmegeed, M. A., & Song, B. J. (2014). Functional roles of protein nitration in acute and chronic liver diseases. *Oxidative Medicine and Cellular Longevity*, *2014*, 149627.

Abdelmegeed, M. A., Yoo, S. H., Henderson, L. E., Gonzalez, F. J., Woodcroft, K. J., & Song, B. J. (2011). PPARalpha expression protects male mice from high fat-induced nonalcoholic fatty liver. *Journal of Nutrition*, *141*(4), 603–610.

Addolorato, G., Capristo, E., Greco, A. V., Caputo, F., Stefanini, G. F., & Gasbarrini, G. (1998). Three months of abstinence from alcohol normalizes energy expenditure and substrate oxidation in alcoholics: A longitudinal study. *American Journal of Gastroenterology*, *93*(12), 2476–2481.

Adkins, Y., Schie, I. W., Fedor, D., Reddy, A., Nguyen, S., Zhou, P., et al. (2013). A novel mouse model of nonalcoholic steatohepatitis with significant insulin resistance. *Laboratory Investigation*, *93*(12), 1313–1322.

Agarwal, D. P. (2001). Genetic polymorphisms of alcohol metabolizing enzymes. *Pathological Biology (Paris)*, *49*(9), 703–709.

Alary, J., Gueraud, F., & Cravedi, J. P. (2003). Fate of 4-hydroxynonenal in vivo: Disposition and metabolic pathways. *Molecular Aspects of Medicine*, *24*(4–5), 177–187.

Albano, E., Clot, P., Morimoto, M., Tomasi, A., Ingelman-Sundberg, M., & French, S. W. (1996). Role of cytochrome P4502E1-dependent formation of hydroxyethyl free radical in the development of liver damage in rats intragastrically fed with ethanol. *Hepatology*, *23*(1), 155–163.

Alexanian, A., Miller, B., Roman, R. J., & Sorokin, A. (2012). 20-HETE-producing enzymes are up-regulated in human cancers. *Cancer Genomics Proteomics*, *9*(4), 163–169.

Amir, M., & Czaja, M. J. (2011). Autophagy in nonalcoholic steatohepatitis. *Expert Reviews of Gastroenterology and Hepatology*, *5*(2), 159–166.

Anderson, D. R., Duryee, M. J., Shurmur, S. W., Um, J. Y., Bussey, W. D., Hunter, C. D., et al. (2014). Unique antibody responses to malondialdehyde-acetaldehyde (MAA)-protein adducts predict coronary artery disease. *PLoS One*, *9*(9), e107440.

Andringa, K. K., King, A. L., Eccleston, H. B., Mantena, S. K., Landar, A., Jhala, N. C., et al. (2010). Analysis of the liver mitochondrial proteome in response to ethanol and S-adenosylmethionine treatments: Novel molecular targets of disease and hepatoprotection. *American Journal of Physiology. Gastrointestinal and Liver Physiology*, *298*(5), G732–G745.

Andringa, K. K., Udoh, U. S., Landar, A., & Bailey, S. M. (2014). Proteomic analysis of 4-hydroxynonenal (4-HNE) modified proteins in liver mitochondria from chronic ethanol-fed rats. *Redox Biology*, *2C*, 1038–1047.

Arteel, G. E., Iimuro, Y., Yin, M., Raleigh, J. A., & Thurman, R. G. (1997). Chronic enteral ethanol treatment causes hypoxia in rat liver tissue in vivo. *Hepatology*, *25*(4), 920–926.

Arteel, G. E., Raleigh, J. A., Bradford, B. U., & Thurman, R. G. (1996). Acute alcohol produces hypoxia directly in rat liver tissue in vivo: Role of Kupffer cells. *American Journal of Physiology*, *271*(3 Pt. 1), G494–G500.

Athinarayanan, S., Wei, R., Zhang, M., Bai, S., Traber, M. G., Yates, K., et al. (2014). Genetic polymorphism of cytochrome P450 4 F2, vitamin E level and histological response in adults and children with nonalcoholic fatty liver disease who participated in PIVENS and TONIC clinical trials. *PLoS One*, *9*(4), e95366.

Aubert, J., Begriche, K., Knockarert, L., Robin, M. A., & Fromenty, B. (2011). Increased expression of cytochrome P450 2E1 in nonalcoholic fatty live disease: Mechanism and pathophysiological role. *Clinical Research in Hepatology and Gastroenterology*, *35*(10), 630–637.

Auld, D. S., & Bergman, T. (2008). Medium- and short-chain dehydrogenase/reductase gene and protein families: The role of zinc for alcohol dehydrogenase structure and function. *Cellular and Molecular Life Sciences*, *65*(24), 3961–3970.

Badger, T. M., Ronis, M. J., Seitz, H. K., Albano, E., Ingelman-Sundberg, M., & Lieber, C. S. (2003). Alcohol metabolism: Role in toxicity and carcinogenesis. *Alcoholism: Clinical and Experimental Research*, *27*, 336e347.

Bae, M. A., Pie, J. E., & Song, B. J. (2001). Acetaminophen induces apoptosis of C6 glioma cells by activating the c-Jun NH(2)-terminal protein kinase-related cell death pathway. *Molecular Pharmacology*, *60*(4), 847–856.

Bae, M. A., & Song, B. J. (2003). Critical role of c-Jun N-terminal protein kinase activation in troglitazone-induced apoptosis of human HepG2 hepatoma cells. *Molecular Pharmacology*, *63*(2), 401–408.

Bailey, S. M., & Cunningham, C. C. (1999). Effect of dietary fat on chronic ethanol-induced oxidative stress in hepatocytes. *Alcoholism: Clinical and Experimental Research*, *23*(7), 1210–1218.

Bailey, S. M., Mantena, S. K., Millender-Swain, T., Cakir, Y., Jhala, N. C., Chhieng, D., et al. (2009). Ethanol and tobacco smoke increase hepatic steatosis and hypoxia in the hypercholesterolemic apoE(−/−) mouse: Implications for a "multihit" hypothesis of fatty liver disease. *Free Radical Biology and Medicine*, *46*(7), 928–938.

Bailey, S. M., Robinson, G., Pinner, A., Chamlee, L., Ulasova, E., Pompilius, M., et al. (2006). S-adenosylmethionine prevents chronic alcohol-induced mitochondrial dysfunction in the rat liver. *American Journal of Physiology Gastrointestinal and Liver Physiology*, *291*(5), G857–G867.

Bala, S., Marcos, M., Gattu, A., Catalano, D., & Szabo, G. (2014). Acute binge drinking increases serum endotoxin and bacterial DNA levels in healthy individuals. *PLoS One*, *9*(5), e96864.

Banfi, P., Lanzi, C., Falvella, F. S., Gariboldi, M., Gambetta, R. A., & Dragani, T. A. (1994). The daunorubicin-binding protein of Mr 54,000 is an aldehyde dehydrogenase and is down-regulated in mouse liver tumors and in tumor cell lines. *Molecular Pharmacology*, *46*(5), 896–900.

Bansal, S., Anandatheerthavarada, H. K., Prabu, G. K., Milne, G. L., Martin, M. V., Guengerich, F. P., et al. (2013). Human cytochrome P450 2E1 mutations that alter mitochondrial targeting efficiency and susceptibility to ethanol-induced toxicity in cellular models. *Journal of Biological Chemistry*, *288*(18), 12627–12644.

Bansal, S., Liu, C. P., Sepuri, N. B., Anandatheerthavarada, H. K., Selvaraj, V., Hoek, J., et al. (2010). Mitochondria-targeted cytochrome P450 2E1 induces oxidative damage and augments alcohol-mediated oxidative stress. *Journal of Biological Chemistry*, *285*(32), 24609–24619.

Bansal, S., Srinivasan, S., Anandasadagopan, S., Chowdhury, A. R., Selvaraj, V., Kalyanaraman, B., et al. (2012). Additive effects of mitochondrion-targeted cytochrome CYP2E1 and alcohol toxicity on cytochrome c oxidase function and stability of respirosome complexes. *Journal of Biological Chemistry*, *287*(19), 15284–15297.

Barak, A. J., Beckenhauer, H. C., Kharbanda, K. K., & Tuma, D. J. (2001). Chronic ethanol consumption increases homocysteine accumulation in hepatocytes. *Alcohol*, *25*(2), 77–81.

Bardag-Gorce, F., Li, J., French, B. A., & French, S. W. (2005). The effect of ethanol-induced CYP2E1 on proteasome activity: The role of 4-hydroxynonenal. *Experimental and Molecular Pathology*, *78*(2), 109–115.
Bardag-Gorce, F., Wilson, L., Nan, L., Li, J., French, B. A., Morgan, T. R., et al. (2005). CYP2E1 inhibition enhances mallory body formation. *Experimental and Molecular Pathology*, *78*(3), 207–211.
Barnes, M. A., Roychowdhury, S., & Nagy, L. E. (2014). Innate immunity and cell death in alcoholic liver disease: Role of cytochrome P4502E1. *Redox Biology*, *2*, 929–935.
Basuroy, S., Sheth, P., Mansbach, C. M., & Rao, R. K. (2005). Acetaldehyde disrupts tight junctions and adherens junctions in human colonic mucosa: Protection by EGF and L-glutamine. *American Journal of Physiology. Gastrointestinal and Liver Physiology*, *289*(2), G367–G375.
Battelli, M. G., Abbondanza, A., & Stirpe, F. (1992). Effects of hypoxia and ethanol on xanthine oxidase of isolated rat hepatocytes: Conversion from D to O form and leakage from cells. *Chemico-Biological Interactions*, *83*(1), 73–84.
Beck, A., & Heinz, A. (2013). Alcohol-related aggression-social and neurobiological factors. *Deutsches Ärzteblatt International*, *110*(42), 711–715.
Bergheim, I., Weber, S., Vos, M., Krämer, S., Volynets, V., Kaserouni, S., et al. (2008). Antibiotics protect against fructose-induced hepatic lipid accumulation in mice: Role of endotoxin. *Journal of Hepatology*, *48*(6), 983–992.
Beyer, C., Stearns, N. A., Giessl, A., Distler, J. H., Schett, G., & Pisetsky, D. S. (2012). The extracellular release of DNA and HMGB1 from Jurkat T cells during in vitro necrotic cell death. *Innate Immunity*, *18*(5), 727–737.
Biemond, P., Swaak, A. J., Beindorff, C. M., & Koster, J. F. (1986). Superoxide-dependent and -independent mechanisms of iron mobilization from ferritin by xanthine oxidase. Implications for oxygen-free-radical-induced tissue destruction during ischaemia and inflammation. *Biochemical Journal*, *239*(1), 169–173.
Bode, C., Kugler, V., & Bode, J. C. (1987). Endotoxemia in patients with alcoholic and non-alcoholic cirrhosis and in subjects with no evidence of chronic liver disease following acute alcohol excess. *Journal of Hepatology*, *4*(1), 8–14.
Boelsterli, U. A., & Lee, K. K. (2014). Mechanisms of isoniazid-induced idiosyncratic liver injury: Emerging role of mitochondrial stress. *Journal of Gastroenterology and Hepatology*, *29*(4), 678–687.
Bora, P. S., Spilburg, C. A., & Lange, L. G. (1989). Identification of a satellite fatty acid ethyl ester synthase from human myocardium as a glutathione S-transferase. *Journal of Clinical Investigation*, *84*(6), 1942–1946.
Bosron, W. F., Ehrig, T., & Li, T. K. (1993). Genetic factors in alcohol metabolism and alcoholism. *Seminars in Liver Diseases*, *13*, 126–135.
Bradford, B. U., Kono, H., Isayama, F., Kosyk, O., Wheeler, M. D., Akiyama, T. E., et al. (2005). Cytochrome P450 CYP2E1, but not nicotinamide adenine dinucleotide phosphate oxidase, is required for ethanol-induced oxidative DNA damage in rodent liver. *Hepatology*, *41*(2), 336–344.
Brenner, C., Galluzzi, L., Kepp, O., & Kroemer, G. (2013). Decoding cell death signals in liver inflammation. *Journal of Hepatology*, *59*(3), 583–594.
Brooks, P. J., Enoch, M. A., Goldman, D., Li, T. K., & Yokoyama, A. (2009). The alcohol flushing response: An unrecognized risk factor for esophageal cancer from alcohol consumption. *PLoS Medicine*, *6*, e50.
Budas, G. R., Disatnik, M. H., Chen, C. H., & Mochly-Rosen, D. (2010). Activation of aldehyde dehydrogenase 2 (ALDH2) confers cardioprotection in protein kinase C epsilon (PKCvarepsilon) knockout mice. *Journal of Molecular and Cellular Cardiology*, *48*(4), 757–764.

Cai, C., Huang, H., Whelan, S., Liu, L., Kautza, B., Luciano, J., et al. (2014). Benzyl alcohol attenuates acetaminophen-induced acute liver injury in a Toll-like receptor-4-dependent pattern in mice. *Hepatology, 60*(3), 990–1002.

Caldwell, S. H., Swerdlow, R. H., Khan, E. M., Iezzoni, J. C., Hespenheide, E. E., Parks, J. K., et al. (1999). Mitochondrial abnormalities in non-alcoholic steatohepatitis. *Journal of Hepatology, 31*(3), 430–434.

Cao, X., Chen, A., Yang, P., Song, X., Liu, Y., Li, Z., et al. (2013). Alpha-lipoic acid protects cardiomyocytes against hypoxia/reoxygenation injury by inhibiting autophagy. *Biochemical and Biophysical Research Communications, 441*(4), 935–940.

Cao, Q., Mak, K. M., & Lieber, C. S. (2005). Cytochrome P4502E1 primes macrophages to increase TNF-alpha production in response to lipopolysaccharide. *American Journal of Physiology. Gastrointestinal and Liver Physiology, 289*(1), G95–G107.

Cappelle, D., Neels, H., Yegles, M., Paulus, J., van Nuijs, A. L., Covaci, A., et al. (2015). Gas chromatographic determination of ethyl glucuronide in hair: Comparison between tandem mass spectrometry and single quadrupole mass spectrometry. *Forensic Science International, 249C*, 20–24.

Carmiel-Haggai, M., Cederbaum, A. I., & Nieto, N. (2005). A high-fat diet leads to the progression of non-alcoholic fatty liver disease in obese rats. *FASEB Journal, 19*(1), 136–138.

Caro, A. A., & Cederbaum, A. I. (2004). Oxidative stress, toxicology, and pharmacology of CYP2E1. *Annual Review of Pharmacology and Toxicology, 44*, 27–42.

Carr, R. M., Dhir, R., Yin, X., Agarwal, B., & Ahima, R. S. (2013). Temporal effects of ethanol consumption on energy homeostasis, hepatic steatosis, and insulin sensitivity in mice. *Alcoholism: Clinical and Experimental Research, 37*(7), 1091–1099.

Casazza, J. P., Felver, M. E., & Veech, R. L. (1984). The metabolism of acetone in rat. *Journal of Biological Chemistry, 259*(1), 231–236.

Cederbaum, A. I. (2010). Hepatoprotective effects of S-adenosyl-L-methionine against alcohol- and cytochrome P450 2E1-induced liver injury. *World Journal of Gastroenterology, 16*(11), 1366–1376.

Cederbaum, A. I. (2012a). CYP2E1 potentiates toxicity in obesity and after chronic ethanol treatment. *Drug Metabolism and Drug Interactions, 27*(3), 125–144.

Cederbaum, A. I. (2012b). Alcohol metabolism. *Clinics in Liver Disease, 16*(4), 667–685.

Cederbaum, A. I. (2014). Methodology to assay CYP2E1 mixed function oxidase catalytic activity and its induction. *Redox Biology, 2C*, 1048–1054.

Cederbaum, A. I., Lu, Y., Wang, X., & Wu, D. (2015). Synergistic toxic interactions between CYP2E1, LPS/TNFα, and JNK/p38 MAP Kinase and their implications in alcohol-induced liver injury. *Advances in Experimental Medicine and Biology, 815*, 145–172.

Cederbaum, A. I., Lu, Y., & Wu, D. (2009). Role of oxidative stress in alcohol-induced liver injury. *Archives of Toxicology, 83*(6), 519–548.

Cederbaum, A. I., Yang, L., Wang, X., & Wu, D. (2012). CYP2E1 Sensitizes the liver to LPS- and TNF α-induced toxicity via elevated oxidative and nitrosative stress and activation of ASK-1 and JNK mitogen-activated kinases. *International Journal of Hepatology, 2012*, 582790.

Ceni, E., Mello, T., & Galli, A. (2014). Pathogenesis of alcoholic liver disease: Role of oxidative metabolism. *World Journal of Gastroenterology, 20*(47), 17756–17772.

Chacko, B. K., Srivastava, A., Johnson, M. S., Benavides, G. A., Chang, M. J., Ye, Y., et al. (2011). Mitochondria-targeted ubiquinone (MitoQ) decreases ethanol-dependent micro and macro hepatosteatosis. *Hepatology, 54*(1), 153–163.

Chalasani, N., Gorski, J. C., Asghar, M. S., Asghar, A., Foresman, B., Hall, S. D., et al. (2003). Hepatic cytochrome P450 2E1 activity in nondiabetic patients with nonalcoholic steatohepatitis. *Hepatology, 37*(3), 544–550.

Charbonneau, A., & Marette, A. (2010). Inducible nitric oxide synthase induction underlies lipid-induced hepatic insulin resistance in mice: Potential role of tyrosine nitration of insulin signaling proteins. *Diabetes*, *59*(4), 861–871.

Chatterjee, S., Ganini, D., Tokar, E. J., Kumar, A., Das, S., Corbett, J., et al. (2013). Leptin is key to peroxynitrite-mediated oxidative stress and Kupffer cell activation in experimental non-alcoholic steatohepatitis. *Journal of Hepatology*, *58*(4), 778–784.

Chen, C.-H., Budas, G. R., Churchill, E. N., Disatnik, M. H., Hurley, T. D., & Mochly-Rosen, D. (2008). Activation of aldehyde dehydrogenase-2 reduces ischemic damage to the heart. *Science*, *321*(5895), 1493–1495.

Chen, C.-H., Gray, M. O., & Mochly-Rosen, D. (1999). Cardioprotection from ischemia by a brief exposure to physiological levels of ethanol: Role of epsilon protein kinase C. *Proceedings of the National Academy of Sciences of the United States of America*, *96*(22), 12784–12789.

Chen, G., Ma, C., Bower, K. A., Shi, X., Ke, Z., & Luo, J. (2008). Ethanol promotes endoplasmic reticulum stress-induced neuronal death: Involvement of oxidative stress. *Journal of Neuroscience Research*, *86*(4), 937–946.

Chen, Y. C., Peng, G. S., Wang, M. F., Tsao, T. P., & Yin, S. J. (2009). Polymorphism of ethanol-metabolism genes and alcoholism: Correlation of allelic variations with the pharmacokinetic and pharmacodynamic consequences. *Chemical and Biological Interactions*, *178*(1–3), 2–7.

Chen, J., Robinson, N. C., Schenker, S., Frosto, T. A., & Henderson, G. I. (1999). Formation of 4-hydroxynonenal adducts with cytochrome c oxidase in rats following short-term ethanol intake. *Hepatology*, *29*(6), 1792–1798.

Choi, M. K., Han, J. M., Kim, H. G., Lee, J. S., Lee, J. S., Wang, J. H., et al. (2013). Aqueous extract of Artemisia capillaris exerts hepatoprotective action in alcohol-pyrazole-fed rat model. *Journal of Ethnopharmacology*, *147*(3), 662–670.

Chung, M. Y., Park, H. J., Manautou, J. E., Koo, S. I., & Bruno, R. S. (2012). Green tea extract protects against nonalcoholic steatohepatitis in ob/ob mice by decreasing oxidative and nitrative stress responses induced by proinflammatory enzymes. *Journal of Nutritional Biochemistry*, *23*(4), 361–367.

Churchill, E. N., Disatnik, M. H., & Mochly-Rosen, D. (2009). Time-dependent and ethanol-induced cardiac protection from ischemia mediated by mitochondrial translocation of varepsilonPKC and activation of aldehyde dehydrogenase 2. *Journal of Molecular and Cellular Cardiology*, *46*(2), 278–284.

Clot, P., Albano, E., Eliasson, E., Tabone, M., Aricò, S., Israel, Y., et al. (1996). Cytochrome P4502E1 hydroxyethyl radical adducts as the major antigen in autoantibody formation among alcoholics. *Gastroenterology*, *111*(1), 206–216.

Cnop, M., Foufelle, F., & Velloso, L. A. (2012). Endoplasmic reticulum stress, obesity and diabetes. *Trends in Molecular Medicine*, *18*(1), 59–68.

Collins, M. A., Neafsey, E. J., Mukamal, K. J., Gray, M. O., Parks, D. A., Das, D. K., et al. (2009). Alcohol in moderation, cardioprotection, and neuroprotection: Epidemiological considerations and mechanistic studies. *Alcoholism: Clinical and Experimental Research*, *33*(2), 206–219.

Cremers, C. M., & Jakob, U. (2013). Oxidant sensing by reversible disulfide bond formation. *Journal of Biological Chemistry*, *288*(37), 26489–26496.

Criddle, D. N., Murphy, J., Fistetto, G., Barrow, S., Tepikin, A. V., Neoptolemos, J. P., et al. (2006). Fatty acid ethyl esters cause pancreatic calcium toxicity via inositol trisphosphate receptors and loss of ATP synthesis. *Gastroenterology*, *130*(3), 781–793.

Criddle, D. N., Raraty, M. G., Neoptolemos, J. P., Tepikin, A. V., Petersen, O. H., & Sutton, R. (2004). Ethanol toxicity in pancreatic acinar cells: Mediation by nonoxidative fatty acid metabolites. *Proceedings of the National Academy of Sciences of the United States of America*, *101*(29), 10738–10743.

Czaja, M. J. (2010). JNK regulation of hepatic manifestations of the metabolic syndrome. *Trends in Endocrinology and Metabolism*, *21*(12), 707–713.

Czaja, M. J. (2011). Functions of autophagy in hepatic and pancreatic physiology and disease. *Gastroenterology*, *140*(7), 1895–1908.

Dai, Z., Wu, Z., Yang, Y., Wang, J., Satterfield, M. C., Meininger, C. J., et al. (2013). Nitric oxide and energy metabolism in mammals. *Biofactors*, *39*(4), 383–391.

Daly, A. K. (2013). Relevance of CYP2E1 to non-alcoholic fatty liver disease. *Subcellular Biochemistry*, *67*, 165–175.

Day, C. P., Bashir, R., James, O. F., Bassendine, M. F., Crabb, D. W., Thomasson, H. R., et al. (1991). Investigation of the role of polymorphisms at the alcohol and aldehyde dehydrogenase loci in genetic predisposition to alcohol-related end-organ damage. *Hepatology*, *14*(5), 798–801.

Day, C. P., & James, O. F. (1998). Steatohepatitis: A tale of two "hits"? *Gastroenterology*, *114*(4), 842–845.

Deitrich, R. (2011). Ethanol as a prodrug: Brain metabolism of ethanol mediates its reinforcing effects—A commentary. *Alcoholism: Clinical and Experimental Research*, *35*(4), 581–583.

Del Ben, M., Polimeni, L., Baratta, F., Pastori, D., Loffredo, L., & Angelico, F. (2014). Modern approach to the clinical management of non-alcoholic fatty liver disease. *World Journal of Gastroenterology*, *20*(26), 8341–8350.

dela Peña, A., Leclercq, I. A., Williams, J., & Farrell, G. C. (2007). NADPH oxidase is not an essential mediator of oxidative stress or liver injury in murine MCD diet-induced steatohepatitis. *Journal of Hepatology*, *46*(2), 304–313.

Deng, X. S., & Deitrich, R. A. (2007). Ethanol metabolism and effects: Nitric oxide and its interaction. *Current Clinical Pharmacology*, *2*(2), 145–153.

Depew, W., Boyer, T., Omata, M., Redeker, A., & Reynolds, T. (1980). Double-blind controlled trial of prednisolone therapy in patients with severe acute alcoholic hepatitis and spontaneous encephalopathy. *Gastroenterology*, *78*(3), 524–529.

Depner, C. M., Philbrick, K. A., & Jump, D. B. (2013). Docosahexaenoic acid attenuates hepatic inflammation, oxidative stress, and fibrosis without decreasing hepatosteatosis in a Ldlr(−/−) mouse model of western diet-induced nonalcoholic steatohepatitis. *Journal of Nutrition*, *143*(3), 315–323.

Dey, A., & Kumar, S. M. (2011). Cytochrome P450 2E1 and hyperglycemia-induced liver injury. *Cell Biology and Toxicology*, *27*(40), 285–310.

Dey, A., & Swaminathan, K. (2010). Hyperglycemia-induced mitochondrial alterations in liver. *Life Sciences*, *87*(7–8), 197–214.

Ding, W. X., Manley, S., & Ni, H. M. (2011). The emerging role of autophagy in alcoholic liver disease. *Experimental Biology and Medicine (Maywood)*, *236*(5), 546–556.

Doggett, T. M., & Breslin, J. W. (2014). Acute alcohol intoxication-induced microvascular leakage. *Alcoholism: Clinical and Experimental Research*, *38*(9), 2414–2426.

Doorn, J. A., Hurley, T. D., & Petersen, D. R. (2006). Inhibition of human mitochondrial aldehyde dehydrogenase by 4-hydroxynon-2-enal and 4-oxonon-2-enal. *Chemical Research in Toxicology*, *19*(1), 102–110.

Doser, T. A., Turdi, S., Thomas, D. P., Epstein, P. N., Li, S. Y., & Ren, J. (2009). Transgenic overexpression of aldehyde dehydrogenase-2 rescues chronic alcohol intake-induced myocardial hypertrophy and contractile dysfunction. *Circulation*, *119*(14), 1941–1949.

Dunagan, M., Chaudhry, K., Samak, G., & Rao, R. K. (2012). Acetaldehyde disrupts tight junctions in Caco-2 cell monolayers by a protein phosphatase 2A-dependent mechanism. *American Journal of Physiology. Gastrointestinal and Liver Physiology*, *303*(12), G1356–G1364.

Duryee, M. J., Freeman, T. L., Willis, M. S., Hunter, C. D., Hamilton, B. C., 3rd, Suzuki, H., et al. (2005). Scavenger receptors on sinusoidal liver endothelial cells are involved in the uptake of aldehyde-modified proteins. *Molecular Pharmacology, 68*(5), 1423–1430.

Eccleston, H. B., Andringa, K. K., Betancourt, A. M., King, A. L., Mantena, S. K., Swain, T. M., et al. (2011). Chronic exposure to a high-fat diet induces hepatic steatosis, impairs nitric oxide bioavailability, and modifies the mitochondrial proteome in mice. *Antioxidants and Redox Signaling, 15*(2), 447–459.

Edson, K. Z., & Rettie, A. E. (2013). CYP4 Enzymes as potential drug targets: Focus on enzyme multiplicity, inducers and inhibitors, and therapeutic modulation of 20-hydroxyeicosatetraenoic acid (20-HETE) synthase and fatty acid ω-hydroxylase activities. *Current Topics in Medicinal Chemistry, 13*(12), 1429–1440.

Eguchi, A., Wree, A., & Feldstein, A. E. (2014). Biomarkers of liver cell death. *Journal of Hepatology, 60*(5), 1063–1074.

Ekström, G., & Ingelman-Sundberg, M. (1989). Rat liver microsomal NADPH-supported oxidase activity and lipid peroxidation dependent on ethanol-inducible cytochrome P-450 (P-450IIE1). *Biochemical Pharmacology, 38*(8), 1313–1319.

Elamin, E. E., Masclee, A. A., Dekker, J., & Jonkers, D. M. (2013). Ethanol metabolism and its effects on the intestinal epithelial barrier. *Nutrition Reviews, 71*(7), 483–499.

Elamin, E., Masclee, A., Juuti-Uusitalo, K., van Ijzendoorn, S., Troost, F., Pieters, H. J., et al. (2013). Fatty acid ethyl esters induce intestinal epithelial barrier dysfunction via a reactive oxygen species-dependent mechanism in a three-dimensional cell culture model. *PLoS One, 8*(3), e58561.

Elamin, E., Masclee, A., Troost, F., Dekker, J., & Jonkers, D. (2014). Activation of the epithelial-to-mesenchymal transition factor snail mediates acetaldehyde-induced intestinal epithelial barrier disruption. *Alcoholism: Clinical and Experimental Research, 38*(2), 344–353.

Enriquez, A., Leclercq, I., Farrell, G. C., & Robertson, G. (1999). Altered expression of hepatic CYP2E1 and CYP4A in obese, diabetic ob/ob mice, and fa/fa Zucker rats. *Biochemical and Biophysical Research Communications, 255*(2), 300–306.

Esfandiari, F., You, M., Villanueva, J. A., Wong, D. H., French, S. W., & Halsted, C. H. (2007). S-adenosylmethionine attenuates hepatic lipid synthesis in micropigs fed ethanol with a folate-deficient diet. *Alcoholism: Clinical and Experimental Research, 31*(7), 1231–1239.

Fan, X., Joshi, P. C., Koval, M., & Guidot, D. M. (2011). Chronic alcohol ingestion exacerbates lung epithelial barrier dysfunction in HIV-1 transgenic rats. *Alcoholism: Clinical and Experimental Research, 35*(10), 1866–1875.

Farres, J., Wang, X., Takahashi, K., Cunningham, S. J., Wang, T. T., & Weiner, H. (1994). Effects of changing glutamate 487 to Lysine in rat and human liver mitochondrial aldehyde dehydrogenase. A model to study human (Oriental type) class 2 aldehyde dehydrogenase. *Journal of Biological Chemistry, 269*(19), 13854–13860.

Feillet-Coudray, C., Fouret, G., Ebabe, E. R., Rieusset, J., Bonafo, B., Chabi, B., et al. (2014). The mitochondrial-targeted antioxidant MitoQ ameliorates metabolic syndrome features in obesogenic diet-fed rats better than Apocynin or Allopurinol. *Free Radical Research, 48*(10), 1232–1246.

Feldstein, A. E., & Bailey, S. M. (2011). Emerging role of redox dysregulation in alcoholic and nonalcoholic fatty liver disease. *Antioxidants and Redox Signaling, 15*(2), 421–424.

Feltes, B. C., de Faria Poloni, J., Nunes, I. J., & Bonatto, D. (2014). Fetal alcohol syndrome, chemo-biology and OMICS: Ethanol effects on vitamin metabolism during neurodevelopment as measured by systems biology analysis. *OMICS, 18*(6), 344–363.

Fischer, M., You, M., Matsumoto, M., & Crabb, D. W. (2003). Peroxisome proliferator-activated receptor alpha (PPARalpha) agonist treatment reverses PPARalpha dysfunction and abnormalities in hepatic lipid metabolism in ethanol-fed mice. *Journal of Biological Chemistry, 278*(30), 27997–28004.

Fisher, C. D., Lickteig, A. J., Augustine, L. M., Ranger-Moore, J., Jackson, J. P., Ferguson, S. S., et al. (2009). Hepatic cytochrome P450 enzyme alterations in humans with progressive stages of nonalcoholic fatty liver disease. *Drug Metabolism and Disposition*, *37*(10), 2087–2094.

Forsyth, C. B., Voigt, R. M., & Keshavarzian, A. (2014). Intestinal CYP2E1: A mediator of alcohol-induced gut leakiness. *Redox Biology*, *3*, 40–46.

Forsyth, C. B., Voigt, R. M., Shaikh, M., Tang, Y., Cederbaum, A. I., Turek, F. W., et al. (2013). Role for intestinal CYP2E1 in alcohol-induced circadian gene-mediated intestinal hyperpermeability. *American Journal of Physiology. Gastrointestinal and Liver Physiology*, *305*(2), G185–G195.

Frazier, T. H. (2011). Treatment of alcoholic liver disease. *Therapeutic Advances in Gastroenterology*, *4*(1), 63–81.

French, S. W., Wong, K., Jui, L., Albano, E., Hagbjork, A. L., & Ingelman-Sundberg, M. (1993). Effect of ethanol on cytochrome P450 2E1 (CYP2E1), lipid peroxidation, and serum protein adduct formation in relation to liver pathology pathogenesis. *Experimental and Molecular Pathology*, *58*(1), 61–75.

Fritz, K. S., Galligan, J. J., Smathers, R. L., Roede, J. R., Shearn, C. T., Reigan, P., et al. (2011). 4-Hydroxynonenal inhibits SIRT3 via thiol-specific modification. *Chemical Research in Toxicology*, *24*(5), 651–662.

Fritz, K. S., & Petersen, D. R. (2013). An overview of the chemistry and biology of reactive aldehydes. *Free Radical Biology and Medicine*, *59*, 85–91.

Fromenty, B., Grimbert, S., Mansouri, A., Beaugrand, M., Erlinger, S., Rötig, A., et al. (1995). Hepatic mitochondrial DNA deletion in alcoholics: Association with microvesicular steatosis. *Gastroenterology*, *108*(1), 193–200.

Fujimoto, M., Shimizu, N., Kunii, K., Martyn, J. A., Ueki, K., & Kaneki, M. (2005). A role for iNOS in fasting hyperglycemia and impaired insulin signaling in the liver of obese diabetic mice. *Diabetes*, *54*(5), 1340–1348.

Galli, A., Pinaire, J., Fischer, M., Dorris, D., & Crabb, D. R. (2001). The transcriptional and DNA binding activity of peroxisome proliferator-activated receptor α is inhibited by ethanol metabolism. A novel mechanism for the development for the development of ethanol-induced fatty liver. *Journal of Biological Chemistry*, *276*(1), 68–75.

Galligan, J. J., Fritz, K. S., Tipney, H., Smathers, R. L., Roede, J. R., Shearn, C. T., et al. (2012). Profiling impaired hepatic endoplasmic reticulum glycosylation as a consequence of ethanol ingestion. *Journal of Proteome Research*, *10*(4), 1837–1847.

Galligan, J. J., Smathers, R. L., Fritz, K. S., Epperson, L. E., Hunter, L. E., & Petersen, D. R. (2012). Protein carbonylation in a murine model for early alcoholic liver disease. *Chemical Research in Toxicology*, *25*(5), 1012–1021.

Galligan, J. J., Smathers, R. L., Shearn, C. T., Fritz, K. S., Backos, D. S., Jiang, H., et al. (2012). Oxidative stress and the ER stress response in a murine model for early-stage alcoholic liver disease. *Journal of Toxicology*, *2012*, 207594.

Gandhi, A. V., Saxena, S., Relles, D., Sarosiek, K., Kang, C. Y., Chipitsyna, G., et al. (2013). Differential expression of cytochrome P450 omega-hydroxylase isoforms and their association with clinicopathological features in pancreatic ductal adenocarcinoma. *Annals of Surgical Oncology*, *20*(Suppl. 3), S636–S643.

Gao, B., & Bataller, R. (2011). Alcoholic liver disease: Pathogenesis and new therapeutic targets. *Gastroenterology*, *141*(5), 1572–1585.

Gao, D., Nong, S., Huang, X., Lu, Y., Zhao, H., Lin, Y., et al. (2010). The effects of palmitate on hepatic insulin resistance are mediated by NADPH Oxidase 3-derived reactive oxygen species through JNK and p38MAPK pathways. *Journal of Biological Chemistry*, *285*(39), 29965–29973.

Gao, Z., Zhang, J., Kheterpal, I., Kennedy, N., Davis, R. J., & Ye, J. (2011). Sirtuin 1 (SIRT1) protein degradation in response to persistent c-Jun N-terminal kinase 1

(JNK1) activation contributes to hepatic steatosis in obesity. *Journal of Biological Chemistry, 286*, 22227–22234.
Gao, W., Zhou, P., Ma, X., Tschudy-Seney, B., Chen, J., Magner, N. L., et al. (2014). Ethanol negatively regulates hepatic differentiation of hESC by inhibition of the MAPK/ERK signaling pathway in vitro. *PLoS One, 9*(11), e112698.
Ge, X., Antoine, D. J., Lu, Y., Arriazu, E., Leung, T. M., Klepper, A. L., et al. (2014). High mobility group box-1 (HMGB1) participates in the pathogenesis of alcoholic liver disease (ALD). *Journal of Biological Chemistry, 289*(33), 22672–22691.
Gentile, C. L., Frye, M., & Pagliassotti, M. J. (2011). Endoplasmic reticulum stress and the unfolded protein response in nonalcoholic fatty liver disease. *Antioxidants and Redox Signaling, 15*(2), 505–521.
Gomes, K. M., Bechara, L. R., Lima, V. M., Ribeiro, M. A., Campos, J. C., Dourado, P. M., et al. (2015). Aldehydic load and aldehyde dehydrogenase 2 profile during the progression of post-myocardial infarction cardiomyopathy: Benefits of Alda-1. *International Journal of Cardiology, 179*, 129–138.
Guengerich, F. P., Kim, D. H., & Iwasaki, M. (1991). Role of human cytochrome P-450 IIE1 in the oxidation of many low molecular weight cancer suspects. *Chemical Research in Toxicology, 4*(2), 168–179.
Gunzerath, L., Faden, V., Zakhari, S., & Warren, K. (2004). National Institute on Alcohol Abuse and Alcoholism report on moderate drinking. *Alcoholism: Clinical and Experimental Research, 28*(6), 829–847.
Guo, M., Roman, R. J., Falck, J. R., Edwards, P. A., & Scicli, A. G. (2005). Human U251 glioma cell proliferation is suppressed by HET0016 [N-hydroxy-N′-(4-butyl-2-methylphenyl)formamidine], a selective inhibitor of CYP4A. *The Journal of Pharmacology and Experimental Therapeutics, 315*(2), 526–533.
Guo, M., Roman, R. J., Fenstermacher, J. D., Brown, S. L., Falck, J. R., Arbab, A. S., et al. (2006). 9 L gliosarcoma cell proliferation and tumor growth in rats are suppressed by N-hydroxy-N′-(4-butyl-2-methylphenol) formamidine (HET0016): A selective inhibitor of CYP4A. *The Journal of Pharmacology and Experimental Therapeutics, 317*(1), 97–108.
Hardwick, J. P. (2008). Cytochrome P450 omega hydroxylase (CYP4) function in fatty acid metabolism and metabolic diseases. *Biochemical Pharmacology, 75*(12), 2263–2275.
Hardwick, J. P., Eckman, K., Lee, Y. K., Abdelmegeed, M. A., Esterle, A., Chilian, W. M., et al. (2013). Eicosanoids in metabolic syndrome. *Advances in Pharmacology, 66*, 157–266.
Hart, C. L., Morrison, D. S., Batty, G. D., Mitchell, R. J., & Davey, S. G. (2010). Effect of body mass index and alcohol consumption on liver disease: Analysis of data from two prospective cohort studies. *British Medical Journal, 340*, c1240.
Hartley, D. P., Ruth, J. A., & Petersen, D. R. (1995). The hepatocellular metabolism of 4-hydroxynonenal by alcohol dehydrogenase, aldehyde dehydrogenase, and glutathione S-transferase. *Archives of Biochemistry and Biophysics, 316*(1), 197–205.
Hassan, M. M., Hwang, L. Y., Hatten, C. J., Swaim, M., Li, D., Abbruzzese, J. L., et al. (2002). Risk factors for hepatocellular carcinoma: Synergism of alcohol with viral hepatitis and diabetes mellitus. *Hepatology, 36*(5), 1206–1213.
He, J., Yang, Z., Yang, H., Wang, L., Wu, H., Fan, Y., et al. (2015). Regulation of insulin sensitivity, insulin production, and pancreatic β cell survival by angiotensin-(1–7) in a rat model of streptozotocin-induced diabetes mellitus. *Peptides, 64C*, 49–54.
Heneberg, P., & Dráber, P. (2005). Regulation of cys-based protein tyrosine phosphatases via reactive oxygen and nitrogen species in mast cells and basophils. *Current Medicinal Chemistry, 12*(16), 1859–1871.
Hernández-Gea, V., Hilscher, M., Rozenfeld, R., Lim, M. P., Nieto, N., Werner, S., et al. (2013). Endoplasmic reticulum stress induces fibrogenic activity in hepatic stellate cells through autophagy. *Journal of Hepatology, 59*(1), 98–104.

Hetz, C., Chevet, E., & Harding, H. P. (2013). Targeting the unfolded protein response in disease. *Nature Reviews Drug Discovery*, *12*(9), 703–719.

Higuchi, S., Matsushita, S., Imazaki, H., Kinoshita, T., Takagi, S., & Kono, H. (1994). Aldehyde dehydrogenase genotypes in Japanese alcoholics. *Lancet*, *343*(8899), 741–742.

Hirschey, M. D., Shimazu, T., Jing, E., Grueter, C. A., Collins, A. M., Aouizerat, B., et al. (2011). SIRT3 deficiency and mitochondrial protein hyperacetylation accelerate the development of the metabolic syndrome. *Molecular Cell*, *44*(2), 177–190.

Hoek, J. B., Cahill, A., & Pastorino, J. G. (2002). Alcohol and mitochondria: A dysfunctional relationship. *Gastroenterology*, *122*(7), 2049–2063.

Hoek, J., Thiele, G. M., Klassen, L. W., Mandrekar, P., Zakhari, S., Cook, R. T., et al. (2004). RSA 2004: Combined basic research satellite symposium-mechanisms of alcohol-mediated organ and tissue damage: Inflammation and immunity and alcohol and mitochondrial metabolism: At the crossroads of life and death session one: Alcohol, cellular and organ damage. *Alcoholism: Clinical and Experimental Research*, *29*(9), 1735–1743.

Hong, F., Radaeva, S., Pan, H. N., Tian, Z., Veech, R., & Gao, B. (2004). Interleukin 6 alleviates hepatic steatosis and ischemia/reperfusion injury in mice with fatty liver disease. *Hepatology*, *40*(4), 933–941.

Horton, J. D., Shah, N. A., Warrington, J. A., Anderson, N. N., Park, S. W., Brown, M. S., et al. (2003). Combined analysis of oligonucleotide microarray data from transgenic and knockout mice identifies direct SREBP target genes. *Proceedings of the National Academy of Sciences of the United States of America*, *100*(21), 12027–12032.

Hosomi, H., Akai, S., Minami, K., Yoshikawa, Y., Fukami, T., Nakajima, M., et al. (2010). An in vitro drug-induced hepatotoxicity screening system using CYP3A4-expressing and gamma-glutamylcysteine synthetase knockdown cells. *Toxicology In Vitro*, *24*(3), 1032–1038.

Hosomi, H., Fukami, T., Iwamura, A., Nakajima, M., & Yokoi, T. (2011). Development of a highly sensitive cytotoxicity assay system for CYP3A4-mediated metabolic activation. *Drug Metabolism and Disposition*, *39*(8), 1388–1395.

Hu, Y., Mishin, V., Johansson, I., von Bahr, C., Cross, A., Ronis, M. J., et al. (1994). Chlormethiazole as an efficient inhibitor of cytochrome P450 2E1 expression in rat liver. *Journal of Pharmacology and Experimental Therapeutics*, *269*(3), 1286–1291.

Huang, H., Nace, G. W., McDonald, K. A., Tai, S., Klune, J. R., Rosborough, B. R., et al. (2014). Hepatocyte-specific high-mobility group box 1 deletion worsens the injury in liver ischemia/reperfusion: A role for intracellular high-mobility group box 1 in cellular protection. *Hepatology*, *59*(5), 1984–1997.

Ip, E., Farrell, G., Hall, P., Robertson, G., & Leclercq, I. (2004). Administration of the potent PPARalpha agonist, Wy-14,643, reverses nutritional fibrosis and steatohepatitis in mice. *Hepatology*, *39*(5), 1286–1296.

Israel, Y. (1997). Antibodies against ethanol-derived protein adducts: Pathogenic implications. *Gastroenterology*, *113*(1), 353–355.

Israel, Y., Kalant, H., Orrego, H., Khanna, J. M., Videla, L., & Phillips, J. M. (1975). Experimental alcohol-induced hepatic necrosis: Suppression by propylthiouracil. *Proceedings of the National Academy of Sciences of the United States of America*, *72*(3), 1137–1141.

Israel, Y., Khanna, J. M., & Lin, R. (1970). Effect of 2,4-dinitrophenol on the rate of ethanol elimination in the rat in vivo. *Biochemical Journal*, *120*(2), 447–448.

Israel, Y., Orrego, H., & Carmichael, F. J. (1994). Acetate-mediated effects of ethanol. *Alcoholism: Clinical and Experimental Research*, *18*(1), 144–148.

Isse, T., Matsuno, K., Oyama, T., Kitagawa, K., & Kawamoto, T. (2005). Aldehyde dehydrogenase 2 gene targeting mouse lacking enzyme activity shows high acetaldehyde level in blood, brain, and liver after ethanol gavages. *Alcoholism: Clinical and Experimental Research*, *29*(11), 1959–1964.

Ito, Y., Abril, E. R., Bethea, N. W., & McCuskey, R. S. (2004). Role of nitric oxide in hepatic microvascular injury elicited by acetaminophen in mice. *American Journal of Physiology. Gastrointestinal and Liver Physiology*, *286*(1), G60–G67.

Jaeschke, H., Gores, G. J., Cederbaum, A. I., Hinson, J. A., Pessayre, D., & Lemasters, J. J. (2002). Mechanisms of hepatotoxicity. *Toxicological Sciences*, *65*(2), 166–176.

Jaeschke, H., McGill, M. R., & Ramachandran, A. (2012). Oxidant stress, mitochondria, and cell death mechanisms in drug-induced liver injury: Lessons learned from acetaminophen hepatotoxicity. *Drug Metabolism Reviews*, *44*(1), 88–106.

James, L. P., Donahower, B., Burke, A. S., McCullough, S., & Hinson, J. A. (2006). Induction of the nuclear factor HIF-1alpha in acetaminophen toxicity: Evidence for oxidative stress. *Biochemical and Biophysical Research Communications*, *343*(1), 171–176.

Jennett, R. B., Sorrell, M. F., Saffari-Fard, A., Ockner, J. L., & Tuma, D. J. (1989). Preferential covalent binding of acetaldehyde to the alpha-chain of purified rat liver tubulin. *Hepatology*, *9*(1), 57–62.

Jeong, K. S., Lee, I. J., Roberts, B. J., Soh, Y., Yoo, J. K., Lee, J. W., et al. (1996). Transcriptional inhibition of cytochrome P4502E1 by a synthetic compound, YH439. *Archives of Biochemistry and Biophysics*, *326*(1), 137–144.

Jeong, K. S., Soh, Y., Jeng, J., Felder, M. R., Hardwick, J. P., & Song, B. J. (2000). Cytochrome P450 2E1 (CYP2E1)-dependent production of a 37-kDa acetaldehyde-protein adduct in the rat liver. *Archives of Biochemistry and Biophysics*, *384*(1), 81–87.

Jerrells, T. R., Peritt, D., Marietta, C., & Eckardt, M. J. (1989). Mechanisms of suppression of cellular immunity induced by ethanol. *Alcoholism: Clinical and Experimental Research*, *13*(4), 490–493.

Ji, C. (2008). Dissection of endoplasmic reticulum stress signaling in alcoholic and nonalcoholic liver injury. *Journal of Gastroenterology and Hepatology*, *23*(Suppl. 1), S16–S24.

Ji, C., & Kaplowitz, N. (2003). Betaine decreases hyperhomocysteinemia, endoplasmic reticulum stress, and liver injury in alcohol-fed mice. *Gastroenterology*, *124*(5), 1488–1499.

Ji, S., Lemasters, J. J., Christenson, V., & Thurman, R. G. (1982). Periportal and pericentral pyridine nucleotide fluorescence from the surface of the perfused liver: Evaluation of the hypothesis that chronic treatment with ethanol produces pericentral hypoxia. *Proceedings of the National Academy of Sciences of the United States of America*, *79*(17), 5415–5419.

Jia, G., Habibi, J., Bostick, B. P., Ma, L., DeMarco, V. G., Aroor, A. R., et al. (2015). Uric Acid promotes left ventricular diastolic dysfunction in mice fed a Western diet. *Hypertension*, *65*(3), 531–539.

Jin, Y., & Penning, T. M. (2007). Aldo-keto reductases and bioactivation/detoxication. *Annual Review of Pharmacology and Toxicology*, *47*, 263–292.

Jo, S. A., Kim, E. K., Park, M. H., Han, C., Park, H. Y., Jang, Y., et al. (2007). A Glu487Lys polymorphism in the gene for mitochondrial aldehyde dehydrogenase 2 is associated with myocardial infarction in elderly Korean men. *Clinica Chimica Acta*, *382*(1–2), 43–47.

Johansson, I., Ekström, G., Scholte, B., Puzycki, D., Jörnvall, H., & Ingelman-Sundberg, M. (1988). Ethanol-, fasting-, and acetone-inducible cytochromes P-450 in rat liver: Regulation and characteristics of enzymes belonging to the IIB and IIE gene subfamilies. *Biochemistry*, *27*(6), 1925–1934.

Ka, T., Moriwaki, Y., Inokuchi, T., Yamamoto, A., Takahashi, S., Tsutsumi, Z., et al. (2006). Effects of allopurinol on beer-induced increases in plasma concentrations and urinary excretion of purine bases (uric acid, hypoxanthine, and xanthine). *Hormone and Metabolic Research*, *38*(3), 188–192.

Kalf, G. F., Post, G. B., & Snyder, R. (1987). Solvent toxicology: Recent advances in the toxicology of benzene, the glycol ethers, and carbon tetrachloride. *Annual Review of Pharmacology and Toxicology*, *27*, 399–427.

Kamanaka, Y., Kawabata, A., Matsuya, H., Taga, C., Sekiguchi, F., & Kawao, N. (2003). Effect of a potent iNOS inhibitor (ONO-1714) on acetaminophen-induced hepatotoxicity in the rat. *Life Sciences*, *74*(6), 793–802.

Kamo, N., Ke, B., Ghaffari, A. A., Shen, X. D., Busuttil, R. W., Cheng, G., et al. (2013). ASC/caspase-1/IL-1β signaling triggers inflammatory responses by promoting HMGB1 induction in liver ischemia/reperfusion injury. *Hepatology*, *58*(1), 351–362.

Kane, M. A., Folias, A. E., Wang, C., & Napoli, J. L. (2010). Ethanol elevates physiological all-trans-retinoic acid levels in select loci through altering retinoid metabolism in multiple loci: A potential mechanism of ethanol toxicity. *FASEB Journal*, *24*(3), 823–832.

Kaphalia, B. S., Bhopale, K. K., Kondraganti, S., Wu, H., Boor, P. J., & Ansari, G. A. (2010). Pancreatic injury in hepatic alcohol dehydrogenase-deficient deer mice after subchronic exposure to ethanol. *Toxicology and Applied Pharmacology*, *246*(3), 154–162.

Kaplowitz, N., & Ji, C. (2006). Unfolding new mechanisms of alcoholic liver disease in the endoplasmic reticulum. *Journal of Gastroenterology and Hepatology*, *21*(Suppl. 3), S7–S9.

Karahanian, E., Quintanilla, M. E., Tampier, L., Rivera-Meza, M., Bustamante, D., Gonzalez-Lira, V., et al. (2011). Ethanol as a prodrug: Brain metabolism of ethanol mediates its reinforcing effects. *Alcoholism: Clinical and Experimental Research*, *35*(4), 606–612.

Karahanian, E., Rivera-Meza, M., Tampier, L., Quintanilla, M. E., Herrera-Marschitz, M., & Israel, Y. (2015). Long-term inhibition of ethanol intake by the administration of an aldehyde dehydrogenase-2 (ALDH2)-coding lentiviral vector into the ventral tegmental area of rats. *Addiction Biology*, *20*(2), 336–344.

Kathirvel, E., Chen, P., Morgan, K., French, S. W., & Morgan, T. R. (2010). Oxidative stress and regulation of anti-oxidant enzymes in cytochrome P4502E1 transgenic mouse model of non-alcoholic fatty liver. *Journal of Gastroenterology and Hepatology*, *25*(6), 1136–1143.

Kathirvel, E., Morgan, K., French, S. W., & Morgan, T. R. (2009). Overexpression of liver-specific cytochrome P4502E1 impairs hepatic insulin signaling in a transgenic mouse model of nonalcoholic fatty liver disease. *European Journal of Gastroenterology and Hepatology*, *21*(9), 973–983.

Kato, S., Kawase, T., Alderman, J., Inatomi, N., & Lieber, C. S. (1990). Role of xanthine oxidase in ethanol-induced lipid peroxidation in rats. *Gastroenterology*, *98*(1), 203–210.

Katsiki, N., Tziomalos, K., & Mikhailidis, D. P. (2014). Alcohol and the cardiovascular system: A double-edged sword. *Current Pharmaceutical Design*, *20*(40), 6276–6288.

Keshavarzian, A., Jacyno, M., Urban, G., Winship, D., & Fields, J. Z. (1996). The role of nitric oxide in ethanol-induced gastrointestinal dysfunction. *Alcoholism: Clinical and Experimental Research*, *20*(9), 1618–1624.

Keyte, A., & Hutson, M. R. (2012). The neural crest in cardiac congenital anomalies. *Differentiation*, *84*(1), 25–40.

Kharbanda, K. K., Todero, S. L., King, A. L., Osna, N. A., McVicker, B. L., Tuma, D. J., et al. (2012). Betaine treatment attenuates chronic ethanol-induced hepatic steatosis and alterations to the mitochondrial respiratory chain proteome. *International Journal of Hepatology*, *2012*(2012), 962183.

Kim, B. J., Hood, B. L., Aragon, R. A., Hardwick, J. P., Conrads, T. P., Veenstra, T. D., et al. (2006). Increased oxidation and degradation of cytosolic proteins in alcohol-exposed mouse liver and hepatoma cells. *Proteomics*, *6*(4), 1250–1260.

Kim, J., Kim, S. H., Lee, S. U., Ha, G. H., Kang, D. G., Ha, N. Y., et al. (2002). Proteome analysis of human liver tumor tissue by two-dimensional gel electrophoresis and matrix assisted laser desorption/ionization-mass spectrometry for identification of disease-related proteins. *Electrophoresis*, *23*(24), 4142–4156.

Kim, M. J., Nagy, L. E., & Park, P. H. (2014). Globular adiponectin inhibits ethanol-induced reactive oxygen species production through modulation of NADPH oxidase in macrophages: Involvement of liver kinase B1/AMP-activated protein kinase pathway. *Molecular Pharmacology*, *86*(3), 284–296.

Kim, H. S., Quon, M. J., & Kim, J. A. (2014). New insights into the mechanisms of polyphenols beyond antioxidant properties; lessons from the green tea polyphenol, epigallocatechin 3-gallate. *Redox Biology*, *2*, 187–195.

Kim, B. J., Ryu, S. W., & Song, B. J. (2006). JNK- and p38 kinase mediated phosphorylation of Bax leads to its activation, mitochondrial translocation and apoptosis of human hepatoma HepG2 cells. *Journal of Biological Chemistry*, *281*(30), 21256–21265.

Klyosov, A. A., Rashkovetsky, L. G., Tahir, M. K., & Keung, W. M. (1996). Possible role of liver cytosolic and mitochondrial aldehyde dehydrogenases in acetaldehyde metabolism. *Biochemistry*, *35*(14), 4445–4456.

Knockaert, L., Fromenty, B., & Robin, M. A. (2011). Mechanisms of mitochondrial targeting of cytochrome P450 2E1: Physiopathological role in liver injury and obesity. *FEBS Journal*, *278*(22), 4252–4260.

Kodde, I. F., van der Stok, J., Smolenski, R. T., & de Jong, J. W. (2007). Metabolic and genetic regulation of cardiac energy substrate preference. *Comparative Biochemistry and Physiology Part A: Molecular & Integrative Physiology*, *146*(1), 26–39.

Kojima, H., Sakurai, S., Uemura, M., Fukui, H., Morimoto, H., & Tamagawa, Y. (2007). Mitochondrial abnormality and oxidative stress in nonalcoholic steatohepatitis. *Alcoholism: Clinical and Experimental Research*, *31*(1 Suppl.), S61–S66.

Kono, H., Bradford, B. U., Yin, M., Sulik, K. K., Koop, D. R., Peters, J. M., et al. (1999). CYP2E1 is not involved in early alcohol-induced liver injury. *American Journal of Physiology*, *277*(6 Pt. 1), G1259–G1267.

Koop, D. R. (1992). Oxidative and reductive metabolism by cytochrome P4502E1. *FASEB Journal*, *6*(2), 724–730.

Kozlov, A. V., Duvigneau, J. C., Miller, I., Nürnberger, S., Gesslbauer, B., Kungl, A., et al. (2009). Endotoxin causes functional endoplasmic reticulum failure, possibly mediated by mitochondria. *Biochimica et Biophysica Acta*, *1792*(6), 521–530.

Kubes, P., & Mehal, W. Z. (2012). Sterile inflammation in the liver. *Gastroenterology*, *143*(5), 1158–1172.

Kuper, H., Tzonou, A., Kaklamani, E., Hsieh, C. C., Lagiou, P., Adami, H. O., et al. (2000). Tobacco smoking, alcohol consumption and their interaction in the causation of hepatocellular carcinoma. *International Journal of Cancer*, *85*(4), 498–502.

Kurkivuori, J., Salaspuro, V., Kaihovaara, P., Kari, K., Rautemaa, R., Grönroos, L., et al. (2007). Acetaldehyde production from ethanol by oral streptococci. *Oral Oncology*, *43*(2), 181–186.

Kwon, H. J., Won, Y. S., Park, O., Chang, B., Duryee, M. J., Thiele, G. E., et al. (2014). Aldehyde dehydrogenase 2 deficiency ameliorates alcoholic fatty liver but worsens liver inflammation and fibrosis in mice. *Hepatology*, *60*(1), 146–157.

Lakshman, M. R., Garige, M., Gong, M. A., Leckey, L., Varatharajalu, R., Redman, R. S., et al. (2013). CYP2E1, oxidative stress, post-translational modifications and lipid metabolism. *Subcellular Biochemistry*, *67*, 199–233.

Lambert, J. C., Zhou, Z., Wang, L., Song, Z., McClain, C. J., & Kang, Y. J. (2003). Prevention of alterations in intestinal permeability is involved in zinc inhibition of acute ethanol-induced liver damage in mice. *Journal of Pharmacology and Experimental Therapeutics*, *305*(3), 880–886.

Lambert, J. C., Zhou, Z., Wang, L., Song, Z., McClain, C. J., & Kang, Y. J. (2004). Preservation of intestinal structural integrity by zinc is independent of metallothionein in alcohol-intoxicated mice. *American Journal of Pathology*, *164*(6), 1959–1966.

Landin, J. S., Cohen, S. D., & Khairallah, E. A. (1996). Identification of a 54-kDa mitochondrial acetaminophen-binding protein as aldehyde dehydrogenase. *Toxicology and Applied Pharmacology*, *141*(1), 299–307.

Lange, L. G. (1982). Nonoxidative ethanol metabolism: Formation of fatty acid ethyl esters by cholesterol esterase. *Proceedings of the National Academy of Sciences of the United States of America, 79*(13), 3954–3957.
Lange, L. G., & Sobel, B. E. (1983). Mitochondrial dysfunction induced by fatty acid ethyl esters, myocardial metabolites of ethanol. *Journal of Clinical Investigation, 72*(2), 724–731.
Laposata, M. (1998). Fatty acid ethyl esters: Ethanol metabolites which mediate ethanol-induced organ damage and serve as markers of ethanol intake. *Progress in Lipid Research, 37*(5), 307–316.
Laposata, E. A., & Lange, L. G. (1986). Presence of nonoxidative ethanol metabolism in human organs commonly damaged by ethanol abuse. *Science, 231*(4737), 497–499.
Lavandera, J., Ruspini, S., Batlle, A., & Buzaleh, A. M. (2015). Cytochrome P450 expression in mouse brain: Specific isoenzymes involved in Phase I metabolizing system of porphyrinogenic agents in both microsomes and mitochondria. *Biochemistry and Cell Biology, 93*(1), 102–107.
Leclercq, I. A., Farrell, G. C., Field, J., Bell, D. R., Gonzalez, F. J., & Robertson, G. R. (2000). CYP2E1 and CYP4A as microsomal catalysts of lipid peroxides in murine non-alcoholic steatohepatitis. *Journal of Clinical Investigation, 105*(8), 1067–1075.
Leclercq, I. A., Field, J., Enriquez, A., Farrell, G. C., & Robertson, G. R. (2000). Constitutive and inducible expression of hepatic CYP2E1 in leptin-deficient ob/ob mice. *Biochemical and Biophysical Research Communications, 268*(2), 337–344.
Lee, S. S., Buters, J. T., Pineau, T., Fernandez-Salguero, P., & Gonzalez, F. J. (1996). Role of CYP2E1 in the hepatotoxicity of acetaminophen. *Journal of Biological Chemistry, 271*(20), 12063–12067.
Lee, J. H., Jeong, J., Jeong, E. M., Cho, S. Y., Kang, J. W., Lim, J., et al. (2014). Endoplasmic reticulum stress activates transglutaminase 2 leading to protein aggregation. *International Journal of Molecular Medicine, 33*(4), 849–855.
Lee, Y. P., Liao, J. T., Cheng, Y. W., Wu, T. L., Lee, S. L., Liu, J. K., et al. (2013). Inhibition of human alcohol and aldehyde dehydrogenases by acetaminophen: Assessment of the effects on first-pass metabolism of ethanol. *Alcohol, 47*(7), 559–565.
Lee, H. I., McGregor, R. A., Choi, M. S., Seo, K. I., Jung, U. J., Yeo, J., et al. (2013). Low doses of curcumin protect alcohol-induced liver damage by modulation of the alcohol metabolic pathway, CYP2E1 and AMPK. *Life Sciences, 93*(18–19), 693–699.
Lee, Y. J., & Shukla, S. D. (2005). Pro- and anti-apoptotic roles of c-Jun N-terminal kinase (JNK) in ethanol and acetaldehyde exposed rat hepatocytes. *European Journal of Pharmacology, 508*(1–3), 31–45.
Lee, H. I., Yun, K. W., Seo, K. I., Kim, M. J., & Lee, M. K. (2014). Scopoletin prevents alcohol-induced hepatic lipid accumulation by modulating the AMPK-SREBP pathway in diet-induced obese mice. *Metabolism, 63*(4), 593–601.
Leo, M. A., & Lieber, C. S. (1999). Alcohol, vitamin A, and beta-carotene: Adverse interactions, including hepatotoxicity and carcinogenicity. *American Journal of Clinical Nutrition, 69*(6), 1071–1085.
Leung, T. M., & Nieto, N. (2013). CYP2E1 and oxidant stress in alcoholic and non-alcoholic fatty liver disease. *Journal of Hepatology, 58*(2), 395–398.
Li, T. K. (2000). Pharmacogenetics of responses to alcohol and genes that influence alcohol drinking. *Journal of Studies on Alcohol, 61*, 5–12.
Li, Y., Wong, K., Giles, A., Jiang, J., Lee, J. W., Adams, A. C., et al. (2014). Hepatic SIRT1 attenuates hepatic steatosis and controls energy balance in mice by inducing fibroblast growth factor 21. *Gastroenterology, 146*(2), 539–549.
Li, T. K., Yin, S. J., Crabb, D. W., O'Connor, S., & Ramchandani, V. A. (2001). Genetic and environmental influences on alcohol metabolism in humans. *Alcoholism: Clinical and Experimental Research, 25*(1), 136–144.

Li, X., Zhao, G., Ma, B., Li, R., Hong, J., Liu, S., et al. (2014). 20-Hydroxyeicosatetraenoic acid impairs endothelial insulin signaling by inducing phosphorylation of the insulin receptor substrate-1 at Ser616. *PLoS One*, *9*(4), e95841.

Lieber, C. S. (1997). Role of oxidative stress and antioxidant therapy in alcoholic and non-alcoholic liver diseases. *Advances in Pharmacology*, *38*, 601–628.

Lieber, C. S. (2002). S-Adenosyl-L-methionine and alcoholic liver disease in animal models: Implications for early intervention in human beings. *Alcohol*, *27*(3), 173–177.

Lieber, C. S. (2004a). Alcoholic fatty liver: Its pathogenesis and mechanism of progression to inflammation and fibrosis. *Alcohol*, *34*(1), 9–19.

Lieber, C. S. (2004b). CYP2E1: From ASH to NASH. *Hepatology Research*, *28*, 1–11.

Lieber, C. S. (2005). Metabolism of alcohol. *Clinics in Liver Disease*, *9*(1), 1–35.

Lieber, C. S., Leo, M. A., Wang, X., & Decarli, L. M. (2008a). Effect of chronic alcohol consumption on Hepatic SIRT1 and PGC-1alpha in rats. *Biochemical and Biophysical Research Communications*, *370*(1), 44–48.

Lieber, C. S., Leo, M. A., Wang, X., & Decarli, L. M. (2008b). Alcohol alters hepatic FoxO1, p53, and mitochondrial SIRT5 deacetylation function. *Biochemical and Biophysical Research Communications*, *373*(2), 246–252.

Lindblad, B., & Olsson, R. (1976). Unusually high levels of blood alcohol? *JAMA*, *236*(14), 1600–1602.

Linhart, K., Bartsch, H., & Seitz, H. K. (2014). The role of reactive oxygen species (ROS) and cytochrome P-450 2E1 in the generation of carcinogenic etheno-DNA adducts. *Redox Biology*, *3*, 56–62.

Liu, L. Q., Fan, Z. Q., Tang, Y. F., & Ke, Z. J. (2014). The resveratrol attenuates ethanol-induced hepatocyte apoptosis via inhibiting ER-related caspase-12 activation and PDE activity in vitro. *Alcoholism: Clinical and Experimental Research*, *38*(3), 683–693.

Liu, C., Russell, R. M., Seitz, H. K., & Wang, X. D. (2001). Ethanol enhances retinoic acid metabolism into polar metabolites in rat liver via induction of cytochrome P4502E1. *Gastroenterology*, *120*(1), 179–189.

Lluis, J. M., Colell, A., García-Ruiz, C., Kaplowitz, N., & Fernández-Checa, J. C. (2003). Acetaldehyde impairs mitochondrial glutathione transport in HepG2 cells through endoplasmic reticulum stress. *Gastroenterology*, *124*(3), 708–724.

Lomonaco, R., Sunny, N. E., Bril, F., & Cusi, K. (2013). Nonalcoholic fatty liver disease: Current issues and novel treatment approaches. *Drugs*, *73*(1), 1–14.

Loomba, R., Yang, H. I., Su, J., Brenner, D., Barrett-Connor, E., Iloeje, U., et al. (2013). Synergism between obesity and alcohol in increasing the risk of hepatocellular carcinoma: A prospective cohort study. *American Journal of Epidemiology*, *177*(4), 333–342.

Loomba, R., Yang, H. I., Su, J., Brenner, D., Iloeje, U., & Chen, C. J. (2010). Obesity and alcohol synergize to increase the risk of incident hepatocellular carcinoma in men. *Clinical Gastroenterology and Hepatology*, *8*(10), 891–898, e1-2.

Lu, Y., Cheng, J., Chen, L., Li, C., Chen, G., Gui, L., et al. (2015). Endoplasmic reticulum stress involved in high-fat diet and palmitic acid-induced vascular damages and fenofibrate intervention. *Biochemical and Biophysical Research Communications*, *458*(1), 1–7.

Lu, Y., Ward, S. C., & Cederbaum, A. I. (2013). Nicotine enhances ethanol-induced fat accumulation and collagen deposition but not inflammation in mouse liver. *Alcohol*, *47*(5), 353–357.

Lu, Y., Zhang, X. H., & Cederbaum, A. I. (2012). Ethanol induction of CYP2A5: Role of CYP2E1-ROS-Nrf2 pathway. *Toxicological Sciences*, *128*(2), 427–438.

Lu, Y., Zhuge, J., Wu, D., & Cederbaum, A. I. (2011). Ethanol induction of CYP2A5: Permissive role for CYP2E1. *Drug Metabolism and Disposition*, *39*(2), 330–336.

Lucey, M. R., Mathurin, P., & Morgan, T. R. (2009). Alcoholic hepatitis. *New England Journal of Medicine*, *360*(26), 2758–2769.

Lukivskaya, O. Y., & Buko, V. U. (1993). Utilization of ketone bodies by the rat liver, brain and heart in chronic alcohol intoxication. *Alcohol and Alcoholism, 28*(4), 431–436.
Luo, X. J., Liu, B., Ma, Q. L., & Peng, J. (2014). Mitochondrial aldehyde dehydrogenase, a potential drug target for protection of heart and brain from ischemia/reperfusion injury. *Current Drug Targets, 15*(10), 948–955.
Maher, J. J. (2009). DAMPs ramp up drug toxicity. *Journal of Clinical Investigation, 119*(2), 246–249.
Malhi, H., & Kaufman, R. J. (2014). Endoplasmic reticulum stress in liver disease. *Journal of Hepatology, 54*(4), 795–809.
Mali, V. R., Ning, R., Chen, J., Yang, X. P., Xu, J., & Palaniyandi, S. S. (2014). Impairment of aldehyde dehydrogenase-2 by 4-hydroxy-2-nonenal adduct formation and cardiomyocyte hypertrophy in mice fed a high-fat diet and injected with low-dose streptozotocin. *Experimental Biology and Medicine (Maywood), 239*(5), 610–618.
Mantena, S. K., Vaughn, D. P., Andringa, K. K., Eccleston, H. B., King, A. L., Abrams, G. A., et al. (2009). High fat diet induces dysregulation of hepatic oxygen gradients and mitochondrial function in vivo. *Biochemical Journal, 417*(1), 183–193.
Marchitti, S. A., Brocker, C., Stagos, D., & Vasiliou, V. (2008). Non-P450 aldehyde oxidizing enzymes: The aldehyde dehydrogenase superfamily. *Expert Opinion on Drug Metabolism & Toxicology, 4*(6), 697–720.
Marcolin, E., San-Miguel, B., Vallejo, D., Tieppo, J., Marroni, N., González-Gallego, J., et al. (2012). Quercetin treatment ameliorates inflammation and fibrosis in mice with nonalcoholic steatohepatitis. *Journal of Nutrition, 142*(10), 1821–1828.
Marí, M., Morales, A., Colell, A., García-Ruiz, C., & Fernández-Checa, J. C. (2009). Mitochondrial glutathione, a key survival antioxidant. *Antioxidants and Redox Signaling, 11*(11), 2685–2700.
Marietta, C. A., Jerrells, T. R., Meagher, R. C., Karanian, J. W., Weight, F. F., & Eckardt, M. J. (1988). Effects of long-term ethanol inhalation on the immune and hematopoietic systems of the rat. *Alcoholism: Clinical and Experimental Research, 12*(2), 211–214.
Marseglia, L., Manti, S., D'Angelo, G., Nicotera, A., Parisi, E., Di Rosa, G., et al. (2014). Oxidative stress in obesity: A critical component in human diseases. *International Journal of Molecular Sciences, 16*(1), 378–400.
Martin-Murphy, B. V., Holt, M. P., & Ju, C. (2010). The role of damage associated molecular pattern molecules in acetaminophen-induced liver injury in mice. *Toxicology Letters, 192*(3), 387–394.
McClain, C. J., Kromhout, J. P., Peterson, F. J., & Holtzman, J. L. (1980). Potentiation of acetaminophen hepatotoxicity by alcohol. *JAMA, 244*(3), 251–253.
McClain, C. J., Mokshagundam, S. P., Barve, S. S., Song, Z., Hill, D. B., Chen, T., et al. (2004). Mechanisms of non-alcoholic steatohepatitis. *Alcohol, 34*(1), 67–79.
McCuskey, R. S., Ito, Y., Robertson, G. R., McCuskey, M. K., Perry, M., & Farrell, G. C. (2004). Hepatic microvascular dysfunction during evolution of dietary steatohepatitis in mice. *Hepatology, 40*(2), 386–393.
McGill, M. R., & Jaeschke, H. (2014). Mechanistic biomarkers in acetaminophen-induced hepatotoxicity and acute liver failure: From preclinical models to patients. *Expert Opinion on Drug Metabolism & Toxicology, 10*(7), 1005–1017.
Mello, T., Ceni, E., Surrenti, C., & Galli, A. (2008). Alcohol induced hepatic fibrosis: Role of acetaldehyde. *Molecular Aspects of Medicine, 29*(1–2), 17–21.
Mells, J. E., Fu, P. P., Kumar, P., Smith, T., Karpen, S. J., & Anania, F. A. (2015). Saturated fat and cholesterol are critical to inducing murine metabolic syndrome with robust nonalcoholic steatohepatitis. *Journal of Nutritional Biochemistry, 26*(3), 285–292.
Melov, S., Doctrow, S. R., Schneider, J. A., Haberson, J., Patel, M., Coskun, P. E., et al. (2001). Lifespan extension and rescue of spongiform encephalopathy in superoxide dismutase 2 nullizygous mice treated with superoxide dismutase-catalase mimetics. *Journal of Neurosciences, 21*(21), 8348–8353.

Méndez, L., Pazos, M., Molinar-Toribio, E., Sánchez-Martos, V., Gallardo, J. M., Rosa, N. M., et al. (2014). Protein carbonylation associated to high-fat, high-sucrose diet and its metabolic effects. *Journal of Nutritional Biochemistry*, *25*(12), 1243–1253.

Michaut, A., Moreau, C., Robin, M. A., & Fromenty, B. (2014). Acetaminophen-induced liver injury in obesity and nonalcoholic fatty liver disease. *Liver International*, *34*(7), e171–e179.

Minko, I. G., Kozekov, I. D., Harris, T. M., Rizzo, C. J., Lloyd, R. S., & Stone, M. P. (2009). Chemistry and biology of DNA containing 1, N(2)-deoxyguanosine adducts of the alpha, beta-unsaturated aldehydes acrolein, crotonaldehyde, and 4-hydroxynonenal. *Chemical Research in Toxicology*, *22*(5), 759–778.

Mitchell, D. Y., & Petersen, D. R. (1988). Inhibition of rat liver aldehyde dehydrogenases by acrolein. *Drug Metabolism and Disposition*, *16*, 37–42.

Miyata, N., Taniguchi, K., Seki, T., Ishimoto, T., Sato-Watanabe, M., Yasuda, Y., et al. (2001). HET0016, a potent and selective inhibitor of 20-HETE synthesizing enzyme. *British Journal of Pharmacology*, *133*(3), 325–329.

Mogelson, S., & Lange, L. G. (1984). Nonoxidative ethanol metabolism in rabbit myocardium: Purification to homogeneity of fatty acyl ethyl ester synthase. *Biochemistry*, *23*(18), 4075–4081.

Moncada, C., Torres, V., Varghese, G., Albano, E., & Israel, Y. (1994). Ethanol-derived immunoreactive species formed by free radical mechanisms. *Molecular Pharmacology*, *46*(4), 786–791.

Moon, K. H., Hood, B. L., Kim, B. J., Hardwick, J. P., Conrads, T. P., Veenstra, T. D., et al. (2006). Inactivation of oxidized and S-nitrosylated mitochondrial proteins in alcoholic fatty liver of rats. *Hepatology*, *44*(5), 1218–1230.

Moon, K. H., Hood, B. L., Mukhopadhyay, P., Rajesh, M., Abdelmegeed, M. A., Kwon, Y. I., et al. (2008). Oxidative inactivation of key mitochondrial proteins leads to dysfunction and injury in hepatic ischemia reperfusion. *Gastroenterology*, *135*(4), 1344–1357.

Moon, K. H., Kim, B. J., & Song, B. J. (2005). Inhibition of mitochondrial aldehyde dehydrogenase by nitric oxide-mediated S-nitrosylation. *FEBS Letters*, *579*(27), 6115–6120.

Moon, K. H., Lee, Y. M., & Song, B. J. (2010). Inhibition of hepatic mitochondrial aldehyde dehydrogenase by carbon tetrachloride through JNK-mediated phosphorylation. *Free Radical Biology and Medicine*, *48*(3), 391–398.

Moon, K. H., Upreti, V. V., Yu, L. R., Lee, I. J., Ye, X., Eddington, N. D., et al. (2008). Mechanism of 3,4-methylenedioxymethamphetamine (MDMA, ecstasy)-mediated mitochondrial dysfunction in rat liver. *Proteomics*, *8*(18), 3906–3918.

Moraes, L. A., Piqueras, L., & Bishop-Bailey, D. (2006). Peroxisome proliferator-activated receptors and inflammation. *Pharmacology & Therapeutics*, *110*(3), 371–385.

Morgan, K., French, S. W., & Morgan, T. R. (2002). Production of a cytochrome P450 2E1 transgenic mouse and initial evaluation of alcoholic liver damage. *Hepatology*, *36*(1), 122–134.

Morgan, T. R., Mandayam, S., & Jamal, M. M. (2004). Alcohol and hepatocellular carcinoma. *Gastroenterology*, *127*(5 Suppl. 1), S87–S96.

Mori, Y., Murakawa, Y., Yokoyama, J., Tajima, N., Ikeda, Y., Nobukata, H., et al. (1999). Effect of highly purified eicosapentaenoic acid ethyl ester on insulin resistance and hypertension in Dahl salt-sensitive rats. *Metabolism*, *48*(9), 1089–1095.

Mottaran, E., Stewart, S. F., Rolla, R., Vay, D., Cipriani, V., Moretti, M., et al. (2002). Lipid peroxidation contributes to immune reactions associated with alcoholic liver disease. *Free Radical Biology and Medicine*, *32*(1), 38–45.

Mueller, S., Millonig, G., & Seitz, H. K. (2009). Alcoholic liver disease and hepatitis C: A frequently underestimated combination. *World Journal of Gastroenterology*, *15*(28), 3462–3471.

Mukhopadhyay, P., Horváth, B., Zsengellėr, Z., Bátkai, S., Cao, Z., Kechrid, M., et al. (2012). Mitochondrial reactive oxygen species generation triggers inflammatory response and tissue injury associated with hepatic ischemia-reperfusion: Therapeutic potential of mitochondrially targeted antioxidants. *Free Radical Biology and Medicine, 53*(5), 1123–1138.

Murphy, M. P. (2014). Antioxidants as therapies: Can we improve on nature? *Free Radical Biology and Medicine, 66*, 20–23.

Muto, M., Hitomi, Y., Ohtsu, A., Ebihara, S., Yoshida, S., & Esumi, H. (2000). Association of aldehyde dehydrogenase 2 gene polymorphism with multiple oesophageal dysplasia in head and neck cancer patients. *Gut, 47*(2), 256–261.

Nakajima, T., Kamijo, Y., Tanaka, N., Sugiyama, E., Tanaka, E., Kiyosawa, K., et al. (2004). Peroxisome proliferator-activated receptor alpha protects against alcohol-induced liver damage. *Hepatology, 40*(4), 972–980.

Nakamura, M., Higuchi, N., Kato, M., Kotoh, K., Yoshimoto, T., Yada, M., et al. (2008). The significance of differences in fatty acid metabolism between obese and non-obese patients with non-alcoholic fatty liver disease. *International Journal of Molecular Medicine, 22*(5), 663–667.

Nakano, M., Kelly, E. J., & Rettie, A. E. (2009). Expression and characterization of CYP4V2 as a fatty acid omega-hydroxylase. *Drug Metabolism & Disposition, 37*(11), 2119–2122.

Nanji, A. A., Dannenberg, A. J., Jokelainen, K., & Bass, N. M. (2004). Alcoholic liver injury in the rat is associated with reduced expression of peroxisome proliferator-alpha (PPARalpha)-regulated genes and is ameliorated by PPARalpha activation. *Journal of Pharmacology and Experimental Therapeutics, 310*(1), 417–424.

Nanji, A. A., Jokelainen, K., Tipoe, G. L., Rahemtulla, A., Thomas, P., & Dannenberg, A. J. (2003). Curcumin prevents alcohol-induced liver disease in rats by inhibiting the expression of NF-kappa B-dependent genes. *American Journal of Physiology Gastrointestinal and Liver Physiology, 284*(2), G321–G327.

Nath, B., & Szabo, G. (2012). Hypoxia and hypoxia inducible factors: Diverse roles in liver diseases. *Hepatology, 55*(2), 622–633.

Navarro, V. J., Barnhart, H., Bonkovsky, H. L., Davern, T., Fontana, R. J., Grant, L., et al. (2014). Liver injury from herbals and dietary supplements in the U.S. Drug-Induced Liver Injury Network. *Hepatology, 60*(4), 1399–1408.

Neuman, M. G., French, S. W., French, B. A., Seitz, H. K., Cohen, L. B., Mueller, S., et al. (2014). Alcoholic and non-alcoholic steatohepatitis. *Experimental and Molecular Pathology, 97*(3), 492–510.

Nguyen, K. H., Lee, J. H., & Nyomba, B. L. (2012). Ethanol causes endoplasmic reticulum stress and impairment of insulin secretion in pancreatic β-cells. *Alcohol, 46*(1), 89–99.

Niemelä, O., Parkkila, S., Juvonen, R. O., Viitala, K., Gelboin, H. V., & Pasanen, M. (2000). Cytochromes P450 2A6, 2E1, and 3A and production of protein-aldehyde adducts in the liver of patients with alcoholic and non-alcoholic liver diseases. *Journal of Hepatology, 33*(6), 893–901.

Niemelä, O., Parkkila, S., Pasanen, M., Iimuro, Y., Bradford, B., & Thurman, R. G. (1998). Early alcoholic liver injury: Formation of protein adducts with acetaldehyde and lipid peroxidation products, and expression of CYP2E1 and CYP3A. *Alcoholism: Clinical and Experimental Research, 22*(9), 2118–2124.

Niemelä, O., Parkkila, S., Ylä-Herttuala, S., Halsted, C., Witztum, J. L., Lanca, A., et al. (1994). Covalent protein adducts in the liver as a result of ethanol metabolism and lipid peroxidation. *Laboratory Investigation, 70*(4), 537–546.

Nishitani, Y., & Matsumoto, H. (2006). Ethanol rapidly causes activation of JNK associated with ER stress under inhibition of ADH. *FEBS Letters, 580*(1), 9–14.

Noll, T., & De Groot, H. (1984). The critical steady-state hypoxic conditions in carbon tetrachloride-induced lipid peroxidation in rat liver microsomes. *Biochimica et Biophysica Acta*, *795*(2), 356–362.

Nordmann, R., Ribière, C., & Rouach, H. (1990). Ethanol-induced lipid peroxidation and oxidative stress in extrahepatic tissues. *Alcohol and Alcoholism*, *25*(2–3), 231–237.

Ogawa, M., Oyama, T., Isse, T., Saito, K., Tomigahara, Y., Endo, Y., et al. (2007). A comparison of covalent binding of ethanol metabolites to DNA according to Aldh2 genotype. *Toxicology Letters*, *168*(2), 148–154.

Oshita, F., Morita, A., Ito, H., Kameda, Y., Tsuchiya, E., Asai, S., et al. (2010). Proteomic screening of completely resected tumors in relation to survival in patients with stage I non-small cell lung cancer. *Oncology Reports*, *24*(3), 637–645.

Osna, N. A., & Donohue, T. M., Jr. (2013). CYP2E1-catalyzed alcohol metabolism: Role of oxidant generation in interferon signaling, antigen presentation and autophagy. *Subcellular Biochemistry*, *67*, 177–197.

Otani, K., Korenaga, M., Beard, M. R., Li, K., Qian, T., Showalter, L. A., et al. (2005). Hepatitis C virus core protein, cytochrome P450 2E1, and alcohol produce combined mitochondrial injury and cytotoxicity in hepatoma cells. *Gastroenterology*, *128*(1), 96–107.

Pallauf, K., Giller, K., Huebbe, P., & Rimbach, G. (2013). Nutrition and healthy ageing: Calorie restriction or polyphenol-rich "MediterrAsian" diet? *Oxidative Medicine and Cellular Longevity*, *2013*, 707421.

Paradies, G., Paradies, V., Ruggiero, F. M., & Petrosillo, G. (2014). Oxidative stress, cardiolipin and mitochondrial dysfunction in nonalcoholic fatty liver disease. *World Journal of Gastroenterology*, *20*(39), 14205–14218.

Park, K. S., Cho, S. Y., Kim, H., & Paik, Y. K. (2002). Proteomic alterations of the variants of human aldehyde dehydrogenase isozymes correlate with hepatocellular carcinoma. *International Journal of Cancer*, *97*(2), 261–265.

Park, E. C., Kim, S. I., Hong, Y., Hwang, J. W., Cho, G. S., Cha, H. N., et al. (2014). Inhibition of CYP4A reduces hepatic endoplasmic reticulum stress and features of diabetes in mice. *Gastroenterology*, *147*(4), 860–869.

Pawlosky, R. J., Flynn, B. M., & Salem, N., Jr. (1997). The effects of low dietary levels of polyunsaturates on alcohol-induced liver disease in rhesus monkeys. *Hepatology*, *26*(6), 1386–1392.

Pawlosky, R. J., Kashiwaya, Y., Srivastava, S., King, M. T., Crutchfield, C., Volkow, N., et al. (2010). Alterations in brain glucose utilization accompanying elevations in blood ethanol and acetate concentrations in the rat. *Alcoholism: Clinical and Experimental Research*, *34*(2), 375–381.

Peng, G. S., Chen, Y. C., Wang, M. F., Lai, C. L., & Yin, S. J. (2014). ALDH2*2 but not ADH1B*2 is a causative variant gene allele for Asian alcohol flushing after a low-dose challenge: Correlation of the pharmacokinetic and pharmacodynamic findings. *Pharmacogenetics and Genomics*, *24*(12), 607–617.

Pérez-Echarri, N., Pérez-Matute, P., Marcos-Gómez, B., Baena, M. J., Marti, A., Martínez, J. A., et al. (2008). Differential inflammatory status in rats susceptible or resistant to diet-induced obesity: Effects of EPA ethyl ester treatment. *European Journal of Nutrition*, *47*(7), 380–386.

Pérez-Echarri, N., Pérez-Matute, P., Marcos-Gómez, B., Marti, A., Martínez, J. A., & Moreno-Aliaga, M. J. (2009). Down-regulation in muscle and liver lipogenic genes: EPA ethyl ester treatment in lean and overweight (high-fat-fed) rats. *Journal of Nutritional Biochemistry*, *20*(9), 705–714.

Pérez-Matute, P., Pérez-Echarri, N., Martínez, J. A., Marti, A., & Moreno-Aliaga, M. J. (2007). Eicosapentaenoic acid actions on adiposity and insulin resistance in control and high-fat-fed rats: Role of apoptosis, adiponectin and tumour necrosis factor-alpha. *British Journal of Nutrition, 97*(2), 389–398.

Perez-Miller, S., Younus, H., Vanam, R., Chen, C. H., Mochly-Rosen, D., & Hurley, T. D. (2010). Alda-1 is an agonist and chemical chaperone for the common human aldehyde dehydrogenase 2 variant. *Nature Structural and Molecular Biology, 17*(2), 159–164.

Persidsky, Y., Ho, W., Ramirez, S. H., Potula, R., Abood, M. E., Unterwald, E., et al. (2011). HIV-1 infection and alcohol abuse: Neurocognitive impairment, mechanisms of neurodegeneration and therapeutic interventions. *Brain, Behavior, and Immunity, 1*, S61–S70.

Pessayre, D., Fromenty, B., Berson, A., Robin, M. A., Lettéron, P., Moreau, R., et al. (2012). Central role of mitochondria in drug-induced liver injury. *Drug Metabolism Review, 44*(1), 34–87.

Picklo, M. J., Sr. (2008). Ethanol intoxication increases hepatic N-lysyl protein acetylation. *Biochemical and Biophysical Research Communications, 376*(3), 615–619.

Pijls, K. E., Jonkers, D. M., Elamin, E. E., Masclee, A. A., & Koek, G. H. (2013). Intestinal epithelial barrier function in liver cirrhosis: An extensive review of the literature. *Liver International, 33*(10), 1457–1469.

Powell, C. L., Bradford, B. U., Craig, C. P., Tsuchiya, M., Uehara, T., O'Connell, T. M., et al. (2010). Mechanism for prevention of alcohol-induced liver injury by dietary methyl donors. *Toxicological Sciences, 115*(1), 131–139.

Purohit, V., Abdelmalek, M. F., Barve, S., Benevenga, N. J., Halsted, C. H., Kaplowitz, N., et al. (2007). Role of S-adenosylmethionine, folate, and betaine in the treatment of alcoholic liver disease: Summary of a symposium. *American Journal of Clinical Nutrition, 86*(1), 14–24.

Purohit, V., Gao, B., & Song, B. J. (2009). Molecular mechanisms of alcoholic fatty liver. *Alcoholism: Clinical and Experimental Research, 33*, 191–205.

Purohit, V., Rapaka, R., Kwon, O. S., & Song, B. J. (2013). Roles of alcohol and tobacco exposure in the development of hepatocellular carcinoma. *Life Sciences, 92*, 3–9.

Quintanilla, M. E., Israel, Y., Sapag, A., & Tampier, L. (2006). The UChA and UChB rat lines: Metabolic and genetic differences influencing ethanol intake. *Addiction Biology, 11*(3–4), 310–323.

Quintanilla, M. E., Tampier, L., Sapag, A., & Israel, Y. (2005). Polymorphisms in the mitochondrial aldehyde dehydrogenase gene (Aldh2) determine peak blood acetaldehyde levels and voluntary ethanol consumption in rats. *Pharmacogenetics and Genomics, 15*(6), 427–431.

Raetsch, C., Jia, J. D., Boigk, G., Bauer, M., Hahn, E. G., Riecken, E. O., et al. (2002). Pentoxifylline downregulates profibrogenic cytokines and procollagen I expression in rat secondary biliary fibrosis. *Gut, 50*(2), 241–247.

Ramirez, T., Longato, L., Dostalek, M., Tong, M., Wands, J. R., & de la Monte, S. M. (2013). Insulin resistance, ceramide accumulation and endoplasmic reticulum stress in experimental chronic alcohol-induced steatohepatitis. *Alcohol and Alcoholism, 48*(1), 39–52.

Reily, C., Mitchell, T., Chacko, B. K., Benavides, G., Murphy, M. P., & Darley-Usmar, V. (2013). Mitochondrially targeted compounds and their impact on cellular bioenergetics. *Redox Biology, 1*(1), 86–93.

Rigamonti, C., Mottaran, E., Reale, E., Rolla, R., Cipriani, V., Capelli, F., et al. (2003). Moderate alcohol consumption increases oxidative stress in patients with chronic hepatitis C. *Hepatology, 38*(1), 42–49.

Roberts, B. J., Shoaf, S. E., Jeong, K. S., & Song, B. J. (1994). Induction of CYP2E1 in liver, kidney, brain and intestine during chronic ethanol administration and withdrawal: Evidence that CYP2E1 possesses a rapid phase half-life of 6 hours or less. *Biochemical and Biophysical Research Communications, 205*(2), 1064–1071.

Roberts, B. J., Song, B. J., Soh, Y., Park, S. S., & Shoaf, S. E. (1995). Ethanol induces CYP2E1 by protein stabilization. Role of ubiquitin conjugation in the rapid degradation of CYP2E1. *Journal of Biological Chemistry, 270*(50), 29632–29635.

Robertson, G., Leclercq, I., Farrell, G. C., & Robertson, G. (2001). Nonalcoholic steatosis and steatohepatitis. II. Cytochrome P-450 enzymes and oxidative stress. *American Journal of Physiology. Gastrointestinal and Liver Physiology, 281*(5), G1135–G1139.

Robin, M. A., Anandatheerthavarada, H. K., Biswas, G., Sepuri, N. B., Gordon, D. M., Pain, D., et al. (2002). Bimodal targeting of microsomal CYP2E1 to mitochondria through activation of an N-terminal chimeric signal by cAMP-mediated phosphorylation. *Journal of Biological Chemistry, 277*(43), 40583–40593.

Robin, M. A., Sauvage, I., Grandperret, T., Descatoire, V., Pessayre, D., & Fromenty, B. (2005). Ethanol increases mitochondrial cytochrome P450 2E1 in mouse liver and rat hepatocytes. *FEBS Letters, 579*(30), 6895–6902.

Rodd-Henricks, Z. A., Melendez, R. I., Zaffaroni, A., Goldstein, A., McBride, W. J., & Li, T. K. (2002). The reinforcing effects of acetaldehyde in the posterior ventral tegmental area of alcohol-preferring rats. *Pharmacology, Biochemistry and Behavior, 72*(1–2), 55–64.

Rodrigues, C. F., Ascenção, K., Silva, F. A., Sarmento, B., Oliveira, M. B., & Andrade, J. C. (2013). Drug-delivery systems of green tea catechins for improved stability and bioavailability. *Current Medicinal Chemistry, 20*(37), 4744–4757.

Rodriguez, J., Gilson, H., Jamart, C., Naslain, D., Pierre, N., Deldicque, L., et al. (2015). Pomegranate and green tea extracts protect against ER stress induced by a high-fat diet in skeletal muscle of mice. *European Journal of Nutrition, 54*, 377–389.

Roychowdhury, S., McMullen, M. R., Pritchard, M. T., Li, W., Salomon, R. G., & Nagy, L. E. (2009). Formation of gamma-ketoaldehyde-protein adducts during ethanol-induced liver injury in mice. *Free Radical Biology & Medicine, 47*(11), 1526–1538.

Saberi, B., Ybanez, M. D., Johnson, H. S., Gaarde, W. A., Han, D., & Kaplowitz, N. (2014). Protein kinase C (PKC) participates in acetaminophen hepatotoxicity through c-jun-N-terminal kinase (JNK)-dependent and -independent signaling pathways. *Hepatology, 59*(4), 1543–1554.

Salaspuro, V., & Salaspuro, M. (2004). Synergistic effect of alcohol drinking and smoking on in vivo acetaldehyde concentration in saliva. *International Journal of Cancer, 111*(4), 480–483.

Salaspuro, M. P., Shaw, S., Jayatilleke, E., Ross, W. A., & Lieber, C. S. (1981). Attenuation of the ethanol-induced hepatic redox change after chronic alcohol consumption in baboons: Metabolic consequences in vivo and in vitro. *Hepatology, 1*, 33–38.

Samak, G., Suzuki, T., Bhargava, A., & Rao, R. K. (2010). c-Jun NH2-terminal kinase-2 mediates osmotic stress-induced tight junction disruption in the intestinal epithelium. *American Journal of Physiology. Gastrointestinal and Liver Physiology, 299*(3), G572–G584.

Sano, R., & Reed, J. C. (2013). ER stress-induced cell death mechanisms. *Biochimica et Biophysica Acta, 1833*(12), 3460–3470.

Santamaria, E., Avila, M. A., Latasa, M. U., Rubio, A., Martin-Duce, A., Lu, S. C., et al. (2003). Functional proteomics of nonalcoholic steatohepatitis: Mitochondrial proteins as targets of S-adenosylmethionine. *Proceedings of the National Academy of Sciences of the United States of America, 100*(6), 3065–3070.

Sanyal, A. J., Campbell-Sargent, C., Mirshahi, F., Rizzo, W. B., Contos, M. J., Sterling, R. K., et al. (2001). Nonalcoholic steatohepatitis: Association of insulin resistance and mitochondrial abnormalities. *Gastroenterology, 120*(5), 1183–1192.

Sarna, L. K., Wu, N., Wang, P., Hwang, S. Y., & Siow, Y. L. (2012). Folic acid supplementation attenuates high fat diet induced hepatic oxidative stress via regulation of NADPH oxidase. *Canadian Journal of Physiological Pharmacology, 90*(2), 155–165.

Sato, M., Ishii, T., Kobayashi-Matsunaga, Y., Amada, H., Taniguchi, K., Miyata, N., et al. (2001). Discovery of a N′-hydroxyphenylformamidine derivative HET0016 as a potent

and selective 20-HETE synthase inhibitor. *Bioorganic & Medicinal Chemistry Letters, 11*(23), 2993–29942.
Schattenberg, J. M., & Czaja, M. J. (2014). Regulation of the effects of CYP2E1-induced oxidative stress by JNK signaling. *Redox Biology, 3*, 7–15.
Schug, Z. T., Peck, B., Jones, D. T., Zhang, Q., Grosskurth, S., Alam, I. S., et al. (2015). Acetyl-CoA synthetase 2 promotes acetate utilization and maintains cancer cell growth under metabolic stress. *Cancer Cell, 27*(1), 57–71.
Schwab, N., & Skopp, G. (2014). Identification and preliminary characterization of UDP-glucuronosyltransferases catalyzing formation of ethyl glucuronide. *Analytical and Bioanalytical Chemistry, 406*(9–10), 2325–2332.
Scorletti, E., Bhatia, L., McCormick, K. G., Clough, G. F., Nash, K., Hodson, L., et al. (2014). Effects of purified eicosapentaenoic and docosahexaenoic acids in nonalcoholic fatty liver disease: Results from the Welcome* study. *Hepatology, 60*(4), 1211–1221.
Seeff, L. B., Cuccherini, B. A., Zimmerman, H. J., Adler, E., & Benjamin, S. B. (1986). Acetaminophen hepatotoxicity in alcoholics. A therapeutic misadventure. *Annals of Internal Medicine, 104*(3), 399–404.
Seitz, H. K., & Cho, C. H. (2009). Contribution of alcohol and tobacco use in gastrointestinal cancer development. *Methods in Molecular Biology, 472*, 217–241.
Seitz, H. K., & Stickel, F. (2010). Acetaldehyde as an underestimated risk factor for cancer development: Role of genetics in ethanol metabolism. *Genes and Nutrition, 5*(2), 121–128.
Seitz, H. K., & Wang, X. D. (2013). The role of cytochrome P450 2E1 in ethanol-mediated carcinogenesis. *Subcellular Biochemistry, 67*, 131–143.
Seki, E., Brenner, D. A., & Karin, M. (2012). A liver full of JNK: Signaling in regulation of cell function and disease pathogenesis, and clinical approaches. *Gastroenterology, 143*(2), 307–320.
Seo, Y. S., Kwon, J. H., Yaqoob, U., Yang, L., De Assuncao, T. M., Simonetto, D. A., et al. (2013). HMGB1 recruits hepatic stellate cells and liver endothelial cells to sites of ethanol-induced parenchymal cell injury. *American Journal of Physiology Gastrointestinal and Liver Physiology, 305*(11), G838–G848.
Seth, R. K., Kumar, A., Das, S., Kadiiska, M. B., Michelotti, G., Diehl, A. M., et al. (2013). Environmental toxin-linked nonalcoholic steatohepatitis and hepatic metabolic reprogramming in obese mice. *Toxicological Sciences, 134*(2), 291–303.
Setshedi, M., Wands, J. R., & de la Monte, S. M. (2010). Acetaldehyde adducts in alcoholic liver disease. *Oxidative Medicine and Cellular Longevity, 3*(3), 178–185.
Sheikh, S., Ni, L., Hurley, T. D., & Weiner, H. (1997). The potential roles of the conserved amino acids in human liver mitochondrial aldehyde dehydrogenase. *Journal of Biological Chemistry, 272*(30), 18817–18822.
Shin, S., Park, J., Li, Y., Min, K. N., Kong, G., Hur, G. M., et al. (2014). β-Lapachone alleviates alcoholic fatty liver disease in rats. *Cell Signaling, 26*(2), 295–305.
Shukla, S. D., & Lim, R. W. (2013). Epigenetic effects of ethanol on the liver and gastrointestinal system. *Alcohol Research, 35*(1), 47–55.
Shukla, S. D., Velazquez, J., French, S. W., Lu, S. C., Ticku, M. K., & Zakhari, S. (2008). Emerging role of epigenetics in the actions of alcohol. *Alcoholism: Clinical and Experimental Research, 32*(9), 1525–1534.
Singh, R., Wang, Y., Xiang, Y., Tanaka, K. E., Gaarde, W. A., & Czaja, M. J. (2009). Differential effects of JNK1 and JNK2 inhibition on murine steatohepatitis and insulin resistance. *Hepatology, 49*(1), 87–96.
Soderberg, B. L., Salem, R. O., Best, C. A., Cluette-Brown, J. E., & Laposata, M. (2003). Fatty acid ethyl esters. Ethanol metabolites that reflect ethanol intake. *American Journal of Clinical Pathology, 119*(Suppl.), S94–S99.

Soh, Y., Jeong, K. S., Lee, I. J., Bae, M. A., Kim, Y. C., & Song, B. J. (2000). Selective activation of the c-Jun N-terminal protein kinase pathway during 4-hydroxynonenal induced apoptosis of PC12 cells. *Molecular Pharmacology*, *58*(3), 535–541.

Sohn, D. H., Yun, Y. P., Park, K. S., Veech, R. L., & Song, B. J. (1991). Post-translational reduction of cytochrome P450IIE by CCl4, its substrate. *Biochemical and Biophysical Research Communications*, *179*(1), 449–454.

Son, Y., Kim, S., Chung, H. T., & Pae, H. O. (2013). Reactive oxygen species in the activation of MAP kinases. *Methods in Enzymology*, *528*, 27–48.

Song, B. J., Abdelmegeed, M. A., Henderson, L. E., Yoo, S. H., Wan, J., Purohit, V., et al. (2013). Increased nitroxidative stress promotes mitochondrial dysfunction in alcoholic and nonalcoholic fatty liver disease. *Oxidative Medicine and Cellular Longevity*, *2013*, 781050.

Song, B. J., Abdelmegeed, M. A., Yoo, S. H., Kim, B. J., Jo, S. A., Jo, I., et al. (2011). Post-translational modifications of mitochondrial aldehyde dehydrogenase and biomedical implications. *Journal of Proteomics*, *74*(12), 2691–2702.

Song, B. J., Akbar, M., Abdelmegeed, M. A., Byun, K., Lee, B., Yoon, S. K., et al. (2014). Mitochondrial dysfunction and tissue injury by alcohol, high fat, nonalcoholic substances and pathological conditions through post-translational protein modifications. *Redox Biology*, *3*, 109–123.

Song, E., Fu, J., Xia, X., Su, C., & Song, Y. (2014). Bazhen decoction protects against acetaminophen induced acute liver injury by inhibiting oxidative stress, inflammation and apoptosis in mice. *PLoS One*, *9*(9), e107405.

Song, B. J., Gelboin, H. V., Park, S. S., Yang, C. S., & Gonzalez, F. J. (1986). Complementary DNA and protein sequences of ethanol-inducible rat and human cytochrome P-450s. Transcriptional and post-transcriptional regulation of the rat enzyme. *Journal of Biological Chemistry*, *261*(35), 16689–16697.

Song, B. J., Koop, D. R., Ingelman-Sundberg, M., Nanji, A. A., & Cederbaum, A. I. (1996). Ethanol-inducible cytochrome P450 (CYP2E1): Biochemistry, molecular biology and clinical relevance: 1996 update. *Alcoholism: Clinical and Experimental Research*, *20*(8 Suppl.), 138A–146A.

Song, B. J., Matsunaga, T., Hardwick, J. P., Park, S. S., Veech, R. L., Yang, C. S., et al. (1987). Stabilization of cytochrome P450j messenger ribonucleic acid in the diabetic rat. *Molecular Endocrinology*, *1*(8), 542–547.

Song, B. J., Moon, K. H., Olsson, N., & Salem, N., Jr. (2008). Prevention of alcoholic fatty liver and mitochondrial dysfunction in the rat by long-chain polyunsaturated fatty acids. *Journal of Hepatology*, *49*, 262–273.

Song, B. J., Veech, R. L., Park, S. S., Gelboin, H. V., & Gonzalez, F. J. (1989). Induction of rat hepatic N-nitrosodimethylamine demethylase by acetone is due to protein stabilization. *Journal of Biological Chemistry*, *264*(6), 3568–3572.

Song, B. J., Veech, R. L., & Saenger, P. (1990). Cytochrome P450IIE1 is elevated in lymphocytes from poorly controlled insulin-dependent diabetics. *Journal of Clinical Endocrinology and Metabolism*, *71*(4), 1036–1040.

Sparkenbaugh, E. M., Saini, Y., Greenwood, K. K., LaPres, J. J., Luyendyk, J. P., Copple, B. L., et al. (2011). The role of hypoxia-inducible factor-1α in acetaminophen hepatotoxicity. *Journal of Pharmacology and Experimental Therapeutics*, *338*(2), 492–502.

Spencer, M., Finlin, B. S., Unal, R., Zhu, B., Morris, A. J., Shipp, L. R., et al. (2013). Omega-3 fatty acids reduce adipose tissue macrophages in human subjects with insulin resistance. *Diabetes*, *62*(5), 1709–1717.

Spruss, A., Kanuri, G., Stahl, C., Bischoff, S. C., & Bergheim, I. (2012). Metformin protects against the development of fructose-induced steatosis in mice: Role of the intestinal barrier function. *Laboratory Investigation*, *92*(7), 1020–1032.

Spruss, A., Kanuri, G., Uebel, K., Bischoff, S. C., & Bergheim, I. (2011). Role of the inducible nitric oxide synthase in the onset of fructose-induced steatosis in mice. *Antioxidants and Redox Signaling*, *14*(11), 2121–2135.

Spruss, A., Kanuri, G., Wagnerberger, S., Haub, S., Bischoff, S. C., & Bergheim, I. (2009). Toll-like receptor 4 is involved in the development of fructose-induced hepatic steatosis in mice. *Hepatology*, *50*(4), 1094–1104.

Stachlewitz, R. F., Arteel, G. E., Raleigh, J. A., Connor, H. D., Mason, R. P., & Thurman, R. G. (1997). Development and characterization of a new model of tacrine-induced hepatotoxicity: Role of the sympathetic nervous system and hypoxia-reoxygenation. *Journal of Pharmacology and Experimental Therapeutics*, *282*(3), 1591–1599.

Stice, C. P., Liu, C., Aizawa, K., Greenberg, A. S., Ausman, L. M., & Wang, X. D. (2015). Dietary tomato powder inhibits alcohol-induced hepatic injury by suppressing cytochrome p450 2E1 induction in rodent models. *Archives of Biochemistry and Biophysics*, *572*, 81–88.

Stickel, F., Kessebohm, K., Weimann, R., & Seitz, H. K. (2011). Review of liver injury associated with dietary supplements. *Liver International*, *31*(5), 595–605.

Stickel, F., & Seitz, H. K. (2010). Alcoholic steatohepatitis. *Best Practice & Research Clinical Gastroenterology*, *24*(5), 683–693.

Stickel, F., & Shouval, D. (2015). Hepatotoxicity of herbal and dietary supplements: An update. *Archives of Toxicology*, *89*(6), 851–865.

Suh, S. K., Hood, B. L., Kim, B. J., Conrads, T. P., Veenstra, T. D., & Song, B. J. (2004). Identification of oxidized mitochondrial proteins in alcohol-exposed human hepatoma cells and mouse liver. *Proteomics*, *4*(11), 3401–3412.

Sultatos, L. G. (1988). Effects of acute ethanol administration on the hepatic xanthine dehydrogenase/oxidase system in the rat. *Journal of Pharmacology and Experimental Therapeutics*, *246*(3), 946–949.

Sun, A., Cheng, Y., Zhang, Y., Zhang, Q., Wang, S., Tian, S., et al. (2014). Aldehyde dehydrogenase 2 ameliorates doxorubicin-induced myocardial dysfunction through detoxification of 4-HNE and suppression of autophagy. *Journal of Molecular and Cellular Cardiology*, *71*, 92–104.

Surapaneni, K. M., Priya, V. V., & Mallika, J. (2014). Pioglitazone, quercetin and hydroxy citric acid effect on cytochrome P450 2E1 (CYP2E1) enzyme levels in experimentally induced non alcoholic steatohepatitis (NASH). *European Review for Medical and Pharmacological Sciences*, *18*(18), 2736–2741.

Sutti, S., Rigamonti, C., Vidali, M., & Albano, E. (2014). CYP2E1 autoantibodies in liver diseases. *Redox Biology*, *3*, 72–78.

Swaminathan, K., Clemens, D. L., & Dey, A. (2013). Inhibition of CYP2E1 leads to decreased malondialdehyde-acetaldehyde adduct formation in VL-17A cells under chronic alcohol exposure. *Life Sciences*, *92*(6–7), 325–336.

Swaminathan, K., Kumar, S. M., Clemens, D. L., & Dey, A. (2013). Inhibition of CYP2E1 leads to decreased advanced glycated end product formation in high glucose treated ADH and CYP2E1 over-expressing VL-17A cells. *Biochimica et Biophysica Acta*, *1830*(10), 4407–4416.

Szabo, G. (2015). Gut-liver axis in alcoholic liver disease. *Gastroenterology*, *148*(1), 30–36.

Szabo, G., & Lippai, D. (2012). Molecular hepatic carcinogenesis: Impact of inflammation. *Digestive Diseases*, *30*(3), 243–248.

Szabo, G., Saha, B., & Bukong, T. N. (2015). Alcohol and HCV: Implications for liver cancer. *Advances in Experimental Medicine and Biology*, *815*, 197–216.

Szabo, G., Wands, J. R., Eken, A., Osna, N. A., Weinman, S. A., Machida, K., et al. (2010). Alcohol and hepatitis C virus-interactions in immune dysfunctions and liver damage. *Alcoholism: Clinical and Experimental Research*, *34*(10), 1675–1686.

Takagi, S., Iwai, N., Yamauchi, R., Kojima, S., Yasuno, S., Baba, T., et al. (2002). Aldehyde dehydrogenase 2 gene is a risk factor for myocardial infarction in Japanese men. *Hypertension Research*, *25*, 677–681.

Tang, Y., Forsyth, C. B., Farhadi, A., Rangan, J., Jakate, S., Shaikh, M., et al. (2009). Nitric oxide-mediated intestinal injury is required for alcohol-induced gut leakiness and liver damage. *Alcoholism: Clinical and Experimental Research*, *33*, 1220–1230.

Tarnawski, A. S., Ahluwalia, A., & Jones, M. K. (2012). The mechanisms of gastric mucosal injury: Focus on microvascular endothelium as a key target. *Current Medicinal Chemistry*, *19*(1), 4–15.

Terelius, Y., & Ingelman-Sundberg, M. (1988). Cytochrome P-450-dependent oxidase activity and hydroxyl radical production in micellar and membranous types of reconstituted systems. *Biochemical Pharmacology*, *37*(7), 1383–1389.

Thiele, G. M., Worrall, S., Tuma, D. J., Klassen, L. W., Wyatt, T. A., & Nagata, N. (2001). The chemistry and biological effects of malondialdehyde-acetaldehyde adducts. *Alcoholism: Clinical and Experimental Research*, *25*(5 Suppl. ISBRA), 218S–224S.

Thuy, S., Ladurner, R., Volynets, V., Wagner, S., Strahl, S., Königsrainer, A., et al. (2008). Nonalcoholic fatty liver disease in humans is associated with increased plasma endotoxin and plasminogen activator inhibitor 1 concentrations and with fructose intake. *Journal of Nutrition*, *138*(8), 1452–1455.

Tipoe, G. L., Leung, T. M., Liong, E., So, H., Leung, K. M., Lau, T. Y., et al. (2006). Inhibitors of inducible nitric oxide (NO) synthase are more effective than an NO donor in reducing carbon-tetrachloride induced acute liver injury. *Histology and Histopathology*, *21*(11), 1157–1165.

Ugarte, G., Pino, M. E., & Insunza, I. (1967). Hepatic alcohol dehydrogenase in alcoholic addicts with and without hepatic damage. *American Journal of Digestive Diseases*, *94*, 589–592.

Valdecantos, M. P., Pérez-Matute, P., González-Muniesa, P., Prieto-Hontoria, P. L., Moreno-Aliaga, M. J., & Martínez, J. A. (2012). Lipoic acid improves mitochondrial function in nonalcoholic steatosis through the stimulation of sirtuin 1 and sirtuin 3. *Obesity (Silver Spring)*, *20*(10), 1974–1983.

Vavilin, V. A., Nepomnyashchikh, D. L., Shchepotina, E. G., Karavaeva, Y. Y., Makarova, S. I., Vinogradova, E. V., et al. (2013). Cytochrome P450 4 F2 polymorphism in patients with liver cirrhosis. *Bulletin of Experimental Biology and Medicine*, *156*(2), 181–184.

Venkatraman, A., Landar, A., Davis, A. J., Chamlee, L., Sanderson, T., Kim, H., et al. (2004). Modification of the mitochondrial proteome in response to the stress of ethanol-dependent hepatotoxicity. *Journal of Biological Chemistry*, *279*(21), 22092–22101.

Venkatraman, A., Landar, A., Davis, A. J., Ulasova, E., Page, G., Murphy, M. P., et al. (2004). Oxidative modification of hepatic mitochondria protein thiols: Effect of chronic alcohol consumption. *American Journal of Physiology. Gastrointestinal and Liver Physiology*, *286*(4), G521–G527.

Videla, L., & Israel, Y. (1970). Factors that modify the metabolism of ethanol in rat liver and adaptive changes produced by its chronic administration. *Biochemical Journal*, *118*(2), 275–281.

Volkow, N. D., Kim, S. W., Wang, G. J., Alexoff, D., Logan, J., Muench, L., et al. (2013). Acute alcohol intoxication decreases glucose metabolism but increases acetate uptake in the human brain. *NeuroImage*, *64*, 277–283.

Volkow, N. D., Ma, Y., Zhu, W., Fowler, J. S., Li, J., Rao, M., et al. (2008). Moderate doses of alcohol disrupt the functional organization of the human brain. *Psychiatry Research*, *162*(3), 205–213.

Volkow, N. D., Wang, G. J., Franceschi, D., Fowler, J. S., Thanos, P. P., Maynard, L., et al. (2006). Low doses of alcohol substantially decrease glucose metabolism in the human brain. *NeuroImage*, *29*(1), 295–301.

Vos, M. B., & Lavine, J. E. (2013). Dietary fructose in nonalcoholic fatty liver disease. *Hepatology, 57*(6), 2525–2531.
Walter, P., & Ron, D. (2011). The unfolded protein response: From stress pathway to homeostatic regulation. *Science, 334*(6059), 1081–1086.
Wang, K. (2014). Molecular mechanisms of hepatic apoptosis. *Cell Death & Disease, 5*, e996.
Wang, Y., Millonig, G., Nair, J., Patsenker, E., Stickel, F., Mueller, S., et al. (2009). Ethanol-induced cytochrome P4502E1 causes carcinogenic etheno-DNA lesions in alcoholic liver disease. *Hepatology, 50*(2), 453–461.
Wang, S., Moustaid-Moussa, N., Chen, L., Mo, H., Shastri, A., Su, R., et al. (2014). Novel insights of dietary polyphenols and obesity. *Journal of Nutritional Biochemistry, 25*(1), 1–18.
Wang, X., Wu, D., Yang, L., Gan, L., & Cederbaum, A. I. (2013). Cytochrome P450 2E1 potentiates ethanol induction of hypoxia and HIF-1α in vivo. *Free Radical Biology and Medicine, 63*, 175–186.
Watanabe, A., Hobara, N., Nakatsukasa, H., Shiota, T., Kobayashi, M., & Nagashima, H. (1985). Impaired acetaldehyde metabolism in partially hepatectomized rats. *Research in Experimental Medicine, 185*, 13–20.
Weltman, M. D., Farrell, G. C., Hall, P., Ingelman-Sundberg, M., & Liddle, C. (1998). Hepatic cytochrome P450 2E1 is increased in patients with nonalcoholic steatohepatitis. *Hepatology, 27*(1), 128–133.
Weltman, M. D., Farrell, G. C., & Liddle, C. (1996). Increased hepatocyte CYP2E1 expression in a rat nutritional model of hepatic steatosis with inflammation. *Gastroenterology, 111*(6), 1645–1653.
Wenzel, P., Hink, U., Oelze, M., Schuppan, S., Schaeuble, K., Schildknecht, S., et al. (2007). Role of reduced lipoic acid in the redox regulation of mitochondrial aldehyde dehydrogenase (ALDH-2) activity. Implications for mitochondrial oxidative stress and nitrate tolerance. *Journal of Biological Chemistry, 282*(1), 792–799.
Wenzel, P., Müller, J., Zurmeyer, S., Schuhmacher, S., Schulz, E., Oelze, M., et al. (2008). ALDH-2 deficiency increases cardiovascular oxidative stress–evidence for indirect antioxidative properties. *Biochemical and Biophysical Research Communications, 367*(1), 137–143.
Werner, J., Laposata, M., Fernandez-Del Castillo, C., Saghir, M., Iozzo, R. V., Lewandrowski, K. B., et al. (1997). Pancreatic injury in rats induced by fatty acid ethyl ester, a nonoxidative metabolite of alcohol. *Gastroenterology, 113*(1), 286–294.
Witschi, A., Mossi, S., Meyer, B., Junker, E., & Lauterburg, B. H. (1994). Mitochondrial function reflected by the decarboxylation of [13C]ketoisocaproate is impaired in alcoholics. *Alcoholism: Clinical and Experimental Research, 18*(4), 951–955.
Wolf, K. K., Wood, S. G., Allard, J. L., Hunt, J. A., Gorman, N., Walton-Strong, B. W., et al. (2007). Role of CYP3A and CYP2E1 in alcohol-mediated increases in acetaminophen hepatotoxicity: Comparison of wild-type and Cyp2e1(−/−) mice. *Drug Metabolism and Disposition, 35*(7), 1223–1231.
Wong, F. W., Chan, W. Y., & Lee, S. S. (1998). Resistance to carbon tetrachloride-induced hepatotoxicity in mice which lack CYP2E1 expression. *Toxicology and Applied Pharmacology, 153*(1), 109–118.
Wu, D., & Cederbaum, A. I. (2013). Inhibition of autophagy promotes CYP2E1-dependent toxicity in HepG2 cells via elevated oxidative stress, mitochondria dysfunction and activation of p38 and JNK MAPK. *Redox Biology, 1*, 552–565.
Wu, D., Wang, X., Zhou, R., Yang, L., & Cederbaum, A. I. (2012). Alcohol steatosis and cytotoxicity: The role of cytochrome P4502E1 and autophagy. *Free Radical Biology and Medicine, 53*(6), 1346–1357.
Wu, D., Xu, C., & Cederbaum, A. I. (2009). Role of nitric oxide and nuclear factor-kappaB in the CYP2E1 potentiation of tumor necrosis factor alpha hepatotoxicity in mice. *Free Radical Biology and Medicine, 46*(4), 480–491.

Xiao, J., Ching, Y. P., Liong, E. C., Nanji, A. A., Fung, M. L., & Tipoe, G. L. (2013). Garlic-derived S-allylmercaptocysteine is a hepato-protective agent in non-alcoholic fatty liver disease in vivo animal model. *European Journal of Nutrition*, *52*(1), 179–191.

Xie, Y., McGill, M. R., Dorko, K., Kumer, S. C., Schmitt, T. M., Forster, J., et al. (2014). Mechanisms of acetaminophen-induced cell death in primary human hepatocytes. *Toxicology and Applied Pharmacology*, *279*(3), 266–274.

Xu, C., Wan, X., Xu, L., Weng, H., Yan, M., Miao, M., et al. (2015). Xanthine oxidase in nonalcoholic fatty liver disease and hyperuricemia: One stone hits two birds. *Journal of Hepatology*, *62*(6), 1412–1419.

Xu, A., Wang, Y., Keshaw, H., Xu, L. Y., Lam, K. S., & Cooper, G. J. (2003). The fat-derived hormone adiponectin alleviates alcoholic and nonalcoholic fatty liver diseases in mice. *Journal of Clinical Investigation*, *112*(1), 91–100.

Xu, T., Zheng, L., Xu, L., Yin, L., Qi, Y., Xu, Y., et al. (2014). Protective effects of dioscin against alcohol-induced liver injury. *Archives of Toxicology*, *88*(3), 739–753.

Yamamoto, T., Moriwaki, Y., & Takahashi, S. (2005). Effect of ethanol on metabolism of purine bases (hypoxanthine, xanthine, and uric acid). *Clinica Chimica Acta*, *356*(1–2), 35–57.

Yamauchi, T., Kamon, J., Minokoshi, Y., Ito, Y., Waki, H., Uchida, S., et al. (2002). Adiponectin stimulates glucose utilization and fatty-acid oxidation by activating AMP-activated protein kinase. *Nature Medicine*, *8*(11), 1288–1295.

Yamauchi, T., Kamon, J., Waki, H., Terauchi, Y., Kubota, N., Hara, K., et al. (2001). The fat-derived hormone adiponectin reverses insulin resistance associated with both lipoatrophy and obesity. *Nature Medicine*, 7(8), 941–946.

Yang, M., Dai, J., Jia, Y., Suo, L., Li, S., Guo, Y., et al. (2014). Overexpression of juxtaposed with another zinc finger gene 1 reduces proinflammatory cytokine release via inhibition of stress-activated protein kinases and nuclear factor-κB. *FEBS Journal*, *281*(14), 3193–3205.

Yang, S. J., & Lim, Y. (2014). Resveratrol ameliorates hepatic metaflammation and inhibits NLRP3 inflammasome activation. *Metabolism*, *63*(5), 693–701.

Yang, L., Wu, D., Wang, X., & Cederbaum, A. I. (2012). Cytochrome P4502E1, oxidative stress, JNK, and autophagy in acute alcohol-induced fatty liver. *Free Radical Biology and Medicine*, *53*(5), 1170–1180.

Yin, O. Q., Tomlinson, B., & Chow, M. S. (2010). CYP3A5 but not CYP2D6 polymorphism contributes significantly to the variability in dextropropoxyphene disposition. *Journal of Clinical Pharmacology*, *50*(10), 1136–1141.

Yokoyama, A., Muramatsu, T., Ohmori, T., Yokoyama, T., Okuyama, K., Takahashi, H., et al. (1998). Alcohol-related cancers and aldehyde dehydrogenase-2 in Japanese alcoholics. *Carcinogenesis*, *19*(8), 1383–1387.

Yokoyama, T., Muramatsu, T., Omori, T., Yokoyama, S., Matsushita, S., Higuchi, K., et al. (2001). Alcohol and aldehyde dehydrogenase gene polymorphisms and oropharyngolaryngeal, esophageal and stomach cancers in Japanese alcoholics. *Carcinogenesis*, *22*(3), 433–439.

Yoo, S. H., Abdelmegeed, M. A., & Song, B. J. (2013). Activation of PPARα by Wy-14643 ameliorates systemic lipopolysaccharide-induced acute lung injury. *Biochemical and Biophysical Research Communications*, *436*(3), 366–371.

Yoshida, A., Huang, I. Y., & Ikawa, M. (1984). Molecular abnormality of an inactive aldehyde dehydrogenase variant commonly found in Orientals. *Proceedings of the National Academy of Sciences of the United States of America*, *81*(1), 258–261.

Yoshigae, Y., Sridar, C., Kent, U. M., & Hollenberg, P. F. (2013). The inactivation of human CYP2E1 by phenethyl isothiocyanate, a naturally occurring chemopreventive agent, and its oxidative bioactivation. *Drug Metabolism & Disposition*, *41*(4), 858–869.

You, M., Fischer, M., Deeg, M. A., & Crabb, D. W. (2002). Ethanol induces fatty acid synthesis pathways by activation of sterol regulatory element-binding protein (SREBP). *Journal of Biological Chemistry*, *277*(32), 29342–29347.

You, M., Liang, X., Ajmo, J. M., & Ness, G. C. (2008). Involvement of mammalian sirtuin 1 in the action of ethanol in the liver. *American Journal of Physiology. Gastrointestinal and Liver Physiology*, *294*(4), G892–G898.

You, M., Matsumoto, M., Pacold, C. M., Cho, W. K., & Crabb, D. W. (2004). The role of AMP-activated protein kinase in the action of ethanol in the liver. *Gastroenterology*, *127*(6), 1798–1808.

Younes, M., & Strubelt, O. (1987). Enhancement of hypoxic liver damage by ethanol. Involvement of xanthine oxidase and the role of glycolysis. *Biochemical Pharmacology*, *36*(18), 2973–2977.

Yu, W., Chen, L., Yang, Y., Falck, J. R., Guo, A. M., Li, Y., et al. (2011). Cytochrome P450 omega hydroxylase promotes angiogenesis and metastasis by upregulation of VEGF and MMP-9 in non-small cell lung cancer. *Cancer Chemotherapy and Pharmacology*, *68*(3), 619–629.

Yu, H. S., Oyama, T., Isse, T., Kitagawa, K., Pham, T. T., Tanaka, M., et al. (2010). Formation of acetaldehyde-derived DNA adducts due to alcohol exposure. *Chemico-Biological Interactions*, *188*(3), 367–375.

Yu, H. S., Oyama, T., Matsuda, T., Isse, T., Yamaguchi, T., Tanaka, M., et al. (2012). The effect of ethanol on the formation of N2-ethylidene-dG adducts in mice: Implications for alcohol-related carcinogenicity of the oral cavity and esophagus. *Biomarkers*, *17*(3), 269–274.

Yuan, L., & Kaplowitz, N. (2013). Mechanisms of drug-induced liver injury. *Clinics in Liver Disease*, *17*(4), 507–518.

Yuan, F., Lei, Y., Wang, Q., Esberg, L. B., Huang, Z., Scott, G. I., et al. (2014). Moderate ethanol administration accentuates cardiomyocyte contractile dysfunction and mitochondrial injury in high fat diet-induced obesity. *Toxicology Letters*, *233*(3), 267–277.

Yukawa, Y., Ohashi, S., Amanuma, Y., Nakai, Y., Tsurumaki, M., Kikuchi, O., et al. (2014). Impairment of aldehyde dehydrogenase 2 increases accumulation of acetaldehyde-derived DNA damage in the esophagus after ethanol ingestion. *American Journal of Cancer Research*, *4*(3), 279–284.

Yun, Y. P., Casazza, J. P., Sohn, D. H., Veech, R. L., & Song, B. J. (1992). Pretranslational activation of cytochrome P450IIE during ketosis induced by a high fat diet. *Molecular Pharmacology*, *41*(3), 474–479.

Yun, J. W., Son, M. J., Abdelmegeed, M. A., Banerjee, A., Morgan, T. R., Yoo, S. H., et al. (2014). Binge alcohol promotes hypoxic liver injury through CYP2E1-HIF1a-dependent apoptosis pathway in mice and humans. *Free Radical Biology and Medicine*, *77*, 183–194.

Zakhari, S. (2013). Bermuda Triangle for the liver: Alcohol, obesity, and viral hepatitis. *Journal of Gastroenterology & Hepatology*, *28*(Suppl. 1), 18–25.

Zakhari, S., & Li, T.-K. (2007). Determinants of alcohol use and abuse: Impact of quantity and frequency patterns on liver disease. *Hepatology*, *46*(6), 2032–2039.

Zhang, R. H., Gao, J. Y., Guo, H. T., Scott, G. I., Eason, A. R., Wang, X. M., et al. (2013). Inhibition of CYP2E1 attenuates chronic alcohol intake-induced myocardial contractile dysfunction and apoptosis. *Biochimica et Biophysica Acta*, *1832*(1), 128–141.

Zhang, Y., Mi, S. L., Hu, N., Doser, T. A., Sun, S., Ge, J., et al. (2014). Mitochondrial aldehyde dehydrogenase 2 accentuates aging-induced cardiac remodeling and contractile dysfunction: Role of AMPK, Sirt1, and mitochondrial function. *Free Radical Biology & Medicine*, *71*, 208–220.

Zhang, Y., & Ren, J. (2011). ALDH2 in alcoholic heart diseases: Molecular mechanism and clinical implications. *Pharmacology & Therapeutics*, *132*, 86–95.

Zhang, W., Wang, L. W., Wang, L. K., Li, X., Zhang, H., Luo, L. P., et al. (2013). Betaine protects against high-fat-diet-induced liver injury by inhibition of high-mobility group box 1 and Toll-like receptor 4 expression in rats. *Digestive Diseases and Sciences*, *58*(11), 3198–3206.

Zhang, X., Xu, A., Chung, S. K., Cresser, J. H., Sweeney, G., Wong, R. L., et al. (2011). Selective inactivation of c-Jun NH2-terminal kinase in adipose tissue protects against diet-induced obesity and improves insulin sensitivity in both liver and skeletal muscle in mice. *Diabetes*, *60*(2), 486–495.

Zhong, W., Zhang, W., Li, Q., Xie, G., Sun, Q., Sun, X., et al. (2014). Pharmacological activation of aldehyde dehydrogenase 2 by Alda-1 reverses alcohol-induced hepatic steatosis and cell death in mice. *Journal of Hepatology*, *62*(6), 1375–1381.

Zong, H., Armoni, M., Harel, C., Karnieli, E., & Pessin, J. E. (2012). Cytochrome P-450 CYP2E1 knockout mice are protected against high-fat diet-induced obesity and insulin resistance. *American Journal of Physiology, Endocrinology and Metabolism*, *302*(5), E532–E539.

CHAPTER ELEVEN

Role of Retinoic Acid-Metabolizing Cytochrome P450s, CYP26, in Inflammation and Cancer

Faith Stevison, Jing Jing, Sasmita Tripathy, Nina Isoherranen[1]
Department of Pharmaceutics, School of Pharmacy, University of Washington, Seattle, Washington, USA
[1]Corresponding author: e-mail address: ni2@uw.edu

Contents

Abstract

Vitamin A (retinol) and its active metabolite, *all-trans*-retinoic acid (*at*RA), play critical roles in regulating the differentiation, growth, and migration of immune cells. Similarly, as critical signaling molecules in the regulation of the cell cycle, retinoids are important in cancers. Concentrations of *at*RA are tightly regulated in tissues, predominantly by the availability of retinol, synthesis of *at*RA by ALDH1A enzymes and metabolism and clearance of *at*RA by CYP26 enzymes. The ALDH1A and CYP26 enzymes are expressed in several cell types in the immune system and in cancer cells. In the immune system, the ALDH1A and CYP26 enzymes appear to modulate RA concentrations. Consequently, alterations in the activity of ALDH1A and CYP26 enzymes are expected to change disease outcomes in inflammation. There is increasing evidence from various disease models of intestinal and skin inflammation that treatment with *at*RA has a positive effect on disease markers. However, whether aberrant *at*RA concentrations or *at*RA synthesis

Advances in Pharmacology, Volume 74
ISSN 1054-3589
http://dx.doi.org/10.1016/bs.apha.2015.04.006

and metabolism play a role in inflammatory disease development and progression is not well understood. In cancers, especially in acute promyelocytic leukemia and neuroblastoma, increasing intracellular concentrations of *at*RA appears to provide clinical benefit. Inhibition of the CYP26 enzymes to increase *at*RA concentrations and combat therapy resistance has been pursued as a drug target in these cancers. This chapter covers the current knowledge of how *at*RA and retinol regulate the immune system and inflammation, how retinol and *at*RA metabolism is altered in inflammation and cancer, and what roles *at*RA-metabolizing enzymes have in immune responses and cancers.

ABBREVIATIONS

9-*cis*RA 9-*cis*-retinoic acid
13-*cis*RA 13-*cis*-retinoic acid
ALDH aldehyde dehydrogenase
APL acute promyelocytic leukemia
***at*RA** *all-trans*-retinoic acid
BCDO2 β,β-carotene-9,10-dioxygenase 2
BCMO1 β,β-carotene-15,15′-monooxygenase 1
CD Crohn's disease
CRABP cellular retinoic acid-binding protein
CRBP cellular retinol binding protein
CYP26 cytochrome P450 family 26
DC dendritic cell
ER endoplasmic reticulum
FABP fatty acid binding protein
GALT gut-associated lymphoid tissue
IBD inflammatory bowel disease
LRAT lecithin retinol acyltransferase
NAFLD non-alcoholic fatty liver disease
NASH non-alcoholic steatohepatitis
PML promyelocytic leukemia gene
PPAR peroxisome proliferator-activated receptor
RA retinoic acid
RAMBA retinoic acid metabolism blocking agent
RAR retinoic acid receptor
RBP retinol binding protein
RDH1 retinol dehydrogenase 1
REH retinyl ester hydrolase
RXR retinoid X receptor
UC ulcerative colitis

1. INTRODUCTION

Vitamin A is one of the fat-soluble vitamins critical for health in humans and other mammals. Typically, the term vitamin A is used to

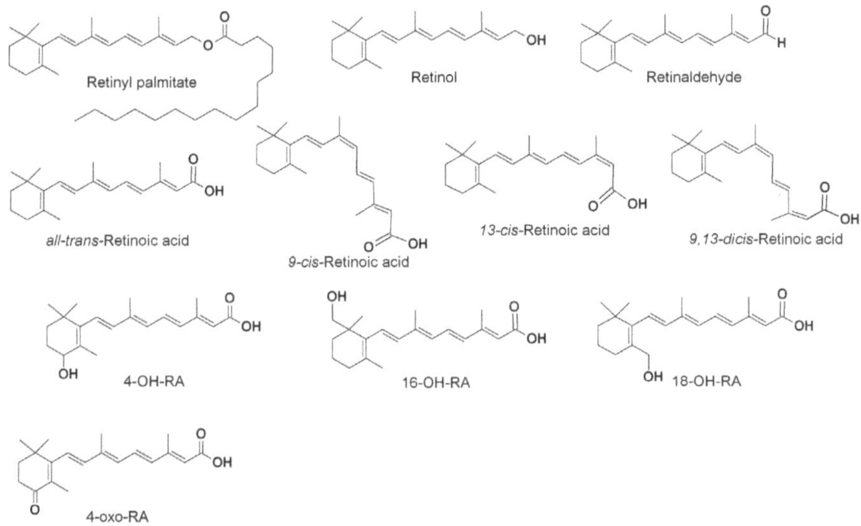

Figure 1 Chemical structures of retinyl palmitate, retinol, retinaldehyde, and the retinoic acid isomers and metabolites.

describe retinol and retinyl esters (Fig. 1), including retinyl palmitate, that are the main dietary forms of vitamin A obtained from animal products (Schoeff, 1983; Sklan, 1987). However, retinol and retinyl esters are not biologically active but require oxidation by several enzymes to generate retinoic acid (RA), the main compound responsible for the biological activity of vitamin A. Retinyl esters serve as the reservoir of vitamin A in an organism and are stored in the body primarily in the liver stellate cells and in the lungs and adipose (Chytil, 1992; Sauvant, Cansell, & Atgié, 2011). The esterification of retinol and the hydrolysis and release of the stored esters as retinol are regulated by the activity of two enzymes, lecithin retinol acyltransferase (LRAT) and retinyl ester hydrolase (REH), respectively (Fig. 2). The circulating retinol is taken up into target cells by an uptake transporter STRA6, but due to its high lipophilicity retinol may also cross membranes by passive diffusion. In target cells, retinol is oxidized to retinaldehyde by several alcohol dehydrogenase enzymes including retinol dehydrogenase (RDH) 1, RDH10, RDH-E, RDHE2, and DHRS9 (D'Ambrosio, Clugston, & Blaner, 2011; Napoli, 2012; Ross, 2003; Fig. 2). Alternatively, retinaldehyde can be obtained via direct cleavage of β-carotene by carotenoid cleavage enzymes (Lobo, Amengual, Palczewski, Babino, & von Lintig, 2012). Two enzymes have been indicated as the primary enzymes generating retinaldehyde, mammalian β,β-carotene-15,15′-monooxygenase 1 (BCMO1), and β,β-carotene-9,10-dioxygenase 2 (BCDO2) (Lobo et al., 2012).

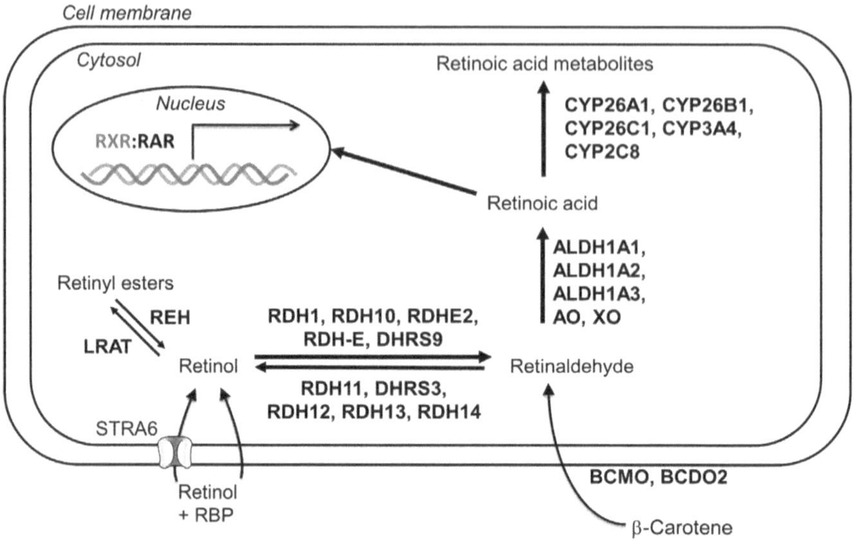

Figure 2 Enzymatic processes that regulate retinoic acid homeostasis in humans. The retinyl ester hydrolase (REH) enzymes hydrolyze retinyl esters, acquired from the diet, to retinol, and lecithin retinol acyltransferase (LRAT) esterifies retinol for storage. STRA6 transports retinol bound to the retinol binding protein (RBP) into the target cell where several enzymes control the reversible oxidation of retinol to retinaldehyde. Retinol can also cross cell membranes via passive diffusion. Retinaldehyde, derived via oxidation of retinol by RDH enzymes or from β-carotene precursors, is irreversibly oxidized to the active metabolite retinoic acid. In the cell retinoic acid will traverse to the nucleus and bind to RXR or RAR to initiate gene transcription. Retinoic acid is cleared through oxidation to polar metabolites by several P450s, predominantly the CYP26 family.

To generate vitamin A from dietary carotenoids, retinaldehyde has to be reduced to retinol, and retinol is subsequently esterified and stored as esters (Fig. 2). The reduction of retinaldehyde to retinol in humans is catalyzed by several enzymes including RDH11, RDH12, RDH13, RDH14, and DHRS3 (Kedishvili, 2013). The relative contribution of these reductive enzymes to retinol synthesis from retinaldehyde is not well characterized, and the importance may vary between species, tissues, developmental stages, and age after birth. While the conversion of retinol to retinaldehyde is reversible, oxidation of retinaldehyde to RA appears to be irreversible. Four aldehyde dehydrogenase (ALDH) enzymes and aldehyde and xanthine oxidases have been reported to catalyze the oxidation of retinaldehyde to RA (Fig. 2; Ambroziak, Izaguirre, & Pietruszko, 1999; Kumar, Sandell, Trainor, Koentgen, & Duester, 2012; Taibi & Nicotra, 2007; Terao et al., 2009). In addition, several cytochrome P450 enzymes in humans and rats oxidize retinaldehyde to RA (Raner, Vaz, & Coon, 1996; Tomita, Okuyama, Ohnishi, & Ichikawa, 1996; Zhang, Dunbar, & Kaminsky, 2000). It is generally believed that ALDH1A2 and ALDH1A3 are the predominant enzymes that form RA during fetal and embryonic development, as neither the Aldh1a2$^{-/-}$ nor the Aldh1a3$^{-/-}$ mice are viable. However, the importance of ALDH1A1 in RA synthesis is unclear as Aldh1a1$^{-/-}$ mice are viable and fertile (Kumar et al., 2012). Very little is known of the potential importance of ALDH8A1 in RA synthesis. Recent studies have shown that the relative importance of ALDH1A enzymes in RA synthesis is tissue specific. For example, ALDH1A1 appears to be responsible for RA synthesis in the liver while ALDH1A2 and ALDH1A1 make the main contributions to RA synthesis in the testes (Arnold, Kent, Hogarth, Griswold, et al., 2015; Arnold, Kent, Hogarth, Schlatt, et al., 2015). This tissue-specific expression is important, as locally generated RA gradients are believed to be a key factor in regulating biological processes. Yet, the overall quantitative importance of the different enzymes in retinaldehyde oxidation in different species and organs is not well characterized.

RA has five chemical isomers (*all-trans*-RA, 9-*cis*-retinoic acid (9-*cis*RA), 13-*cis*-retinoic acid (13-*cis*RA), 11-*cis*RA, and 9,13-*dicis*RA) (Fig. 1). Of these isomers, *all-trans*-RA (*at*RA) is considered to be the biologically active isomer and the primary enzymatic product of retinaldehyde oxidation. However, in various species including humans, the 9-*cis*RA, 13-*cis*RA, and 9,13-*dicis*RA isomers have also been detected in different organs (Arnold, Amory, Walsh, & Isoherranen, 2012; Kane, Folias, Wang, & Napoli, 2008). The *in vivo* source of these isomers is not clear. Although isomerization between *all-trans* and *cis* retinols has been characterized (McBee et al., 2000), the stereochemistry of the retinaldehyde precursor *in vivo* has not been determined. 9-*cis*RA binds to retinoid X receptors (RXRs) and this isomer may be important in some tissues such as the pancreas where it has been detected (Kane et al., 2010). Yet, 9-*cis*RA concentrations *in vivo* are typically very low or undetectable, and the biological role of this isomer has not been confirmed. In contrast, 13-*cis*RA is detected in all tissues and the concentrations of 13-*cis*RA are often higher than *at*RA. Yet, 13-*cis*RA is considered to be devoid of biological activity due to its low affinity to nuclear retinoic acid receptors (RARs) in comparison to *at*RA (Aström, Pettersson, Krust, Chambon, & Voorhees, 1990). *In vivo*, 13-*cis*RA isomerizes to *at*RA both via glutathione-*S*-transferase-mediated and thermodynamic processes (Chen & Juchau, 1997; Sass, Forster,

Bock, & Nau, 1994). Hence, the activity of 13-*cis*RA is believed to be a result of isomerization to *at*RA. In many species, including human, 9,13-*dicis*RA is also found at concentrations as high as *at*RA and 13-*cis*RA. However, this isomer appears to be devoid of biological activity (Chen et al., 2000; Nonnecke, Horst, Dubeski, & Reinhardt, 1997; Zile, Emerick, & DeLuca, 1967). Throughout this text, the terms *at*RA and 13-*cis*RA are used when a specific isomer has been identified in the research, and the generic term RA is employed when the identity of the isomer has not been specified.

1.1 RA Signaling and Biological Activity

Retinoid signaling and *at*RA have been shown to be important in various biological processes and in treatment of several diseases. For example, endogenous *at*RA is critical in the maintenance of healthy skin, epithelia, and the immune system (Napoli, 2012), in regulation of bone growth and homeostasis, and in regulating continuous, asynchronous spermatogenesis (Chung & Wolgemuth, 2004; Hogarth & Griswold, 2013). In addition, *at*RA is a classic morphogen that regulates embryonic development (Duester, 2008), organogenesis, stem cell differentiation (Gudas & Wagner, 2011), and body patterning in all chordates. Retinoid signaling also plays a central role in neuronal differentiation (Maden, 2007) and in tissue repair and regeneration (Gudas, 2012). In addition, retinoid signaling has been linked to regulation of insulin-stimulated glucose secretion (Chung & Wolgemuth, 2004; Kane et al., 2010, 2008) and to lipid homeostasis and adiposity (Bonet, Ribot, & Palou, 2012). Recently, *at*RA has been found to also play a role in modulating the inflammation that accompanies many human diseases. Exogenous *at*RA and 13-*cis*RA have been shown to be beneficial in the treatment of chronic hand eczema (Ruzicka et al., 2004) and in a variety of cancers including acute promyelocytic leukemia (APL) (Tang & Gudas, 2011), neuroblastoma in children (Veal et al., 2007), and Kaposi's sarcoma (Altucci, Leibowitz, Ogilvie, de Lera, & Gronemeyer, 2007). Whether exogenous RA functions via similar pathways and receptors as endogenous *at*RA is not well defined.

The biological activity of *at*RA is predominantly mediated by binding to RARα, RARβ, and RARγ, but *at*RA has also been shown to bind to peroxisome proliferator-activated receptors (PPARβ/δ) (Schug, Berry, Shaw, Travis, & Noy, 2007). Binding of retinoids to either RARs or PPARs results in increased transcription of respective target genes, and these regulatory mechanisms have been thoroughly reviewed elsewhere (Gudas, 2012; Noy, 2010). The observed effects of *at*RA on gene transcription are dependent on the cellular concentrations of *at*RA as well as the expression levels of RAR and PPAR isoforms. In addition, the expression of cellular binding proteins that bind *at*RA, most notably cellular retinoic acid-binding proteins (CRABPs) and fatty acid binding proteins (FABPs), has been shown to direct the activity of *at*RA isomers toward the different nuclear receptors (Noy, 2000; Schug et al., 2007). Overall, *at*RA signaling in various tissues is tightly controlled by the interplay between expression levels of the nuclear receptors and binding proteins, and the expression and activity of the *at*RA synthesizing and eliminating enzymes that control *at*RA concentrations in target tissues.

1.2 Biochemistry, Expression, and Function of CYP26 Enzymes

The clearance of *at*RA is mediated predominantly by the cytochrome P450 family 26 enzymes (CYP26) in all chordates, although many other CYP enzymes including human CYP3A4 and CYP2C8 and rat Cyp2c22 also oxidize *at*RA and its 9-*cis*RA and 13-*cis*RA isomers (Marill, Cresteil, Lanotte, & Chabot, 2000; McSorley & Daly, 2000; Nadin & Murray, 1999). It is widely believed that oxidation of *at*RA in the four-position of the β-ionone ring to generate 4-OH-RA (Fig. 1) is the predominant route of elimination of *at*RA but no unequivocal mass balance studies confirm this. Determining the major route of *at*RA elimination in humans or animal species is challenging as 4-OH-RA is extensively metabolized both *in vitro* and *in vivo*. 4-OH-RA is glucuronidated predominantly by UGT2B7 and oxidized by CYP26s and microsomal ADHs to 4-oxo-RA and other products (Samokyszyn et al., 2000; Topletz et al., 2015). The CYP26s and other CYP enzymes also generate several other hydroxylation products from *at*RA (Thatcher et al., 2011). The best characterized of these additional hydroxylation products is the 18-OH-RA (Fig. 1), which has been synthesized (Rosenberger & Neukom, 1982) and shown to be a metabolite formed by CYP26A1, CYP26B1, and other CYPs (Topletz et al., 2012). In addition, 16-OH-RA (Fig. 1) has been proposed as a metabolite of *at*RA, but this metabolite has never been synthesized, and therefore the identification of the metabolite is based on interpretation of mass spectrometry fragmentation patterns (Thatcher et al., 2011).

The CYP26 family consists of three highly conserved enzymes CYP26A1, CYP26B1, and CYP26C1. Despite the fact that RA appears to be the primary substrate of all three CYP26 enzymes and all CYP26s metabolize *at*RA predominantly to 4-OH-RA, these three enzymes share only 40–50% sequence similarity in a given species (Thatcher & Isoherranen, 2009). Yet, the individual isoforms are highly conserved across chordates which all have three CYP26 enzymes. Overall, from genomic and phylogenetic analysis it appears that CYP26 enzymes evolved very early together with RA signaling and play critical roles in regulating developmental processes. As such, it is not surprising that both *Cyp26a1*$^{-/-}$ and *Cyp26b1*$^{-/-}$ mice suffer from multiple developmental defects. The *Cyp26a1*$^{-/-}$ mice die during gestation, while *Cyp26b1*$^{-/-}$ mice survive till birth but die shortly after and have severe, but specific, malformations (Abu-Abed et al., 2001; MacLean et al., 2001; Uehara et al., 2007; Yashiro et al., 2004). Notably, the pattern of developmental defects in the two mouse models is different, pointing to the unique roles the two CYP26 enzymes play during organogenesis and body axis formation. In contrast, *Cyp26c1*$^{-/-}$ mice are viable and free of malformations although the simultaneous knockout of *Cyp26c1* with *Cyp26a1* aggravates the phenotype of the *Cyp26a1*$^{-/-}$ mice (Uehara et al., 2007).

Functionally, CYP26 enzymes are membrane-anchored P450 proteins that require NADPH and P450 reductase for their function. While it is theoretically possible that the CYP26 enzymes reside in the mitochondria instead of the endoplasmic reticulum (ER), the fact that the CYP26 enzymes require P450 reductase from the ER membrane to function suggests that these enzymes are ER membrane bound (Lutz et al., 2009; Topletz et al., 2012). However, the localization of the CYP26 enzymes within a cell has not been unequivocally determined. Interestingly, unlike other P450 enzymes such as CYP17, cytochrome *b*5 does not affect the catalytic activity of the CYP26 enzymes, and P450 reductase seems to solely support the catalytic activity of these enzymes (Lutz et al., 2009). Overall, the affinity of *at*RA to each of the CYP26 enzymes is high and the K_m values for CYP26A1 and CYP26B1 are <100 nM for *at*RA. Interestingly, the K_m value for *at*RA with the CYP26s approximates the concentrations of *at*RA in various tissues (Topletz et al., 2012). While the affinity of the substrate is high, the k_{cat} values for *at*RA with CYP26 enzymes are similar to other mammalian P450 enzymes (approximately 1–10 pmol/min/pmol) (Thatcher, Zelter, & Isoherranen, 2010; Topletz et al., 2012). Since the affinity of *at*RA to CYP26 is approximately 1000-fold higher than that to other CYP enzymes, the overall intrinsic clearance of *at*RA by CYP26 enzymes is 1000- to 10,000-fold higher than other CYP enzymes such as

CYP3A4 and CYP2C8 (Thatcher et al., 2010). Therefore, even in tissues such as the human liver which have high expression levels of CYP3A4 and CYP2C8, the CYP26s are expected to be the main contributors to RA clearance even if they are expressed at low levels (Thatcher et al., 2011, 2010; Thatcher & Isoherranen, 2009). Indeed, in a study of individual donors in a human liver bank, CYP26A1 was shown to be the main enzyme contributing to *at*RA clearance, but the expression of CYP26A1 was subject to considerable interindividual variability (Thatcher et al., 2010). Interestingly, in the human liver CYP26B1 protein was completely absent pointing to tissue- and cell-type-specific roles of the CYP26 enzymes in adult tissues and during fetal development.

It is generally believed that RA concentrations in target tissues are regulated locally by RA synthesis and clearance, and that systemic RA concentrations do not reflect RA homeostasis in specific tissues. Therefore, extrahepatic expression of CYP26 enzymes is expected and likely plays a role in tissue-specific clearance and regulation of RA concentrations. Indeed, CYP26A1 and CYP26B1 mRNA and protein are found in multiple extrahepatic human tissues (Topletz et al., 2012; Xi & Yang, 2008) with CYP26B1 appearing to have a higher expression in extrahepatic tissues than CYP26A1. In agreement with the human tissue data of CYP26 expression, metabolism of *at*RA was shown in rat testes, kidney, and lung microsomes (Fiorella & Napoli, 1994). In addition, CYP26 enzymes have been suggested to constitute a tissue barrier for RA distribution from circulation to specific organs. This hypothesis originated from the observation that uptake of administered radiolabeled RA to some tissues, such as the testes, pancreas, and spleen, was limited (Ahluwalia, Gambhir, & Sekhon, 1975; Kurlandsky, Gamble, Ramakrishnan, & Blaner, 1995; McCormick, Kroll, & Napoli, 1983; Smith, Milch, Muto, & Goodman, 1973), and that CYP26 enzymes were detected in several barrier cells (Heise et al., 2006; Vernet et al., 2006; Xia, Ma, Sun, Yang, & Peng, 2010). The metabolic barrier provided by CYP26 enzymes in some tissues such as the testes likely ensures that RA gradients are regulated by enzyme expression and activity within the tissue and not by circulating concentrations. Therefore, understanding the activity and expression of CYP26 enzymes in individual tissues and cell types is critically important for defining the relationship between *at*RA concentrations and biological outcomes within a tissue. Based on tissue-specific expression and activity of CYP26 enzymes, it can be hypothesized that tissue-specific alterations in CYP26 activity will alter *at*RA concentrations in the given tissue and result in altered *at*RA signaling. For example, inhibition of CYP26s or genetic polymorphisms that decrease CYP26 activity would be expected to increase *at*RA concentrations within the tissue in which CYP26s are expressed. On the other hand, induction of CYP26s or increased activity of CYP26 enzymes within an organ is expected to decrease the overall exposure to *at*RA and possibly increase exposure to *at*RA metabolites that may also have pharmacological activity. Yet, a direct effect of decreased or increased CYP26 activity and *at*RA concentrations in a tissue has never been demonstrated, even though P450 inhibitors such as talarozole and liarozole that also inhibit CYP26 do increase RA concentrations in various tissues and plasma (Nelson, Buttrick, & Isoherranen, 2013; Stoppie et al., 2000; Van Wauwe, Coene, Goossens, Cools, & Monbaliu, 1990). Likewise, specific effects of CYP26A1 or CYP26B1 inhibition have never been shown to have different effects in different tissues.

In addition to metabolic enzymes, small binding proteins of the class of lipid binding proteins such as CRABP-I and CRABP-II that bind *at*RA with high affinity ($K_d = 0.06$ nM for CRABP-I and $K_d = 0.13$ nM for CRABP-II) (Dong, Ruuska, Levinthal, & Noy, 1999) are believed to alter *at*RA homeostasis and CYP26 activity. However, CRABP$^{-/-}$ mice are viable and devoid of any malformations or negative health effects (De Bruijn et al., 1994; Lampron et al., 1995) putting the biological significance of the binding proteins to question. CRABP-II has been shown to deliver *at*RA to RAR in the nucleus whereas CRABP-I may serve a different function such as delivering *at*RA to the CYP26 enzymes. Whether these functions affect overall *at*RA homeostasis is still unknown.

2. ROLE OF RA IN THE IMMUNE SYSTEM

For nearly 100 years, vitamin A has been recognized as a critical nutritional component needed to support immunity. It is well established that vitamin A deficiency results in increased morbidity and mortality from certain pathogens, and several clinical trials have shown that vitamin A supplementation in at risk populations improves overall outcomes (Ross, 2012). In support of human studies, vitamin A-deficient mice were shown to be completely deficient in CD4+ and CD8+ T-cells in the submucosal lamina propria in the intestine (Iwata et al., 2004). This suggests that gut mucosal immunity is impaired in vitamin A deficiency, and potentially explains the specific link between poor outcomes in diarrheal disease and vitamin A deficiency. Yet, it is possible that retinol and retinol binding protein (RBP) also contribute to regulation of gut mucosal immunity either together with RA or independent of RA as retinol and RBP concentrations are significantly decreased in vitamin A deficiency (De Pee & Dary, 2002; Gamble et al., 2001) and retinol and RBP have been shown to have activity in the absence of RA via binding to STRA6 (Berry, Levi, & Noy, 2014; Marwarha, Berry, Croniger, & Noy, 2014). However, current evidence suggests that *at*RA itself is critical in maintaining immune homeostasis (Hall, Grainger, Spencer, & Belkaid, 2011; Raverdeau & Mills, 2014), is the required signaling molecule in the regulation of T-cell, B-cell, and dendritic cell (DC) differentiation and function, and in the production of both pro- and anti-inflammatory cytokines (Hansen, Westendorf, & Buer, 2008; Raverdeau & Mills, 2014).

Vitamin A deficiency, which presumably results in lower concentrations of *at*RA, promotes naive CD4+ T-cell differentiation into Th1 cells, which secrete interferon-γ (Cantorna, Nashold, & Hayes, 1995). High concentrations of *at*RA, on the other hand, generally promote T-cell differentiation into Th2 cells producing IL-4 and IL-5 (Fig. 3). Hence, it appears that

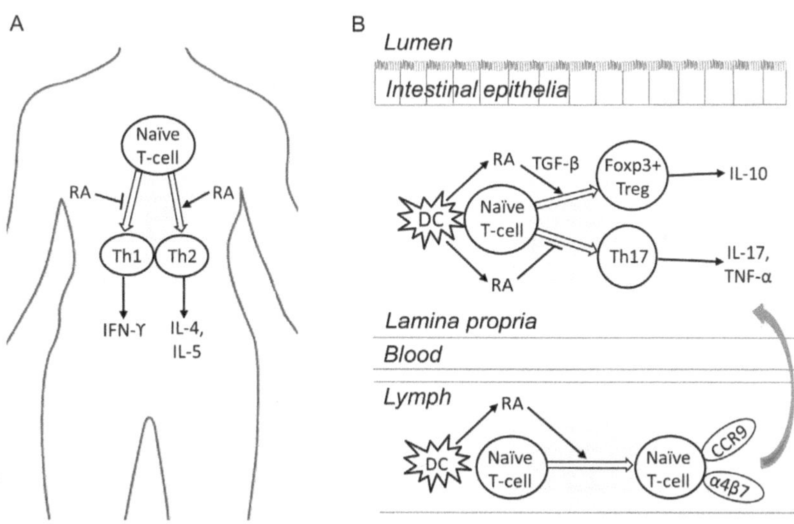

Figure 3 Retinoic acid in steady-state immune homoestasis. (A) Throughout the body, RA plays an important role in the ratio of Th1:Th2 cells by reducing differentiation of naïve T-cells into IFN-γ producing Th1 cells and promoting differentiation of Th2 cells, which secrete IL-4 and IL-5. (B) In gut-associated lymphoid tissue, dendritic cells (DCs) synthesize RA, which acts synergistically with TGF-β to initiate differentiation of Foxp3+ Treg cells and suppress T-cell differentiation into Th17 cells. RA produced by DCs also induces the expression of α4β7 and CCR9 on naïve T-cells to prompt gut homing to the lamina propria over peripheral lymph nodes. Cytokines shown are not a complete list but were chosen based on relevance to the text.

increasing concentrations of *at*RA in the precursor T-cell environment will increase the Th2 to Th1 cell ratio (Ross, 2012). In addition, of interest especially to gut mucosal immunity, *at*RA also regulates Th17 cell function by opposing Th17 cell commitment and inhibiting T-cell differentiation into Th17 cells (Mucida et al., 2007; Fig. 3). RA synthesis by the DCs in the gut-associated lymphoid tissue (GALT) is believed to initiate Treg cell differentiation. RA produced by DCs in GALT also induces the expression of α4β7 and CCR9 on T-cells upon antigen presentation (Iwata et al., 2004) inducing gut homing of the T-cells to the lamina propria over peripheral lymph nodes (Fig. 3). Activation of naïve T-cells in the presence of RA induced CCR9 and the α4 integrin as well (Kang, Park, Cho, Ulrich, & Kim, 2011). In Treg cell activation, RA acts synergistically with TGF-β to enhance the expression of Foxp3, a critical transcription factor for Treg cells (Hall et al., 2011; Mucida et al., 2007). The effects of *at*RA on differentiation and homing of Treg cells to the gut mucosa and inhibition of Th17 cell formation appear to be regulated via RARα (Mucida et al., 2007; Raverdeau et al., 2014). Hence, increasing RA concentrations is expected to prevent the induction of autoimmune T-cells and modulate inflammatory responses. Yet, many of the responses observed with RA appear to be dependent on the cytokine environment. Overall, RA plays a critical role in modulating immune responses and inflammation, and regulation of RA concentrations is likely important in inflammatory and autoimmune diseases. However, how RA concentrations are controlled within the immune system is not well established, and it is possible that the roles of the individual enzymes are divergent in different inflammatory conditions and in different tissues.

Based on current data, the DCs in the GALT express RDH and ALDH1A enzymes and produce RA (Iwata et al., 2004). The elimination of *at*RA also appears to be regulated within the immune system as CYP26 enzymes are expressed and inducible by *at*RA in T- and B-cells (Takeuchi, Yokota, Ohoka, & Iwata, 2011; Wen et al., 2013). Interestingly, in naïve T-cells *in vitro*, TGF-β inhibits and TNF-α enhances CYP26B1 induction by *at*RA (Takeuchi et al., 2011). CYP26B1 mRNA was found in T-cells from mouse GALT, including mesenteric lymph nodes and Payer's patches, but not from peripheral lymph nodes (Takeuchi et al., 2011). CYP26A1 and CYP26B1 mRNA expression and RA-dependent induction have also been demonstrated in B-cells of healthy humans (Wen et al., 2013). *In vitro*, when T-cells from mice were stimulated under conditions promoting T-cell differentiation, CYP26B1 expression was significantly increased in differentiated Th17 cells but was significantly lower in differentiated Treg cells and lowest in undifferentiated T-cells suggesting that depletion of RA by CYP26B1 regulates RA concentrations within the T-cells (Chenery et al., 2013). A higher expression of CYP26B1 should result in lower RA concentrations within the cells removing the RA signaling. However, somewhat surprisingly when CYP26B1 was knocked out in the T-cells, the frequency of both IL-17 producing T-cells (Th17 cells) and FoxP3+ Treg cells was increased (Chenery et al., 2013). Based on these data, it is likely that alterations in the activity of ALDH1A and CYP26 enzymes within the immune system are a key factor modulating *at*RA concentrations and signaling that subsequently drives immune and inflammatory responses.

3. CYP26 AND RA IN CHRONIC INFLAMMATORY DISEASES

A large portion of human diseases are associated with chronic inflammation. In some of these conditions inflammation is related to the primary pathology of the disease, whereas in others only clinical associations between disease progression and inflammatory markers have been observed. For example, non-alcoholic fatty liver disease (NAFLD) is defined as an excess accumulation of lipids in the liver but progression to non-alcoholic steatohepatitis (NASH) is thought to be a result of triglyceride and cholesterol accumulation in the liver leading to oxidative stress, insulin resistance, and inflammation (Depner, Philbrick, & Jump, 2013). As RA and RAR activation is known to play an important role in the liver and RA is an important regulator of inflammation, it is possible that development of NASH and associated inflammation involves altered *at*RA signaling. In the case of NAFLD and NASH, analysis of mRNA from patient liver biopsies showed increased mRNA levels of a variety of enzymes involved in regulation of *at*RA and vitamin A concentrations including CRBP1, LRAT, RDH5, RDH11, DHRS3, ALDH1, ALDH3, and CYP26A1 (Ashla et al., 2010). However, whether any of these changes are associated with altered liver *at*RA concentrations, play a role in disease development and progression, or are a result of liver disease is not known.

Perhaps the best-demonstrated connections between chronic inflammatory markers and disease progression have emerged from autoimmune diseases. Autoimmune diseases arise from an abnormal immune response of the body against its own tissues or substances (Abella et al., 2014; Feldmann, Brennan, & Maini, 1996). Treatment of these diseases generally focuses on regulation of immune system activity. While autoimmune diseases as well as inflammatory conditions are a large group of diseases, this section focuses on those conditions in which *at*RA signaling and CYP26 activity have been shown to be altered or potentially play a role in modulating the disease progression.

3.1 Inflammatory Bowel Disease

Inflammatory bowel disease (IBD) is an inflammatory disease of the gastrointestinal tract and a link may exist between regulation of intestinal immunity by RA and IBD progression. The principal forms of IBD are Crohn's disease (CD) and ulcerative colitis (UC). The pathology of these diseases is complicated and thought to arise as a result of environmental, genetic, and immunologic factors. Both CD and UC are associated with an excessive response of mucosal adaptive and innate immunity (Brown & Mayer, 2007; Talley et al., 2011). In active IBD, increased levels of macrophages and pro-inflammatory cytokines are found in intestinal mucosa (Mahida, 2000). *Ex vivo* studies in cells from UC patients and *in vivo* mouse studies have found that increasing *at*RA concentrations could overcome the dysfunction of the immune system and ameliorate inflammation. For example, in an *in vitro* colitis model using murine macrophages, *at*RA treatment decreased the activation of NFκB, a macrophage nuclear transcription factor, and production of the pro-inflammatory cytokine TNF-α. In a mouse model of colitis, *at*RA treatment reduced symptoms of inflammation including reduced infiltration of neutrophils, decreased levels of NFκB, and lower levels of TNF-α in colon tissues in comparison to control mice (Hong et al., 2014). Similar to mouse models of colitis, treatment of cultured colonic mucosa cells from UC patients with RA decreased concentrations of TNF-α, upregulated Foxp3 and downregulated IL-17, returning immunologic balance. The nuclear receptor RARα was implicated in regulating these RA effects (Bai et al., 2009). These studies did not evaluate whether RA was *a priori* decreased in the disease setting due to either decreased synthesis or increased metabolism of RA. However, these studies suggest that increasing levels of *at*RA may be effective in halting the acute inflammatory response in IBD patients.

Further evidence of the potential benefit of increased RA in IBD comes from studies in T-cell-specific CYP26B1$^{-/-}$ mice (Chenery et al., 2013). Although RA concentrations were not measured, it is expected that CYP26B1$^{-/-}$ T-cells have increased RA concentrations compared to wild-type mouse T-cells. When naïve CD4+ T-cells from T-cell-specific CYP26B1$^{-/-}$ mice and control mice were implanted into a murine model of T-cell-dependent intestinal inflammation, inflammation and disease progression were attenuated in the presence of CYP26B1$^{-/-}$ T-cells versus control (Chenery et al., 2013). Still, T-cell-specific CYP26B1$^{-/-}$ mice had equivalent T-cell development as controls. In addition, IL17a and TNF-α mRNA levels were decreased in the colon of mice with CYP26B1$^{-/-}$ T-cells but Foxp3 mRNA was unaffected by the knockout. Assuming that CYP26B1 metabolizes RA within the T-cells these findings are consistent with decreased Th17 cells and unaltered Treg cells. The findings in the T-cell-specific CYP26B1$^{-/-}$ mice suggest that CYP26B1 plays an important role in modulation of T-cell homeostasis. Surprisingly, the expression of RA-inducible homing signals α4β7 and CCR9 was not affected in the T-cell-specific CYP26B1$^{-/-}$ mice (Chenery et al., 2013).

In addition to the mouse T-cell-specific CYP26B1$^{-/-}$ model, the effect of CYP26B1 polymorphisms on CD and IBD risk was studied (Fransén et al., 2013). The CYP26B1 T to C polymorphism (rs2241057) that results in L264S amino acid change displays enhanced catabolic activity compared to the T allele suggesting concentrations of RA may be higher in patients with the T allele (Fransén et al., 2013). The frequency of the lower activity T allele was not significantly higher in CD patients compared UC patients or healthy controls. However, *at*RA concentrations were not measured in this study and no firm relationship between the polymorphism, RA levels and risk of developing CD could be made.

Despite a large volume of data supporting the concept that increasing RA levels would combat IBD, it has also been proposed that increased gut RA levels contribute to IBD pathogenesis. This suggestion is based on the observation that ALDH1A mRNA was increased in intestinal mucosa of CD patients compared to controls. Specifically, ALDH1A1 was predominantly expressed in macrophages from patients and ALDH1A2 was found in DCs (Sanders et al., 2014). While increased ALDH1A mRNA in CD patients suggests that formation of *at*RA in some or all subsets of antigen-presenting cells in intestinal mucosa is increased, altered RA concentrations, or increased RA synthesis was not measured in this study. Hence, it is possible that ALDH1A expression is increased in the intestinal mucosa as a compensatory mechanism to low *at*RA or vitamin A concentrations. Overall further studies that directly measure RA concentrations in IBD and CD are needed to determine the role of *at*RA signaling in these diseases.

3.2 Psoriasis

Psoriasis is characterized by an abnormally excessive growth and aberrant differentiation of the epidermal layer of skin. In psoriasis, premature maturation of keratinocytes is induced by an inflammatory cascade originating from DCs, macrophages, and T-cells (Nestle, Kaplan, & Barker, 2009). The beneficial effect of oral and topical retinoids in psoriasis is well established. In general, retinoids have an anti-proliferative effect in epidermal cells if the microenvironment is hyperproliferative and a proliferative effect if the microenvironment is normal (Harper, 1988; Harper & Burgoon, 1982). In normal cells, RA stimulates keratinocyte growth and differentiation, while in psoriasis RA is anti-proliferative. In psoriasis, retinoid treatment decreases cytotoxic T-cell induction, inhibits the chemotactic responses of activated lymphocytes, and inhibits expression of migration inhibitory factor-related protein-8 (a marker for hyperproliferation of keratinocytes) resulting in improvement in patient symptoms (DiSepio, Malhotra, Chandraratna, & Nagpal, 1997; Dupuy, Bagot, Heslan, & Dubertret, 1989; Van de Kerkhof, 2006).

Similar to RA signaling in gut mucosal immunity, RA signaling is important in the DCs and T-cells in the skin. Similar to DCs in the GALT, DCs in the skin have been shown to express ALDH1A enzymes and to produce RA (Guilliams et al., 2010). Analogously to the gut mucosa, DCs in the skin were shown to induce Treg cell differentiation and Foxp3 expression in ALDH1A expression-dependent manner (Guilliams et al., 2010). Interestingly, in contrast to the DCs in the GALT, DCs in the skin do not induce the gut homing signals such as CCR9 in CD4+ T-cells despite the fact that skin DCs also synthesize RA, demonstrating that gut homing signal generation is specific to GALT DCs (Guilliams et al., 2010; McCully et al., 2012).

As in most other cells of the body, retinoid levels in human skin are under tight control. Retinoids are stored in the skin as retinyl esters and RA is produced locally in the epidermis (Fisher & Voorhees, 1996; Pavez Loriè, Li, Vahlquist, & Törmä, 2009). Expression of CYP26A1 has been confirmed in the basal layer of normal and psoriatic human epidermal keratinocytes with immunofluorescence and immunohistochemical staining (Heise et al., 2006). In addition to CYP26A1, another RA-metabolizing enzyme, CYP2S1, has been identified in keratinocytes in psoriatic plaques and normal human skin (Smith et al., 2006) but the quantitative role of CYP2S1 in *at*RA clearance is not known. Still, the expression of two RA-metabolizing enzymes is indicative of the importance of RA homeostasis in skin.

Natural and synthetic retinoids have been used in the oral and topical treatment of psoriasis since the 1970s (Dogra & Yadav, 2014; Roos, Jugert, Merk, & Bickers, 1998). While clinical results have been positive, severe side effects such as teratogenicity, hypertriglyceridemia, and dermatitis can be dose limiting. Thus, the use of inhibitors of RA metabolism, that is, RA metabolism blocking agents (RAMBAs) has also been investigated as a means to increase endogenous RA levels in psoriasis and avoid complications from exogenous administration (Verfaille, Borgers, & van Steensel, 2008). Although RAMBAs were originally developed before the identification of CYP26 enzymes, it is now assumed that CYP26s are the primary target of these compounds (Nelson et al., 2013). Treatment of keratinocyte cultures with CYP26 inhibitors, such as talarozole (also known as R115866 or rambazole), has been shown to potentiate cell differentiation effects of RA (Giltaire et al., 2009). At the same time, treatment with RAMBAs induced only CYP26 expression, whereas treatment with RA altered multiple enzymes in the retinoid synthesis pathway (Pavez Loriè, Chamcheu, Vahlquist, & Törmä, 2009; Smith et al., 2006). These results indicate that inhibition of CYP26 may achieve the same therapeutic benefits as exogenous RA administration without the side effects observed with *at*RA administration. However, it is not clear how CYP26 inhibitors will affect the immune system and inflammation in the skin.

Clinical studies with CYP26 inhibitors have also been reported in psoriasis. Patients treated with oral ketoconazole, a broad CYP inhibitor that also inhibits CYP26s, had some improvement in their scalp psoriasis. However, the trial was discontinued due to concerns of drug toxicity (Farr, Krause, Marks, & Shuster, 1985). A second-generation RAMBA, liarozole, was developed with more specific activity towards CYP26, and it was effective in treating psoriasis in early clinical trials (Bhushan et al., 2001; Dockx, Decree, & Degreef, 1995). Further improvement in CYP26 selectivity and potency led to the development of talarozole (Pavez Loriè, Li, et al., 2009). Oral administration of talarozole significantly improved psoriasis symptoms (noted as a reduction in the Psoriasis Area Severity Index or plaque severity) in Phase I and II clinical trials. Increases in plasma concentrations of *at*RA were observed in a single-dose escalation trial of talarozole, but RA concentrations were within or slightly above normal physiological concentrations. Talarozole was well tolerated with no severe side effects noted in Phase I or II studies (Bovenschen et al., 2007; Verfaille et al., 2007). Additionally, topical treatment with talarozole was shown to increase CRABP2 and keratin 4 and decrease keratin 2 (markers of increased RA activity) in epidermis in healthy volunteers (Pavez Loriè, Cools, et al., 2009). Collectively, these data demonstrate the important role of CYP26 in regulation of RA metabolism in the skin and in psoriasis. Unfortunately, none of the studies with RAMBAs have assessed RA concentrations in the skin or inflammatory markers. Hence, it is not possible to determine whether modulation of inflammatory responses also contributes to clinical improvement.

3.3 Atherosclerosis

Atherosclerosis is a slowly progressing chronic inflammatory disease of blood vessels characterized by accumulation of lipids in the arterial wall (Hansson & Hermansson, 2011; Hansson, Robertson, & Soderberg-Naucler, 2006). Atherosclerosis is a complex disease resulting from multiple risk factors such as excess lipid accumulation, coagulation and fibrosis, cell proliferation, and inflammatory cell migration to the inflamed sites. During early stages of atherosclerosis, vascular smooth muscle cells migrate to the intima, promote extracellular matrix deposition, and reduce the lumen size (Rhee, Nallamshetty, & Plutzky, 2012). However, inflammation, including innate and adaptive immune responses, is involved in all stages of the development of atherosclerosis (Libby, Ridker, & Hansson, 2009; Wiedermann et al., 1996).

RA has diverse biological effects on arterial wall function and plays an important role in the coagulation system. Treatment of human aortic smooth muscle cells with *at*RA decreased cell proliferation, cell migration to the lumen, and extracellular matrix formation in the lumen (Johst et al., 2003). Additionally, administration of *at*RA attenuated pro-coagulant properties, promoted fibrinolysis, reduced the induction of monocytes and tissue factor-dependent arterial thrombus formation and suppressed the expression of vasoconstrictor thromboxane A2 in vascular smooth muscle cells (Barstad, Hamers, Stephens, & Sakariassen, 1995; Bulens et al., 1995; Van Giezen, Boon, Jansen, & Bouma, 1993; Watanabe et al., 2002).

Several enzymes involved in retinoid metabolism and action are expressed in the vasculature including alcohol dehydrogenases and RARs (Allali-Hassani et al., 1997; Miano, Topouzis, Majesky, & Olson, 1996), and RAR activity in human smooth muscle cells was confirmed in transient transfected reporter gene assays (Ocaya et al., 2011). Of the three CYP26 enzymes, CYP26B1 was found to be constitutively expressed in human vascular smooth cells and play a major role in *at*RA clearance in the arterial wall (Elmabsout et al., 2012; Krivospitskaya et al., 2012; Ocaya et al., 2011).

Based on this, RA may also play a role in regulating other functions in the vasculature. In human atherosclerotic arteries CYP26B1 mRNA expression was increased suggesting that RA concentrations are decreased in atherosclerosis. Murine atherosclerotic model arteries also expressed higher CYP26B1 mRNA than normal arteries (Krivospitskaya et al., 2012). Additionally, based on immunohistochemistry of arterial biopsies from atherosclerotic patients CYP26B1 colocalizes with CD68, a marker for activated macrophages, in atherosclerotic lesions (Krivospitskaya et al., 2012). The accumulation of CYP26B1-expressing macrophages may be the reason for increased CYP26B1 mRNA in the atherosclerotic arteries. When the frequency of the CYP26B1 SNP (L264S) with higher catalytic activity than wild type was tested in atherosclerotic patients and matched controls, no association of the variant allele expression between patients and matched controls and basic clinical characteristics was detected. However, patients who carried the minor allele variant and may have lower levels of *at*RA had a small increase in atherosclerotic lesion size (Krivospitskaya et al., 2012). Expression of CYP26B1 and RARβ was enhanced in human aortic smooth muscle cells treated with *at*RA alone or with talarozole and *at*RA (Ocaya et al., 2011). Talarozole together with *at*RA treatment also increased cellular *at*RA concentrations and decreased DNA synthesis and cell proliferation compared to cells treated with *at*RA alone. In addition, transfection with CYP26B1 siRNA led to increased *at*RA-mediated RAR/RXR signaling.

4. CYP26 AND RA IN CANCER

Vitamin A and RA play a crucial role in regulating cell proliferation, differentiation, and apoptosis, and as early as 1925 an association between vitamin A deficiency and carcinogenesis was suggested (Wolbach & Howe, 1925). Subsequent epidemiological, preclinical, and clinical studies established a relationship between vitamin A deficiency and a higher incidence of carcinogenesis in both humans and rodents (Basu, Temple, & Hodgson, 1988; Chaudhy, Jafarey, & Ibrahim, 1980; Dogra, Khanduja, & Gupta, 1985; Ito, Gajalakshmi, Sasaki, Suzuki, & Shanta, 1999; Stehr et al., 1985). In addition, abnormal RA metabolism potentially leading to altered RA levels has also been implicated in tumorigenesis (Moulas, Gerogianni, Papadopoulos, & Gourgoulianis, 2006; Osanai & Petkovich, 2005).

Altered expression of proteins controlling RA concentrations has been associated with multiple cancers but changes in RA concentrations have not been directly measured in cancers. Significantly, higher activity of alcohol dehydrogenase enzymes has been shown in breast, ovarian, colorectal, and cervical cancer tissues compared to healthy human tissues (Jelski, Chrostek, Szmitkowski, & Markiewicz, 2006; Jelski, Zalewski, Chrostek, & Szmitkowski, 2004; Orywal, Jelski, Zdrodowski, & Szmitkowski, 2011, 2013). ALDH1A3 expression was also high in breast cancer tissues and ALDH1A3 levels significantly correlated with distant metastasis (Qiu et al., 2014). Further, ALDH1A1 expression was observed in 63.0% (92 of 146) of human breast invasive ductal carcinoma tissues (Zhong et al., 2013). It is possible that increases in alcohol dehydrogenase and ALDH enzymes will lead to increased RA concentrations. However, it is also possible that low RA concentrations in cancer prevent RA-mediated regulation of these enzymes or that alcohol dehydrogenase and ALDH1A enzymes are increased in an attempt to correct low RA concentrations in cancer.

The role of the retinol transport proteins RBP and STRA6 in cancer development and progression is still unclear. In human breast and colon cancers STRA6 and RBP mRNA levels were elevated in comparison to normal breast and colon tissue, and higher expression of STRA6 and RBP correlated with oncogenic properties and tumor progression suggesting an oncogenic role of STRA6 (Berry et al., 2014). In agreement, binding of holo-RBP4 (RBP4-ROH) to STRA6 was shown to trigger phosphorylation of STAT3, resulting in recruitment and activation of the JAK/STAT pathway and increased oncogenic properties in tumor cell lines (Berry et al., 2014). In contrast, in human keratinocytes, knockdown of STRA6 was shown to increase keratinocyte proliferation and lead to thickening of the epidermal layer and increased proliferation (Skazik et al., 2014). Similarly, increased expression of STRA6 increased apoptosis in normal and cancer cells and influenced p53-mediated apoptosis in bladder, ovarian, and colon cancer cell lines (Carrera et al., 2013). Overall, the findings are controversial and further work is needed to determine how STRA6 expression and RA signaling affect cancer development and oncogenesis. Additionally, downregulation or loss of CRBP1, which regulates retinoid uptake, intracellular availability, and metabolism, has been shown in both human breast and ovarian cancer (Cvetković, Williams, & Hamilton, 2003; Kuppumbatti, Bleiweiss, Mandeli, Waxman, & Mira-Y-Lopez, 2000). Together, these studies show that proteins controlling RA signaling play an important role in many cancers.

Increased expression of CYP26 enzymes in cancers, such as breast, colorectal, and head and neck cancers has recently been shown in several studies (Brown et al., 2014; Osanai & Lee, 2014; Osanai, Sawada, & Lee, 2009). Elevated CYP26 levels in cancer tissue resulting in RA deficiency have also been linked to carcinogenesis in epidemiological studies (Berlin Grace, Niranjali Devaraj, Radhakrishnan Pillai, & Devaraj, 2006; Moulas et al., 2006). CYP26A1 was shown to be highly expressed in 42% (27/65) of surgically resected breast cancer specimens (Osanai & Lee, 2014). CYP26A1 expression was significantly higher in metastatic carcinomas ($n=10$) than in normal breast tissue suggesting that high expression of CYP26A1 is associated with decreased disease-free and overall survival in breast cancer (Osanai et al., 2014). Similarly, CYP26A1 and CYP26B1 expression was significantly increased in primary colorectal cancer compared with normal colonic mucosa (Brown et al., 2014), and CYP26A1 expression was significantly higher in sporadic human colon tumor samples than in normal tissues (Shelton et al., 2006). In addition to breast and colorectal cancer, CYP26A1 overexpression might also contribute to the development and progression of cervical malignancies and squamous neoplasia of the head and neck. For example, strong expression of CYP26A1 was observed in 19 of 50 (38%) cervical squamous cell carcinoma cases but no expression was found in normal cervical epithelium. In addition, strong expression of CYP26A1 was shown in 52 of 128 (41%) head and neck cancer cases with no CYP26A1 expressed in non-neoplastic tissues of the head and neck (Osanai et al., 2014). In Barrett's carcinogenesis, a significant upregulation of CYP26A1 was observed in adenocarcinoma compared to non-dysplastic Barrett's tissues (Chang, Hong, Lao-Sirieix, & Fitzgerald, 2008). Finally, CYP26A1 expression was significantly enhanced in primary ovarian cancers in comparison to normal ovarian tissue (Downie et al., 2005). While these studies show that CYP26 expression is elevated in various types of cancer, the underlying mechanism resulting in elevated CYP26 expression in human cancers and the exact role of CYP26 in carcinogenesis has not been fully clarified.

In vitro studies with CYP26 transfection to induce overexpression of CYP26 have been conducted and offer insights into the role of CYP26 in carcinogenesis (Chang et al., 2008; Xia et al., 2010). Increases in anti-apoptotic signals and oncogenes (e.g., epidermal growth factor receptor and c-Myc) were observed in dysplastic Barrett's cells transfected with CYP26A1 in comparison to control cells. Cells overexpressing CYP26A1 also exhibited increased proliferation (Chang et al., 2008). In addition, downregulation of p53, an apoptotic signal and tumor suppressor gene was found following overexpression of CYP26A1 in endometrial stromal cells (Xia et al., 2010). Based on these studies, use of RAMBAs to inhibit CYP26A1 and increase tumor RA levels is being pursued as a drug target in cancer (Nelson et al., 2013).

Efforts to synthesize RAMBAs with increased potency and favorable pharmacokinetics that selectively inhibit CYP26s are ongoing in several laboratories and details of these compounds can be found in several recent reviews (Nelson et al., 2013; Njar et al., 2006). These CYP26 inhibitors could potentially be used alone to increase RA concentrations or in combination with RA treatment to combat resistance to treatment and increase potency of RA. Preclinical and early clinical data for CYP26 inhibitors are promising (Godbole, Purushottamachar, Martin, Daskalakis, & Njar, 2012; Gomaa et al., 2011). Of the well-studied RAMBAs, liarozole, an inhibitor of CYP26 (IC_{50} 2.2–6.0 μM), reduced tumor growth in androgen-dependent and androgen-independent rat prostate carcinoma models. Liarozole also decreased risk of death compared to cyproterone acetate in patients with metastatic prostate cancer in relapse after first-line endocrine therapy (Njar et al., 2006; Stearns, Wang, & Fudge, 1993). Another CYP26 inhibitor R116010, a benzothiazolamine, inhibited *at*RA metabolism in intact T47D human breast cancer cells with an IC_{50} value of 8.7 nM (Van Heusden et al., 2002).

4.1 Acute Promyelocytic Leukemia

Leukemia results from an uncontrolled increase in the number of immature white blood cells. A rare subset of leukemia, APL, is definitively marked by a chromosomal translocation affecting the RARα gene (Martens et al., 2010). The predominant translocation, occurring in 90–95% of APL patients, is *t*(15; 17) where chromosome 15 maps to the promyelocytic leukemia gene (PML) and chromosome 17 to the RARα gene (Kakizuka et al., 1991; Orfali, McKenna, Cahill, Gudas, & Mongan, 2014). The *t*(15;17) translocation ultimately results in a PML–RARα fusion protein (Guidez et al., 1998; Kakizuka et al., 1991; Pandolfi et al., 1991), which binds with RXR to the RA response element of target genes (Martens et al., 2010) ultimately blocking myeloid differentiation and leading to leukemogenesis (Collins, 2002; Guidez et al., 1998; Fig. 4). With wild-type RARα, physiological levels of *at*RA are sufficient to dissociate negative corepressors but PML–RARα

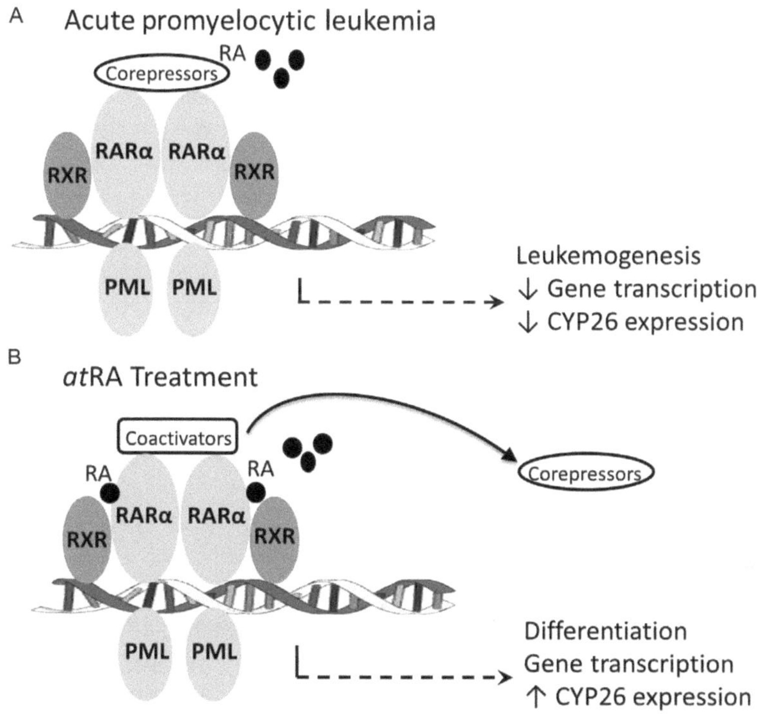

Figure 4 Retinoic acid and CYP26 signaling in acute promyelocytic leukemia. In acute promyelocytic leukemia, a chromosomal translocation forms the PML–RARα fusion protein which binds DNA as either a homodimer or multimeric complex with RXR. (A) Physiological levels of the signaling molecule, RA, are insufficient to bind and promote release of corepressors from the complex which blocks gene transcription (including the gene for CYP26) and promotes leukemogenesis. (B) Pharmacological levels of *at*RA bind the receptor, cause release of corepressors, and recruit activators for gene transcription ultimately resulting in terminal myeloid differentiation.

fusion proteins are less sensitive to *at*RA and pharmacological levels of *at*RA are required to dissociate negative corepressors from the fusion protein for gene transcription and terminal myeloid differentiation (Guidez et al., 1998).

Treatment of APL with *at*RA began in the 1980s and remains standard of care to date (Huang et al., 1988; Watts & Tallman, 2014). *at*RA treatment causes terminal myeloid differentiation, degradation of the fusion protein and autophagy of APL blasts. APL is highly curable with approximately 90% complete remission following front-line therapy with *at*RA in combination with either anthracycline-based chemotherapy or arsenic trioxide (Tallman & Altman, 2009; Watts & Tallman, 2014). However, despite the high remission rates with front-line treatment of APL, relapse occurs in 10–30% of patients (Idres, Marill, & Chabot, 2005; Kelaidi et al., 2006). Several factors have been proposed to contribute to the relapse in APL patients, including mutations of the PML–RARα fusion protein blocking transcriptional activation, increased binding affinity of corepressors to the fusion protein, and increased *at*RA metabolism due to induction of CYP26 (Guidez et al., 1998; Idres et al., 2005). Autoinduction of *at*RA metabolism and resulting resistance to *at*RA is the best-characterized mechanism of APL relapse. In clinical pharmacokinetic studies, in APL patients exposure to *at*RA was significantly decreased at relapse when compared to exposure after the first dose (Muindi et al., 1992). Doubling the clinical dose of *at*RA at relapse was insufficient to match initial *at*RA exposures. Blasts cells were collected from patients at relapse and were sensitive *in vitro* to *at*RA differentiation and maintained the PML–RARα fusion protein, indicating that decreased systemic exposure to *at*RA was the root cause of relapse. The authors proposed that autoinduction of a CYP enzyme responsible for *at*RA metabolism occurred with prolonged treatment with *at*RA (Muindi et al., 1992). This proposal was further confirmed by the observation that pretreatment with fluconazole, an inhibitor of CYP enzymes, caused over fourfold increase in *at*RA AUC in APL patients (Schwartz, Hallam, Gallagher, & Wiernik, 1995). After identification of CYP26 as the *at*RA hydroxylase, *at*RA treatment was shown to induce CYP26 mRNA in APL patient derived model cell lines (NB4 and HL-60) (Idres et al., 2005; Ozpolat, Mehta, & Lopez-Berestein, 2005; Ozpolat, Mehta, Tari, & Lopez-Berestein, 2002). Basal CYP26 expression in NB4 cells is very low but mRNA levels can be induced up to 22-fold following 24–96 h of treatment with pharmacologically relevant concentrations of *at*RA. Furthermore, *at*RA binding to the RARα nuclear receptor was shown to mediate CYP26 induction (Idres et al., 2005; Ozpolat et al., 2002; Roussel & Lanotte, 2001). These findings support the notion that therapeutic CYP26 inhibition may circumvent APL relapse following *at*RA treatment by preventing rapid metabolism of *at*RA due to increased expression of CYP26.

An alternative hypothesis for APL relapse was investigated by Quere et al., in blasts from *at*RA treatment-naïve APL patients. Blast cells were harvested from patients and treated *in vitro* with *at*RA for 3 days. Blasts with delayed maturation, marked as fewer than 50% differentiated cells on day 3, were associated with poor clinical prognosis and categorized as having low-sensitivity to *at*RA treatment. Blasts with rapid maturation, defined as greater than 50% differentiated cells following 3 days of *in vitro* incubation with *at*RA, were associated with significantly improved disease-free survival clinically and therefore categorized as highly sensitive to *at*RA treatment. In the two classes of blasts, higher transcription of *at*RA-target genes, transcriptional regulators, and chromatin remodelers was observed in low-sensitive cells. Conversely, CYP26 gene expression was limited to highly sensitive blasts with little to no expression observed in low-sensitive blasts. The authors proposed that low expression of CYP26 in APL blasts with low-sensitivity to *at*RA *in vitro* led to increased nuclear *at*RA concentrations and corresponding increased *at*RA-target gene transcription. Further, when NB4 cells were incubated with either *at*RA or the CYP26 inhibitor 4-azolyl retinoic acid, accumulation of *at*RA and development of subclones resistant to *at*RA-induced differentiation were observed following NB4 cell treatment with 4-azolyl retinoic acid (Quere et al., 2007). While this conflicting data of mechanisms of resistance remain to be resolved, it appears that the activity of CYP26 in regulating nuclear *at*RA levels is important in preventing APL relapse. Indeed, CYP26-specific RAMBAs have been developed and investigated preclinically but no clinical studies have yet shown the effects of coadministration of *at*RA and a RAMBA in APL (Li et al., 2013; Njar et al., 2006). Several other strategies have been proposed to address *at*RA resistance and APL relapse. Intermittent dosing of *at*RA combined with chemotherapy is used in standard APL therapy during consolidation and maintenance. A significantly lower 2-year relapse rate was reported in patients receiving intermittent *at*RA compared to patients without *at*RA treatment during maintenance (Fenaux, Wang, & Degos, 2007).

4.2 Neuroblastoma

Neuroblastoma is a malignant tumor comprised of neuroblasts. It develops most commonly in the abdomen but also appears in the neck, chest, and pelvis. Neuroblastoma is the most common extracranial solid tumor in childhood and accounts for 15% of oncology deaths in patients younger than 15 years (Maris, Hogarty, Bagatell, & Cohn, 2007). Treatment of neuroblastoma is based on disease risk stratification and can include surgery alone or chemotherapy and autologous bone marrow transplant. Treatment with 13-*cis*RA significantly improves overall survival of high risk patients when used following transplantation or chemotherapy (Maris et al., 2007; Matthay et al., 1999, 2009). In neuroblastoma cell lines, 13-*cis*RA and *at*RA have similar potency on cellular differentiation and growth arrest (Reynolds et al., 1994) but 13-*cis*RA is superior to *at*RA in the treatment of neuroblastoma due to its favorable pharmacokinetic and toxicological properties (Reynolds, Matthay, Villablanca, & Maurer, 2003).

Disease relapse is common in neuroblastoma. One approach to improve clinical outcomes is identifying an optimal dosing regimen of 13-*cis*RA. Intermittent high-dose administration of 13-*cis*RA improved clinical outcomes when compared to chronic low-dose administration but significant variability in 13-*cis*RA exposures was still observed causing suboptimal concentrations in some patients (Matthay, 2013; Matthay et al., 2009; Villablanca et al., 1995). Polymorphisms in several enzymes that clear 13-*cis*RA such as CYP2C8, CYP3A5, CYP3A7, and UGT2B1 have been studied as the cause of variability in 13-*cis*RA exposures. 4-oxo-13-*cis*RA is the major 13-*cis*RA metabolite observed in neuroblastoma patients. While 4-oxo-13-*cis*RA exposure was associated with CYP2C8 and CYP3A7 polymorphisms, variability in 13-*cis*RA exposure and 13-*cis*RA:4-oxo-13-*cis*RA ratio could not be attributed to any of the polymorphisms investigated (Veal et al., 2013). It is reasonable to suggest that CYP26 plays a role in the metabolism of 13-*cis*RA since CYP26 is the major catabolic enzyme of *at*RA and isomerization between *at*RA and 13-*cis*RA has been observed (Tsukada et al., 2000).

Given that 13-*cis*RA is generally considered to be devoid of biological activity due to its low affinity for RARs, neuroblastoma cellular differentiation, and growth arrest in response to 13-*cis*RA treatment (Reynolds et al., 2003) can be attributed to isomerization of 13-*cis*RA to *at*RA. However, this has not been directly proven. Determination of retinoid levels following treatment of neuroblastoma cell lines with 13-*cis*RA revealed little extracellular *at*RA but increased intracellular *at*RA due to either intracellular isomerization or selective uptake of *at*RA (Veal, Errington, Redfern, Pearson, & Boddy, 2002). Furthermore, incubation with 13-*cis*RA alone in neuroblastoma cell lines resulted in *at*RA accounting for 15–31% of the total RA. In contrast, after incubation with *at*RA, 13-*cis*RA accounted for <10% of total RA (Armstrong et al., 2007). Intracellular *at*RA concentrations but not 13-*cis*RA concentrations were increased in neuroblastoma cell lines following administration of the CYP26 inhibitor R116010 following either *at*RA or 13-*cis*RA treatment (Armstrong et al., 2005, 2007). This suggests that coadministration of a CYP26 inhibitor may improve clinical outcomes in neuroblastoma patients with low, subtherapeutic 13-*cis*RA concentrations. Surprisingly and in contrast to preclinical results, overexpression of the CYP26A1 gene in neuroblastoma tumors from over 100 patients was found to be significantly associated with favorable clinical outcomes (Kamei et al., 2009). However, no concentrations of *at*RA or 13-*cis*RA were reported in this study. It is possible that higher CYP26A1 expression in the neuroblastoma tumors in clinical samples is indicative of higher RA exposure upon treatment as CYP26A1 is induced by RA. Collectively, these findings highlight the need for more investigation of the concentrations of RA in neuroblastoma and the effects of altered RA exposures to disease outcomes.

5. CONCLUSION

There is considerable evidence that RA plays an important role in modulating innate and adaptive immunity and inflammation, and studies in various disease models suggest that increasing *at*RA concentrations will be beneficial in inflammatory diseases. The most important role of *at*RA in the immune system is perhaps inducing T-cell differentiation to Treg cells and suppressing T-cell differentiation to Th17 cells. This activity is also likely important for the anti-inflammatory actions of *at*RA. In addition, RA has a critical role in inducing the gut homing signals in T-cells. It is not clear how altered expression of RA-synthesizing and -metabolizing enzymes modulates RA concentrations in various immune cells, but there is emerging evidence that CYP26B1 is the predominant enzyme clearing RA in the T- and B-cells and ALDH1A2 appears to be mainly responsible for synthesizing RA in these cells and in DCs. Therefore, alterations in the activity of these enzymes are expected to affect immune homeostasis. However, whether the expression of these enzymes varies dynamically with immune responses and in inflammation is not known. Thus, the endogenous roles of CYP26s and ALDHs in immune modulation are not yet established. It is important to note that endogenous RA is generally synthesized within target cells and RA signaling appears to be modulated by highly specific gradients of RA concentrations (Hogarth et al., 2015). In contrast, when RA is administered exogenously for pharmacological studies, RA gradients are not necessarily generated which may confound the mechanistic information obtained regarding the endogenous regulatory systems and the role of RA in inflammatory processes.

*at*RA and its stereoisomer 13-*cis*RA are useful therapeutic agents in the treatment of APL and neuroblastoma, respectively. However, resistance to RA treatment is observed in both diseases and appears to involve induction of the CYP26 enzymes, likely CYP26A1, and increased clearance of RA. Therefore, inhibition of the CYP26s in these cancers is predicted to improve therapy outcomes. However, some of the mechanistic data of resistance development is controversial and challenging to interpret, as RA induces CYP26 expression but CYP26s also clear RA and likely reduce its concentrations in target cells. Several studies have also suggested that RA signaling and aberrant synthesis and metabolism of RA play a role in other cancers such as breast, prostate, head and neck, colorectal, and ovarian cancers. In many *in vitro* models of these cancers, increased expression of CYP26A1 appears to promote tumorigenesis while increased concentrations of RA induce apoptotic pathways. Yet, evidence from animal and human studies is incomplete to demonstrate that altered CYP26A1 expression will cause cancer or contribute to cancer development. In addition, it has not been directly shown that RA concentrations are altered in cancer simultaneously with increased CYP26 expression in malignant cells.

Overall, it appears that increasing RA concentrations via inhibition of CYP26 enzymes will have many positive effects in inflammatory diseases such as IBD, psoriasis, and atherosclerosis and in various cancers by promoting apoptosis and cell cycle arrest. However, potent inhibitors of CYP26 enzymes that have suitable *in vivo* pharmacokinetics to use in disease models or humans have not yet been developed or approved. Therefore, the effect of altered RA homeostasis and the magnitude of required change in RA concentrations to have therapeutic benefit are not well defined. While many of the beneficial effects on the immune system and cancer appear to be associated with increasing RA concentrations, in many other situations, such as fetal development, both too high and too low RA concentrations are detrimental and manifest themselves with similar phenotypes. This may in part explain some of the controversies observed in different inflammatory diseases and cancers and may affect future strategies for development of therapies relating to retinoids.

CONFLICT OF INTEREST

The authors have no conflicts of interest to declare.

ACKNOWLEDGMENT

This work was supported in part by an NIH NIGMS grant R01 GM111772.

REFERENCES

Abella, V., Scotece, M., Conde, J., López, V., Lazzaro, V., Pino, J., et al. (2014). Adipokines, metabolic syndrome and rheumatic diseases. *Journal of Immunology Research*, *2014*, 343746.

Abu-Abed, S., Dollé, P., Metzger, D., Beckett, B., Chambon, P., & Petkovich, M. (2001). The retinoic acid-metabolizing enzyme, CYP26A1, is essential for normal hindbrain patterning, vertebral identity, and development of posterior structures. *Genes & Development, 15*, 226–240.

Ahluwalia, B., Gambhir, K., & Sekhon, H. (1975). Distribution of labeled retinyl acetate and retinoic acid in rat and human testes. A possible site of retinyl acetate incorporation in rat testes. *The Journal of Nutrition, 105*, 467–474.

Allali-Hassani, A., Martinez, S. E., Peralba, J. M., Vaglenova, J., Vidal, F., Richart, C., et al. (1997). Alcohol dehydrogenase of human and rat blood vessels. Role in ethanol metabolism. *FEBS Letters, 405*, 26–30.

Altucci, L., Leibowitz, M. D., Ogilvie, K. M., de Lera, A. R., & Gronemeyer, H. (2007). RAR and RXR modulation in cancer and metabolic disease. *Nature Reviews. Drug Discovery, 6*, 793–810.

Ambroziak, W., Izaguirre, G., & Pietruszko, R. (1999). Metabolism of retinaldehyde and other aldehydes in soluble extracts of human liver and kidney. *Journal of Biological Chemistry, 274*, 33366–33373.

Armstrong, J. L., Ruiz, M., Boddy, A. V., Redfern, C. P. F., Pearson, A. D. J., & Veal, G. J. (2005). Increasing the intracellular availability of all-trans retinoic acid in neuroblastoma cells. *British Journal of Cancer, 92*, 696–704.

Armstrong, J. L., Taylor, G. A., Thomas, H. D., Boddy, A. V., Redfern, C. P. F., & Veal, G. J. (2007). Molecular targeting of retinoic acid metabolism in neuroblastoma: The role of the CYP26 inhibitor R116010 in vitro and in vivo. *British Journal of Cancer, 96*, 1675–1683.

Arnold, S. L. M., Amory, J. K., Walsh, T. J., & Isoherranen, N. (2012). A sensitive and specific method for measurement of multiple retinoids in human serum with UHPLC-MS/MS. *Journal of Lipid Research, 53*, 587–598.

Arnold, S. L. M., Kent, T., Hogarth, C. A., Griswold, M. D., Amory, J. K., & Isoherranen, N. (2015). Pharmacological inhibition of ALDH1A in mice decreases all-trans retinoic acid concentrations in a tissue specific manner. *Biochemical Pharmacology*, http://dx.doi.org/10.1016/j.bcp.2015.03.001.

Arnold, S. L., Kent, T., Hogarth, C. A., Schlatt, S., Prasad, B., Haenisch, M., et al. (2015). Importance of ALDH1A enzymes in determining human testicular retinoic acid concentrations. *Journal of Lipid Research, 56*, 342–357.

Ashla, A. A., Hoshikawa, Y., Tsuchiya, H., Hashiguchi, K., Enjoji, M., Nakamuta, M., et al. (2010). Genetic analysis of expression profile involved in retinoid metabolism in non-alcoholic fatty liver disease. *Hepatology Research, 40*, 594–604.

Aström, A., Pettersson, U., Krust, A., Chambon, P., & Voorhees, J. J. (1990). Retinoic acid and synthetic analogs differentially activate retinoic acid receptor dependent transcription. *Biochemical and Biophysical Research Communications, 173*, 339–345.

Bai, A., Lu, N., Guo, Y., Liu, Z., Chen, J., & Peng, Z. (2009). All-trans retinoic acid down-regulates inflammatory responses by shifting the Treg/Th17 profile in human ulcerative and murine colitis. *Journal of Leukocyte Biology, 86*, 959–969.

Barstad, R. M., Hamers, M. J., Stephens, R. W., & Sakariassen, K. S. (1995). Retinoic acid reduces induction of monocyte tissue factor and tissue factor/factor VIIa-dependent arterial thrombus formation. *Blood, 86*, 212–218.

Basu, T. K., Temple, N. J., & Hodgson, A. M. (1988). Vitamin A, beta-carotene and cancer. *Progress in Clinical and Biological Research, 259*, 217–228.

Berlin Grace, V. M., Niranjali Devaraj, S., Radhakrishnan Pillai, M., & Devaraj, H. (2006). HPV-induced carcinogenesis of the uterine cervix is associated with reduced serum ATRA level. *Gynecologic Oncology, 103*, 113–119.

Berry, D. C., Levi, L., & Noy, N. (2014). Holo-retinol-binding protein and its receptor STRA6 drive oncogenic transformation. *Cancer Research, 74*, 6341–6351.

Bhushan, M., Burden, A. D., McElhone, K., James, R., Vanhoutte, F. P., & Griffiths, C. E. (2001). Oral liarozole in the treatment of palmoplantar pustular psoriasis: A randomized, double-blind, placebo-controlled study. *The British Journal of Dermatology*, *145*, 546–553.
Bonet, M. L., Ribot, J., & Palou, A. (2012). Lipid metabolism in mammalian tissues and its control by retinoic acid. *Biochimica et Biophysica Acta*, *1821*, 177–189.
Bovenschen, H. J., Otero, M. E., Langewouters, A. M. G., van Vlijmen-Willems, I. M. J. J., van Rens, D. W. A., Seyger, M. M. B., et al. (2007). Oral retinoic acid metabolism blocking agent Rambazole for plaque psoriasis: An immunohistochemical study. *The British Journal of Dermatology*, *156*, 263–270.
Brown, G. T., Cash, B. G., Blihoghe, D., Johansson, P., Alnabulsi, A., & Murray, G. I. (2014). The expression and prognostic significance of retinoic acid metabolising enzymes in colorectal cancer. *PLoS One*, *9*, e90776.
Brown, S. J., & Mayer, L. (2007). The immune response in inflammatory bowel disease. *The American Journal of Gastroenterology*, *102*, 2058–2069.
Bulens, F., Ibanez-Tallon, I., Van Acker, P., De Vriese, A., Nelles, L., Belayew, A., et al. (1995). Retinoic acid induction of human tissue-type plasminogen activator gene expression via a direct repeat element (DR5) located at -7 kilobases. *Journal of Biological Chemistry*, *270*, 7167–7175.
Cantorna, M. T., Nashold, F. E., & Hayes, C. E. (1995). Vitamin A deficiency results in a priming environment conducive for Th1 cell development. *European Journal of Immunology*, *25*, 1673–1679.
Carrera, S., Cuadrado-Castano, S., Samuel, J., Jones, G. D. D., Villar, E., Lee, S. W., et al. (2013). Stra6, a retinoic acid-responsive gene, participates in p53-induced apoptosis after DNA damage. *Cell Death and Differentiation*, *20*, 910–919.
Chang, C.-L., Hong, E., Lao-Sirieix, P., & Fitzgerald, R. C. (2008). A novel role for the retinoic acid-catabolizing enzyme CYP26A1 in Barrett's associated adenocarcinoma. *Oncogene*, *27*, 2951–2960.
Chaudhy, N. A., Jafarey, N. A., & Ibrahim, K. (1980). Plasma vitamin A and carotene levels in relation to the clinical stages of carcinoma of the oral cavity and oropharynx. *JPMA. The Journal of the Pakistan Medical Association*, *30*, 221–223.
Chen, H., & Juchau, M. R. (1997). Glutathione S-transferases act as isomerases in isomerization of 13-cis-retinoic acid to all-trans-retinoic acid in vitro. *Biochemical Journal*, *327*, 721–726.
Chen, W. C., Sass, J. O., Seltmann, H., Nau, H., Orfanos, C. E., & Zouboulis, C. C. (2000). Biological effects and metabolism of 9-cis-retinoic acid and its metabolite 9,13-di-cis-retinoic acid in HaCaT keratinocytes in vitro: Comparison with all-trans-retinoic acid. *Archives of Dermatological Research*, *292*, 612–620.
Chenery, A., Burrows, K., Antignano, F., Underhill, T. M., Petkovich, M., & Zaph, C. (2013). The retinoic acid-metabolizing enzyme Cyp26b1 regulates CD4 T cell differentiation and function. *PLoS One*, *8*, e72308.
Chung, S. S. W., & Wolgemuth, D. J. (2004). Role of retinoid signaling in the regulation of spermatogenesis. *Cytogenetic and Genome Research*, *105*, 189–202.
Chytil, F. (1992). The lungs and vitamin A. *The American Journal of Physiology*, *262*, L517–L527.
Collins, S. (2002). The role of retinoids and retinoic acid receptors in normal hematopoiesis. *Leukemia*, *16*, 1896–1905.
Cvetković, D., Williams, S. J., & Hamilton, T. C. (2003). Loss of cellular retinol-binding protein 1 gene expression in microdissected human ovarian cancer. *Clinical Cancer Research: An Official Journal of the American Association for Cancer Research*, *9*, 1013–1020.
D'Ambrosio, D. N., Clugston, R. D., & Blaner, W. S. (2011). Vitamin A metabolism: An update. *Nutrients*, *3*, 63–103.

De Bruijn, D. R., Oerlemans, F., Hendriks, W., Baats, E., Ploemacher, R., Wieringa, B., et al. (1994). Normal development, growth and reproduction in cellular retinoic acid binding protein-I (CRABPI) null mutant mice. *Differentiation; Research in Biological Diversity, 58*, 141–148.
De Pee, S., & Dary, O. (2002). Biochemical indicators of vitamin A deficiency: Serum retinol and serum retinol binding protein. *The Journal of Nutrition, 132*, 2895S–2901S.
Depner, C. M., Philbrick, K. A., & Jump, D. B. (2013). Docosahexaenoic acid attenuates hepatic inflammation, oxidative stress, and fibrosis without decreasing hepatosteatosis in a Ldlr−/− mouse model of western diet-induced nonalcoholic steatohepatitis. *The Journal of Nutrition, 143*, 315–323.
DiSepio, D., Malhotra, M., Chandraratna, R. A., & Nagpal, S. (1997). Retinoic acid receptor-nuclear factor-interleukin 6 antagonism. A novel mechanism of retinoid-dependent inhibition of a keratinocyte hyperproliferative differentiation marker. *The Journal of Biological Chemistry, 272*, 25555–25559.
Dockx, P., Decree, J., & Degreef, H. (1995). Inhibition of the metabolism of endogenous retinoic acid as treatment for severe psoriasis: An open study with oral liarozole. *The British Journal of Dermatology, 133*, 426–432.
Dogra, S. C., Khanduja, K. L., & Gupta, M. P. (1985). The effect of vitamin A deficiency on the initiation and postinitiation phases of benzo(a)pyrene-induced lung tumourigenesis in rats. *British Journal of Cancer, 52*, 931–935.
Dogra, S., & Yadav, S. (2014). Acitretin in psoriasis: An evolving scenario. *International Journal of Dermatology, 53*, 525–538.
Dong, D., Ruuska, S. E., Levinthal, D. J., & Noy, N. (1999). Distinct roles for cellular retinoic acid-binding proteins I and II in regulating signaling by retinoic acid. *The Journal of Biological Chemistry, 274*, 23695–23698.
Downie, D., McFadyen, M. C. E., Rooney, P. H., Cruickshank, M. E., Parkin, D. E., Miller, I. D., et al. (2005). Profiling cytochrome P450 expression in ovarian cancer: Identification of prognostic markers. *Clinical Cancer Research: An Official Journal of the American Association for Cancer Research, 11*, 7369–7375.
Duester, G. (2008). Retinoic acid synthesis and signaling during early organogenesis. *Cell, 134*, 921–931.
Dupuy, P., Bagot, M., Heslan, M., & Dubertret, L. (1989). Synthetic retinoids inhibit the antigen presenting properties of epidermal cells in vitro. *The Journal of Investigative Dermatology, 93*, 455–459.
Elmabsout, A. A., Kumawat, A., Saenz-Méndez, P., Krivospitskaya, O., Sävenstrand, H., Olofsson, P. S., et al. (2012). Cloning and functional studies of a splice variant of CYP26B1 expressed in vascular cells. *PLoS One*, 7, e36839.
Farr, P. M., Krause, L. B., Marks, J. M., & Shuster, S. (1985). Response of scalp psoriasis to oral ketoconazole. *Lancet, 2*, 921–922.
Feldmann, M., Brennan, F. M., & Maini, R. N. (1996). Rheumatoid arthritis. *Cell, 85*, 307–310.
Fenaux, P., Wang, Z. Z., & Degos, L. (2007). Treatment of acute promyelocytic leukemia by retinoids. *Current Topics in Microbiology and Immunology, 313*, 101–128.
Fiorella, P. D., & Napoli, J. L. (1994). Microsomal retinoic acid metabolism. Effects of cellular retinoic acid-binding protein (type I) and C18-hydroxylation as an initial step. *The Journal of Biological Chemistry, 269*, 10538–10544.
Fisher, G. J., & Voorhees, J. J. (1996). Molecular mechanisms of retinoid actions in skin. *FASEB Journal: Official Publication of the Federation of American Societies for Experimental Biology, 10*, 1002–1013.
Fransén, K., Franzén, P., Magnuson, A., Elmabsout, A. A., Nyhlin, N., Wickbom, A., et al. (2013). Polymorphism in the retinoic acid metabolizing enzyme CYP26B1 and the development of Crohn's disease. *PLoS One, 8*, e72739.

Gamble, M. V., Ramakrishnan, R., Palafox, N. A., Briand, K., Berglund, L., & Blaner, W. S. (2001). Retinol binding protein as a surrogate measure for serum retinol: Studies in vitamin A-deficient children from the Republic of the Marshall Islands. *The American Journal of Clinical Nutrition*, *73*, 594–601.

Giltaire, S., Herphelin, F., Frankart, A., Herin, M., Stoppie, P., & Poumay, Y. (2009). The CYP26 inhibitor R115866 potentiates the effects of all-trans retinoic acid on cultured human epidermal keratinocytes. *British Journal of Dermatology*, *160*, 505–513.

Godbole, A. M., Purushottamachar, P., Martin, M. S., Daskalakis, C., & Njar, V. C. O. (2012). Autophagy inhibition synergistically enhances anticancer efficacy of RAMBA, VN/12-1 in SKBR-3 cells, and tumor xenografts. *Molecular Cancer Therapeutics*, *11*, 898–908.

Gomaa, M. S., Bridgens, C. E., Veal, G. J., Redfern, C. P. F., Brancale, A., Armstrong, J. L., et al. (2011). Synthesis and biological evaluation of 3-(1H-imidazol- and triazol-1-yl)-2,2-dimethyl-3-[4-(naphthalen-2-ylamino)phenyl]propyl derivatives as small molecule inhibitors of retinoic acid 4-hydroxylase (CYP26). *Journal of Medicinal Chemistry*, *54*, 6803–6811.

Gudas, L. J. (2012). Emerging roles for retinoids in regeneration and differentiation in normal and disease states. *Biochimica et Biophysica Acta*, *1821*, 213–221.

Gudas, L. J., & Wagner, J. A. (2011). Retinoids regulate stem cell differentiation. *Journal of Cellular Physiology*, *226*, 322–330.

Guidez, F., Ivins, S., Zhu, J., Söderström, M., Waxman, S., & Zelent, A. (1998). Reduced retinoic acid-sensitivities of nuclear receptor corepressor binding to PML- and PLZF-RARα underlie molecular pathogenesis and treatment of acute promyelocytic leukemia. *Blood*, *91*, 2634–2642.

Guilliams, M., Crozat, K., Henri, S., Tamoutounour, S., Grenot, P., Devilard, E., et al. (2010). Skin-draining lymph nodes contain dermis-derived CD103(-) dendritic cells that constitutively produce retinoic acid and induce Foxp3(+) regulatory T cells. *Blood*, *115*, 1958–1968.

Hall, J. A., Grainger, J. R., Spencer, S. P., & Belkaid, Y. (2011). The role of retinoic acid in tolerance and immunity. *Immunity*, *35*, 13–22.

Hansen, W., Westendorf, A. M., & Buer, J. (2008). Regulatory T cells as targets for immunotherapy of autoimmunity and inflammation. *Inflammation & Allergy Drug Targets*, *7*, 217–223.

Hansson, G. K., & Hermansson, A. (2011). The immune system in atherosclerosis. *Nature Immunology*, *12*, 204–212.

Hansson, G. K., Robertson, A. K., & Soderberg-Naucler, C. (2006). Inflammation and atherosclerosis. *Annual Review of Pathology*, *1*, 297–329.

Harper, R. A. (1988). Specificity in the synergism between retinoic acid and EGF on the growth of adult human skin fibroblasts. *Experimental Cell Research*, *178*, 254–263.

Harper, R. A., & Burgoon, T. (1982). Differential effects of retinoic acid on the growth of normal fibroblast-like cells in vitro from human, swine and rabbit skin. *Cell Biology International Reports*, *6*, 163–170.

Heise, R., Mey, J., Neis, M. M., Marquardt, Y., Joussen, S., Ott, H., et al. (2006). Skin retinoid concentrations are modulated by CYP26AI expression restricted to basal keratinocytes in normal human skin and differentiated 3D skin models. *The Journal of Investigative Dermatology*, *126*, 2473–2480.

Hogarth, C. A., Arnold, S., Kent, T., Mitchell, D., Isoherranen, N., & Griswold, M. D. (2015). Processive pulses of retinoic acid propel asynchronous and continuous murine sperm production. *Biology of Reproduction*, *92*, 37.

Hogarth, C. A., & Griswold, M. D. (2013). Retinoic acid regulation of male meiosis. *Current Opinion in Endocrinology, Diabetes, and Obesity*, *20*, 217–223.

Hong, K., Zhang, Y., Guo, Y., Xie, J., Wang, J., He, X., et al. (2014). All-trans retinoic acid attenuates experimental colitis through inhibition of NF-κB signaling. *Immunology Letters, 162*, 34–40.

Huang, M. E., Ye, Y. C., Chen, S. R., Chai, J. R., Lu, J. X., Zhoa, L., et al. (1988). Use of all-trans retinoic acid in the treatment of acute promyelocytic leukemia. *Blood, 72*, 567–572.

Idres, N., Marill, J., & Chabot, G. G. (2005). Regulation of CYP26A1 expression by selective RAR and RXR agonists in human NB4 promyelocytic leukemia cells. *Biochemical Pharmacology, 69*, 1595–1601.

Ito, Y., Gajalakshmi, K. C., Sasaki, R., Suzuki, K., & Shanta, V. (1999). A study on serum carotenoid levels in breast cancer patients of Indian women in Chennai (Madras), India. *Journal of Epidemiology/Japan Epidemiological Association, 9*, 306–314.

Iwata, M., Hirakiyama, A., Eshima, Y., Kagechika, H., Kato, C., & Song, S.-Y. (2004). Retinoic acid imprints gut-homing specificity on T cells. *Immunity, 21*, 527–538.

Jelski, W., Chrostek, L., Szmitkowski, M., & Markiewicz, W. (2006). The activity of class I, II, III and IV alcohol dehydrogenase isoenzymes and aldehyde dehydrogenase in breast cancer. *Clinical and Experimental Medicine, 6*, 89–93.

Jelski, W., Zalewski, B., Chrostek, L., & Szmitkowski, M. (2004). The activity of class I, II, III, and IV alcohol dehydrogenase isoenzymes and aldehyde dehydrogenase in colorectal cancer. *Digestive Diseases and Sciences, 49*, 977–981.

Johst, U., Betsch, A., Wiskirchen, J., Schöber, W., Vonthein, R., Rinkert, N., et al. (2003). All-trans and 9-cis retinoid acids inhibit proliferation, migration, and synthesis of extracellular matrix of human vascular smooth muscle cells by inducing differentiation in vitro. *Journal of Cardiovascular Pharmacology, 41*, 526–535.

Kakizuka, A., Miller, W. H., Jr., Umesono, K., Warrell, R. P., Jr., Frankel, S. R., Murty, V. V. V., et al. (1991). Chromosomal translocation t(15;17) in human acute promyelocytic leukemia fuses RARα with a novel putative transcription factor, PML. *Cell, 66*, 663–674.

Kamei, N., Hiyama, K., Yamaoka, H., Kamimatsuse, A., Onitake, Y., Sueda, T., et al. (2009). Evaluation of genes identified by microarray analysis in favorable neuroblastoma. *Pediatric Surgery International, 25*, 931–937.

Kane, M. A., Folias, A. E., Pingitore, A., Perri, M., Obrochta, K. M., Krois, C. R., et al. (2010). Identification of 9-cis-retinoic acid as a pancreas-specific autacoid that attenuates glucose-stimulated insulin secretion. *Proceedings of the National Academy of Sciences, 107*, 21884–21889.

Kane, M. A., Folias, A. E., Wang, C., & Napoli, J. L. (2008). Quantitative profiling of endogenous retinoic acid in vivo and in vitro by tandem mass spectrometry. *Analytical Chemistry, 80*, 1702–1708.

Kang, S. G., Park, J., Cho, J. Y., Ulrich, B., & Kim, C. H. (2011). Complementary roles of retinoic acid and TGF-β1 in coordinated expression of mucosal integrins by T cells. *Mucosal Immunology, 4*, 66–82.

Kedishvili, N. Y. (2013). Enzymology of retinoic acid biosynthesis and degradation. *Journal of Lipid Research, 54*, 1744–1760.

Kelaidi, C., Ades, L., Chevret, S., Sanz, M., Guerci, A., Thomas, X., et al. (2006). Late first relapses in APL treated with all-trans-retinoic acid- and anthracycline-based chemotherapy: The European APL group experience (APL 91 and APL 93 trials). *Leukemia, 20*, 905–907.

Krivospitskaya, O., Elmabsout, A. A., Sundman, E., Soderstrom, L. A., Ovchinnikova, O., Gidlof, A. C., et al. (2012). A CYP26B1 polymorphism enhances retinoic acid catabolism and may aggravate atherosclerosis. *Molecular Medicine, 18*, 712–718.

Kumar, S., Sandell, L. L., Trainor, P. A., Koentgen, F., & Duester, G. (2012). Alcohol and aldehyde dehydrogenases: Retinoid metabolic effects in mouse knockout models. *Biochimica et Biophysica Acta (BBA): Molecular and Cell Biology of Lipids*, *1821*, 198–205.

Kuppumbatti, Y. S., Bleiweiss, I. J., Mandeli, J. P., Waxman, S., & Mira-Y-Lopez, R. (2000). Cellular retinol-binding protein expression and breast cancer. *Journal of the National Cancer Institute*, *92*, 475–480.

Kurlandsky, S. B., Gamble, M. V., Ramakrishnan, R., & Blaner, W. S. (1995). Plasma delivery of retinoic acid to tissues in the rat. *The Journal of Biological Chemistry*, *270*, 17850–17857.

Lampron, C., Rochette-Egly, C., Gorry, P., Dollé, P., Mark, M., Lufkin, T., et al. (1995). Mice deficient in cellular retinoic acid binding protein II (CRABPII) or in both CRABPI and CRABPII are essentially normal. *Development (Cambridge, England)*, *121*, 539–548.

Li, F., Zhao, D., Ren, J., Hao, F., Liu, G., Jin, S., et al. (2013). 2-(2-Methylfuran-3-carboxamido)-3-phenylpropanoic acid, a potential CYP26A1 inhibitor to enhance all-trans retinoic acid-induced leukemia cell differentiation based on virtual screening and biological evaluation. *Bioorganic & Medicinal Chemistry*, *21*, 3256–3261.

Libby, P., Ridker, P. M., & Hansson, G. K. (2009). Inflammation in atherosclerosis: From pathophysiology to practice. *Journal of the American College of Cardiology*, *54*, 2129–2138.

Lobo, G. P., Amengual, J., Palczewski, G., Babino, D., & von Lintig, J. (2012). Mammalian carotenoid-oxygenases: Key players for carotenoid function and homeostasis. *Biochimica et Biophysica Acta*, *1821*, 78–87.

Lutz, J. D., Dixit, V., Yeung, C. K., Dickmann, L. J., Zelter, A., Thatcher, J. E., et al. (2009). Expression and functional characterization of cytochrome P450 26A1, a retinoic acid hydroxylase. *Biochemical Pharmacology*, *77*, 258–268.

MacLean, G., Abu-Abed, S., Dollé, P., Tahayato, A., Chambon, P., & Petkovich, M. (2001). Cloning of a novel retinoic-acid metabolizing cytochrome P450, Cyp26B1, and comparative expression analysis with Cyp26A1 during early murine development. *Mechanisms of Development*, *107*, 195–201.

Maden, M. (2007). Retinoic acid in the development, regeneration and maintenance of the nervous system. *Nature Reviews. Neuroscience*, *8*, 755–765.

Mahida, Y. R. (2000). The key role of macrophages in the immunopathogenesis of inflammatory bowel disease. *Inflammatory Bowel Diseases*, *6*, 21–33.

Marill, J., Cresteil, T., Lanotte, M., & Chabot, G. G. (2000). Identification of human cytochrome P450s involved in the formation of all-trans-retinoic acid principal metabolites. *Molecular Pharmacology*, *58*, 1341–1348.

Maris, J. M., Hogarty, M. D., Bagatell, R., & Cohn, S. L. (2007). Neuroblastoma. *The Lancet*, *369*, 2106–2120.

Martens, J. H. A., Brinkman, A. B., Simmer, F., Francoijs, K.-J., Nebbioso, A., Ferrara, F., et al. (2010). PML-RARα/RXR alters the epigenetic landscape in acute promyelocytic leukemia. *Cancer Cell*, *17*, 173–185.

Marwarha, G., Berry, D. C., Croniger, C. M., & Noy, N. (2014). The retinol esterifying enzyme LRAT supports cell signaling by retinol-binding protein and its receptor STRA6. *FASEB Journal: Official Publication of the Federation of American Societies for Experimental Biology*, *28*, 26–34.

Matthay, K. K. (2013). Targeted isotretinoin in neuroblastoma: Kinetics, genetics, or absorption. *Clinical Cancer Research: An Official Journal of the American Association for Cancer Research*, *19*, 311–313.

Matthay, K. K., Reynolds, C. P., Seeger, R. C., Shimada, H., Adkins, E. S., Haas-Kogan, D., et al. (2009). Long-term results for children with high-risk neuroblastoma treated on a randomized trial of myeloablative therapy followed by 13-cis-retinoic acid: A children's oncology group study. *Journal of Clinical Oncology*, *27*, 1007–1013.

Matthay, K. K., Villablanca, J. G., Seeger, R. C., Stram, D. O., Harris, R. E., Ramsay, N. K., et al. (1999). Treatment of high-risk neuroblastoma with intensive chemotherapy, radiotherapy, autologous bone marrow transplantation, and 13-cis-retinoic acid. *New England Journal of Medicine*, *341*, 1165–1173.

McBee, J. K., Kuksa, V., Alvarez, R., de Lera, A. R., Prezhdo, O., Haeseleer, F., et al. (2000). Isomerization of all-trans-retinol to cis-retinols in bovine retinal pigment epithelial cells: Dependence on the specificity of retinoid-binding proteins. *Biochemistry*, *39*, 11370–11380.

McCormick, A. M., Kroll, K. D., & Napoli, J. L. (1983). 13-cis-Retinoic acid metabolism in vivo. The major tissue metabolites in the rat have the all-trans configuration. *Biochemistry*, *22*, 3933–3940.

McCully, M. L., Ladell, K., Hakobyan, S., Mansel, R. E., Price, D. A., & Moser, B. (2012). Epidermis instructs skin homing receptor expression in human T cells. *Blood*, *120*, 4591–4598.

McSorley, L. C., & Daly, A. K. (2000). Identification of human cytochrome P450 isoforms that contribute to all-trans-retinoic acid 4-hydroxylation. *Biochemical Pharmacology*, *60*, 517–526.

Miano, J. M., Topouzis, S., Majesky, M. W., & Olson, E. N. (1996). Retinoid receptor expression and all-trans retinoic acid-mediated growth inhibition in vascular smooth muscle cells. *Circulation*, *93*, 1886–1895.

Moulas, A. N., Gerogianni, I. C., Papadopoulos, D., & Gourgoulianis, K. I. (2006). Serum retinoic acid, retinol and retinyl palmitate levels in patients with lung cancer. *Respirology (Carlton, Vic.)*, *11*, 169–174.

Mucida, D., Park, Y., Kim, G., Turovskaya, O., Scott, I., Kronenberg, M., et al. (2007). Reciprocal TH17 and regulatory T cell differentiation mediated by retinoic acid. *Science*, *317*, 256–260.

Muindi, J., Frankel, S. R., Miller, W. J., Jakubowski, A., Scheinberg, D. A., Young, C. W., et al. (1992). Continuous treatment with all-trans retinoic acid causes a progressive reduction in plasma drug concentrations: Implications for relapse and retinoid "resistance" in patients with acute promyelocytic leukemia [published erratum appears in Blood 1992 Aug 1;80(3):855]. *Blood*, *79*, 299–303.

Nadin, L., & Murray, M. (1999). Participation of CYP2C8 in retinoic acid 4-hydroxylation in human hepatic microsomes. *Biochemical Pharmacology*, *58*, 1201–1208.

Napoli, J. L. (2012). Physiological insights into all-trans-retinoic acid biosynthesis. *Biochimica et Biophysica Acta (BBA): Molecular and Cell Biology of Lipids*, *1821*, 152–167.

Nelson, C. H., Buttrick, B. R., & Isoherranen, N. (2013). Therapeutic potential of the inhibition of the retinoic acid hydroxylases CYP26A1 and CYP26B1 by xenobiotics. *Current Topics in Medicinal Chemistry*, *13*, 1402–1428.

Nestle, F. O., Kaplan, D. H., & Barker, J. (2009). Psoriasis. *New England Journal of Medicine*, *361*, 496–509.

Njar, V. C. O., Gediya, L., Purushottamachar, P., Chopra, P., Vasaitis, T. S., Khandelwal, A., et al. (2006). Retinoic acid metabolism blocking agents (RAMBAs) for treatment of cancer and dermatological diseases. *Bioorganic & Medicinal Chemistry*, *14*, 4323–4340.

Nonnecke, B. J., Horst, R. L., Dubeski, P. L., & Reinhardt, T. A. (1997). Reactivity and phenotype of mononuclear leukocytes from nongravid heifers after in vitro exposure to 9,13-di-cis-retinoic acid. *Journal of Dairy Science*, *80*, 2833–2841.

Noy, N. (2000). Retinoid-binding proteins: Mediators of retinoid action. *The Biochemical Journal*, *348*(Pt. 3), 481–495.

Noy, N. (2010). Between death and survival: Retinoic acid in regulation of apoptosis. *Annual Review of Nutrition*, *30*, 201–217.

Ocaya, P. A., Elmabsout, A. A., Olofsson, P. S., Torma, H., Gidlof, A. C., & Sirsjo, A. (2011). CYP26B1 plays a major role in the regulation of all-trans-retinoic acid metabolism and signaling in human aortic smooth muscle cells. *Journal of Vascular Research*, *48*, 23–30.

Orfali, N., McKenna, S. L., Cahill, M. R., Gudas, L. J., & Mongan, N. P. (2014). Retinoid receptor signaling and autophagy in acute promyelocytic leukemia. *Experimental Cell Research*, *324*, 1–12.

Orywal, K., Jelski, W., Zdrodowski, M., & Szmitkowski, M. (2011). The activity of class I, II, III and IV alcohol dehydrogenase isoenzymes and aldehyde dehydrogenase in cervical cancer. *Clinical Biochemistry*, *44*, 1231–1234.

Orywal, K., Jelski, W., Zdrodowski, M., & Szmitkowski, M. (2013). The activity of class I, II, III and IV alcohol dehydrogenase isoenzymes and aldehyde dehydrogenase in ovarian cancer and ovarian cysts. *Advances in Medical Sciences*, *58*, 216–220.

Osanai, M., & Lee, G.-H. (2014). Increased expression of the retinoic acid-metabolizing enzyme CYP26A1 during the progression of cervical squamous neoplasia and head and neck cancer. *BMC Research Notes*, *7*, 697.

Osanai, M., & Petkovich, M. (2005). Expression of the retinoic acid-metabolizing enzyme CYP26A1 limits programmed cell death. *Molecular Pharmacology*, *67*, 1808–1817.

Osanai, M., Sawada, N., & Lee, G.-H. (2009). Oncogenic and cell survival properties of the retinoic acid metabolizing enzyme, CYP26A1. *Oncogene*, *29*, 1135–1144.

Ozpolat, B., Mehta, K., & Lopez-Berestein, G. (2005). Regulation of a highly specific retinoic acid-4-hydroxylase (CYP26A1) enzyme and all-trans-retinoic acid metabolism in human intestinal, liver, endothelial, and acute promyelocytic leukemia cells. *Leukemia & Lymphoma*, *46*, 1497–1506.

Ozpolat, B., Mehta, K., Tari, A. M., & Lopez-Berestein, G. (2002). all-trans-Retinoic acid-induced expression and regulation of retinoic acid 4-hydroxylase (CYP26) in human promyelocytic leukemia. *American Journal of Hematology*, *70*, 39–47.

Pandolfi, P. P., Grignani, F., Alcalay, M., Mencarelli, A., Biondi, A., LoCoco, F., et al. (1991). Structure and origin of the acute promyelocytic leukemia myl/RAR alpha cDNA and characterization of its retinoid-binding and transactivation properties. *Oncogene*, *6*, 1285–1292.

Pavez Loriè, E., Chamcheu, J. C., Vahlquist, A., & Törmä, H. (2009). Both all-trans retinoic acid and cytochrome P450 (CYP26) inhibitors affect the expression of vitamin A metabolizing enzymes and retinoid biomarkers in organotypic epidermis. *Archives for Dermatological Research Archiv Fur Dermatologische Forschung*, *301*, 475–485.

Pavez Loriè, E., Cools, M., Borgers, M., Wouters, L., Shroot, B., Hagforsen, E., et al. (2009). Topical treatment with CYP26 inhibitor talarozole (R115866) dose dependently alters the expression of retinoid-regulated genes in normal human epidermis. *The British Journal of Dermatology*, *160*, 26–36.

Pavez Loriè, E., Li, H., Vahlquist, A., & Törmä, H. (2009). The involvement of cytochrome p450 (CYP) 26 in the retinoic acid metabolism of human epidermal keratinocytes. *Biochimica et Biophysica Acta*, *1791*, 220–228.

Qiu, Y., Pu, T., Li, L., Cheng, F., Lu, C., Sun, L., et al. (2014). The expression of aldehyde dehydrogenase family in breast cancer. *Journal of Breast Cancer*, *17*, 54–60.

Quere, R., Baudet, A., Cassinat, B., Bertrand, G., Marti, J., Manchon, L., et al. (2007). Pharmacogenomic analysis of acute promyelocytic leukemia cells highlights CYP26 cytochrome metabolism in differential all-trans retinoic acid sensitivity. *Blood*, *109*, 4450–4460.

Raner, G. M., Vaz, A. D., & Coon, M. J. (1996). Metabolism of all-trans, 9-cis, and 13-cis isomers of retinal by purified isozymes of microsomal cytochrome P450 and mechanism-based inhibition of retinoid oxidation by citral. *Molecular Pharmacology*, *49*, 515–522.

Raverdeau, M., & Mills, K. H. G. (2014). Modulation of T cell and innate immune responses by retinoic acid. *Journal of Immunology (Baltimore, Md: 1950)*, *192*, 2953–2958.

Reynolds, C. P., Matthay, K. K., Villablanca, J. G., & Maurer, B. J. (2003). Retinoid therapy of high-risk neuroblastoma. *Cancer Letters*, *197*, 185–192.

Reynolds, C. P., Schindler, P. F., Jones, D. M., Gentile, J. L., Proffitt, R. T., & Einhorn, P. A. (1994). Comparison of 13-cis-retinoic acid to trans-retinoic acid using human neuroblastoma cell lines. *Progress in Clinical and Biological Research*, *385*, 237–244.

Rhee, E. J., Nallamshetty, S., & Plutzky, J. (2012). Retinoid metabolism and its effects on the vasculature. *Biochimica et Biophysica Acta*, *1821*, 230–240.

Roos, T. C., Jugert, F. K., Merk, H. F., & Bickers, D. R. (1998). Retinoid metabolism in the skin. *Pharmacological Reviews*, *50*, 315–333.

Rosenberger, M., & Neukom, C. (1982). Retinoic acid metabolites. 3. Total synthesis of (2E,4E,6E,8E)-3,7-dimethyl-9-[6,6-dimethyl-2-(hydroxymethyl)-1-cyclohexen-1-yl]-2,4,6,8-nonatetraenoic acid. *Journal of Organic Chemistry*, *47*, 1782–1785.

Ross, A. C. (2003). Retinoid production and catabolism: Role of diet in regulating retinol esterification and retinoic acid oxidation. *The Journal of Nutrition*, *133*, 291S–296S.

Ross, A. C. (2012). Vitamin A and retinoic acid in T cell-related immunity. *The American Journal of Clinical Nutrition*, *96*, 1166S–1172S.

Roussel, M. J., & Lanotte, M. (2001). Maturation sensitive and resistant t(15;17) NB4 cell lines as tools for APL physiopathology: Nomenclature of cells and repertory of their known genetic alterations and phenotypes. *Oncogene*, *20*, 7287–7291. http://dx.doi.org/10.1038/sj.onc.1204863, Published online: 29 October 2001.

Ruzicka, T., Larsen, F. G., Galewicz, D., Horváth, A., Coenraads, P. J., Thestrup-Pedersen, K., et al. (2004). Oral alitretinoin (9-cis-retinoic acid) therapy for chronic hand dermatitis in patients refractory to standard therapy: Results of a randomized, double-blind, placebo-controlled, multicenter trial. *Archives of Dermatology*, *140*, 1453–1459.

Samokyszyn, V. M., Gall, W. E., Zawada, G., Freyaldenhoven, M. A., Chen, G., Mackenzie, P. I., et al. (2000). 4-Hydroxyretinoic acid, a novel substrate for human liver microsomal UDP-glucuronosyltransferase(s) and recombinant UGT2B7. *The Journal of Biological Chemistry*, *275*, 6908–6914.

Sanders, T. J., McCarthy, N. E., Giles, E. M., Davidson, K. L., Haltalli, M. L., Hazell, S., et al. (2014). Increased production of retinoic acid by intestinal macrophages contributes to their inflammatory phenotype in patients with Crohn's disease. *Gastroenterology*, *146*, 1278.e1–2–1288.e1–2.

Sass, J. O., Forster, A., Bock, K. W., & Nau, H. (1994). Glucuronidation and isomerization of all-trans- and 13-cis-retinoic acid by liver microsomes of phenobarbital- or 3-methylcholanthrene-treated rats. *Biochemical Pharmacology*, *47*, 485–492.

Sauvant, P., Cansell, M., & Atgié, C. (2011). Vitamin A and lipid metabolism: Relationship between hepatic stellate cells (HSCs) and adipocytes. *Journal of Physiology and Biochemistry*, *67*, 487–496.

Schoeff, L. (1983). Vitamin A. *The American Journal of Medical Technology*, *49*, 447–452.

Schug, T. T., Berry, D. C., Shaw, N. S., Travis, S. N., & Noy, N. (2007). Opposing effects of retinoic acid on cell growth result from alternate activation of two different nuclear receptors. *Cell*, *129*, 723–733.

Schwartz, E. L., Hallam, S., Gallagher, R. E., & Wiernik, P. H. (1995). Inhibition of all-trans-retinoic acid metabolism by fluconazole in vitro and in patients with acute promyelocytic leukemia. *Biochemical Pharmacology*, *50*, 923–928.

Shelton, D. N., Sandoval, I. T., Eisinger, A., Chidester, S., Ratnayake, A., Ireland, C. M., et al. (2006). Up-regulation of CYP26A1 in adenomatous polyposis coli-deficient vertebrates via a WNT-dependent mechanism: Implications for intestinal cell differentiation and colon tumor development. *Cancer Research*, *66*, 7571–7577.

Skazik, C., Amann, P. M., Heise, R., Marquardt, Y., Czaja, K., Kim, A., et al. (2014). Downregulation of STRA6 expression in epidermal keratinocytes leads to hyperproliferation-associated differentiation in both in vitro and in vivo skin models. *The Journal of Investigative Dermatology*, *134*, 1579–1588.
Sklan, D. (1987). Vitamin A in human nutrition. *Progress in Food & Nutrition Science*, *11*, 39–55.
Smith, G., Ibbotson, S. H., Comrie, M. M., Dawe, R. S., Bryden, A., Ferguson, J., et al. (2006). Regulation of cutaneous drug-metabolizing enzymes and cytoprotective gene expression by topical drugs in human skin in vivo. *The British Journal of Dermatology*, *155*, 275–281.
Smith, J. E., Milch, P. O., Muto, Y., & Goodman, D. S. (1973). The plasma transport and metabolism of retinoic acid in the rat. *The Biochemical Journal*, *132*, 821–827.
Stearns, M. E., Wang, M., & Fudge, K. (1993). Liarozole and 13-cis-retinoic acid anti-prostatic tumor activity. *Cancer Research*, *53*, 3073–3077.
Stehr, P. A., Gloninger, M. F., Kuller, L. H., Marsh, G. M., Radford, E. P., & Weinberg, G. B. (1985). Dietary vitamin A deficiencies and stomach cancer. *American Journal of Epidemiology*, *121*, 65–70.
Stoppie, P., Borgers, M., Borghgraef, P., Dillen, L., Goossens, J., Sanz, G., et al. (2000). R115866 inhibits all-trans-retinoic acid metabolism and exerts retinoidal effects in rodents. *The Journal of Pharmacology and Experimental Therapeutics*, *293*, 304–312.
Taibi, G., & Nicotra, C. M. A. (2007). Xanthine oxidase catalyzes the oxidation of retinol. *Journal of Enzyme Inhibition and Medicinal Chemistry*, *22*, 471–476.
Takeuchi, H., Yokota, A., Ohoka, Y., & Iwata, M. (2011). Cyp26b1 regulates retinoic acid-dependent signals in T cells and its expression is inhibited by transforming growth factor-β. *PLoS One*, *6*, e16089.
Talley, N. J., Abreu, M. T., Achkar, J.-P., Bernstein, C. N., Dubinsky, M. C., Hanauer, S. B., et al. (2011). An evidence-based systematic review on medical therapies for inflammatory bowel disease. *The American Journal of Gastroenterology*, *106*, S2–S25.
Tallman, M. S., & Altman, J. K. (2009). How I treat acute promyelocytic leukemia. *Blood*, *114*, 5126–5135.
Tang, X.-H., & Gudas, L. J. (2011). Retinoids, retinoic acid receptors, and cancer. *Annual Review of Pathology: Mechanisms of Disease*, *6*, 345–364.
Terao, M., Kurosaki, M., Barzago, M. M., Fratelli, M., Bagnati, R., Bastone, A., et al. (2009). Role of the molybdoflavoenzyme aldehyde oxidase homolog 2 in the biosynthesis of retinoic acid: Generation and characterization of a knockout mouse. *Molecular and Cellular Biology*, *29*, 357–377.
Thatcher, J. E., Buttrick, B., Shaffer, S. A., Shimshoni, J. A., Goodlett, D. R., Nelson, W. L., et al. (2011). Substrate specificity and ligand interactions of CYP26A1, the human liver retinoic acid hydroxylase. *Molecular Pharmacology*, *80*, 228–239.
Thatcher, J. E., & Isoherranen, N. (2009). The role of CYP26 enzymes in retinoic acid clearance. *Expert Opinion on Drug Metabolism & Toxicology*, *5*, 875–886.
Thatcher, J. E., Zelter, A., & Isoherranen, N. (2010). The relative importance of CYP26A1 in hepatic clearance of all-trans retinoic acid. *Biochemical Pharmacology*, *80*, 903–912.
Tomita, S., Okuyama, E., Ohnishi, T., & Ichikawa, Y. (1996). Characteristic properties of a retinoic acid synthetic cytochrome P-450 purified from liver microsomes of 3-methylcholanthrene-induced rats. *Biochimica et Biophysica Acta (BBA): General Subjects*, *1290*, 273–281.
Topletz, A. R., Thatcher, J. E., Zelter, A., Lutz, J. D., Tay, S., Nelson, W. L., et al. (2012). Comparison of the function and expression of CYP26A1 and CYP26B1, the two retinoic acid hydroxylases. *Biochemical Pharmacology*, *83*, 149–163.
Topletz, A. R., Tripathy, S., Foti, R. S., Shimshoni, J. A., Nelson, W. L., & Isoherranen, N. (2015). Induction of CYP26A1 by metabolites of retinoic acid: Evidence that CYP26A1

is an important enzyme in the elimination of active retinoids. *Molecular Pharmacology, 87*, 430–441.

Tsukada, M., Schröder, M., Roos, T. C., Chandraratna, R. A., Reichert, U., Merk, H. F., et al. (2000). 13-cis Retinoic acid exerts its specific activity on human sebocytes through selective intracellular isomerization to all-trans retinoic acid and binding to retinoid acid receptors. *The Journal of Investigative Dermatology, 115*, 321–327.

Uehara, M., Yashiro, K., Mamiya, S., Nishino, J., Chambon, P., Dolle, P., et al. (2007). CYP26A1 and CYP26C1 cooperatively regulate anterior-posterior patterning of the developing brain and the production of migratory cranial neural crest cells in the mouse. *Developmental Biology, 302*, 399–411.

Van de Kerkhof, P. C. M. (2006). Update on retinoid therapy of psoriasis in: An update on the use of retinoids in dermatology. *Dermatologic Therapy, 19*, 252–263.

Van Giezen, J. J., Boon, G. I., Jansen, J. W., & Bouma, B. N. (1993). Retinoic acid enhances fibrinolytic activity in-vivo by enhancing tissue type plasminogen activator (t-PA) activity and inhibits venous thrombosis. *Thrombosis and Haemostasis, 69*, 381–386.

Van Heusden, J., Van Ginckel, R., Bruwiere, H., Moelans, P., Janssen, B., Floren, W., et al. (2002). Inhibition of all-TRANS-retinoic acid metabolism by R116010 induces antitumour activity. *British Journal of Cancer, 86*, 605–611.

Van Wauwe, J. P., Coene, M. C., Goossens, J., Cools, W., & Monbaliu, J. (1990). Effects of cytochrome P-450 inhibitors on the in vivo metabolism of all-trans-retinoic acid in rats. *The Journal of Pharmacology and Experimental Therapeutics, 252*, 365–369.

Veal, G. J., Cole, M., Errington, J., Pearson, A. D. J., Foot, A. B. M., Whyman, G., et al. (2007). Pharmacokinetics and metabolism of 13-cis-retinoic acid (isotretinoin) in children with high-risk neuroblastoma—A study of the United Kingdom Children's Cancer Study Group. *British Journal of Cancer, 96*, 424–431.

Veal, G. J., Errington, J., Redfern, C. P. F., Pearson, A. D. J., & Boddy, A. V. (2002). Influence of isomerisation on the growth inhibitory effects and cellular activity of 13-cis and all-trans retinoic acid in neuroblastoma cells. *Biochemical Pharmacology, 63*, 207–215.

Veal, G. J., Errington, J., Rowbotham, S. E., Illingworth, N. A., Malik, G., Cole, M., et al. (2013). Adaptive dosing approaches to the individualization of 13-cis-retinoic acid (isotretinoin) treatment for children with high-risk neuroblastoma. *Clinical Cancer Research: An Official Journal of the American Association for Cancer Research, 19*, 469–479.

Verfaille, C. J., Borgers, M., & van Steensel, M. A. M. (2008). Retinoic acid metabolism blocking agents (RAMBAs): A new paradigm in the treatment of hyperkeratotic disorders. *Journal Der Deutschen Dermatologischen Gesellschaft = Journal of the German Society of Dermatology: JDDG, 6*, 355–364.

Verfaille, C. J., Thissen, C. A., Bovenschen, H. J., Mertens, J., Steijlen, P. M., & van de Kerkhof, P. C. M. (2007). Oral R115866 in the treatment of moderate to severe plaque-type psoriasis. *Journal of the European Academy of Dermatology and Venereology: JEADV, 21*, 1038–1046.

Vernet, N., Dennefeld, C., Rochette-Egly, C., Oulad-Abdelghani, M., Chambon, P., Ghyselinck, N. B., et al. (2006). Retinoic acid metabolism and signaling pathways in the adult and developing mouse testis. *Endocrinology, 147*, 96–110.

Villablanca, J. G., Khan, A. A., Avramis, V. I., Seeger, R. C., Matthay, K. K., Ramsay, N. K., et al. (1995). Phase I trial of 13-cis-retinoic acid in children with neuroblastoma following bone marrow transplantation. *Journal of Clinical Oncology, 13*, 894–901.

Watanabe, A., Kanai, H., Arai, M., Sekiguchi, K., Uchiyama, T., Nagai, R., et al. (2002). Retinoids induce the PAI-1 gene expression through tyrosine kinase-dependent pathways in vascular smooth muscle cells. *Journal of Cardiovascular Pharmacology, 39*, 503–512.

Watts, J. M., & Tallman, M. S. (2014). Acute promyelocytic leukemia: What is the new standard of care? *Blood Reviews, 28*, 205–212.

Wen, J., Lopes, F., Soares, G., Farrell, S. A., Nelson, C., Qiao, Y., et al. (2013). Phenotypic and functional consequences of haploinsufficiency of genes from exocyst and retinoic acid pathway due to a recurrent microdeletion of 2p13.2. *Orphanet Journal of Rare Diseases*, *8*, 100.

Wiedermann, U., Chen, X. J., Enerbäck, L., Hanson, L. A., Kahu, H., & Dahlgren, U. I. (1996). Vitamin A deficiency increases inflammatory responses. *Scandinavian Journal of Immunology*, *44*, 578–584.

Wolbach, S. B., & Howe, P. R. (1925). Tissue changes following deprivation of fat-soluble a vitamin. *The Journal of Experimental Medicine*, *42*, 753–777.

Xi, J., & Yang, Z. (2008). Expression of RALDHs (ALDH1As) and CYP26s in human tissues and during the neural differentiation of P19 embryonal carcinoma stem cell. *Gene Expression Patterns: GEP*, *8*, 438–442.

Xia, H.-F., Ma, J.-J., Sun, J., Yang, Y., & Peng, J.-P. (2010). Retinoic acid metabolizing enzyme CYP26A1 is implicated in rat embryo implantation. *Human Reproduction (Oxford, England)*, *25*, 2985–2998.

Yashiro, K., Zhao, X., Uehara, M., Yamashita, K., Nishijima, M., Nishino, J., et al. (2004). Regulation of retinoic acid distribution is required for proximodistal patterning and outgrowth of the developing mouse limb. *Developmental Cell*, *6*, 411–422.

Zhang, Q.-Y., Dunbar, D., & Kaminsky, L. (2000). Human cytochrome P-450 metabolism of retinals to retinoic acids. *Drug Metabolism and Disposition*, *28*, 292–297.

Zhong, Y., Lin, Y., Shen, S., Zhou, Y., Mao, F., Guan, J., et al. (2013). Expression of ALDH1 in breast invasive ductal carcinoma: An independent predictor of early tumor relapse. *Cancer Cell International*, *13*, 60.

Zile, M. H., Emerick, R. J., & DeLuca, H. F. (1967). Identification of 13-cis retinoic acid in tissue extracts and its biological activity in rats. *Biochimica et Biophysica Acta*, *141*, 639–641.

CHAPTER TWELVE

Cytochrome P450 Vitamin D Hydroxylases in Inflammation and Cancer

Charlotte Gröschel, Samawansha Tennakoon, Enikö Kállay[1]
Department of Pathophysiology and Allergy Research, Center of Pathophysiology, Infectiology & Immunology, Medical University of Vienna, Währinger Gürtel, Vienna, Austria
[1]Corresponding author: e-mail address: enikoe.kallay@meduniwien.ac.at

Contents

Abstract

Vitamin D insufficiency correlates with increased incidence of inflammatory disorders and cancer of the colon, breast, liver, and prostate. Preclinical studies demonstrated that the hormonally active form of vitamin D, 1,25(OH)$_2$D$_3$, has antiproliferative,

Advances in Pharmacology, Volume 74
ISSN 1054-3589
http://dx.doi.org/10.1016/bs.apha.2015.03.002

proapoptotic, anti-inflammatory, and immunomodulatory effects. Tissue levels of 1,25 $(OH)_2D_3$ are determined by expression and activity of specific vitamin D hydroxylases expressed at renal and extrarenal sites. In order to understand how perturbations in the vitamin D system affect human health, we need to understand the steps involved in the synthesis and catabolism of the active metabolite. This review provides an overview about recent findings on the altered vitamin D metabolism in inflammatory conditions and carcinogenesis. We will summarize existing data on the pathophysiological regulation of vitamin D hydroxylases and outline the role of adequate levels of 1,25 $(OH)_2D_3$ on tissue homeostasis.

ABBREVIATIONS

1α,25$(OH)_2$-3-*epi*-D_3 1α,25-dihydroxy-3-*epi*-vitamin-D_3
1,25$(OH)_2D_3$ 1,25 dihydroxyvitamin D_3
15-PGDH 15 prostaglandin dehydrogenase
25$(OH)D_3$ 25 hydroxyvitamin D_3
COX-2 cyclooxygenase-2
CRC colorectal cancer
CYP cytochrome P450
CYP24A1 1,25 dihydroxyvitamin D_3 24 hydroxylase
CYP27B1 25 hydroxyvitamin D_3 1α hydroxylase
CYP2R1 vitamin D 25-hydroxylase
CYP11A1 (CYPssc) cholesterol side chain cleavage enzyme
DCs dendritic cells
DSS dextran sodium sulfate
EGF epidermal growth factor
FGF23 fibroblast growth factor 23
HCV hepatitis C virus
IBD inflammatory bowel diseases
IFN-γ interferon gamma
IL interleukin
IκBα nuclear factor of kappa light polypeptide gene enhancer in B-cells inhibitor alpha
JAK Janus kinase
K-RAS Kirsten rat sarcoma viral oncogene
LPS lipopolysaccharide
MAPK mitogen-activated protein kinases
NADPH nicotinamide adenine dinucleotide phosphate
NAFLD nonalcoholic fatty liver disease
NFκB nuclear factor kappa B
NOD2 nucleotide-binding oligomerization domain-containing protein 2
NSCLC nonsmall cell lung cancer
OSCCs oral squamous cell carcinomas
PGE_2 prostaglandin E_2
PTH parathyroid hormone
RXR retinoid X receptor
SNPs single nucleotide polymorphisms

STAT3 signal transducer and activator of transcription 3
TGFβ transforming growth factor beta
TLR toll-like receptor
TNF-α tumor necrosis factor alpha
Treg cells T-regulatory cells
VDR vitamin D receptor
VDRE vitamin D response element
Wnt Wingless-type MMTV integration site family member

1. INTRODUCTION

The term vitamin D refers to two highly lipophilic substances: cholecalciferol (vitamin D_3) and ergocalciferol (vitamin D_2). Technically, vitamin D_3 is a secosteroid, officially named 9,10-seco(5Z,7E)-5,7,10(19) cholestatriene-3β-ol. Moreover, vitamin D_3 is not a vitamin, but a prosteroid hormone because it is synthesized from 7-dehydrocholesterol in the skin under the influence of 270–290 nm wavelength solar ultraviolet B radiations. Cholecalciferol is barely active. It needs further activation steps to become the biologically active 1α,25-dihydroxyvitamin D_3 (1,25 $(OH)_2D_3$) or calcitriol (Fig. 1). The first hydroxylated vitamin D metabolites were discovered in 1968 (Blunt, DeLuca, & Schnoes, 1968). All activation and degradation steps are performed by a small number of cytochrome P450 (CYP) enzymes, known as vitamin D hydroxylases (Table 1). The vitamin D hydroxylases are mixed-function oxidases that use nicotinamide adenine dinucleotide phosphate (NADPH), as electron donor and are located either in the mitochondria or in microsomes (Fig. 2) (Henry, 2011b). The main circulating form of vitamin D, 25 hydroxyvitamin D_3 (25(OH)D_3), is the precursor of the active secosteroid hormone 1,25$(OH)_2D_3$. Systemic 1,25 $(OH)_2D_3$ levels are controlled by the kidney, which is the main organ responsible both for its synthesis by 25 hydroxyvitamin D_3 1α hydroxylase (encoded by the gene *CYP27B1*) and degradation by the 1,25 dihydroxyvitamin D_3 24 hydroxylase (encoded by the gene *CYP24A1*). 1,25$(OH)_2D_3$ is the natural ligand of the vitamin D receptor (VDR), a transcription factor belonging to the nuclear receptor superfamily (Carlberg & Molnar, 2012). Vitamin D induces its own catabolism by upregulation of CYP24A1, the enzyme responsible for the degradation of vitamin D and all its metabolites. Ergocalciferol (vitamin D_2) is derived from ergosterol and is synthesized in yeasts and plants and can be taken up with the diet

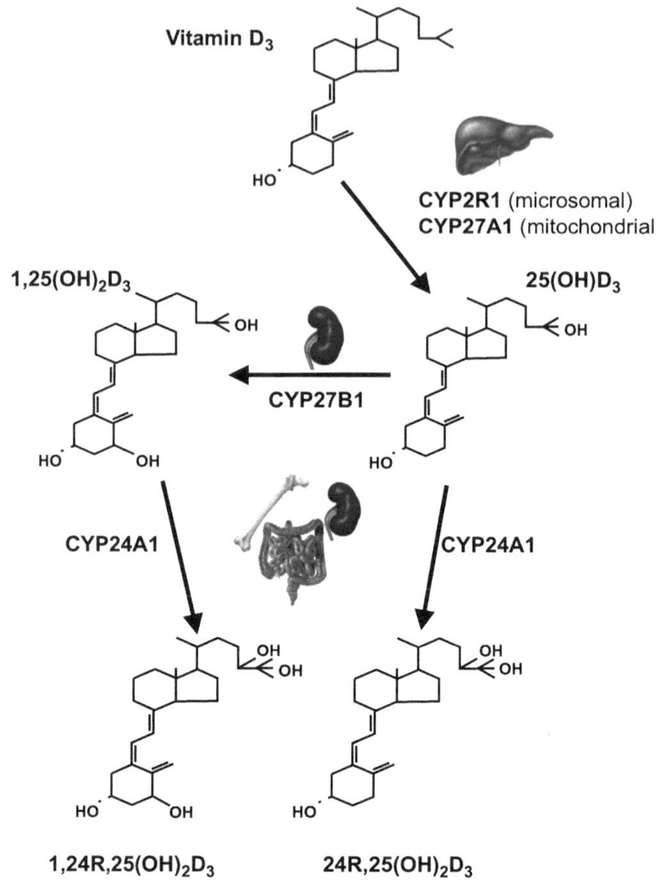

Figure 1 Steps in vitamin D_3 metabolism and catabolism. Hepatic 25-hydroxylation of vitamin D_3 leads to formation of the main circulating metabolite $25(OH)D_3$. 1α-hydroxylation in the kidney generates the active $1{,}25(OH)_2D_3$. Both $25(OH)D_3$ and $1{,}25(OH)_2D_3$ are catabolized by CYP24A1 at position 24.

or supplements (Keegan, Lu, Bogusz, Williams, & Holick, 2013). Similarly to vitamin D_3, vitamin D_2 is hydroxylated to $25(OH)D_2$ and $1{,}25(OH)_2D_2$, respectively (Jones, Schnoes, & DeLuca, 1975). Although they are metabolized similarly, the two vitamin D types differ in their potency. A recent, randomized controlled trial showed that ingested vitamin D_3 was almost twice as potent in increasing and maintaining 25(OH)D serum levels as vitamin D_2. However, discordance about the superiority of vitamin D_3 exists because different dose regimens have been used in previous studies (Heaney, Recker, Grote, Horst, & Armas, 2011; Holick et al., 2008).

Table 1 Main Characteristics of the Major Cytochrome P450 Enzymes Involved in the Metabolism of Vitamin D

CYPs	Main Tissue Localization	Major Substrate	Position of Hydroxylation
CYP2R1	**Liver, testes**, most other tissues	Vitamin D	25
CYP27A1	**Liver**, macrophages	Cholesterol	25
CYP3A4	**Liver**, intestine, other	Xenobiotics	24, 25
CYP2J2	Liver, kidney	1α-D, xenobiotics	25
CYP11A1	**Placenta**, intestine	Cholesterol, vitamin D	17, 20, 22, 23
CYP27B1	**Kidney**, ubiquitous	$25(OH)D_3$	1α
CYP24A1	**Kidney**, ubiquitous	$25(OH)D_3$, $1,25(OH)_2D_3$	23, 24

In bold are the major sites of expression.

1.1 Mechanism of Action of Vitamin D

$1,25(OH)_2D_3$ is the main ligand of the VDR, a ligand-induced transcription factor that plays a central role in the biological actions of vitamin D. VDR binds calcitriol with high specificity and affinity (Rochel, Wurtz, Mitschler, Klaholz, & Moras, 2000). Liganded VDR heterodimerizes with the retinoid X receptor (RXR) (Orlov, Rochel, Moras, & Klaholz, 2012) and binds to vitamin D response elements (VDREs) in the promoter and regulatory region of vitamin D target genes. Binding of VDR to chromatin is observed even in the absence of its ligand, however, at much lower frequency. The binding is increased significantly upon $1,25(OH)_2D_3$ stimulation (Ramagopalan et al., 2010). Liganded VDR–RXR can either activate or repress gene transcription, depending on the presence of coactivators or corepressors (Fujiki et al., 2005; Lonard & O'Malley, 2006). Many target genes possess multiple VDREs, and maximal transcriptional induction is achieved by binding of the VDR–RXR heterodimer to all VDREs. Chromatin looping brings distal VDREs in close proximity to proximal VDREs forming a single platform for gene regulation (Matilainen, Malinen, Turunen, Carlberg, & Vaisanen, 2010).

Genome-wide data on VDR-binding locations suggest that up to 5% of the genome is regulated by $1,25(OH)_2D_3$ and more than 900 genes are direct targets (Carlberg, 2014; Pike et al., 2007; Seuter, Neme, & Carlberg, 2014). The genome-wide actions of $1,25(OH)_2D_3$ result in changes in the expression of thousands of protein-coding mRNAs but also a comparable number

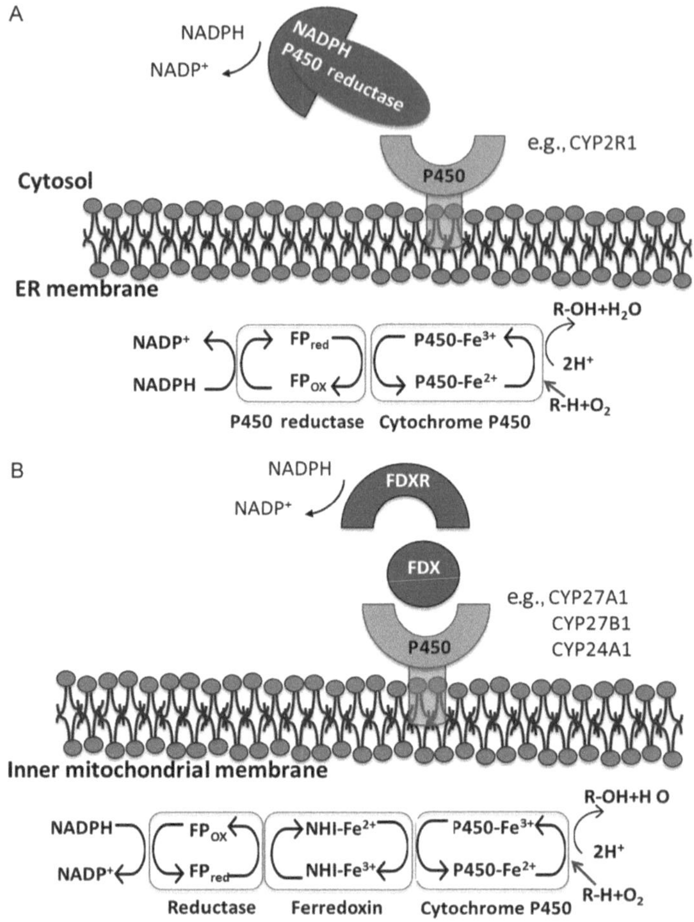

Figure 2 Electron transport chains and protein components of cytochrome p450 vitamin D hydroxylases. (A) Microsomal P450: In the endoplasmic reticulum, electron equivalents are transferred from NADPH via NADPH P450 reductase. The two electrons from NADPH are transferred sequentially to the microsomal P450. (B) Mitochondrial P450: In mitochondria, FDXR oxidizes NADPH. FDXR, a flavoprotein, transfers single electrons through a pool of ferredoxin iron-sulfur proteins to the mitochondrial P450s located on the inner membrane. NADPH, nicotinamide adenine dinucleotide phosphate hydrogen; FDXR, ferredoxin reductase; FDX, ferredoxin; FP, ferroportin.

of noncoding RNAs (Campbell, 2014). VDR is expressed almost ubiquitously, albeit the highest expression was found in tissues associated with maintenance of calcium homeostasis, such as the intestine, kidney, parathyroid, and bone (Wang, Zhu, & DeLuca, 2012).

1.2 Physiological Role of Vitamin D

The first function of vitamin D during evolution may have been detoxification rather than regulation of bone mineralization. This detoxification function is found also in higher vertebrates (Haussler et al., 2013; Kutuzova & DeLuca, 2007). The VDR is able to bind also non-vitamin D-related compounds, such as lithocholic acid, arachidonic acid, curcumin, or polyunsaturated fatty acids, but with a much lower affinity than 1,25 $(OH)_2D_3$ (Haussler et al., 2008).

The main endocrine function of serum $1,25(OH)_2D_3$ is regulation of bone mineralization and calcium/phosphate homeostasis (Holick, 2007). In addition to the kidney, several other tissues, such as the skin, intestine, placenta, immune cells, are able both to activate and degrade 1,25 $(OH)_2D_3$ (Bikle, 2009; Cross & Kallay, 2009; Hobaus, Thiem, Hummel, & Kallay, 2013). Autocrine and paracrine actions of the locally synthesized $1,25(OH)_2D_3$ include boosting of the innate immune system, lowering autoimmune disease by regulating the adaptive immune system, reducing risk of stroke and heart attack, and exerting antitumorigenic activities (Haussler et al., 2013). Association studies of $25(OH)D_3$ levels and chronic diseases led to the conclusion that much higher serum $25(OH)D_3$ levels are needed for these local autocrine/paracrine actions than for the endocrine actions of $1,25(OH)_2D_3$ (Bikle, 2014).

In case of certain diseases, i.e., chronic renal failure, liver failure, tumor-induced osteomalacia, or hyperparathyroidism, that functionally impair the enzymes of the vitamin D system, the endocrine vitamin D metabolism is deregulated leading to vitamin D deficiency (Holick, 2007). On the other hand, meta-analyses from several epidemiological studies show that low serum $25(OH)D_3$ levels cause rickets and osteomalacia and correlate with incidence and severity of various chronic diseases, such as osteoporosis, metabolic syndrome, cardiovascular diseases, multiple sclerosis, cancer of the colon, breast, or prostate (Bouillon et al., 2008; Brondum-Jacobsen, Benn, Jensen, & Nordestgaard, 2012).

1.3 Metabolism of Vitamin D

All steps of the vitamin D metabolism from its synthesis to the degradation are controlled by CYP enzymes (Table 1).

The first activation step is hydroxylation of vitamin D_3 on C-25, followed by hydroxylation at the C-1α position (Fig. 1). Initially, it was thought that these steps are restricted to the liver (25-hydroxylation) and kidney (1α-hydroxylation). Indeed, the kidney is the main source of serum $1{,}25(OH)_2D_3$ for its endocrine role of regulating serum calcium and phosphate levels (DeLuca, 1997).

The production of systemic $1{,}25(OH)_2D_3$ is regulated by serum calcium and phosphate levels (Dusso, Brown, & Slatopolsky, 2005). Low serum-calcium levels induce parathyroid hormone (PTH) secretion, increasing CYP27B1 mRNA expression and activity, reducing CYP24A1 levels. At low serum phosphate concentrations, synthesis of $1{,}25(OH)_2D_3$ is high and independent of PTH. $1{,}25(OH)_2D_3$ regulates its own concentration by a negative feedback mechanism. It inhibits PTH synthesis and secretion from the parathyroid glands. Additionally, $1{,}25(OH)_2D_3$ induces expression of CYP24A1, which degrades both $25(OH)D_3$ and $1{,}25(OH)_2D_3$. Both metabolites are degraded via 24-hydroxylation, 24-oxidation, 23-hydroxylation, and side chain cleavage to the final catabolite, calcitroic acid, which is then secreted into the bile (Beckman et al., 1996) (Fig. 3). $1{,}25(OH)_2D_3$ is catabolized almost 10-fold more efficiently than $25(OH)D_3$ (Burgos-Trinidad & DeLuca, 1991; Jones & Tenenhouse, 2002).

$1{,}25(OH)_2D_3$ levels are regulated also by the fibroblast growth factor 23 (FGF23). Secreted from osteocytes, FGF23 inhibits renal phosphate reabsorption and CYP27B1 expression, in parallel upregulating CYP24A1 expression (Shimada et al., 2004). A negative regulator of CYP24A1 in the kidney is PTH.

$1{,}25(OH)_2D_3$ levels in extrarenal tissues are controlled by local factors such as cytokines and growth factors that adjust $1{,}25(OH)_2D_3$ levels for cell-specific actions (Dusso et al., 2005).

Besides the enzymatic inactivation by the side chain oxidation at C-23 and C-24, a C-3 epimerization pathway has been also observed (Bischof et al., 1998). The C-3 epimer of the active form of vitamin D was detected for the first time in 1994 in human neonatal keratinocytes (Reddy, Muralidharan, Okamura, Tserng, & McLane, 2001). Both $1{,}25(OH)_2D_3$ and $25(OH)D_3$ are sensitive to 3-epimerization in a tissue- and cell-dependent manner, leading to biologically active 3-*epi* metabolites. The enzyme catalyzing the epimerization of the hydroxyl group at the C-3 of the A-ring is still unknown (Kamao et al., 2004; Molnár et al., 2011). Interestingly, the C-3 epimer of $1{,}25(OH)_2D_3$ has less calcaemic effects (Fleet, Bradley, Reddy, Ray, & Wood, 1996), but its clinical relevance remains to be investigated.

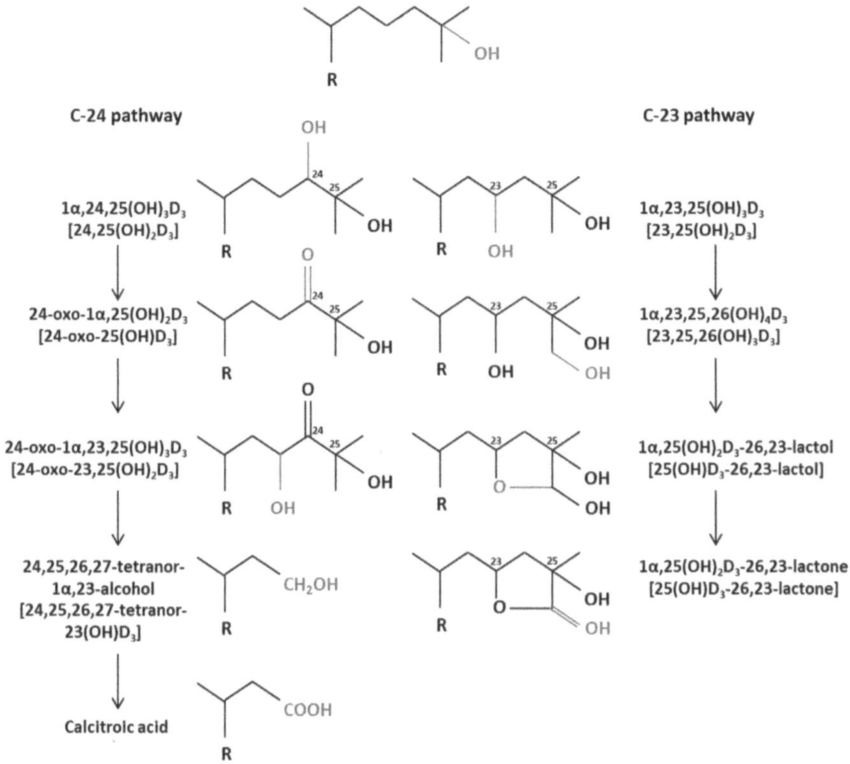

Figure 3 Catabolism of calcitriol and calcidiol by CYP24A1. CYP24A1 catabolizes both $1{,}25(OH)_2D_3$ and $25(OH)D_3$ via the C-24 (left) and C-23 (right) pathway leading to different end products. For simplification, only the side chain structure is shown. The product names are stated for the degradation of $1{,}25(OH)_2D_3$, product names for $25(OH)D_3$ are stated in brackets.

2. VITAMIN D HYDROXYLASES

The vitamin D metabolizing CYP enzymes are mixed-function oxidases as the oxygen bound in their active center is reduced to a hydroxyl group, while the other atom is reduced to H_2O. The energy for this reaction is derived from NAPDH (Annalora et al., 2010). The vitamin D hydroxylases are localized either in mitochondria or in microsomes (Prosser & Jones, 2004). The microsomal enzymes require a general-purpose NADPH–CYP reductase, while the mitochondrial enzymes require two additional electron-transporting proteins: ferredoxin and ferredoxin reductase (Danielson, 2002) (Fig. 2).

2.1 25-Hydroxylase (Cytochrome P450 Family 2 Subfamily R Polypeptide 1, CYP2R1)

The major circulating form of vitamin D_3, $25(OH)D_3$, is synthesized from cholecalciferol through hydroxylation on position C-25 by 25-hydroxylases. The primary site of 25-hydroxylation is the liver. It was only recently that an orphan microsomal CYP, encoded by the gene *CYP2R1*, was identified as the main vitamin D 25-hydroxylase, a highly conserved and substrate-specific microsomal enzyme.

Single nucleotide polymorphisms (SNPs) in or close to the *CYP2R1* gene were associated with circulating $25(OH)D_3$ concentrations (Ahn et al., 2010; Barry et al., 2014; Bu et al., 2010; Ramos-Lopez, Bruck, Jansen, Herwig, & Badenhoop, 2007; Simon et al., 2011; Wang et al., 2010), although others found no association (Signorello et al., 2011). A homozygous mutation in exon 2 of the *CYP2R1* gene caused a hereditary form of rickets, confirming that CYP2R1 is the predominant 25-hydroxylase (Cheng, Levine, Bell, Mangelsdorf, & Russell, 2004). Moreover, the affinity of CYP2R1 to vitamin D is higher than that of any other enzyme with 25-hydroxylase activity. No other substrates have been described for CYP2R1 so far (Schuster, 2011). CYP2R1 was found not only in the liver but also in testis, dermal fibroblasts, and LNCaP prostate cancer cells (Foresta et al., 2011). CYP2R1 is expressed both in activated T cells and in dendritic cells (DCs) but activity was found only in the latter (Sigmundsdottir et al., 2007). CYP2R1 has an extended binding site in which vitamin D is positioned in a 6-s-*trans* conformation (Strushkevich, Usanov, Plotnikov, Jones, & Park, 2008).

2.1.1 Other CYP Enzymes with 25 Hydroxylase Activity

Some residual 25-hydoxylation may be performed by the mitochondrial CYP27A1. However, the main physiological function of this enzyme is the side chain hydroxylation of cholesterol in the process of bile acid synthesis. Several other CYP enzymes are also capable of hydroxylating C-25, such as the two microsomal vitamin D_3 25-hydroxylases CYP2J2 and CYP3A4 (Hobaus, Thiem, et al., 2013; Jones & Prosser, 2011), or the porcine CYP2D25 and rat CYP2C11, the earliest isolated cytochrome enzymes with CYP2R1 activity (Zhu & DeLuca, 2012). Due to the biological redundancy, it is difficult to completely understand the impact of the different enzymes to the physiological process of vitamin D 25-hydroxylation.

Hepatic vitamin D_3 25-hydroxylation seems to be mainly dependent on substrate availability. No significant regulation was observed by PTH, dietary calcium, or phosphorus, the known endocrine regulators of the vitamin D endocrine system.

2.2 1α-Hydroxylase (Cytochrome P450, Family 27, Subfamily B, Polypeptide 1, CYP27B1)

The final activation step of $25(OH)D_3$, hydroxylation on position C-1α is achieved by one single enzyme, the 25-hydroxyvitamin D_3 1α-hydroxylase (CYP27B1) (Gray, Omdahl, Ghazarian, & DeLuca, 1972). The mitochondrial enzyme expressed in proximal renal tubular cells is responsible for the systemic levels of $1,25(OH)_2D_3$ (Henry, 2011a). It has been established that CYP27B1 is additionally present in many different tissues, such as the intestine, pancreas, skin, parathyroid glands, adipose tissue, placenta, macrophages (Adams & Hewison, 2012). The extrarenal CYP27B1 synthesizes $1,25(OH)_2D_3$ for local signaling in an autocrine and paracrine fashion (Feldman, Krishnan, Swami, Giovannucci, & Feldman, 2014).

Regulation of CYP27B1 expression is tissue dependent. In the kidney, the main regulators are PTH, serum calcium and phosphate, FGF-23, and $1,25(OH)_2D_3$ (Armbrecht, Boltz, & Hodam, 2003; Perwad & Portale, 2011). Inhibition of CYP27B1 expression by $1,25(OH)_2D_3$ involves complex epigenetic regulation of its promoter through VDR (Kim et al., 2009). However, there are tissues where $1,25(OH)_2D_3$ either induces CYP27B1 expression or has no effect. In extrarenal tissues, CYP27B1 expression and activity are independent of circulating PTH. In prostate epithelial cells, the epidermal growth factor (EGF) induces, while $1,25(OH)_2D_3$ inhibits CYP27B1 expression (Wang, Flanagan, et al., 2004). In macrophages and keratinocytes, the expression is increased by different cytokines, e.g., interferon gamma (IFN-γ), or tumor necrosis factor α (TNF-α) (Hewison et al., 2007), while nuclear factor kappa B (NFκB) is a potent inhibitor of CYP27B1 expression (Fig. 4).

Inactivating mutations of CYP27B1 cause vitamin D-dependent rickets type I (Sawada, Sakaki, Kitanaka, Kato, & Inouye, 2001; St-Arnaud, Messerlian, Moir, Omdahl, & Glorieux, 1997), confirming the critical role of the enzyme in activation of vitamin D. Genetic linkage analysis has revealed recently that rare inactivating mutations of CYP27B1 seem to play a causative role in the development of multiple sclerosis (Ramagopalan et al., 2011). Several CYP27B1 splice variants are expressed both in normal and transformed tissues. The impact of these on the expression and activity of

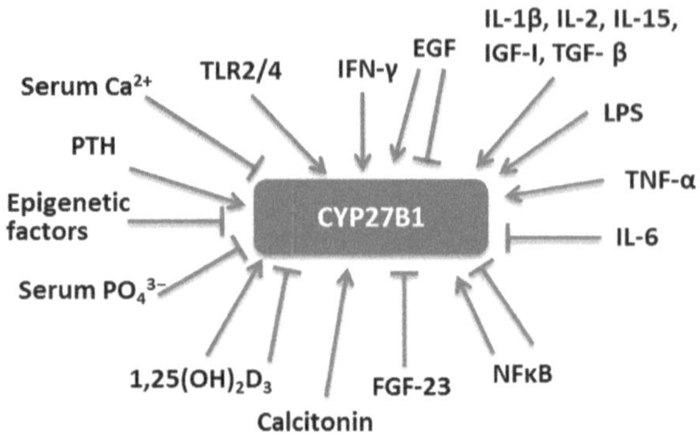

Figure 4 Renal and extrarenal regulation of CYP27B1 expression in inflammation and cancer. Activation (arrows) and inhibition (T lines) by hormones, ions, epigenetic factors, inflammatory mediators, growth factors, and pathways are shown. Renal expression of CYP27B1 is regulated by PTH, serum Ca^{2+} and PO_4^{3-}, FGF23 and $1,25(OH)_2D_3$, extrarenal CYP27B1 is regulated by numerous other factors. PTH, *parathyroid hormone;* TLR, *Toll-like* receptor; IFN-γ, interferon gamma; EGF, *epidermal growth factor;* IL, interleukin; IGF, *insulin*-like growth factor; TGF-β, transforming growth factor beta; LPS, *lipopolysaccharide*; TNF-α, tumor necrosis factor alpha; NFκB, *nuclear factor* kappa B; FGF, *fibroblast* growth factor.

the wild-type enzyme is not yet clear (Wu, Ren, Nguyen, Adams, & Hewison, 2007).

2.3 24-Hydroxylase (Cytochrome P450, Family 24, Subfamily A, Polypeptide 1, CYP24A1)

The $1,25(OH)_2D_3$ 24-hydroxylase, CYP24A1, is a mitochondrial inner membrane CYP enzyme. The tertiary structure of P450 enzyme is conserved and includes 12 α-helices and four β-sheets, while CYP24A1 contains five additional short α-helices. Two of these additional α-helices may have a role in anchoring CYP24A1 into the inner mitochondrial membrane by acting as insertion sequences (Annalora et al., 2010). CYP24A1 is found in all vitamin D target tissues, with the highest expression in the kidneys, where CYP24A1 is essential for the regulation of systemic $25(OH)D_3$ and $1,25(OH)_2D_3$ levels (Dusso et al., 2005; Jones, Strugnell, & DeLuca, 1998).

With the exception of the kidneys, basal expression of CYP24A1 is low. $1,25(OH)_2D_3$ is the primary inducer of CYP24A1 expression both in renal and extrarenal cells (Zierold, Mings, & DeLuca, 2003). Gene expression is rapidly induced after binding of the VDR-$1,25(OH)_2D_3$ complex to two VDRE sites in the proximal *CYP24A1* promoter region (Chen &

DeLuca, 1995). Distal enhancers located approximately 50–69 kB downstream of *CYP24A1* appear to form a chromatin loop and are involved in $1,25(OH)_2D_3$-mediated gene transcription (Meyer, Goetsch, & Pike, 2010). Rapid CYP24A1 expression upon presence of VDR–calcitriol constitutes a negative feedback mechanism to keep calcitriol levels within a narrow range, thus preventing hypercalcaemia. Rats overexpressing CYP24A1 have decreased vitamin D metabolite levels (Kasuga et al., 2002). Mice lacking Cyp24a1 have poor survival rate, only approximately 50% of *cyp24a1*$^{-/-}$ mice survive past week three. The survivors show impaired $1,25(OH)_2D_3$ degradation and completely lack C-24 and C-23 pathway metabolites (Masuda et al., 2005; St-Arnaud et al., 2000).

The main substrates of CYP24A1 are $25(OH)D_3$ and $1,25(OH)_2D_3$ with higher substrate affinity for the latter. Hydroxylation at position C-24 is followed by subsequent degradation steps and results in calcitroic acid. Alternatively, CYP24A1 can hydroxylate both calcidiol and calcitriol at position C-23 which leads to a side chain lactone formation (Fig. 3). In humans, CYP24A1 utilizes both the C-23 and C-24 degradation pathways (Sakaki et al., 2000). While the functional advantage of two independent degradation pathways in humans is debated, evidence suggests that the end product of $1,25(OH)_2D_3$ catalyzed via the C-23 pathway, namely, $1,25(OH)_2D$-26,23-lactone may act as a VDR inhibitor (Ishizuka et al., 2005; Toell et al., 2001), thereby impairing vitamin D signaling. Recombinant CYP24A1 releases its substrate after each step of the C-23 and C-24 pathway (Sakaki et al., 2000); therefore, it is possible that other intermediate substrates might be released into the bloodstream as well.

CYP24A1 is capable to hydroxylate not only $25(OH)D_3$ and $1,25(OH)_2D_3$ but also $20(OH)D_3$, albeit with lower efficiency. However, the $20,24(OH)_2D_3$ metabolites are more potent in inhibiting colony formation of melanoma cells than $20(OH)D_3$ or even $1,25(OH)_2D_3$, suggesting that in this case, CYP24A1 has an activating rather than degrading function (Tieu et al., 2012).

2.4 Other Vitamin D Hydroxylases

Interestingly, the cholesterol side chain cleavage enzyme (CYPscc) (encoded by the *CYP11A1* gene) that catalyzes the conversion of cholesterol to pregnanolone is able to hydroxylate also the side chain of vitamin D_3 on the carbons 17, 20, 22, and 23, producing at least 10 different metabolites (Guryev, Carvalho, Usanov, Gilep, & Estabrook, 2003; Slominski et al.,

2006). CYP11A1 catalyzes only the hydroxylation of the vitamin D_3 side chain without cleaving it. The preferred hydroxylation site is at C-20, with $20(OH)D_3$ being the major metabolite (Tuckey et al., 2008). Tissues with high levels of CYP11A1, such as the placenta, metabolize vitamin D both through the classical pathway and through the CYP11A1-catalyzed pathway (Slominski et al., 2012). Interestingly, this metabolic pathway is active also in the colon cancer cell line Caco-2 that was able to convert exogenous vitamin D_3 mainly to $20(OH)D_3$ beside $22(OH)D_3$ and $25(OH)D_3$ (Slominski et al., 2014). These data suggest that enterocytes also are capable of metabolizing vitamin D_3 through the CYP11A1 pathway.

Several of the CYP11A1-derived vitamin D metabolites were biologically active in both normal and malignant cell lines. In rodents, neither $20(OH)D_3$ nor $20{,}23(OH)_2D_3$ showed toxic calcemic effects (Slominski et al., 2010; Wang, Slominski, et al., 2012).

3. VITAMIN D AND INFLAMMATION

Under physiological conditions, inflammation is a protective, self-limiting response to infectious agents and to tissue injury. Its role is the elimination of pathogens and tissue regeneration (Medzhitov, 2008). The balance of pro- and anti-inflammatory reactions is crucial to avoid transition to chronic inflammation, which would lead to tissue damage, fibrosis, and is a driver for tumorigenesis. Distinct subpopulations of immune cells contribute to acute and chronic inflammatory response. Tissue-resident innate immune cells, namely, macrophages and mast cells recognize pathogens via Toll-like receptors (TLRs) and NOD-nucleotide-binding (oligomerization-domain protein)-like receptors. They release mediators such as proinflammatory cytokines, prostaglandins and attract circulating innate immune cells, neutrophils, and monocytes. If acute inflammatory response fails to eradicate the foreign agent or if it is misdirected against endogenous molecules, it can lead to chronic inflammation in a process involving two cell populations considered as the adaptive immune compartment, the B- and T lymphocytes. Physiologically, it is the second line of defense, based on antigen-specific recognition. The T-lymphocyte lineage comprises effector and regulatory T cells (Tregs). CD4+ effector cells, also referred as T-helper cells, are either proinflammatory (Th1, Th17) or anti-inflammatory (Th2) cells (Crome, Wang, & Levings, 2010). Tregs are an immune-suppressive subgroup of CD4+ cells that reduce proliferation of effector cells (Hewison, 2012). Under pathologic conditions, a shift toward effector cells leads to persistent inflammation and to autoimmune diseases.

Important inflammatory pathways converge at NFκB, STAT3, and p38, major players in transducing signals from inflammatory cytokines such as TNF-α, IL-1β, IFNγ, and IL-6. While NFκB is needed for proper function of the immune system, inappropriately activated NFκB is the major mediator of inflammation (Aggarwal, 2004). It is the key molecule linking inflammation to tumorigenesis (Ben-Neriah & Karin, 2011). Prostaglandin signaling plays an important role in inflammation-associated tumor formation (Nakanishi & Rosenberg, 2013). Prostaglandin E_2 (PGE2) is a lipid mediator that is synthesized by the NFκB target gene cyclooxygenase-2 (COX-2) in all cell types, with higher preference in the cells of the immune system (Agard, Asakrah, & Morici, 2013), in response to inflammatory stimuli (Kalinski, 2012). Inflammation can be considered as the seventh hallmark of cancer (Hanahan & Weinberg, 2011).

3.1 Effect of Vitamin D in Inflammation

$1,25(OH)_2D_3$ inhibits the inflammatory process by suppressing inflammatory pathways in the epithelial cells and by regulating the immune system. The cells of both the innate and adaptive immune system express the VDR, thus are vitamin D targets (Baeke, Takiishi, Korf, Gysemans, & Mathieu, 2010). $1,25(OH)_2D_3$ enhances the antimicrobial response of the innate immune system to pathogens (Baeke et al., 2010) and reduces proinflammatory cytokine expression. Calcitriol promotes tolerogenic effects of the adaptive immune compartment. This is reflected by the observation that VDR is not expressed in resting, but only in activated B- and T-lymphocytes (Adams & Hewison, 2008). $1,25(OH)_2D_3$ limits inflammatory responses of Th1 and Th17 cells, while promoting the Th2 subtype (Cantorna & Waddell, 2014). This requires sufficient tissue levels of the active metabolite, which is provided both systemically and locally, by tissue-specific CYP27B1. Expression of the vitamin D hydroxylases in the inflamed tissue and in cells of the immune system affects the inflammatory process and disease progression. A well-controlled balance between $1,25(OH)_2D_3$ synthesis and degradation is necessary to prevent loss of tissue homeostasis caused by low levels of $1,25(OH)_2D_3$.

3.1.1 Mechanism of Action of Vitamin D in Inflammation

Sustained activity of the monocyte–macrophage system is contributing to chronic inflammation. The active hormone $1,25(OH)_2D_3$ modulates

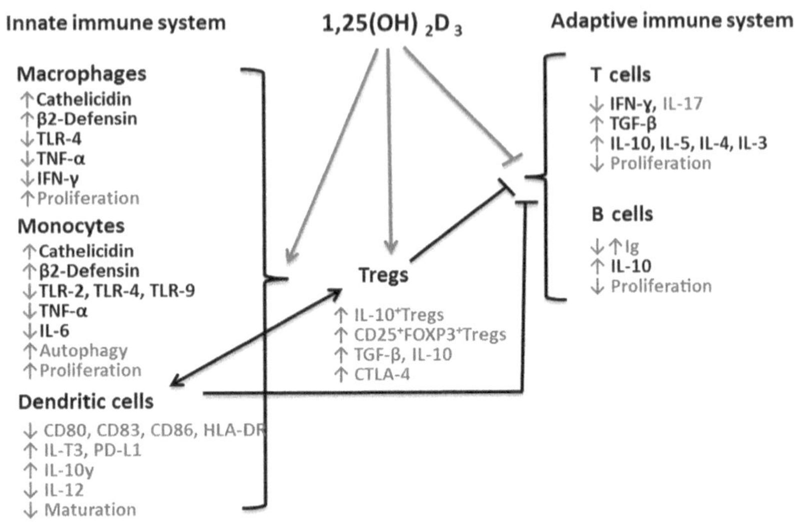

Figure 5 The effects of $1,25(OH)_2D_3$ on cells of the innate and adaptive immune system and on their expression of cytokines and inflammatory mediators. Green (black in the print version) arrows indicate activation, red (black in the print version) arrows indicate inhibition. Effects shown in gray are not further described in this review. CD80, cluster of differentiation 80; HLA-DR, *human leukocyte antigen DR*; PD-L1, programmed death-ligand 1; $FOXP3^+$, forkhead box P3; CTLA-4, cytotoxic T-lymphocyte-associated protein 4.

synthesis of proinflammatory cytokines produced by monocytes, macrophages, fibroblasts, and endothelial cells, thereby protecting the tissue from a hyperinflammatory response (Fig. 5) (Bain & Mowat, 2014). It also inhibits synthesis and secretion of several Th1-specific cytokines (Lemire, Archer, Beck, & Spiegelberg, 1995). Recent studies in human monocytes revealed that $25(OH)D_3$ and $1,25(OH)_2D_3$ inhibited lipopolysaccharide (LPS)-induced synthesis of TNF-α and IL-6 (Murray & Wynn, 2011; Zhang et al., 2012). In human peripheral blood mononuclear cells, $20(OH)D_3$ reduced LPS-induced TNF-α synthesis in a VDR-mediated manner (Slominski et al., 2013).

Vitamin D reduced IL-6 synthesis in prostate epithelial cells (Nonn, Peng, Feldman, & Peehl, 2006). IL-6 is a proinflammatory cytokine with protumorigenic properties and is a crucial key effector in prostate cancer development (Nguyen, Li, & Tewari, 2014; Taniguchi & Karin, 2014).

Distinct subtypes of TLRs are upregulated in chronic inflammatory diseases, e.g., TLR4 is increased during inflammatory bowel diseases (IBD) (Cario, 2010). Several studies have observed an inhibition of TLR expression by 1,25$(OH)_2D_3$. In peripheral human blood monocytes, 1,25 $(OH)_2D_3$ downregulated expression of TLR2, TLR4, and TLR9, leading to reduction of IL-6 synthesis (Dickie et al., 2010). Another mechanism of TL4 inhibition was repression of microRNA-155 by 1,25$(OH)_2D_3$, counteracting indirectly LPS-induced inflammatory response of macrophages (Chen, Liu, et al., 2013). Vitamin D is a potent inhibitor of the NFκB and prostaglandin pathway (Cross, Nittke, & Kallay, 2011). 1,25 $(OH)_2D_3$-induced VDR interaction with the inhibitor of kappa B (IκB) kinase β inhibited NFκB activity in mouse embryonic fibroblasts (Chen, Zhang, et al., 2013). In murine macrophages, 1,25$(OH)_2D_3$ reduced TNF-α secretion and induced IκBα expression (Cohen-Lahav, Shany, Tobvin, Chaimovitz, & Douvdevani, 2006). Similar effects were attributed to 20$(OH)D_3$, which increased IκBα levels in human keratinocytes, leading to a decrease in NFκB activity (Janjetovic, Tuckey, Nguyen, Thorpe, & Slominski, 2010; Janjetovic et al., 2009). 1,25$(OH)_2D_3$ reduced gene expression of the COX-2–PGE2 signaling pathway in prostate epithelial cells (Moreno et al., 2005). In mice, calcitriol inhibited COX-2 expression and reduced subsequent prostaglandin synthesis in macrophages (Wang et al., 2014).

IFN-γ plays a crucial role in host defense against infections but is also a key molecule in chronic inflammatory autoimmune diseases and inflammation-associated cancer. It is usually secreted by proinflammatory Th1 cells together with TNF-α and IL-2. In murine macrophages, 1,25 $(OH)_2D_3$ downregulated expression of IFN-γ (Helming et al., 2005). Similar effect was seen in human peripheral blood monocytes (Reichel, Koeffler, Tobler, & Norman, 1987). Moreover, in microglia cells, the active metabolite suppressed IFN-γ-induced JAK/STAT signaling (Muthian, Raikwar, Rajasingh, & Bright, 2006).

20$(OH)D_3$ suppressed rheumatoid arthritis in a mouse model and decreased expression of several cytokines released by Th1, TH2, and Th17 cells *in vitro* (Slominski et al., 2014). Other studies have shown that 1,25$(OH)_2D_3$ favors a shift of the Th cell population toward Th2 cells and promotes cytokine signature of this subset of $CD4^+$ cells, such as IL-10, IL-5, IL-4, IL-3 (Boonstra et al., 2001; Sloka, Silva, Wang, & Yong, 2011).

Vitamin D stimulates synthesis of negative regulators of inflammation, such as IL-10 and TGFβ. TGFβ plays an important role for the immune suppressive function of T-regulatory cells (Taylor, 2009). IL-10 limits the immune reaction and prevents tissue damage (Iyer & Cheng, 2012).

IL-10 is released not only by Th2 cells but also by B cells, macrophages and tumor cells in order to prevent a hyperinflammatory state of the tissue. In human B cells, its expression is increased by $1{,}25(OH)_2D_3$ (Heine et al., 2008).

3.2 Vitamin D Hydroxylases in Inflammation

Not much is known about the regulation of the vitamin D hydroxylases during inflammation *in vivo*, although there are data suggesting that inflammatory cytokines may increase local synthesis of $1{,}25(OH)_2D_3$ to inhibit the inflammatory process.

There is evidence from *in vitro* studies that inflammatory cytokines influence expression of vitamin D metabolizing enzymes. While renal CYP27B1 expression is tightly controlled by PTH and $1{,}25(OH)_2D_3$, in extra-renal tissues, CYP27B1 is regulated independently of those factors in a tissue-specific manner (Fig. 4). Soluble factors like cytokines and growth factors from the microenvironment affect cellular levels of this enzyme. IL1, IFN-γ, and TNF-α increase CYP27B1 expression in immune cells (van Etten, Stoffels, Gysemans, Mathieu, & Overbergh, 2008). TNF-α stimulated activity of CYP27B1 in untransformed cells such as human keratinocytes, endothelial cells (Zehnder et al., 2002), and peripheral blood monocytes (Gyetko, Hsu, Wilkinson, Patel, & Young, 1993). In cells of the human placenta, TNF-α induced expression of both CYP27B1 and CYP24A1, the latter to a greater extent (Noyola-Martínez et al., 2014).

In the alveolar macrophages of patients with sarcoidosis, an inflammatory granulomatous disorder, CYP27B1 is expressed at a very high level, leading to pathologic increase of systemic $1{,}25(OH)_2D_3$ levels and to hypercalcemia (Inui et al., 2001; Jones, 2008). In addition to the high CYP27B1 expression and activity, $1{,}25(OH)_2D_3$-induced upregulation of CYP24A1 leads to the expression of a splice variant that lacks the mitochondrial targeting sequence thereby blunting the negative feedback regulation (Zehnder et al., 2002), contributing to the high circulating $1{,}25(OH)_2D_3$ levels. CYP27B1 expression in macrophages is controlled by immune signals, such as IFN-γ, LPS, or by viral infections (van Etten et al., 2008).

In DCs, as in other extrarenal tissues, the level of the precursor $25(OH)D_3$ is often the limiting factor in the synthesis of the active metabolite (Jeffery et al., 2012). LPS, the ligand for TLRs, increased gene expression of CYP27B1 in DCs and human macrophages (Liu et al., 2006). $1{,}25(OH)_2D_3$ synthesis in monocyte-derived DC is impaired due to a truncated, less active CYP27B1, while catabolism is not affected. Regulation of VDR targets in these cells seems to occur in a paracrine manner. It is the 1,25 $(OH)_2D_3$ produced by macrophages that induces expression of vitamin D target genes in the neighboring DCs, inhibits maturation of DC, and reduces DC-dependent T-cell responses (Kundu, Chain, Coussens, Khoo, & Noursadeghi, 2014).

T-cell-derived cytokines regulate expression of the activating enzyme CYP27B1 in monocytes via TLR 2/1. IFN-γ increased the activity of CYP27B1 and decreased that of CYP24A1 (Noyola-Martínez et al., 2014). Mechanistic studies using promoter–reporter constructs in monocytes revealed that IFN-γ-induced increase of CYP27B1 is mediated via JAK–STAT, NFκB, and p38-MAPK pathways (Stoffels et al., 2006). Activation of TLR1 and 2 by *Mycobacterium tuberculosis* or IFN-γ increases expression of VDR and CYP27B1, (Adams et al., 2009; Liu et al., 2006). The resultant, locally synthesized 1,25$(OH)_2D_3$ induces the expression of antimicrobial peptides, such as cathelicidin (Gombart, 2009; Hoyer-Hansen, Nordbrandt, & Jaattela, 2010) or defensin 4β (Wang, Nestel, et al., 2004).

3.3 Vitamin D and Vitamin D Hydroxylases in IBD

Vitamin D deficiency increases the severity of several autoimmune diseases including IBD (Froicu et al., 2003). IBD can be classified into Crohn's disease (CD) and ulcerative colitis (UC). It is a chronic, relapsing, and multifactorial disease caused by immunologic, genetic, and environmental factors. In both diseases, defective response of the innate and adaptive immune system leads to intolerance against gut microbiota inducing chronic inflammation (Abraham & Cho, 2009; Sartor, 2006). The incidence of chronic inflammatory disorders such as IBD correlates with low serum vitamin D levels (Ananthakrishnan et al., 2012; Yin & Agrawal, 2014). A multi-institutional cohort study showed that IBD patients with plasma levels of 25(OH)D below 20 ng/ml have an increased risk to develop colorectal cancer (CRC) (Ananthakrishnan et al., 2014). Vitamin D deficiency can be caused by inadequate exposure to sun or dietary vitamin D intake, impaired synthesis, increased catabolism, and genetic variants in vitamin D hydroxylases or transporter (Raftery & O'Sullivan, 2015). However, no associations between polymorphisms of the vitamin D metabolizing enzymes and IBD have been identified so far.

A prospective clinical trial found that raising serum 25(OH)D levels above 40 ng/ml in patients with CD improved disease activity index

(Yang et al., 2013). In a multi-institutional cohort study vitamin D levels $\geq$30 ng/ml reduced the number of surgical interventions in both CD and UC patients (Ananthakrishnan et al., 2013). 1,25$(OH)_2D_3$ was able to upregulate NOD2, a pattern recognition receptor promoting tolerance against commensal bacteria, which is the most common susceptibility gene contributing to CD (Dionne et al., 2014). The protective role of the locally synthesized 1,25$(OH)_2D_3$ is underlined by the observation that in IBD patients the intestinal expression of CYP27B1 is enhanced.

Upregulation of CYP27B1 in chronically inflamed tissue is considered to be a defense mechanism to enhance tissue concentration of 1,25$(OH)_2D_3$ to protect against aggravation of the inflammatory reaction. In a mouse model of colitis, Cyp27b1 was higher expressed in the proximal colon, while *Cyp27b1* null mice developed more severe colitis (Liu et al., 2008) while vitamin D deficiency in otherwise normal mice predisposed to colitis and impaired colonic antibacterial activity (Lagishetty et al., 2010). Overexpression of Cyp27b1 in monocytes improved survival of mice with dextran sodium sulfate (DSS)-induced colitis (Li et al., 2015). The role of vitamin D was intensively studied in several mouse models of IBD. Inflammation in an IL-10 knockout mouse model of DSS-induced colitis was ameliorated in the group fed a high vitamin D diet (Zhu, Mahon, Froicu, & Cantorna, 2005). Mice lacking both *vdr* and *IL-10* have a fulminant intestinal inflammation, and T-cell function is deregulated suggesting a causal relationship between IBD and vitamin D status (Froicu, Zhu, & Cantorna, 2006). It has been shown recently that the knockout of *vdr* exclusively in murine intestinal epithelial cells aggravated DSS-induced colonic inflammation and mucosal damage (Kim et al., 2013). 1,25$(OH)_2D_3$ treatment improved DSS-induced colitis by upregulating IL-10 expression (Froicu & Cantorna, 2007); however, no intestinal inflammation was observed in the animals not exposed to DSS. Tight junctions and desmosomes were severely disrupted in the mucosa of $vdr^{-/-}$ mice compared with wild-type animals, leading to significant lower intestinal transepithelial resistance (Kong et al., 2008). This indicates that the vitamin D system is very important for maintaining the integrity of the intestinal epithelial barrier.

These preclinical studies underline the importance of adequate vitamin D supplementation for IBD patients.

3.4 Hepatic Inflammatory Diseases

Vitamin D suppresses disease progression in patients with nonalcoholic fatty liver disease (NAFLD) and chronic liver disease caused by hepatitis C virus

(HCV), via anti-inflammatory and immune-modulatory effects (Kitson & Roberts, 2012). In a retrospective study of 607 NAFLD patients, 25(OH) D_3 serum levels were decreased compared with controls (Jablonski et al., 2013).

In a cohort study of HCV patients the *CYP27B1*-1260 promoter polymorphism rs10877012 correlated with reduced serum levels of 1,25 $(OH)_2D_3$ and with impaired response to treatment with pegylated interferon and ribavirin (Lange et al., 2011). The anti-inflammatory actions of vitamin D imply that vitamin D has the potential to reduce HCV-mediated liver disease. Therefore, several studies recommend vitamin D supplementation for chronic hepatitis C patients in combination with the conventional interferon and ribavirin therapy (Rahman & Branch, 2013).

3.5 Vitamin D in Autoimmune Diseases

Autoimmune diseases are driven by sustained inflammatory stimuli, caused by a misled reaction of the adaptive immune system against self-antigens. Most of the functions of vitamin D in the adaptive immune system are suppressive and support tolerance, by promoting induction of Tregs (Prietl et al., 2010). Vitamin D deficiency is associated with several autoimmune diseases such as multiple sclerosis, rheumatoid disorders, and type 1 diabetes (Agmon-Levin, Theodor, Segal, & Shoenfeld, 2013). In a large case–control study of type 1 diabetes patients, the common *CYP27B1* polymorphism, rs10877012, was associated with increased disease risk (Bailey et al., 2007).

Several *CYP27B1* polymorphisms are associated also with multiple sclerosis (Sundqvist et al., 2010). The *CYP24A1* SNP rs2248359-C, associated with CYP24A1 overexpression in the frontal cortex, was also identified as a risk allele (Ramasamy et al., 2014). A mechanistic link between vitamin D and multiple sclerosis is provided by a study in a murine model of multiple sclerosis, where vitamin D supplementation in the diet reduced severity of the disease. This effect was prevented in mice lacking STAT6, suggesting that vitamin D exerts its function via inhibition of STAT6 (Sloka et al., 2011).

4. VITAMIN D AND CANCER

The classical hallmarks of cancer include the capability of the tumor cells for sustained proliferative signaling, to evade growth suppressors, resist apoptosis, enable replicative immortality, inducing angiogenesis, and activating invasion and metastasis. Avoiding immune destruction, deregulating cellular energetics, genome instability and mutation, and tumor-promoting

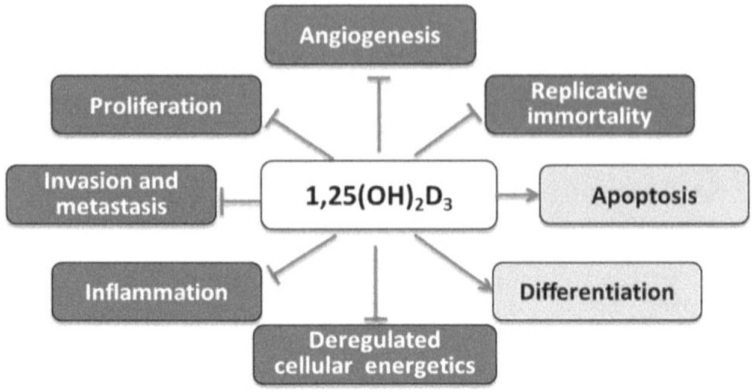

Figure 6 1,25$(OH)_2D_3$ and the hallmarks of cancer.

inflammation are further characteristics of cancer cells (Hanahan & Weinberg, 2011).

In vitro and *in vivo studies* demonstrate that 1,25$(OH)_2D_3$ is able to inhibit most of these hallmarks (Fig. 6). It inhibits proliferation, induces differentiation and activates apoptosis, alters cellular energy metabolism, inhibits angiogenesis of the tumor; in general, it is able to delay tumor formation and prevent spread of metastases (Deeb, Trump, & Johnson, 2007; Feldman et al., 2014; Narvaez et al., 2014; Zheng et al., 2013). The mechanisms behind these effects are complex and tissue dependent. Human clinical data, however, are much less conclusive in showing cancer preventive effects of vitamin D (Chung, Lee, Terasawa, Lau, & Trikalinos, 2011; Hoffmann, Senior, & Mager, 2015; Manson, Mayne, & Clinton, 2011).

High-serum 25$(OH)D_3$ levels were associated with lower incidence of colorectal, breast, and prostate cancer (Feskanich et al., 2004; Garland, Gorham, Mohr, & Garland, 2009; Grant et al., 2009; Leyssens, Verlinden, & Verstuyf, 2013; Song et al., 2015). The most convincing associations were seen in CRC, where several large-scale epidemiological studies suggested that high vitamin D intake decreases cancer risk (Garland et al., 2009; Yin et al., 2009). *In vivo* rodent models show convincingly that vitamin D is protective against this type of cancer (Achkar & Fiocchi, 2009; Hummel et al., 2013; Scaldaferri et al., 2009), while *in vitro* studies unveil the molecular mechanisms behind the growth inhibitory antimitogenic, antiangiogenic, prodifferentiating action of 1,25 $(OH)_2D_3$ (e.g., Pereira, Larriba, & Munoz, 2012; Welsh, 2012).

Besides vitamin D_3 metabolites, vitamin D_2 derivatives and metabolites of the CYP11A1 pathway are also effective in inhibiting proliferation and inducing differentiation of several cell types, such as keratinocytes or melanoma cells. The antiproliferative effect of 20(OH)D_2, a noncalcemic analog is similar to that of 1,25$(OH)_2D_3$ but significantly higher than the effect of 25(OH)D_3. 20(OH)D_2 also displays antitumorigenic effects. In keratinocytes, 20(OH)D_2 inhibits expression of cyclins and stimulates involucrin expression (Slominski et al., 2011).

In addition to the pathways common to most neoplasias, 1,25$(OH)_2D_3$ inhibits pathways specific to distinct tumor types. In sporadic CRC, 1,25 $(OH)_2D_3$ is a potent inhibitor of the Wnt pathway, a major driver of these tumors (Cross & Kallay, 2009; Groschel et al., 2015; Larriba et al., 2011; Pereira et al., 2012). It also regulates the PGE_2 pathway, another pathway deregulated also in inflammation-driven colorectal tumors, by reducing COX-2 expression and increasing 15-PGDH levels (Brozek, Manhardt, Kallay, Peterlik, & Cross, 2012).

In breast cancer cells 1,25$(OH)_2D_3$ prevents carcinogenesis by targeting energy utilization pathways and causing changes in the respiratory capacity of the cells (Narvaez et al., 2014; Zheng et al., 2013). It is also able to inhibit estrogen signaling and synthesis by suppressing aromatase expression (Swami et al., 2011).

4.1 Vitamin D Hydroxylases in Cancer

Evidence is increasing that high tissue levels of 1,25$(OH)_2D_3$ synthesized in the organs prone to sporadic cancers can counteract incipient neoplasia. This underlines the importance of the local, tissue-specific vitamin D system in tumorigenesis. In several malignancies, the vitamin D system is deregulated, i.e., either expression, or function of the VDR, or that of the vitamin D hydroxylases is impaired in a tissue- and neoplasia-specific manner.

Several genetic polymorphisms of the vitamin D pathway correlate with risk, incidence, or outcome of different cancers, although the impact of the SNPs is tissue- and cancer-specific. These polymorphisms affect either the rate of synthesis or degradation of the vitamin D metabolites, thus regulate the availability of the active hormone.

The *CYP2R1* (rs1993116, rs10741657) genotypes were associated with increased risk of HCV-related hepatocellular carcinoma ($OR = 1.56$) (Lange et al., 2013). A recent case–control study of nonsmall cell lung cancer (NSCLC) patients found that the rs3782130 SNP in *CYP27B1* and three SNPs in *CYP24A1* (rs7041, rs6068816, and rs4809957) were associated with higher NSCLC risk. The *CYP27B1* polymorphism (rs3782130) affected gene expression and patient survival (Kong et al., 2014). In a population-based case–control study of incident breast cancer, the *CYP24A1* polymorphism rs6068816-TT was also associated with a 72% reduction in breast cancer risk (Reimers et al., 2015), comparable with its effect in NSCLC patients. In prostate cancer patients, low $25(OH)D_3$ metabolism score was associated with high-grade tumors (OR = 0.76). Similar associations were found for the rs10877012-T allele in *CYP27B1* and the rs6013897-A allele in *CYP24A1* (Gilbert et al., 2015).

Certain types of cancers become resistant to the antitumorigenic actions of vitamin D (Deeb et al., 2007). One major mechanism for vitamin D resistance or reduced sensitivity to $1,25(OH)_2D_3$ in VDR-positive cells might be the loss of the equilibrium between synthesis and degradation of $1,25(OH)_2D_3$ or its precursor $25(OH)D_3$.

4.1.1 Vitamin D_3 25 Hydroxylases (CYP2R1, CYP27A1) in Cancer

Since CYP2R1 was identified as the main 25 hydroxylase only recently, there are only few data reporting the expression level of this enzyme in different tumors. In general, data on expression and activity of 25-hydroxylases during carcinogenesis are limited, due to the compensatory activity of the different hydroxylases.

A recent study showed that CYP2R1 (and CYP27B1) expression was increased in endometrial cancer, as an autocrine/paracrine defense mechanism (Bergada et al., 2014). Similarly, CYP2R1 expression was higher in oral squamous cell carcinomas (OSCCs), compared with the adjacent normal mucosa. OSCC cell lines expressed higher levels of both CYP2R1 and VDR, and $1,25(OH)_2D_3$ further increased their expression (Sundaram, Sambandam, Tsuruga, Wagner, & Reddy, 2014). In renal cell carcinoma, all three vitamin D hydroxylases CYP2R1, CYP27B1, and CYP24A1 were higher expressed in the tumor than in the healthy kidney. However, CYP24A1 levels were significantly higher than those of CYP2R1 and CYP27B1, while in the normal tissue the expression levels were comparable (Urbschat et al., 2013).

In a patient with hepatic carcinoma, the tumor overexpressed CYP27A1 more than 1000-fold compared with the healthy liver tissue surrounding the tumor. In contrast, CYP27A1 levels in intrahepatic colon metastases in two other patients were similar to the surrounding normal areas (Gascon-Barre et al., 2001).

Although hepatic 25-hydroxylases might be important in the activation of vitamin D metabolites and maintenance of serum $25(OH)D_3$ levels, there is no strong evidence suggesting their direct, local involvement in tumor development (Hobaus, Thiem, et al., 2013; Zhu & DeLuca, 2012).

4.1.2 25-Hydroxyvitamin D_3 1-Alpha Hydroxylase (CYP27B1) in Cancer

The renal CYP27B1 controls systemic $1,25(OH)_2D_3$ levels together with CYP24A1. However, both enzymes are almost ubiquitous and are expressed also at various extra-renal sites such as colon, pancreas, brain, skin, lymph nodes, adrenal medulla, placenta, macrophages, and endothelial cells (Zehnder et al., 2001, 2002). CYP27B1 is expressed also in the tumor microenvironment, in tumor-infiltrating macrophages, and in endothelial cells. The availability of local $1,25(OH)_2D_3$ depends on the balance between extra-renal $1,25(OH)_2D_3$ synthesis and catabolism and the tissue-specific level of CYP27B1 and CYP24A1 (Hobaus, Thiem, et al., 2013).

Clinical data on CYP27B1 expression in cancer are inconsistent. During cancer development, expression of CYP27B1 is affected in several tumors, while it remains stable in others, depending on the tissue and the grade of the tumor (Avila et al., 2007; Brozyna, Jozwicki, Janjetovic, & Slominski, 2013; Diesel et al., 2005; Hobaus, Thiem, et al., 2013; Kallay et al., 2002; Lopes et al., 2010; Townsend et al., 2005). In a recent paper, Krishnan et al. reported that feeding mice with 5000 IU vitamin D increased CYP27B1 expression of MCF-7 xenografts and reduced tumor growth. Similar growth inhibition was observed when the mice received $1,25(OH)_2D_3$ intraperitoneally, however, without any effect on CYP27B1 expression in the tumor (Krishnan, Swami, & Feldman, 2013). This experiment demonstrates the importance of the local CYP27B1 expression and activity for restricting tumor growth.

Loss of CYP27B1 expression can be caused by promoter hypermethylation, as it was observed in 41% of breast tumors (Shi et al., 2002). Hypermethylation is one of the causes for loss of CYP27B1 expression also in prostate tumors (Khorchide, Lechner, & Cross, 2005). CYP27B1 expression can be inhibited also by the growth factor independent 1, which represses CYP27B1 transcription by binding to a repressive region in the promoter and recruiting co-repressors with histone deacetylase activity (Dwivedi et al., 2005). A recent study, comparing prostate epithelial and stromal cells from prostate cancer patients, found higher levels of CYP27B1, CYP2R1, and CYP24A1 both in the benign and cancerous epithelia compared with the surrounding stroma (Giangreco et al., 2015).

In ovarian cancer, the lowest level of CYP27B1 protein expression was observed in poorly differentiated primary tumors and metastases. The expression of CYP27B1 correlated negatively with tumor growth (Brozyna, Jozwicki, Jochymski, & Slominski, 2015).

In the skin, CYP27B1 expression was increased in squamous cell carcinomas compared with the normal tissue, and in these cells 1,25$(OH)_2D_3$ inhibited UV-B-induced apoptosis (Reichrath & Reichrath, 2013). CYP27B1 expression was decreased during melanoma progression, and lack of CYP27B1 expression correlated with shorter disease-free survival (Brozyna et al., 2013).

In some cancers, CYP27B1 expression increases in the early phases of tumorigenesis, but then it decreases in advanced, undifferentiated tumors. In well-differentiated colon carcinomas and parathyroid adenomas CYP27B1 levels are high, while in undifferentiated adenocarcinomas the expression is lost (Bises et al., 2004; Kallay et al., 2002; Segersten et al., 2002).

The causes of the upregulation of CYP27B1 expression in early phases of tumorigenesis are not fully understood but it could be a physiological feedback loop coupled to the increased proliferative activity in these tumors, suggesting an autocrine/paracrine antitumor role for 1,25$(OH)_2D_3$ in the early phase. Later, during progression, the tumor seems to develop strategies to overcome the actions of 1,25$(OH)_2D_3$, e.g., by upregulating the expression and/or activity of NFκB, which is able to downregulate the activity of the *CYP27B1* promoter (Ebert et al., 2004; Zordoky & El-Kadi, 2009). A similar strategy could be observed in regulation of CYP27B1 by EGF in CRC cells. In differentiated colon cancer cells, EGF stimulated CYP27B1 expression; however, in cells derived from less differentiated tumors, CYP27B1 expression was inhibited by EGF (Bareis et al., 2002), suggesting that in tumors overexpressing EGF receptor the antitumor role of 1,25 $(OH)_2D_3$ might be impaired. TNF-α and IL-6 downregulated CYP27B1 mRNA expression only in undifferentiated colon cancer cells (Hummel, Fetahu, Gröschel, Manhardt, & Kállay, 2014).

Benign and differentiated malignant thyroid tumors expressed high levels of not only CYP27B1 but also VDR and CYP24A1, however, the expression decreased in metastases (Clinckspoor et al., 2012). In 15–25% of glioblastoma patients, the tumors carried *CYP27B1* gene amplification, which was associated with reduced survival (Fischer et al., 2010). Glioblastoma cell lines expressed several CYP27B1 splice variants. Most of these were nonfunctional (Diesel et al., 2005).

4.1.3 1,25 Hydroxyvitamin D_3 24 Hydroxylase

Expression of the vitamin D catabolizing enzyme CYP24A1 is usually low in extra-renal tissues (Akeno, Saikatsu, Kawane, & Horiuchi, 1997).

CYP24A1 expression is highly inducible by 1,25$(OH)_2D_3$ in a VDR–RXR-dependent manner (Meyer, Zella, Nerenz, & Pike, 2007). In normal tissue, after deprivation of the stimulus 1,25$(OH)_2D_3$, the expression of CYP24A1 rapidly decreases and returns to basal levels. However, an aberrantly high basal expression of CYP24A1 was observed in several tumor types including colon, breast, lung, ovarian, cervical tumors (Anderson, Nakane, Ruan, Kroeger, & Wu-Wong, 2006; Cross & Kallay, 2005; Friedrich et al., 2003; Kallay et al., 2005; Luo, Hershberger, Trump, & Johnson, 2013; Parise et al., 2006). In a colon cancer patient cohort, 69% of the adenocarcinomas overexpressed CYP24A1 protein; while in 60% of the normal and benign tissue, it was barely detectable (Horvath, Lakatos, et al., 2010). High CYP24A1 expression was associated with upregulated expression of key replication licensing factors (Hobaus, Hummel, et al., 2013), suggesting that high degradation rate of 1,25 $(OH)_2D_3$ confers growth advantage to the tumor.

A comparative gene hybridization study of breast cancer patients identified *CYP24A1* as a candidate oncogene (Albertson et al., 2000). In 10–13% of human breast tumors, CYP24A1 expression was increased mainly due to gene amplification (Lopes et al., 2010; Narvaez et al., 2014; Townsend et al., 2005). In esophageal cancer, high CYP24A1 expression correlated with poor clinical outcome (Mimori et al., 2004). While a recent study in melanomas showed inverse correlation between CYP24A1 expression and tumor progression (Brozyna et al., 2014).

Although CYP24A1 is a major target of liganded VDR, very often VDR levels in the tumor do not correlate with CYP24A1, suggesting that high CYP24A1 expression is independent of VDR-mediated transcription (Hobaus, Hummel, et al., 2013; Hummel, Aggarwal, et al., 2014). In a pilot study, we found that during pancreatic ductal adenocarcinoma development, CYP24A1 expression decreases in the islets, weakening the negative feedback regulation of the vitamin D-dependent insulin synthesis/secretion, while in the transformed ducts CYP24A1 expression increases, impairing the antiproliferative effect of vitamin D in these cells (Hummel, Aggarwal, et al., 2014).

Epigenetic alterations, copy number changes, K-RAS mutations and miRNA involvement were suggested as possible causes of CYP24A1 overexpression. In lung adenocarcinomas, CYP24A1 expression is regulated by DNA methylation and by repressive histone modifications (Ramnath et al., 2014). In papillary thyroid carcinoma, CYP24A1 overexpression strongly correlated with the V600E BRAF mutation (Zou et al., 2014). In CRC CYP24A1 overexpression is caused in part by gene amplification and chromosome 20 polysomy (Hobaus, Hummel, et al., 2013), while K-RAS mutations seem to have no impact.

Alternative splicing might be an additional mechanism regulating CYP24A1 activity, or expression also in cancer. CYP24A1 splice variants were identified in colorectal (Horvath, Khabir, et al., 2010; Muindi et al., 2007) and breast cancer cells (Scheible, Thill, Baum, Solomayer, & Friedrich, 2014). Although most variants identified until now seem to be catalytically inactive, some variants retain the substrate-binding domain and thus could act in a dominant negative manner, by competing with CYP27B1 for binding 25(OH)D_3, thus impairing 1,25$(OH)_2D_3$ synthesis (Ren et al., 2005). A SNP located within the VDRE2 in the promoter region of *CYP24A1* was identified, which leads to reduced gene expression; however, there are currently no reports on SNPs associated with increased gene expression (Roff & Wilson, 2008). Better understanding of the mechanisms behind CYP24A1 upregulation, as well as the consequences for the tumor, is necessary in order to develop strategies to restore tissue responsiveness to vitamin D or specifically counteract CYP24A1 activity. Several CYP24A1 inhibitors have been developed in recent years; however, due to the high structural similarities between P450 enzymes, the search for CYP24A1-specific inhibitors continues. Understanding the advantages of CYP24A1 overexpression for the tumors may provide targeting strategies in the future.

5. CONCLUSION

In the last decade, research on the role of vitamin D in human health has intensified. It is considered by numerous groups that vitamin D plays a central regulatory role in human health, protecting against major diseases, including chronic inflammatory diseases and cancer.

However, while preclinical *in vitro studies* and animal models convincingly support the anti-inflammatory and anti-cancer role of vitamin D, large epidemiological and interventional studies are inconclusive, reporting both positive and negative effects. As CYP enzymes play the predominant role in synthesis and degradation of the vitamin D metabolites, reduced levels of the activating enzymes and/or high expression of the degrading enzyme would limit the effectiveness of vitamin D intake or sunlight exposure. Normal expression and activity of the vitamin D metabolizing enzymes are central both to normal immune response and maintenance of tissue homeostasis. Further studies are needed to understand clearly the role of local calcitriol synthesis and of the autocrine/paracrine effect of the locally synthesized 1,25 $(OH)_2D_3$ in mediating the anti-inflammatory and anti-tumorigenic effects of vitamin D.

In addition to the role of $1,25(OH)_2D_3$, it appears possible that there are other active vitamin D metabolites, with anti-inflammatory and cancer preventive properties. Due to their low calcemic effect, the CYP11A1-derived secosteroids could become candidates for treatment of autoimmune diseases such as rheumatoid arthritis, lupus erythomatousus, or IBD. Moreover, they could also be tested as therapeutics or adjuvants for different cancers.

Based on the current information, the authors' opinion is that the main role of vitamin D is maintenance of health and prevention of chronic diseases.

CONFLICT OF INTERESTS

The authors declare that they have no conflict of interest to declare.

ACKNOWLEDGMENTS

This work was supported by Vienna Science and Technology Fund (WWTF) Grant LS12-047 and the European Commission funded FP7 ITN Multifaceted CaSR (#264663). The authors would like to thank Julia Höbaus and Abhishek Aggarwal for their contribution.

REFERENCES

Abraham, C., & Cho, J. H. (2009). Inflammatory bowel disease. *New England Journal of Medicine, 361*(21), 2066–2078. http://dx.doi.org/10.1056/NEJMra0804647.

Achkar, J. P., & Fiocchi, C. (2009). Gene-gene interactions in inflammatory bowel disease: Biological and clinical implications. *American Journal of Gastroenterology, 104*(7), 1734–1736. http://dx.doi.org/10.1038/ajg.2009.179, ajg2009179 [pii].

Adams, J. S., & Hewison, M. (2008). Unexpected actions of vitamin D: New perspectives on the regulation of innate and adaptive immunity. *Nature Clinical Practice. Endocrinology & Metabolism, 4*(2), 80–90. http://dx.doi.org/10.1038/ncpendmet0716.

Adams, J. S., & Hewison, M. (2012). Extrarenal expression of the 25-hydroxyvitamin D-1-hydroxylase. *Archives of Biochemistry and Biophysics, 523*(1), 95–102. http://dx.doi.org/10.1016/j.abb.2012.02.016.

Adams, J. S., Ren, S., Liu, P. T., Chun, R. F., Lagishetty, V., Gombart, A. F., et al. (2009). Vitamin D-directed rheostatic regulation of monocyte antibacterial responses. *Journal of Immunology, 182*(7), 4289–4295. http://dx.doi.org/10.4049/jimmunol.0803736, 182/7/4289 [pii].

Agard, M., Asakrah, S., & Morici, L. A. (2013). PGE(2) suppression of innate immunity during mucosal bacterial infection. *Frontiers in Cellular and Infection Microbiology, 3*, 45. http://dx.doi.org/10.3389/fcimb.2013.00045.

Aggarwal, B. B. (2004). Nuclear factor-kappaB: The enemy within. *Cancer Cell, 6*(3), 203–208. http://dx.doi.org/10.1016/j.ccr.2004.09.003, (Research Support, Non-U.S.

Gov't Research Support, U.S. Gov't, Non-P.H.S. Research Support, U.S. Gov't, P.H.S. Review).
Agmon-Levin, N., Theodor, E., Segal, R. M., & Shoenfeld, Y. (2013). Vitamin D in systemic and organ-specific autoimmune diseases. *Clinical Reviews in Allergy & Immunology*, *45*(2), 256–266. http://dx.doi.org/10.1007/s12016-012-8342-y.
Ahn, J., Yu, K., Stolzenberg-Solomon, R., Simon, K. C., McCullough, M. L., Gallicchio, L., et al. (2010). Genome-wide association study of circulating vitamin D levels. *Human Molecular Genetics*, *19*(13), 2739–2745. http://dx.doi.org/10.1093/hmg/ddq155.
Akeno, N., Saikatsu, S., Kawane, T., & Horiuchi, N. (1997). Mouse vitamin D-24-hydroxylase: Molecular cloning, tissue distribution, and transcriptional regulation by 1alpha,25-dihydroxyvitamin D3. *Endocrinology*, *138*(6), 2233–2240. http://dx.doi.org/10.1210/endo.138.6.5170, (Research Support, Non-U.S. Gov't).
Albertson, D. G., Ylstra, B., Segraves, R., Collins, C., Dairkee, S. H., Kowbel, D., et al. (2000). Quantitative mapping of amplicon structure by array CGH identifies CYP24 as a candidate oncogene. *Nature Genetics*, *25*(2), 144–146. http://dx.doi.org/10.1038/75985.
Ananthakrishnan, A. N., Cagan, A., Gainer, V. S., Cai, T., Cheng, S. C., Savova, G., et al. (2013). Normalization of plasma 25-hydroxy vitamin D is associated with reduced risk of surgery in Crohn's disease. *Inflammatory Bowel Diseases*, *19*(9), 1921–1927. http://dx.doi.org/10.1097/MIB.0b013e3182902ad9.
Ananthakrishnan, A. N., Cheng, S. C., Cai, T., Cagan, A., Gainer, V. S., Szolovits, P., et al. (2014). Association between reduced plasma 25-hydroxy vitamin D and increased risk of cancer in patients with inflammatory bowel diseases. *Clinical Gastroenterology and Hepatology*, *12*(5), 821–827. http://dx.doi.org/10.1016/j.cgh.2013.10.011.
Ananthakrishnan, A. N., Khalili, H., Higuchi, L. M., Bao, Y., Korzenik, J. R., Giovannucci, E. L., et al. (2012). Higher predicted vitamin D status is associated with reduced risk of Crohn's disease. *Gastroenterology*, *142*(3), 482–489. http://dx.doi.org/10.1053/j.gastro.2011.11.040.
Anderson, M. G., Nakane, M., Ruan, X., Kroeger, P. E., & Wu-Wong, J. R. (2006). Expression of VDR and CYP24A1 mRNA in human tumors. *Cancer Chemotherapy and Pharmacology*, *57*(2), 234–240. http://dx.doi.org/10.1007/s00280-005-0059-7.
Annalora, A. J., Goodin, D. B., Hong, W. X., Zhang, Q., Johnson, E. F., & Stout, C. D. (2010). Crystal structure of CYP24A1, a mitochondrial cytochrome P450 involved in vitamin D metabolism. *Journal of Molecular Biology*, *396*(2), 441–451. http://dx.doi.org/10.1016/j.jmb.2009.11.057, (Research Support, N.I.H., Extramural Research Support, U.S. Gov't, Non-P.H.S.).
Armbrecht, H. J., Boltz, M. A., & Hodam, T. L. (2003). PTH increases renal 25(OH)D3-1alpha -hydroxylase (CYP1alpha) mRNA but not renal 1,25(OH)2D3 production in adult rats. *American Journal of Physiology. Renal Physiology*, *284*(5), F1032–F1036. http://dx.doi.org/10.1152/ajprenal.00306.2002, (Research Support, Non-U.S. Gov't Research Support, U.S. Gov't, Non-P.H.S.).
Avila, E., Diaz, L., Barrera, D., Arranz, C., Halhali, A., & Larrea, F. (2007). Metabolism of vitamin D in the human choriocarcinoma cell line JEG-3. *Journal of Steroid Biochemistry and Molecular Biology*, *103*(3–5), 781–785. http://dx.doi.org/10.1016/j.jsbmb.2006.12.047, S0960-0760(06)00425-0 [pii].
Baeke, F., Takiishi, T., Korf, H., Gysemans, C., & Mathieu, C. (2010). Vitamin D: Modulator of the immune system. *Current Opinion in Pharmacology*, *10*(4), 482–496. http://dx.doi.org/10.1016/j.coph.2010.04.001.
Bailey, R., Cooper, J. D., Zeitels, L., Smyth, D. J., Yang, J. H., Walker, N. M., et al. (2007). Association of the vitamin D metabolism gene CYP27B1 with type 1 diabetes. *Diabetes*, *56*(10), 2616–2621. http://dx.doi.org/10.2337/db07-0652.

Bain, C. C., & Mowat, A. M. (2014). The monocyte-macrophage axis in the intestine. *Cellular Immunology*, *291*(1–2), 41–48. http://dx.doi.org/10.1016/j.cellimm.2014.03.012.

Bareis, P., Kallay, E., Bischof, M. G., Bises, G., Hofer, H., Potzi, C., et al. (2002). Clonal differences in expression of 25-hydroxyvitamin D(3)-1alpha-hydroxylase, of 25-hydroxyvitamin D(3)-24-hydroxylase, and of the vitamin D receptor in human colon carcinoma cells: Effects of epidermal growth factor and 1alpha,25-dihydroxyvitamin D (3). *Experimental Cell Research*, *276*(2), 320–327. http://dx.doi.org/10.1006/excr.2002.5528, S001448270295528X [pii].

Barry, E. L., Rees, J. R., Peacock, J. L., Mott, L. A., Amos, C. I., Bostick, R. M., et al. (2014). Genetic variants in CYP2R1, CYP24A1, and VDR modify the efficacy of vitamin D3 supplementation for increasing serum 25-hydroxyvitamin D levels in a randomized controlled trial. *The Journal of Clinical Endocrinology and Metabolism*, *99*(10), E2133–E2137. http://dx.doi.org/10.1210/jc.2014-1389, (Research Support, N.I.H., Extramural).

Beckman, M. J., Tadikonda, P., Werner, E., Prahl, J., Yamada, S., & DeLuca, H. F. (1996). Human 25-hydroxyvitamin D3-24-hydroxylase, a multicatalytic enzyme. *Biochemistry*, *35*(25), 8465–8472. http://dx.doi.org/10.1021/bi960658ibi960658i, (pii).

Ben-Neriah, Y., & Karin, M. (2011). Inflammation meets cancer, with NF-κB as the matchmaker. *Nature Immunology*, *12*(8), 715–723. http://dx.doi.org/10.1038/ni.2060.

Bergada, L., Pallares, J., Maria Vittoria, A., Cardus, A., Santacana, M., Valls, J., et al. (2014). Role of local bioactivation of vitamin D by CYP27A1 and CYP2R1 in the control of cell growth in normal endometrium and endometrial carcinoma. *Laboratory Investigation: A Journal of Technical Methods and Pathology*, *94*(6), 608–622. http://dx.doi.org/10.1038/labinvest.2014.57, (Research Support, Non-U.S. Gov't).

Bikle, D. (2009). Nonclassic actions of vitamin D. *The Journal of Clinical Endocrinology and Metabolism*, *94*(1), 26–34. http://dx.doi.org/10.1210/jc.2008-1454, jc.2008-1454 [pii].

Bikle, D. D. (2014). Vitamin D metabolism, mechanism of action, and clinical applications. *Chemistry & Biology*, *21*(3), 319–329. http://dx.doi.org/10.1016/j.chembiol.2013.12.016, (Research Support, N.I.H., Extramural Research Support, U.S. Gov't, Non-P.H.S.Review).

Bischof, M. G., Siu-Caldera, M. L., Weiskopf, A., Vouros, P., Cross, H. S., Peterlik, M., et al. (1998). Differentiation-related pathways of 1 alpha,25-dihydroxycholecalciferol metabolism in human colon adenocarcinoma-derived Caco-2 cells: Production of 1 alpha,25-dihydroxy-3epi-cholecalciferol. *Experimental Cell Research*, *241*(1), 194–201. http://dx.doi.org/10.1006/excr.1998.4044, S0014-4827(98)94044-7 [pii].

Bises, G., Kallay, E., Weiland, T., Wrba, F., Wenzl, E., Bonner, E., et al. (2004). 25-hydroxyvitamin D3-1alpha-hydroxylase expression in normal and malignant human colon. *Journal of Histochemistry and Cytochemistry*, *52*(7), 985–989. http://dx.doi.org/10.1369/jhc.4B6271.200452/7/985, (pii).

Blunt, J. W., DeLuca, H. F., & Schnoes, H. K. (1968). 25-hydroxycholecalciferol. A biologically active metabolite of vitamin D3. *Biochemistry*, 7(10), 3317–3322.

Boonstra, A., Barrat, F. J., Crain, C., Heath, V. L., Savelkoul, H. F., & O'Garra, A. (2001). 1alpha,25-Dihydroxyvitamin d3 has a direct effect on naive CD4(+) T cells to enhance the development of Th2 cells. *Journal of Immunology*, *167*(9), 4974–4980.

Bouillon, R., Carmeliet, G., Verlinden, L., van Etten, E., Verstuyf, A., Luderer, H. F., et al. (2008). Vitamin D and human health: Lessons from vitamin D receptor null mice. *Endocrine Reviews*, *29*(6), 726–776. http://dx.doi.org/10.1210/er.2008-0004, (Research Support, N.I.H., Extramural Research Support, Non-U.S. Gov't Review).

Brondum-Jacobsen, P., Benn, M., Jensen, G. B., & Nordestgaard, B. G. (2012). 25-hydroxyvitamin d levels and risk of ischemic heart disease, myocardial infarction, and early death: Population-based study and meta-analyses of 18 and 17 studies. *Arteriosclerosis,*

Thrombosis, and Vascular Biology, *32*(11), 2794–2802. http://dx.doi.org/10.1161/ATVBAHA.112.248039, (Meta-Analysis Research Support, Non-U.S. Gov't).
Brozek, W., Manhardt, T., Kallay, E., Peterlik, M., & Cross, H. S. (2012). Relative expression of vitamin D hydroxylases, CYP27B1 and CYP24A1, and of cyclooxygenase-2 and heterogeneity of human colorectal cancer in relation to age, gender, tumor location, and malignancy: Results from factor and cluster analysis. *Cancers (Basel)*, *4*(3), 763–776. http://dx.doi.org/10.3390/cancers4030763, cancers4030763 [pii].
Brozyna, A. A., Jochymski, C., Janjetovic, Z., Jozwicki, W., Tuckey, R. C., & Slominski, A. T. (2014). CYP24A1 expression inversely correlates with melanoma progression: Clinic-pathological studies. *International Journal of Molecular Sciences*, *15*(10), 19000–19017. http://dx.doi.org/10.3390/ijms151019000, (Research Support, N.I.H., Extramural Research Support, Non-U.S. Gov't).
Brozyna, A. A., Jozwicki, W., Janjetovic, Z., & Slominski, A. T. (2013). Expression of the vitamin D-activating enzyme 1alpha-hydroxylase (CYP27B1) decreases during melanoma progression. *Human Pathology*, *44*(3), 374–387. http://dx.doi.org/10.1016/j.humpath.2012.03.031, S0046-8177(12)00205-5 [pii].
Brozyna, A. A., Jozwicki, W., Jochymski, C., & Slominski, A. T. (2015). Decreased expression of CYP27B1 correlates with the increased aggressiveness of ovarian carcinomas. *Oncology Reports*, *33*(2), 599–606. http://dx.doi.org/10.3892/or.2014.3666.
Bu, F. X., Armas, L., Lappe, J., Zhou, Y., Gao, G., Wang, H. W., et al. (2010). Comprehensive association analysis of nine candidate genes with serum 25-hydroxy vitamin D levels among healthy Caucasian subjects. *Human Genetics*, *128*(5), 549–556. http://dx.doi.org/10.1007/s00439-010-0881-9.
Burgos-Trinidad, M., & DeLuca, H. F. (1991). Kinetic properties of 25-hydroxyvitamin D- and 1,25-dihydroxyvitamin D-24-hydroxylase from chick kidney. *Biochimica et Biophysica Acta*, *1078*(2), 226–230.
Campbell, M. J. (2014). Vitamin D and the RNA transcriptome: More than mRNA regulation. *Frontiers in Physiology*, *5*, 181. http://dx.doi.org/10.3389/fphys.2014.00181, (Review).
Cantorna, M. T., & Waddell, A. (2014). The vitamin D receptor turns off chronically activated T cells. *Annals of the New York Academy of Sciences*, *1317*, 70–75. http://dx.doi.org/10.1111/nyas.12408.
Cario, E. (2010). Toll-like receptors in inflammatory bowel diseases: A decade later. *Inflammatory Bowel Diseases*, *16*(9), 1583–1597. http://dx.doi.org/10.1002/ibd.21282.
Carlberg, C. (2014). Genome-wide (over)view on the actions of vitamin D. *Frontiers in Physiology*, *5*, 167. http://dx.doi.org/10.3389/fphys.2014.00167, (Review).
Carlberg, C., & Molnar, F. (2012). Current status of vitamin D signaling and its therapeutic applications. *Current Topics in Medicinal Chemistry*, *12*(6), 528–547, CTMC-EPUB-20120113-004 [pii].
Chen, K. S., & DeLuca, H. F. (1995). Cloning of the human 1 alpha,25-dihydroxyvitamin D-3 24-hydroxylase gene promoter and identification of two vitamin D-responsive elements. *Biochimica et Biophysica Acta*, *1263*(1), 1–9, (Research Support, Non-U.S. Gov'tResearch Support, U.S. Gov't, P.H.S.).
Chen, Y., Liu, W., Sun, T., Huang, Y., Wang, Y., Deb, D. K., et al. (2013). 1,25-Dihydroxyvitamin D promotes negative feedback regulation of TLR signaling via targeting microRNA-155-SOCS1 in macrophages. *Journal of Immunology*, *190*(7), 3687–3695. http://dx.doi.org/10.4049/jimmunol.1203273.
Chen, Y., Zhang, J., Ge, X., Du, J., Deb, D. K., & Li, Y. C. (2013). Vitamin D receptor inhibits nuclear factor κB activation by interacting with IκB kinase β protein. *Journal of Biological Chemistry*, *288*(27), 19450–19458. http://dx.doi.org/10.1074/jbc.M113.467670.
Cheng, J. B., Levine, M. A., Bell, N. H., Mangelsdorf, D. J., & Russell, D. W. (2004). Genetic evidence that the human CYP2R1 enzyme is a key vitamin

D 25-hydroxylase. *Proceedings of the National Academy of Sciences of the United States of America, 101*(20), 7711–7715. http://dx.doi.org/10.1073/pnas.0402490101.

Chung, M., Lee, J., Terasawa, T., Lau, J., & Trikalinos, T. A. (2011). Vitamin D with or without calcium supplementation for prevention of cancer and fractures: An updated meta-analysis for the U.S. Preventive Services Task Force. *Annals of Internal Medicine, 155*(12), 827–838. http://dx.doi.org/10.7326/0003-4819-155-12-201112200-00005, (Meta-Analysis Research Support, U.S. Gov't, P.H.S. Review).

Clinckspoor, I., Hauben, E., Verlinden, L., Van den Bruel, A., Vanwalleghem, L., Vander Poorten, V., et al. (2012). Altered expression of key players in vitamin D metabolism and signaling in malignant and benign thyroid tumors. *The Journal of Histochemistry and Cytochemistry: Official Journal of the Histochemistry Society, 60*(7), 502–511. http://dx.doi.org/10.1369/0022155412447296, (Research Support, Non-U.S. Gov't).

Cohen-Lahav, M., Shany, S., Tobvin, D., Chaimovitz, C., & Douvdevani, A. (2006). Vitamin D decreases NFkappaB activity by increasing IkappaBalpha levels. *Nephrology, Dialysis, Transplantation: Official Publication of the European Dialysis and Transplant Association—European Renal Association, 21*(4), 889–897. http://dx.doi.org/10.1093/ndt/gfi254, (Research Support, Non-U.S. Gov't).

Crome, S. Q., Wang, A. Y., & Levings, M. K. (2010). Translational mini-review series on Th17 cells: Function and regulation of human T helper 17 cells in health and disease. *Clinical and Experimental Immunology, 159*(2), 109–119. http://dx.doi.org/10.1111/j.1365-2249.2009.04037.x.

Cross, H. S., & Kallay, E. (2005). Nutritional regulation of extrarenal vitamin D hydroxylase expression—Potential application in tumor prevention and therapy. *Future Oncology, 1*(3), 415–424. http://dx.doi.org/10.1517/14796694.1.3.415.

Cross, H. S., & Kallay, E. (2009). Regulation of the colonic vitamin D system for prevention of tumor progression: An update. *Future Oncology, 5*(4), 493–507. http://dx.doi.org/10.2217/fon.09.22.

Cross, H. S., Nittke, T., & Kallay, E. (2011). Colonic vitamin D metabolism: Implications for the pathogenesis of inflammatory bowel disease and colorectal cancer. *Molecular and Cellular Endocrinology, 347*(1–2), 70–79. http://dx.doi.org/10.1016/j.mce.2011.07.022, S0303-7207(11)00402-3 [pii].

Danielson, P. B. (2002). The cytochrome P450 superfamily: Biochemistry, evolution and drug metabolism in humans. *Current Drug Metabolism, 3*(6), 561–597.

Deeb, K. K., Trump, D. L., & Johnson, C. S. (2007). Vitamin D signalling pathways in cancer: Potential for anticancer therapeutics. *Nature Reviews. Cancer*, 7(9), 684–700. http://dx.doi.org/10.1038/nrc2196, nrc2196 [pii].

DeLuca, H. F. (1997). Discovery of the vitamin D endocrine system. *Journal of Nutrition, 127*(5 Suppl.), 1050S–1052S.

Dickie, L. J., Church, L. D., Coulthard, L. R., Mathews, R. J., Emery, P., & McDermott, M. F. (2010). Vitamin D3 down-regulates intracellular Toll-like receptor 9 expression and Toll-like receptor 9-induced IL-6 production in human monocytes. *Rheumatology (Oxford), 49*(8), 1466–1471. http://dx.doi.org/10.1093/rheumatology/keq124.

Diesel, B., Radermacher, J., Bureik, M., Bernhardt, R., Seifert, M., Reichrath, J., et al. (2005). Vitamin D(3) metabolism in human glioblastoma multiforme: Functionality of CYP27B1 splice variants, metabolism of calcidiol, and effect of calcitriol. *Clinical Cancer Research, 11*(15), 5370–5380. http://dx.doi.org/10.1158/1078-0432.CCR-04-1968, 11/15/5370 [pii].

Dionne, S., Calderon, M. R., White, J. H., Memari, B., Elimrani, I., Adelson, B., et al. (2014). Differential effect of vitamin D on NOD2- and TLR-induced cytokines in Crohn's disease. *Mucosal Immunology*, 7(6), 1405–1415. http://dx.doi.org/10.1038/mi.2014.30.

Dusso, A. S., Brown, A. J., & Slatopolsky, E. (2005). Vitamin D. *American Journal of Physiology. Renal Physiology, 289*(1), F8–F28. http://dx.doi.org/10.1152/ajprenal.00336.2004, 289/1/F8 [pii].

Dwivedi, P. P., Anderson, P. H., Omdahl, J. L., Grimes, H. L., Morris, H. A., & May, B. K. (2005). Identification of growth factor independent-1 (GFI1) as a repressor of 25-hydroxyvitamin D 1-alpha hydroxylase (CYP27B1) gene expression in human prostate cancer cells. *Endocrine-Related Cancer, 12*(2), 351–365. http://dx.doi.org/10.1677/erc.1.00920, 12/2/351 [pii].

Ebert, R., Jovanovic, M., Ulmer, M., Schneider, D., Meissner-Weigl, J., Adamski, J., et al. (2004). Down-regulation by nuclear factor kappaB of human 25-hydroxyvitamin D3 1alpha-hydroxylase promoter. *Molecular Endocrinology, 18*(10), 2440–2450. http://dx.doi.org/10.1210/me.2002-0441, (Research Support, Non-U.S. Gov't).

Feldman, D., Krishnan, A. V., Swami, S., Giovannucci, E., & Feldman, B. J. (2014). The role of vitamin D in reducing cancer risk and progression. *Nature Reviews. Cancer, 14*(5), 342–357. http://dx.doi.org/10.1038/nrc3691, nrc3691 [pii].

Feskanich, D., Ma, J., Fuchs, C. S., Kirkner, G. J., Hankinson, S. E., Hollis, B. W., et al. (2004). Plasma vitamin D metabolites and risk of colorectal cancer in women. *Cancer Epidemiology, Biomarkers and Prevention, 13*(9), 1502–1508, 13/9/1502 [pii].

Fischer, U., Leidinger, P., Keller, A., Folarin, A., Ketter, R., Graf, N., et al. (2010). Amplicons on chromosome 12q13-21 in glioblastoma recurrences. *International Journal of Cancer, 126*(11), 2594–2602. http://dx.doi.org/10.1002/ijc.24971.

Fleet, J. C., Bradley, J., Reddy, G. S., Ray, R., & Wood, R. J. (1996). 1 alpha,25-(OH)2-vitamin D3 analogs with minimal in vivo calcemic activity can stimulate significant transepithelial calcium transport and mRNA expression in vitro. *Archives of Biochemistry and Biophysics, 329*(2), 228–234. http://dx.doi.org/10.1006/abbi.1996.0213.

Foresta, C., Strapazzon, G., De Toni, L., Perilli, L., Di Mambro, A., Muciaccia, B., et al. (2011). Bone mineral density and testicular failure: Evidence for a role of vitamin D 25-hydroxylase in human testis. *The Journal of Clinical Endocrinology and Metabolism, 96*(4), E646–E652. http://dx.doi.org/10.1210/jc.2010-1628, jc.2010-1628 [pii].

Friedrich, M., Rafi, L., Mitschele, T., Tilgen, W., Schmidt, W., & Reichrath, J. (2003). Analysis of the vitamin D system in cervical carcinomas, breast cancer and ovarian cancer. *Recent Results in Cancer Research. Fortschritte der Krebsforschung. Progres dans les Recherches sur le Cancer, 164*, 239–246.

Froicu, M., & Cantorna, M. T. (2007). Vitamin D and the vitamin D receptor are critical for control of the innate immune response to colonic injury. *BMC Immunology, 8*, 5. http://dx.doi.org/10.1186/1471-2172-8-5, 1471-2172-8-5 [pii].

Froicu, M., Weaver, V., Wynn, T. A., McDowell, M. A., Welsh, J. E., & Cantorna, M. T. (2003). A crucial role for the vitamin D receptor in experimental inflammatory bowel diseases. *Molecular Endocrinology, 17*(12), 2386–2392. http://dx.doi.org/10.1210/me.2003-0281, me.2003-0281 [pii].

Froicu, M., Zhu, Y., & Cantorna, M. T. (2006). Vitamin D receptor is required to control gastrointestinal immunity in IL-10 knockout mice. *Immunology, 117*(3), 310–318. http://dx.doi.org/10.1111/j.1365-2567.2005.02290.x.

Fujiki, R., Kim, M. S., Sasaki, Y., Yoshimura, K., Kitagawa, H., & Kato, S. (2005). Ligand-induced transrepression by VDR through association of WSTF with acetylated histones. *The EMBO Journal, 24*(22), 3881–3894. http://dx.doi.org/10.1038/sj.emboj.7600853, (Research Support, Non-U.S. Gov't Retracted Publication).

Garland, C. F., Gorham, E. D., Mohr, S. B., & Garland, F. C. (2009). Vitamin D for cancer prevention: Global perspective. *Annals of Epidemiology, 19*(7), 468–483. http://dx.doi.org/10.1016/j.annepidem.2009.03.021, S1047-2797(09)00105-7 [pii].

Gascon-Barre, M., Demers, C., Ghrab, O., Theodoropoulos, C., Lapointe, R., Jones, G., et al. (2001). Expression of CYP27A, a gene encoding a vitamin D-25 hydroxylase in human liver and kidney. *Clinical Endocrinology*, *54*(1), 107–115, cen1160 [pii].

Giangreco, A. A., Dambal, S., Wagner, D., Van der Kwast, T., Vieth, R., Prins, G. S., et al. (2015). Differential expression and regulation of vitamin D hydroxylases and inflammatory genes in prostate stroma and epithelium by 1,25-dihydroxyvitamin D in men with prostate cancer and an in vitro model. *Journal of Steroid Biochemistry and Molecular Biology*, *148*, 156–165. http://dx.doi.org/10.1016/j.jsbmb.2014.10.004, S0960-0760(14)00237-4 [pii].

Gilbert, R., Bonilla, C., Metcalfe, C., Lewis, S., Evans, D. M., Fraser, W. D., et al. (2015). Associations of vitamin D pathway genes with circulating 25-hydroxyvitamin-D, 1,25-dihydroxyvitamin-D, and prostate cancer: A nested case–control study. *Cancer Causes and Control*, *26*(2), 205–218. http://dx.doi.org/10.1007/s10552-014-0500-5.

Gombart, A. F. (2009). The vitamin D-antimicrobial peptide pathway and its role in protection against infection. *Future Microbiology*, *4*(9), 1151–1165.

Grant, W. B., Cross, H. S., Garland, C. F., Gorham, E. D., Moan, J., Peterlik, M., et al. (2009). Estimated benefit of increased vitamin D status in reducing the economic burden of disease in western Europe. *Progress in Biophysics and Molecular Biology*, *99*(2–3), 104–113. http://dx.doi.org/10.1016/j.pbiomolbio.2009.02.003, S0079-107(09)00040-6 [pii].

Gray, R. W., Omdahl, J. L., Ghazarian, J. G., & DeLuca, H. F. (1972). 25-Hydroxycholecalciferol-1-hydroxylase. Subcellular location and properties. *The Journal of Biological Chemistry*, *247*(23), 7528–7532.

Groschel, C., Aggarwal, A., Tennakoon, S., Hobaus, J., Prinz-Wohlgenannt, M., Marian, B., et al. (2015). Effect of 1,25-dihydroxyvitamin D on the Wnt pathway in non-malignant colonic cells. *Journal of Steroid Biochemistry and Molecular Biology*. http://dx.doi.org/10.1016/j.jsbmb.2015.02.011, S0960-0760(15)00049-7 [pii].

Guryev, O., Carvalho, R. A., Usanov, S., Gilep, A., & Estabrook, R. W. (2003). A pathway for the metabolism of vitamin D3: Unique hydroxylated metabolites formed during catalysis with cytochrome P450scc (CYP11A1). *Proceedings of the National Academy of Sciences of the United States of America*, *100*(25), 14754–14759. http://dx.doi.org/10.1073/pnas.2336107100, (Research Support, Non-U.S. Gov't Research Support, U.S. Gov't, P.H.S.).

Gyetko, M. R., Hsu, C. H., Wilkinson, C. C., Patel, S., & Young, E. (1993). Monocyte 1 alpha-hydroxylase regulation: Induction by inflammatory cytokines and suppression by dexamethasone and uremia toxin. *Journal of Leukocyte Biology*, *54*(1), 17–22.

Hanahan, D., & Weinberg, R. A. (2011). Hallmarks of cancer: The next generation. *Cell*, *144*(5), 646–674. http://dx.doi.org/10.1016/j.cell.2011.02.013, S0092-8674(11)00127-9 [pii].

Haussler, M. R., Haussler, C. A., Bartik, L., Whitfield, G. K., Hsieh, J. C., Slater, S., et al. (2008). Vitamin D receptor: Molecular signaling and actions of nutritional ligands in disease prevention. *Nutrition Reviews*, *66*(10 Suppl. 2), S98–S112. http://dx.doi.org/10.1111/j.1753-4887.2008.00093.x, (Research Support, N.I.H., Extramural Review).

Haussler, M. R., Whitfield, G. K., Kaneko, I., Haussler, C. A., Hsieh, D., Hsieh, J. C., et al. (2013). Molecular mechanisms of vitamin D action. *Calcified Tissue International*, *92*(2), 77–98. http://dx.doi.org/10.1007/s00223-012-9619-0, (Review).

Heaney, R. P., Recker, R. R., Grote, J., Horst, R. L., & Armas, L. A. (2011). Vitamin D(3) is more potent than vitamin D(2) in humans. *The Journal of Clinical Endocrinology and Metabolism*, *96*(3), E447–E452. http://dx.doi.org/10.1210/jc.2010-2230, jc.2010-2230 [pii].

Heine, G., Niesner, U., Chang, H. D., Steinmeyer, A., Zügel, U., Zuberbier, T., et al. (2008). 1,25-dihydroxyvitamin D(3) promotes IL-10 production in human B cells.

European Journal of Immunology, 38(8), 2210–2218. http://dx.doi.org/10.1002/eji.200838216.

Helming, L., Böse, J., Ehrchen, J., Schiebe, S., Frahm, T., Geffers, R., et al. (2005). 1alpha,25-Dihydroxyvitamin D3 is a potent suppressor of interferon gamma-mediated macrophage activation. *Blood, 106*(13), 4351–4358. http://dx.doi.org/10.1182/blood-2005-03-1029.

Henry, H. L. (2011). Regulation of vitamin D metabolism. *Best Practice & Research. Clinical Endocrinology & Metabolism, 25*(4), 531–541. http://dx.doi.org/10.1016/j.beem.2011.05.003.

Hewison, M. (2012). Vitamin D and immune function: An overview. *Proceedings of the Nutrition Society, 71*(1), 50–61. http://dx.doi.org/10.1017/S0029665111001650.

Hewison, M., Burke, F., Evans, K. N., Lammas, D. A., Sansom, D. M., Liu, P., et al. (2007). Extra-renal 25-hydroxyvitamin D3-1alpha-hydroxylase in human health and disease. *The Journal of Steroid Biochemistry and Molecular Biology, 103*(3–5), 316–321. http://dx.doi.org/10.1016/j.jsbmb.2006.12.078.

Hobaus, J., Hummel, D. M., Thiem, U., Fetahu, I. S., Aggarwal, A., Mullauer, L., et al. (2013). Increased copy-number and not DNA hypomethylation causes overexpression of the candidate proto-oncogene CYP24A1 in colorectal cancer. *International Journal of Cancer. Journal International du Cancer, 133*(6), 1380–1388. http://dx.doi.org/10.1002/ijc.28143, (Research Support, Non-U.S. Gov't).

Hobaus, J., Thiem, U., Hummel, D. M., & Kallay, E. (2013). Role of calcium, vitamin D, and the extrarenal vitamin D hydroxylases in carcinogenesis. *Anti-Cancer Agents in Medicinal Chemistry, 13*(1), 20–35, CMCACA-EPUB-20121012-8 [pii].

Hoffmann, M. R., Senior, P. A., & Mager, D. R. (2015). Vitamin D supplementation and health-related quality of life: a systematic review of the literature. *Journal of the Academy of Nutrition and Dietetics, 115*, 406–418. http://dx.doi.org/10.1016/j.jand.2014.10.023.

Holick, M. F. (2007). Vitamin D deficiency. *New England Journal of Medicine, 357*(3), 266–281. http://dx.doi.org/10.1056/NEJMra070553, 357/3/266 [pii].

Holick, M. F., Biancuzzo, R. M., Chen, T. C., Klein, E. K., Young, A., Bibuld, D., et al. (2008). Vitamin D2 is as effective as vitamin D3 in maintaining circulating concentrations of 25-hydroxyvitamin D. *The Journal of Clinical Endocrinology and Metabolism, 93*(3), 677–681. http://dx.doi.org/10.1210/jc.2007-2308, jc.2007-2308 [pii].

Horvath, H. C., Khabir, Z., Nittke, T., Gruber, S., Speer, G., Manhardt, T., et al. (2010). CYP24A1 splice variants—Implications for the antitumorigenic actions of 1,25-(OH)2D3 in colorectal cancer. *Journal of Steroid Biochemistry and Molecular Biology, 121*(1–2), 76–79. http://dx.doi.org/10.1016/j.jsbmb.2010.03.080, S0960-0760(10)00179-2 [pii].

Horvath, H. C., Lakatos, P., Kosa, J. P., Bacsi, K., Borka, K., Bises, G., et al. (2010). The candidate oncogene CYP24A1: A potential biomarker for colorectal tumorigenesis. *Journal of Histochemistry and Cytochemistry, 58*(3), 277–285. http://dx.doi.org/10.1369/jhc.2009.954339, jhc.2009.954339 [pii].

Hoyer-Hansen, M., Nordbrandt, S. P., & Jaattela, M. (2010). Autophagy as a basis for the health-promoting effects of vitamin D. *Trends in Molecular Medicine, 16*(7), 295–302.

Hummel, D., Aggarwal, A., Borka, K., Bajna, E., Kallay, E., & Horvath, H. C. (2014). The vitamin D system is deregulated in pancreatic diseases. *The Journal of Steroid Biochemistry and Molecular Biology, 144*(Pt B), 402–409. http://dx.doi.org/10.1016/j.jsbmb.2014.07.011, (Research Support, Non-U.S. Gov't).

Hummel, D. M., Fetahu, I. S., Gröschel, C., Manhardt, T., & Kállay, E. (2014). Role of proinflammatory cytokines on expression of vitamin D metabolism and target genes in colon cancer cells. *Journal of Steroid Biochemistry and Molecular Biology, 144*, 91–95. http://dx.doi.org/10.1016/j.jsbmb.2013.09.017.

Hummel, D. M., Thiem, U., Höbaus, J., Mesteri, I., Gober, L., Stremnitzer, C., et al. (2013). Prevention of preneoplastic lesions by dietary vitamin D in a mouse model of colorectal carcinogenesis. *Journal of Steroid Biochemistry and Molecular Biology, 136*, 284–288. http://dx.doi.org/10.1016/j.jsbmb.2012.09.003.

Inui, N., Murayama, A., Sasaki, S., Suda, T., Chida, K., Kato, S., et al. (2001). Correlation between 25-hydroxyvitamin D3 1 alpha-hydroxylase gene expression in alveolar macrophages and the activity of sarcoidosis. *American Journal of Medicine, 110*(9), 687–693.

Ishizuka, S., Kurihara, N., Reddy, S. V., Cornish, J., Cundy, T., & Roodman, G. D. (2005). (23S)-25-Dehydro-1{alpha}-hydroxyvitamin D3-26,23-lactone, a vitamin D receptor antagonist that inhibits osteoclast formation and bone resorption in bone marrow cultures from patients with Paget's disease. *Endocrinology, 146*(4), 2023–2030. http://dx.doi.org/10.1210/en.2004-1140.

Iyer, S. S., & Cheng, G. (2012). Role of interleukin 10 transcriptional regulation in inflammation and autoimmune disease. *Critical Reviews in Immunology, 32*(1), 23–63.

Jablonski, K. L., Jovanovich, A., Holmen, J., Targher, G., McFann, K., Kendrick, J., et al. (2013). Low 25-hydroxyvitamin D level is independently associated with non-alcoholic fatty liver disease. *Nutrition, Metabolism, and Cardiovascular Diseases, 23*(8), 792–798. http://dx.doi.org/10.1016/j.numecd.2012.12.006.

Janjetovic, Z., Tuckey, R. C., Nguyen, M. N., Thorpe, E. M., Jr., & Slominski, A. T. (2010). 20,23-dihydroxyvitamin D3, novel P450scc product, stimulates differentiation and inhibits proliferation and NF-kappaB activity in human keratinocytes. *Journal of Cellular Physiology, 223*(1), 36–48. http://dx.doi.org/10.1002/jcp.21992.

Janjetovic, Z., Zmijewski, M. A., Tuckey, R. C., DeLeon, D. A., Nguyen, M. N., Pfeffer, L. M., et al. (2009). 20-Hydroxycholecalciferol, product of vitamin D3 hydroxylation by P450scc, decreases NF-kappaB activity by increasing IkappaB alpha levels in human keratinocytes. *PLoS One, 4*(6), e5988. http://dx.doi.org/10.1371/journal.pone.0005988.

Jeffery, L. E., Wood, A. M., Qureshi, O. S., Hou, T. Z., Gardner, D., Briggs, Z., et al. (2012). Availability of 25-hydroxyvitamin D(3) to APCs controls the balance between regulatory and inflammatory T cell responses. *Journal of Immunology, 189*(11), 5155–5164. http://dx.doi.org/10.4049/jimmunol.1200786.

Jones, G. (2008). Pharmacokinetics of vitamin D toxicity. *American Journal of Clinical Nutrition, 88*(2), 582S–586S.

Jones, G., & Prosser, D. E. (2011). The activating enzymes of vitamin d metabolism (25- and 1alpha-hydroxylases). In D. Feldman, W. J. Pike, & J. S. Adams (Eds.), *Vitamin D* (pp. 23–42). vol. 1. San Francisco: Elsevier.

Jones, G., Schnoes, H. K., & DeLuca, H. F. (1975). Isolation and identification of 1,25-dihydroxyvitamin D2. *Biochemistry, 14*(6), 1250–1256.

Jones, G., Strugnell, S. A., & DeLuca, H. F. (1998). Current understanding of the molecular actions of vitamin D. *Physiological Reviews, 78*(4), 1193–1231, (Research Support, Non-U.S. Gov't Research Support, U.S. Gov't, P.H.S. Review).

Jones, G., & Tenenhouse, H. S. (2002). 1,25(OH)2D, the preferred substrate for CYP24. *Journal of Bone and Mineral Research, 17*(1), 179–181. http://dx.doi.org/10.1359/jbmr.2002.17.1.179.

Kalinski, P. (2012). Regulation of immune responses by prostaglandin E2. *Journal of Immunology, 188*(1), 21–28. http://dx.doi.org/10.4049/jimmunol.1101029.

Kallay, E., Bareis, P., Bajna, E., Kriwanek, S., Bonner, E., Toyokuni, S., et al. (2002). Vitamin D receptor activity and prevention of colonic hyperproliferation and oxidative stress. *Food and Chemical Toxicology, 40*(8), 1191–1196, S0278691502000303 [pii].

Kallay, E., Bises, G., Bajna, E., Bieglmayer, C., Gerdenitsch, W., Steffan, I., et al. (2005). Colon-specific regulation of vitamin D hydroxylases—A possible approach for tumor prevention. *Carcinogenesis, 26*(9), 1581–1589. http://dx.doi.org/10.1093/carcin/bgi124, bgi124 [pii].

Kamao, M., Tatematsu, S., Sawada, N., Sakaki, T., Hatakeyama, S., Kubodera, N., et al. (2004). Cell specificity and properties of the C-3 epimerization of vitamin D3 metabolites. *Journal of Steroid Biochemistry and Molecular Biology*, *89–90*(1–5), 39–42. http://dx.doi.org/10.1016/j.jsbmb.2004.03.048.

Kasuga, H., Hosogane, N., Matsuoka, K., Mori, I., Sakura, Y., Shimakawa, K., et al. (2002). Characterization of transgenic rats constitutively expressing vitamin D-24-hydroxylase gene. *Biochemical and Biophysical Research Communications*, *297*(5), 1332–1338.

Keegan, R. J., Lu, Z., Bogusz, J. M., Williams, J. E., & Holick, M. F. (2013). Photobiology of vitamin D in mushrooms and its bioavailability in humans. *Dermato-endocrinology*, *5*(1), 165–176. http://dx.doi.org/10.4161/derm.23321.

Khorchide, M., Lechner, D., & Cross, H. S. (2005). Epigenetic regulation of vitamin D hydroxylase expression and activity in normal and malignant human prostate cells. *Journal of Steroid Biochemistry and Molecular Biology*, *93*(2–5), 167–172. http://dx.doi.org/10.1016/j.jsbmb.2004.12.022, S0960-0760(04)00437-6 [pii].

Kim, M. S., Kondo, T., Takada, I., Youn, M. Y., Yamamoto, Y., Takahashi, S., et al. (2009). DNA demethylation in hormone-induced transcriptional derepression. *Nature*, *461*(7266), 1007–1012. http://dx.doi.org/10.1038/nature08456, (Research Support, Non-U.S. Gov't Retracted Publication).

Kim, J. H., Yamaori, S., Tanabe, T., Johnson, C. H., Krausz, K. W., Kato, S., et al. (2013). Implication of intestinal VDR deficiency in inflammatory bowel disease. *Biochimica et Biophysica Acta*, *1830*(1), 2118–2128. http://dx.doi.org/10.1016/j.bbagen.2012.09.020.

Kitson, M. T., & Roberts, S. K. (2012). D-livering the message: The importance of vitamin D status in chronic liver disease. *Journal of Hepatology*, *57*(4), 897–909. http://dx.doi.org/10.1016/j.jhep.2012.04.033.

Kong, J., Xu, F., Qu, J., Wang, Y., Gao, M., Yu, H., et al. (2014). Genetic polymorphisms in the vitamin D pathway in relation to lung cancer risk and survival. *Oncotarget*, *6*, 2573–2582, 2951 [pii].

Kong, J., Zhang, Z., Musch, M. W., Ning, G., Sun, J., Hart, J., et al. (2008). Novel role of the vitamin D receptor in maintaining the integrity of the intestinal mucosal barrier. *American Journal of Physiology Gastrointestinal and Liver Physiology*, *294*(1), G208–G216. http://dx.doi.org/10.1152/ajpgi.00398.2007, 00398.2007 [pii].

Krishnan, A. V., Swami, S., & Feldman, D. (2013). Equivalent anticancer activities of dietary vitamin D and calcitriol in an animal model of breast cancer: Importance of mammary CYP27B1 for treatment and prevention. *The Journal of Steroid Biochemistry and Molecular Biology*, *136*, 289–295. http://dx.doi.org/10.1016/j.jsbmb.2012.08.005, (Research Support, N.I.H., Extramural Review).

Kundu, R., Chain, B. M., Coussens, A. K., Khoo, B., & Noursadeghi, M. (2014). Regulation of CYP27B1 and CYP24A1 hydroxylases limits cell-autonomous activation of vitamin D in dendritic cells. *European Journal of Immunology*, *44*(6), 1781–1790. http://dx.doi.org/10.1002/eji.201344157.

Kutuzova, G. D., & DeLuca, H. F. (2007). 1,25-Dihydroxyvitamin D3 regulates genes responsible for detoxification in intestine. *Toxicology and Applied Pharmacology*, *218*(1), 37–44. http://dx.doi.org/10.1016/j.taap.2006.10.005.

Lagishetty, V., Misharin, A. V., Liu, N. Q., Lisse, T. S., Chun, R. F., Ouyang, Y., et al. (2010). Vitamin D deficiency in mice impairs colonic antibacterial activity and predisposes to colitis. *Endocrinology*, *151*(6), 2423–2432. http://dx.doi.org/10.1210/en.2010-0089, en.2010-0089 [pii].

Lange, C. M., Bojunga, J., Ramos-Lopez, E., von Wagner, M., Hassler, A., Vermehren, J., et al. (2011). Vitamin D deficiency and a CYP27B1-1260 promoter polymorphism are associated with chronic hepatitis C and poor response to interferon-alfa based therapy. *Journal of Hepatology*, *54*(5), 887–893. http://dx.doi.org/10.1016/j.jhep.2010.08.036.

Lange, C. M., Miki, D., Ochi, H., Nischalke, H. D., Bojunga, J., Bibert, S., et al. (2013). Genetic analyses reveal a role for vitamin D insufficiency in HCV-associated hepatocellular carcinoma development. *PLoS One*, *8*(5), e64053. http://dx.doi.org/10.1371/journal.pone.0064053, PONE-D-13-04732 [pii].
Larriba, M. J., Ordonez-Moran, P., Chicote, I., Martin-Fernandez, G., Puig, I., Munoz, A., et al. (2011). Vitamin D receptor deficiency enhances Wnt/beta-catenin signaling and tumor burden in colon cancer. *PLoS One*, *6*(8), e23524. http://dx.doi.org/10.1371/journal.pone.0023524, PONE-D-11-05092 [pii].
Lemire, J. M., Archer, D. C., Beck, L., & Spiegelberg, H. L. (1995). Immunosuppressive actions of 1,25-dihydroxyvitamin D3: Preferential inhibition of Th1 functions. *Journal of Nutrition*, *125*(6 Suppl.), 1704S–1708S.
Leyssens, C., Verlinden, L., & Verstuyf, A. (2013). Antineoplastic effects of 1,25(OH)2D3 and its analogs in breast, prostate and colorectal cancer. *Endocrine-Related Cancer*, *20*(2), R31–R47. http://dx.doi.org/10.1530/ERC-12-0381, ERC-12-0381 [pii].
Li, B., Baylink, D. J., Walter, M. H., Lau, K. H., Meng, X., Wang, J., et al. (2015). Targeted 25-hydroxyvitamin D3 1α-hydroxylase Adoptive gene therapy ameliorates DSS-induced colitis without causing hypercalcemia in mice. *Molecular Therapy*, *23*, 339–351. http://dx.doi.org/10.1038/mt.2014.201.
Liu, N., Nguyen, L., Chun, R. F., Lagishetty, V., Ren, S., Wu, S., et al. (2008). Altered endocrine and autocrine metabolism of vitamin D in a mouse model of gastrointestinal inflammation. *Endocrinology*, *149*(10), 4799–4808. http://dx.doi.org/10.1210/en.2008-0060, en.2008-0060 [pii].
Liu, P. T., Stenger, S., Li, H., Wenzel, L., Tan, B. H., Krutzik, S. R., et al. (2006). Toll-like receptor triggering of a vitamin D-mediated human antimicrobial response. *Science*, *311*(5768), 1770–1773. http://dx.doi.org/10.1126/science.1123933.
Lonard, D. M., & O'Malley, B. W. (2006). The expanding cosmos of nuclear receptor coactivators. *Cell*, *125*(3), 411–414. http://dx.doi.org/10.1016/j.cell.2006.04.021, (Research Support, N.I.H., Extramural Review).
Lopes, N., Sousa, B., Martins, D., Gomes, M., Vieira, D., Veronese, L. A., et al. (2010). Alterations in vitamin D signalling and metabolic pathways in breast cancer progression: A study of VDR, CYP27B1 and CYP24A1 expression in benign and malignant breast lesions. *BMC Cancer*, *10*, 483. http://dx.doi.org/10.1186/1471-2407-10-483, 1471-2407-10-483 [pii].
Luo, W., Hershberger, P. A., Trump, D. L., & Johnson, C. S. (2013). 24-Hydroxylase in cancer: Impact on vitamin D-based anticancer therapeutics. *The Journal of Steroid Biochemistry and Molecular Biology*, *136*, 252–257. http://dx.doi.org/10.1016/j.jsbmb.2012.09.031, (Research Support, N.I.H., Extramural Review).
Manson, J. E., Mayne, S. T., & Clinton, S. K. (2011). Vitamin D and prevention of cancer—Ready for prime time? *The New England Journal of Medicine*, *364*(15), 1385–1387. http://dx.doi.org/10.1056/NEJMp1102022.
Masuda, S., Byford, V., Arabian, A., Sakai, Y., Demay, M. B., St-Arnaud, R., et al. (2005). Altered pharmacokinetics of 1alpha,25-dihydroxyvitamin D3 and 25-hydroxyvitamin D3 in the blood and tissues of the 25-hydroxyvitamin D-24-hydroxylase (Cyp24a1) null mouse. *Endocrinology*, *146*(2), 825–834. http://dx.doi.org/10.1210/en.2004-1116, (Research Support, Non-U.S. Gov't Research Support, U.S. Gov't, P.H.S.).
Matilainen, J. M., Malinen, M., Turunen, M. M., Carlberg, C., & Vaisanen, S. (2010). The number of vitamin D receptor binding sites defines the different vitamin D responsiveness of the CYP24 gene in malignant and normal mammary cells. *The Journal of Biological Chemistry*, *285*(31), 24174–24183. http://dx.doi.org/10.1074/jbc.M110.124073, (Research Support, Non-U.S. Gov't).
Medzhitov, R. (2008). Origin and physiological roles of inflammation. *Nature*, *454*(7203), 428–435. http://dx.doi.org/10.1038/nature07201.

Meyer, M. B., Goetsch, P. D., & Pike, J. W. (2010). A downstream intergenic cluster of regulatory enhancers contributes to the induction of CYP24A1 expression by 1alpha,25-dihydroxyvitamin D3. *The Journal of Biological Chemistry*, *285*(20), 15599–15610. http://dx.doi.org/10.1074/jbc.M110.119958, (Research Support, N.I.H., Extramural).

Meyer, M. B., Zella, L. A., Nerenz, R. D., & Pike, J. W. (2007). Characterizing early events associated with the activation of target genes by 1,25-dihydroxyvitamin D3 in mouse kidney and intestine in vivo. *Journal of Biological Chemistry*, *282*(31), 22344–22352. http://dx.doi.org/10.1074/jbc.M703475200, M703475200 [pii].

Mimori, K., Tanaka, Y., Yoshinaga, K., Masuda, T., Yamashita, K., Okamoto, M., et al. (2004). Clinical significance of the overexpression of the candidate oncogene CYP24 in esophageal cancer. *Annals of Oncology*, *15*(2), 236–241.

Molnár, F., Sigüeiro, R., Sato, Y., Araujo, C., Schuster, I., Antony, P., et al. (2011). 1α,25 (OH)2-3-epi-vitamin D3, a natural physiological metabolite of vitamin D3: Its synthesis, biological activity and crystal structure with its receptor. *PLoS One*, *6*(3), e18124. http://dx.doi.org/10.1371/journal.pone.0018124.

Moreno, J., Krishnan, A. V., Swami, S., Nonn, L., Peehl, D. M., & Feldman, D. (2005). Regulation of prostaglandin metabolism by calcitriol attenuates growth stimulation in prostate cancer cells. *Cancer Research*, *65*(17), 7917–7925. http://dx.doi.org/10.1158/0008-5472.CAN-05-1435.

Muindi, J. R., Nganga, A., Engler, K. L., Coignet, L. J., Johnson, C. S., & Trump, D. L. (2007). CYP24 splicing variants are associated with different patterns of constitutive and calcitriol-inducible CYP24 activity in human prostate cancer cell lines. *Journal of Steroid Biochemistry and Molecular Biology*, *103*(3–5), 334–337. http://dx.doi.org/10.1016/j.jsbmb.2006.12.060, S0960-0760(06)00439-0 [pii].

Murray, P. J., & Wynn, T. A. (2011). Protective and pathogenic functions of macrophage subsets. *Nature Reviews. Immunology*, *11*(11), 723–737. http://dx.doi.org/10.1038/nri3073.

Muthian, G., Raikwar, H. P., Rajasingh, J., & Bright, J. J. (2006). 1,25 Dihydroxyvitamin-D3 modulates JAK-STAT pathway in IL-12/IFNgamma axis leading to Th1 response in experimental allergic encephalomyelitis. *Journal of Neuroscience Research*, *83*(7), 1299–1309. http://dx.doi.org/10.1002/jnr.20826.

Nakanishi, M., & Rosenberg, D. W. (2013). Multifaceted roles of PGE2 in inflammation and cancer. *Seminars in Immunopathology*, *35*(2), 123–137. http://dx.doi.org/10.1007/s00281-012-0342-8.

Narvaez, C. J., Matthews, D., LaPorta, E., Simmons, K. M., Beaudin, S., & Welsh, J. (2014). The impact of vitamin D in breast cancer: Genomics, pathways, metabolism. *Frontiers in Physiology*, *5*, 213. http://dx.doi.org/10.3389/fphys.2014.00213.

Nguyen, D. P., Li, J., & Tewari, A. K. (2014). Inflammation and prostate cancer: The role of interleukin 6 (IL-6). *BJU International*, *113*(6), 986–992. http://dx.doi.org/10.1111/bju.12452.

Nonn, L., Peng, L., Feldman, D., & Peehl, D. M. (2006). Inhibition of p38 by vitamin D reduces interleukin-6 production in normal prostate cells via mitogen-activated protein kinase phosphatase 5: Implications for prostate cancer prevention by vitamin D. *Cancer Research*, *66*(8), 4516–4524. http://dx.doi.org/10.1158/0008-5472.CAN-05-3796.

Noyola-Martínez, N., Díaz, L., Zaga-Clavellina, V., Avila, E., Halhali, A., Larrea, F., et al. (2014). Regulation of CYP27B1 and CYP24A1 gene expression by recombinant pro-inflammatory cytokines in cultured human trophoblasts. *Journal of Steroid Biochemistry and Molecular Biology*, *144*(Pt. A), 106–109. http://dx.doi.org/10.1016/j.jsbmb.2013.12.007.

Orlov, I., Rochel, N., Moras, D., & Klaholz, B. P. (2012). Structure of the full human RXR/VDR nuclear receptor heterodimer complex with its DR3 target DNA. *The*

EMBO Journal, *31*(2), 291–300. http://dx.doi.org/10.1038/emboj.2011.445, (Research Support, Non-U.S. Gov't).

Parise, R. A., Egorin, M. J., Kanterewicz, B., Taimi, M., Petkovich, M., Lew, A. M., et al. (2006). CYP24, the enzyme that catabolizes the antiproliferative agent vitamin D, is increased in lung cancer. *International Journal of Cancer. Journal International du Cancer*, *119*(8), 1819–1828. http://dx.doi.org/10.1002/ijc.22058, (Research Support, N.I.H., Extramural).

Pereira, F., Larriba, M. J., & Munoz, A. (2012). Vitamin D and colon cancer. *Endocrine-Related Cancer*, *19*(3), R51–R71. http://dx.doi.org/10.1530/ERC-11-0388, ERC-11-0388 [pii].

Perwad, F., & Portale, A. A. (2011). Vitamin D metabolism in the kidney: Regulation by phosphorus and fibroblast growth factor 23. *Molecular and Cellular Endocrinology*, *347*(1–2), 17–24. http://dx.doi.org/10.1016/j.mce.2011.08.030, (Review).

Pike, J. W., Meyer, M. B., Watanuki, M., Kim, S., Zella, L. A., Fretz, J. A., et al. (2007). Perspectives on mechanisms of gene regulation by 1,25-dihydroxyvitamin D3 and its receptor. *The Journal of Steroid Biochemistry and Molecular Biology*, *103*(3–5), 389–395. http://dx.doi.org/10.1016/j.jsbmb.2006.12.050, (Research Support, N.I.H., Extramural).

Prietl, B., Pilz, S., Wolf, M., Tomaschitz, A., Obermayer-Pietsch, B., Graninger, W., et al. (2010). Vitamin D supplementation and regulatory T cells in apparently healthy subjects: Vitamin D treatment for autoimmune diseases? *The Israel Medical Association Journal*, *12*(3), 136–139.

Prosser, D. E., & Jones, G. (2004). Enzymes involved in the activation and inactivation of vitamin D. *Trends in Biochemical Sciences*, *29*(12), 664–673. http://dx.doi.org/10.1016/j.tibs.2004.10.005.

Raftery, T., & O'Sullivan, M. (2015). Optimal vitamin D levels in Crohn's disease: A review. *Proceedings of the Nutrition Society*, *74*(1), 56–66. http://dx.doi.org/10.1017/S0029665114001591, S0029665114001591 [pii].

Rahman, A. H., & Branch, A. D. (2013). Vitamin D for your patients with chronic hepatitis C? *Journal of Hepatology*, *58*(1), 184–189. http://dx.doi.org/10.1016/j.jhep.2012.07.026.

Ramagopalan, S. V., Dyment, D. A., Cader, M. Z., Morrison, K. M., Disanto, G., Morahan, J. M., et al. (2011). Rare variants in the CYP27B1 gene are associated with multiple sclerosis. *Annals of Neurology*, *70*(6), 881–886. http://dx.doi.org/10.1002/ana.22678, (Research Support, Non-U.S. Gov't).

Ramagopalan, S. V., Heger, A., Berlanga, A. J., Maugeri, N. J., Lincoln, M. R., Burrell, A., et al. (2010). A ChIP-seq defined genome-wide map of vitamin D receptor binding: Associations with disease and evolution. *Genome Research*, *20*(10), 1352–1360. http://dx.doi.org/10.1101/gr.107920.110, (Research Support, Non-U.S. Gov't).

Ramasamy, A., Trabzuni, D., Forabosco, P., Smith, C., Walker, R., Dillman, A., et al. (2014). Genetic evidence for a pathogenic role for the vitamin D3 metabolizing enzyme CYP24A1 in multiple sclerosis. *Multiple Sclerosis and Related Disorders*, *3*(2), 211–219. http://dx.doi.org/10.1016/j.msard.2013.08.009.

Ramnath, N., Nadal, E., Jeon, C. K., Sandoval, J., Colacino, J., Rozek, L. S., et al. (2014). Epigenetic regulation of vitamin D metabolism in human lung adenocarcinoma. *Journal of Thoracic Oncology*, *9*(4), 473–482. http://dx.doi.org/10.1097/JTO.0000000000000114, 01243894-201404000-00012 [pii].

Ramos-Lopez, E., Bruck, P., Jansen, T., Herwig, J., & Badenhoop, K. (2007). CYP2R1 (vitamin D 25-hydroxylase) gene is associated with susceptibility to type 1 diabetes and vitamin D levels in Germans. *Diabetes/Metabolism Research and Reviews*, *23*(8), 631–636. http://dx.doi.org/10.1002/dmrr.719.

Reddy, G. S., Muralidharan, K. R., Okamura, W. H., Tserng, K. Y., & McLane, J. A. (2001). Metabolism of 1alpha,25-dihydroxyvitamin D(3) and its C-3 epimer

1alpha,25-dihydroxy-3-epi-vitamin D(3) in neonatal human keratinocytes. *Steroids, 66*(3–5), 441–450.

Reichel, H., Koeffler, H. P., Tobler, A., & Norman, A. W. (1987). 1 alpha,25-Dihydroxyvitamin D3 inhibits gamma-interferon synthesis by normal human peripheral blood lymphocytes. *Proceedings of the National Academy of Sciences of the United States of America, 84*(10), 3385–3389.

Reichrath, J., & Reichrath, S. (2013). The relevance of the vitamin D endocrine system (VDES) for tumorigenesis, prevention, and treatment of non-melanoma skin cancer (NMSC): Present concepts and future perspectives. *Dermato-Endocrinology, 5*(1), 38–50. http://dx.doi.org/10.4161/derm.24156, (Review).

Reimers, L. L., Crew, K. D., Bradshaw, P. T., Santella, R. M., Steck, S. E., Sirosh, I., et al. (2015). Vitamin D-related gene polymorphisms, plasma 25-hydroxyvitamin D, and breast cancer risk. *Cancer Causes and Control, 26*(2), 187–203. http://dx.doi.org/10.1007/s10552-014-0497-9.

Ren, S., Nguyen, L., Wu, S., Encinas, C., Adams, J. S., & Hewison, M. (2005). Alternative splicing of vitamin D-24-hydroxylase: A novel mechanism for the regulation of extrarenal 1,25-dihydroxyvitamin D synthesis. *The Journal of Biological Chemistry, 280*(21), 20604–20611. http://dx.doi.org/10.1074/jbc.M414522200, M414522200 [pii].

Rochel, N., Wurtz, J. M., Mitschler, A., Klaholz, B., & Moras, D. (2000). The crystal structure of the nuclear receptor for vitamin D bound to its natural ligand. *Molecular Cell, 5*(1), 173–179, (Research Support, Non-U.S. Gov't).

Roff, A., & Wilson, R. T. (2008). A novel SNP in a vitamin D response element of the CYP24A1 promoter reduces protein binding, transactivation, and gene expression. *Journal of Steroid Biochemistry and Molecular Biology, 112*(1–3), 47–54. http://dx.doi.org/10.1016/j.jsbmb.2008.08.009, S0960-0760(08)00218-5 [pii].

Sakaki, T., Sawada, N., Komai, K., Shiozawa, S., Yamada, S., Yamamoto, K., et al. (2000). Dual metabolic pathway of 25-hydroxyvitamin D3 catalyzed by human CYP24. *European Journal of Biochemistry/FEBS, 267*(20), 6158–6165, (Research Support, Non-U.S. Gov't).

Sartor, R. B. (2006). Mechanisms of disease: Pathogenesis of Crohn's disease and ulcerative colitis. *Nature Clinical Practice. Gastroenterology & Hepatology, 3*(7), 390–407. http://dx.doi.org/10.1038/ncpgasthep0528.

Sawada, N., Sakaki, T., Kitanaka, S., Kato, S., & Inouye, K. (2001). Structure-function analysis of CYP27B1 and CYP27A1. Studies on mutants from patients with vitamin D-dependent rickets type I (VDDR-I) and cerebrotendinous xanthomatosis (CTX). *European Journal of Biochemistry/FEBS, 268*(24), 6607–6615, (Research Support, Non-U.S. Gov't).

Scaldaferri, F., Vetrano, S., Sans, M., Arena, V., Straface, G., Stigliano, E., et al. (2009). VEGF-A links angiogenesis and inflammation in inflammatory bowel disease pathogenesis. *Gastroenterology, 136*(2), 585–595. http://dx.doi.org/10.1053/j.gastro.2008.09.064, S0016-5085(08)01731-9 [pii].

Scheible, C., Thill, M., Baum, S., Solomayer, E., & Friedrich, M. (2014). Implication of CYP24A1 splicing in breast cancer. *Anti-Cancer Agents in Medicinal Chemistry, 14*(1), 109–114, ACAMC-EPUB-54158 [pii].

Schuster, I. (2011). Cytochromes P450 are essential players in the vitamin D signaling system. *Biochimica et Biophysica Acta, 1814*(1), 186–199. http://dx.doi.org/10.1016/j.bbapap.2010.06.022, S1570-9639(10)00181-0 [pii].

Segersten, U., Correa, P., Hewison, M., Hellman, P., Dralle, H., Carling, T., et al. (2002). 25-hydroxyvitamin D(3)-1alpha-hydroxylase expression in normal and pathological parathyroid glands. *The Journal of Clinical Endocrinology and Metabolism, 87*(6), 2967–2972. http://dx.doi.org/10.1210/jcem.87.6.8604, (Research Support, Non-U.S. Gov't).

Seuter, S., Neme, A., & Carlberg, C. (2014). Characterization of genomic vitamin D receptor binding sites through chromatin looping and opening. *PLoS One*, *9*(4), e96184. http://dx.doi.org/10.1371/journal.pone.0096184, (Research Support, Non-U.S. Gov't).

Shi, H., Yan, P. S., Chen, C. M., Rahmatpanah, F., Lofton-Day, C., Caldwell, C. W., et al. (2002). Expressed CpG island sequence tag microarray for dual screening of DNA hypermethylation and gene silencing in cancer cells. *Cancer Research*, *62*(11), 3214–3220.

Shimada, T., Hasegawa, H., Yamazaki, Y., Muto, T., Hino, R., Takeuchi, Y., et al. (2004). FGF-23 is a potent regulator of vitamin D metabolism and phosphate homeostasis. *Journal of Bone and Mineral Research*, *19*(3), 429–435. http://dx.doi.org/10.1359/JBMR.0301264.

Sigmundsdottir, H., Pan, J., Debes, G. F., Alt, C., Habtezion, A., Soler, D., et al. (2007). DCs metabolize sunlight-induced vitamin D3 to 'program' T cell attraction to the epidermal chemokine CCL27. *Nature Immunology*, *8*(3), 285–293. http://dx.doi.org/10.1038/ni1433, ni1433 [pii].

Signorello, L. B., Shi, J., Cai, Q., Zheng, W., Williams, S. M., Long, J., et al. (2011). Common variation in vitamin D pathway genes predicts circulating 25-hydroxyvitamin D levels among African Americans. *PLoS One*, *6*(12), e28623. http://dx.doi.org/10.1371/journal.pone.0028623.

Simon, K. C., Munger, K. L., Kraft, P., Hunter, D. J., De Jager, P. L., & Ascherio, A. (2011). Genetic predictors of 25-hydroxyvitamin D levels and risk of multiple sclerosis. *Journal of Neurology*, *258*(9), 1676–1682.

Sloka, S., Silva, C., Wang, J., & Yong, V. W. (2011). Predominance of Th2 polarization by vitamin D through a STAT6-dependent mechanism. *Journal of Neuroinflammation*, *8*, 56. http://dx.doi.org/10.1186/1742-2094-8-56.

Slominski, A. T., Janjetovic, Z., Fuller, B. E., Zmijewski, M. A., Tuckey, R. C., Nguyen, M. N., et al. (2010). Products of vitamin D3 or 7-dehydrocholesterol metabolism by cytochrome P450scc show anti-leukemia effects, having low or absent calcemic activity. *PLoS One*, *5*(3), e9907. http://dx.doi.org/10.1371/journal.pone.0009907.

Slominski, A. T., Kim, T. K., Janjetovic, Z., Tuckey, R. C., Bieniek, R., Yue, J., et al. (2011). 20-Hydroxyvitamin D2 is a noncalcemic analog of vitamin D with potent antiproliferative and prodifferentiation activities in normal and malignant cells. *American Journal of Physiology. Cell Physiology*, *300*(3), C526–C541. http://dx.doi.org/10.1152/ajpcell.00203.2010, ajpcell.00203.2010 [pii].

Slominski, A. T., Kim, T. K., Li, W., Yi, A. K., Postlethwaite, A., & Tuckey, R. C. (2014). The role of CYP11A1 in the production of vitamin D metabolites and their role in the regulation of epidermal functions. *Journal of Steroid Biochemistry and Molecular Biology*, *144*(Pt A), 28–39. http://dx.doi.org/10.1016/j.jsbmb.2013.10.012, S0960-0760(13)00205-7 [pii].

Slominski, A. T., Kim, T. K., Shehabi, H. Z., Semak, I., Tang, E. K., Nguyen, M. N., et al. (2012). In vivo evidence for a novel pathway of vitamin D(3) metabolism initiated by P450scc and modified by CYP27B1. *FASEB Journal : Official Publication of the Federation of American Societies for Experimental Biology*, *26*(9), 3901–3915. http://dx.doi.org/10.1096/fj.12-208975, (Research Support, N.I.H., Extramural Research Support, Non-U.S. Gov't).

Slominski, A., Kim, T. K., Zmijewski, M. A., Janjetovic, Z., Li, W., Chen, J., et al. (2013). Novel vitamin D photoproducts and their precursors in the skin. *Dermato-endocrinology*, *5*(1), 7–19. http://dx.doi.org/10.4161/derm.23938, 2012DE0223R [pii].

Slominski, A., Semak, I., Wortsman, J., Zjawiony, J., Li, W., Zbytek, B., et al. (2006). An alternative pathway of vitamin D metabolism. Cytochrome P450scc (CYP11A1)-mediated conversion to 20-hydroxyvitamin D2 and 17,20-dihydroxyvitamin D2.

The FEBS Journal, 273(13), 2891–2901. http://dx.doi.org/10.1111/j.1742-4658.2006.05302.x, (Research Support, N.I.H., Extramural).

Song, M., Nishihara, R., Wang, M., Chan, A. T., Qian, Z. R., Inamura, K., et al. (2015). Plasma 25-hydroxyvitamin D and colorectal cancer risk according to tumour immunity status. *Gut*. http://dx.doi.org/10.1136/gutjnl-2014-308852.

St-Arnaud, R., Arabian, A., Travers, R., Barletta, F., Raval-Pandya, M., Chapin, K., et al. (2000). Deficient mineralization of intramembranous bone in vitamin D-24-hydroxylase-ablated mice is due to elevated 1,25-dihydroxyvitamin D and not to the absence of 24,25-dihydroxyvitamin D. *Endocrinology, 141*(7), 2658–2666. http://dx.doi.org/10.1210/endo.141.7.7579, (Research Support, Non-U.S. Gov't).

St-Arnaud, R., Messerlian, S., Moir, J. M., Omdahl, J. L., & Glorieux, F. H. (1997). The 25-hydroxyvitamin D 1-alpha-hydroxylase gene maps to the pseudovitamin D-deficiency rickets (PDDR) disease locus. *Journal of Bone and Mineral Research: The Official Journal of the American Society for Bone and Mineral Research, 12*(10), 1552–1559. http://dx.doi.org/10.1359/jbmr.1997.12.10.1552, (Research Support, Non-U.S. Gov't).

Stoffels, K., Overbergh, L., Giulietti, A., Verlinden, L., Bouillon, R., & Mathieu, C. (2006). Immune regulation of 25-hydroxyvitamin-D3-1alpha-hydroxylase in human monocytes. *Journal of Bone and Mineral Research, 21*(1), 37–47. http://dx.doi.org/10.1359/JBMR.050908.

Strushkevich, N., Usanov, S. A., Plotnikov, A. N., Jones, G., & Park, H. W. (2008). Structural analysis of CYP2R1 in complex with vitamin D3. *Journal of Molecular Biology, 380*(1), 95–106. http://dx.doi.org/10.1016/j.jmb.2008.03.065, S0022-2836(08)00396-3 [pii].

Sundaram, K., Sambandam, Y., Tsuruga, E., Wagner, C. L., & Reddy, S. V. (2014). 1alpha,25-dihydroxyvitamin D3 modulates CYP2R1 gene expression in human oral squamous cell carcinoma tumor cells. *Hormones and Cancer, 5*(2), 90–97. http://dx.doi.org/10.1007/s12672-014-0170-5.

Sundqvist, E., Bäärnhielm, M., Alfredsson, L., Hillert, J., Olsson, T., & Kockum, I. (2010). Confirmation of association between multiple sclerosis and CYP27B1. *European Journal of Human Genetics, 18*(12), 1349–1352. http://dx.doi.org/10.1038/ejhg.2010.113.

Swami, S., Krishnan, A. V., Wang, J. Y., Jensen, K., Peng, L., Albertelli, M. A., et al. (2011). Inhibitory effects of calcitriol on the growth of MCF-7 breast cancer xenografts in nude mice: Selective modulation of aromatase expression in vivo. *Hormones & Cancer, 2*(3), 190–202. http://dx.doi.org/10.1007/s12672-011-0073-7.

Taniguchi, K., & Karin, M. (2014). IL-6 and related cytokines as the critical lynchpins between inflammation and cancer. *Seminars in Immunology, 26*(1), 54–74. http://dx.doi.org/10.1016/j.smim.2014.01.001.

Taylor, A. W. (2009). Review of the activation of TGF-beta in immunity. *Journal of Leukocyte Biology, 85*(1), 29–33. http://dx.doi.org/10.1189/jlb.0708415.

Tieu, E. W., Tang, E. K., Chen, J., Li, W., Nguyen, M. N., Janjetovic, Z., et al. (2012). Rat CYP24A1 acts on 20-hydroxyvitamin D(3) producing hydroxylated products with increased biological activity. *Biochemical Pharmacology, 84*(12), 1696–1704. http://dx.doi.org/10.1016/j.bcp.2012.09.032, S0006-2952(12)00663-6 [pii].

Toell, A., Gonzalez, M. M., Ruf, D., Steinmeyer, A., Ishizuka, S., & Carlberg, C. (2001). Different molecular mechanisms of vitamin D(3) receptor antagonists. *Molecular Pharmacology, 59*(6), 1478–1485, (Research Support, Non-U.S. Gov't).

Townsend, K., Banwell, C. M., Guy, M., Colston, K. W., Mansi, J. L., Stewart, P. M., et al. (2005). Autocrine metabolism of vitamin D in normal and malignant breast tissue. *Clinical Cancer Research, 11*(9), 3579–3586. http://dx.doi.org/10.1158/1078-0432.CCR-04-2359, 11/9/3579 [pii].

Tuckey, R. C., Li, W., Zjawiony, J. K., Zmijewski, M. A., Nguyen, M. N., Sweatman, T., et al. (2008). Pathways and products for the metabolism of vitamin D3 by cytochrome P450scc. *The FEBS Journal, 275*(10), 2585–2596. http://dx.doi.org/10.1111/j.1742-

4658.2008.06406.x, (Research Support, N.I.H., Extramural. Research Support, Non-U.S. Gov't).

Urbschat, A., Paulus, P., von Quernheim, Q. F., Bruck, P., Badenhoop, K., Zeuzem, S., et al. (2013). Vitamin D hydroxylases CYP2R1, CYP27B1 and CYP24A1 in renal cell carcinoma. *European Journal of Clinical Investigation, 43*(12), 1282–1290. http://dx.doi.org/10.1111/eci.12176.

van Etten, E., Stoffels, K., Gysemans, C., Mathieu, C., & Overbergh, L. (2008). Regulation of vitamin D homeostasis: Implications for the immune system. *Nutrition Reviews, 66*(10 Suppl. 2), S125–S134. http://dx.doi.org/10.1111/j.1753-4887.2008.00096.x.

Wang, L., Flanagan, J. N., Whitlatch, L. W., Jamieson, D. P., Holick, M. F., & Chen, T. C. (2004). Regulation of 25-hydroxyvitamin D-1alpha-hydroxylase by epidermal growth factor in prostate cells. *Journal of Steroid Biochemistry and Molecular Biology, 89-90*(1–5), 127–130. http://dx.doi.org/10.1016/j.jsbmb.2004.03.091, S0960076004001505 [pii].

Wang, Q., He, Y., Shen, Y., Zhang, Q., Chen, D., Zuo, C., et al. (2014). Vitamin D inhibits COX-2 expression and inflammatory response by targeting thioesterase superfamily member 4. *Journal of Biological Chemistry, 289*(17), 11681–11694. http://dx.doi.org/10.1074/jbc.M113.517581.

Wang, T. T., Nestel, F. P., Bourdeau, V., Nagai, Y., Wang, Q., Liao, J., et al. (2004). Cutting edge: 1,25-Dihydroxyvitamin D3 is a direct inducer of antimicrobial peptide gene expression. *Journal of Immunology, 173*(5), 2909–2912.

Wang, J., Slominski, A., Tuckey, R. C., Janjetovic, Z., Kulkarni, A., Chen, J., et al. (2012). 20-hydroxyvitamin D(3) inhibits proliferation of cancer cells with high efficacy while being non-toxic. *Anticancer Research, 32*(3), 739–746, 32/3/739 [pii].

Wang, T. J., Zhang, F., Richards, J. B., Kestenbaum, B., van Meurs, J. B., Berry, D., et al. (2010). Common genetic determinants of vitamin D insufficiency: A genome-wide association study. *Lancet, 376*(9736), 180–188. http://dx.doi.org/10.1016/s0140-6736(10)60588-0.

Wang, Y., Zhu, J., & DeLuca, H. F. (2012). Where is the vitamin D receptor? *Archives of Biochemistry and Biophysics, 523*(1), 123–133. http://dx.doi.org/10.1016/j.abb.2012.04.001, (Research Support, Non-U.S. Gov't Review).

Welsh, J. (2012). Vitamin D and cancer: Integration of cellular biology, molecular mechanisms and animal models. *Scandinavian Journal of Clinical and Laboratory Investigation. Supplementum, 243*, 103–111. http://dx.doi.org/10.3109/00365513.2012.682870.

Wu, S., Ren, S., Nguyen, L., Adams, J. S., & Hewison, M. (2007). Splice variants of the CYP27b1 gene and the regulation of 1,25-dihydroxyvitamin D3 production. *Endocrinology, 148*(7), 3410–3418. http://dx.doi.org/10.1210/en.2006-1388, (Research Support, N.I.H., Extramural).

Yang, L., Weaver, V., Smith, J. P., Bingaman, S., Hartman, T. J., & Cantorna, M. T. (2013). Therapeutic effect of vitamin d supplementation in a pilot study of Crohn's patients. *Clinical and Translational Gastroenterology, 4*, e33. http://dx.doi.org/10.1038/ctg.2013.1, ctg20131 [pii].

Yin, K., & Agrawal, D. K. (2014). Vitamin D and inflammatory diseases. *Journal of Inflammation Research, 7*, 69–87. http://dx.doi.org/10.2147/JIR.S63898.

Yin, L., Grandi, N., Raum, E., Haug, U., Arndt, V., & Brenner, H. (2009). Meta-analysis: Longitudinal studies of serum vitamin D and colorectal cancer risk. *Alimentary Pharmacology and Therapeutics, 30*(2), 113–125. http://dx.doi.org/10.1111/j.1365-2036.2009.04022.x, APT4022 [pii].

Zehnder, D., Bland, R., Chana, R. S., Wheeler, D. C., Howie, A. J., Williams, M. C., et al. (2002). Synthesis of 1,25-dihydroxyvitamin D(3) by human endothelial cells is regulated by inflammatory cytokines: A novel autocrine determinant of vascular cell adhesion. *Journal of the American Society of Nephrology, 13*(3), 621–629.

Zehnder, D., Bland, R., Williams, M. C., McNinch, R. W., Howie, A. J., Stewart, P. M., et al. (2001). Extrarenal expression of 25-hydroxyvitamin d(3)-1 alpha-hydroxylase. *The*

Journal of Clinical Endocrinology and Metabolism, *86*(2), 888–894. http://dx.doi.org/10.1210/jcem.86.2.7220.
Zhang, Y., Leung, D. Y., Richers, B. N., Liu, Y., Remigio, L. K., Riches, D. W., et al. (2012). Vitamin D inhibits monocyte/macrophage proinflammatory cytokine production by targeting MAPK phosphatase-1. *Journal of Immunology*, *188*(5), 2127–2135. http://dx.doi.org/10.4049/jimmunol.1102412.
Zheng, W., Tayyari, F., Gowda, G. A., Raftery, D., McLamore, E. S., Shi, J., et al. (2013). 1,25-dihydroxyvitamin D regulation of glucose metabolism in Harvey-ras transformed MCF10A human breast epithelial cells. *Journal of Steroid Biochemistry and Molecular Biology*, *138*, 81–89. http://dx.doi.org/10.1016/j.jsbmb.2013.03.012, S0960-0760(13)00061-7 [pii].
Zhu, J., & DeLuca, H. F. (2012). Vitamin D 25-hydroxylase—Four decades of searching, are we there yet? *Archives of Biochemistry and Biophysics*, *523*(1), 30–36. http://dx.doi.org/10.1016/j.abb.2012.01.013, S0003-9861(12)00028-8 [pii].
Zhu, Y., Mahon, B. D., Froicu, M., & Cantorna, M. T. (2005). Calcium and 1 alpha,25-dihydroxyvitamin D3 target the TNF-alpha pathway to suppress experimental inflammatory bowel disease. *European Journal of Immunology*, *35*(1), 217–224. http://dx.doi.org/10.1002/eji.200425491.
Zierold, C., Mings, J. A., & DeLuca, H. F. (2003). Regulation of 25-hydroxyvitamin D3-24-hydroxylase mRNA by 1,25-dihydroxyvitamin D3 and parathyroid hormone. *Journal of Cellular Biochemistry*, *88*(2), 234–237. http://dx.doi.org/10.1002/jcb.10341.
Zordoky, B. N., & El-Kadi, A. O. (2009). Role of NF-kappaB in the regulation of cytochrome P450 enzymes. *Current Drug Metabolism*, *10*(2), 164–178, (Research Support, Non-U.S. Gov't Review).
Zou, M., BinHumaid, F. S., Alzahrani, A. S., Baitei, E. Y., Al-Mohanna, F. A., Meyer, B. F., et al. (2014). Increased CYP24A1 expression is associated with BRAF(V600E) mutation and advanced stages in papillary thyroid carcinoma. *Clinical Endocrinology*, *81*(1), 109–116. http://dx.doi.org/10.1111/cen.12396, (Research Support, Non-U.S. Gov't).

INDEX

Note: Page numbers followed by "*f*" indicate figures and "*t*" indicate tables.

I

K

L

R

W

X

FSC
www.fsc.org
MIX
Papier | Fördert
gute Waldnutzung
FSC® C083411

Kolibri 360 GmbH
Ferdinand-Jühlke-Straße 7
99095 Erfurt, Deutschland
produktsicherheit@kolibri360.de